DVD DEMYSTIFIED

DVD
Demystified

Jim Taylor

Second Edition

McGraw-Hill
New York   San Francisco   Washington, D.C.
Auckland   Bogotá   Caracas   Lisbon   London
Madrid   Mexico City   Milan   Montreal   New Delhi
San Juan   Singapore   Sydney   Tokyo   Toronto

Cataloging-in-Publication Data is on file with the Library of Congress

McGraw-Hill

A Division of The McGraw·Hill Companies

Copyright © 2001, 1998 by The McGraw-Hill Companies, Inc. All rights reserved. Printed in the United States of America. Except as permitted under the United States Copyright Act of 1976, no part of this publication may be reproduced or distributed in any form or by any means, or stored in a data base or retrieval system, without the prior written permission of the publisher.

2 3 4 5 6 7 8 9 0   DOC/DOC   0 6 5 4 3 2 1

P/N 135027-6
PART OF
ISBN 0-07-135026-8

The sponsoring editor for this book was Stephen S. Chapman and the production supervisor was Pamela A. Pelton. It was set in Century Schoolbook by MacAllister Publishing Services, LLC.

Printed and bound R. R. Donnelley & Sons Company.

 This book is printed on recycled, acid-free paper containing a minimum of 50 percent recycled, de-inked fiber.

McGraw-Hill books are available at special quantity discounts to use as premiums and sales promotions, or for use in corporate training programs. For more information, please write to the Director of Special Sales, Professional Publishing, McGraw-Hill, Two Penn Plaza, New York, NY 10121-2298. Or contact your local bookstore.

To my wife, Julia, who would have preferred more me and less book, but who was amazingly supportive through the entire long endeavor.

PRAISE FOR DVD DEMYSTIFIED

It was on Aril 17, 1992, when four bottles of Napa red wine were emptied between Warran Lieberfarb (President, Warner Home Video) and myself, that the very first shape of the DVD concept was originated. I was very excited but nevertheless seriously anxious about whether this new technology would ever see the light of the multimedia age.

Then, in early 1998, when I learned that DVD Demystified had been published, was the very first time I began to feel there was proof that DVD was heading for success.

I met with Jim in San Francisco later and got his book with his autograph. It sits nicely on my shelf as a monumental book that gave me the confidence that we are doing the right thing. His new edition, as a reference to executives and engineers, will surely add stability for the ever-growing DVD market.

DVD will continue to penetrate into every corner of our lives. It will be a pervasive technology for probably a couple of decades to come. I have the feeling that DVD Demystified will continue in several more editions over the years as a benchmark reference of this technology.

Koji Hase
Acting Chairperson of the DVD Forum
Vice President, Strategic Alliance Division,
Digital Media Network Company, Toshiba Corporation

A clear and intelligent guide to DVD, and a valuable reference for anyone who wants to take full advantage of the technology.

Kamer Davis, Senior Vice President,
Ogilvy Public Relations Worldwide

Required reading for both novices and professionals interested in the fundamentals of DVD, and required reading for all InterActual employees.

Todd Collart, President & CEO,
InterActual Technologies, Inc.

CONTENTS

Contents

Contents

Contents

FOREWORD

It may seem hard to believe, but more than three years have passed since the first edition of Jim Taylor's *DVD Demystified* was written. In those dark, early days, few people knew much about DVD, and its future was by no means assured. At industry conferences, somber war stories about the challenges and high cost of producing theatrical DVD titles were presented. Promises of 17-GB DVD-18 discs seemed more like science fiction than reality. Predictions of home video DVD recorders entering the market seemed preposterous, since the equivalent technology required to author and record DVD programs cost more than $200,000 at the time. Several well-publicized compatibility problems existed between early players and titles, and there were real concerns about whether all of the major motion picture studios would ever commit to the format and release titles on DVD.

These were shaky times for the format, and some cynics joked that DVD stood for "doubtful, very doubtful". In early 1997, I stood in my company's booth at an industry trade show trying to provide a simple description of DVD to groups of skeptical video producers and engineers. After my attempts to explain the format's many complexities were greeted with quizzical looks, I began using a more succinct description: "DVD is movies on little discs." Although many people, even the technical ones, were surprisingly satisfied with this desperate explanation, I remember worrying that if it was so difficult just to explain what DVD is, how would we ever sell it to actual users?

Obviously, there was nothing to worry about.

Today, chances are excellent that most of your neighbors know what DVD is. Everywhere you look, evidence can be found of DVD's astounding success: discount stores prominently display racks of video titles; brand name DVD players sell at low prices that were unthinkable not long ago; and some industry analysts now believe that home DVD video recorders will replace VCRs within a few years. Countless thousands of titles are available worldwide—some on 17-GB DVD-18 discs—and many more arrive every day. The format has in fact become so synonymous with high quality and value that many people happily buy movies on DVD that they already own on tape or laserdisc.

In addition to its success in the consumer video business, a promising industrial market for DVD has also formed. Corporations, schools, hospitals, the military, and other government agencies have all put DVD to use for a variety of applications. That cool video you see playing on a videowall at the mall may in fact be coming from a DVD disc. Kids at a U.S. high school are learning about Newtonian physics from a detailed frame-by-frame study of scenes in an action-adventure movie. Computer images of your heart might be stored in an automated DVD library system that is growing in a number of hospitals. Some military officers are now learning leadership skills through the use of interactive DVD training systems. DVD has, without a doubt, hit the big time, and it's clearly here to stay.

DVD has, of course, always been far more than just "movies on little discs." It includes four physical formats: DVD-ROM, DVD-R, DVD-RAM, and DVD-RW. DVD also has three application formats: DVD Video, DVD Audio Recording, and DVD Video Recording, with more on the way. The richness of these offerings, however, is a mixed bag of goods. Although DVD has a solution for just about any storage problem, making sense of it all is still a challenge. This, of course, is why this book was written.

For the past three years, Jim Taylor's *DVD Demystified* has become the authoritative bible of the industry, bar none. I am certain that every self-respecting DVD professional in the business owns a well-worn, dog-eared copy. It's a book that smart people *never* loan out because there's a good chance it won't be returned. Why? Because it is the most complete, user-friendly source of information available about this very complex format. *DVD Demystified* is aimed at individuals who don't enjoy reading the massive, arcane (and expensive) DVD specification books. In other words, this book targets just about everyone. In fact, I'll let you in on a little secret. I know at least one person who *does* have access to the spec books and always reaches for Jim's book first.

The only thing that improves on *DVD Demystified* is . . . the *updated* version of *DVD Demystified*. As you can tell by now, the world of DVD moves very fast, and no one follows the details as closely—and as accurately—as Jim Taylor. As in the first edition, you can count on an independent, thought-provoking analysis of all that is DVD, and you will find it written in an approachable style that both technical and non-technical readers will enjoy.

So if you're interested in DVD and haven't bought this book yet, what are you waiting for? But a word from the wise: don't let *anyone* borrow it.

ANDY PARSONS
Senior Vice President
Product Development & Technical Support
Pioneer New Media Technologies

ACKNOWLEDGMENTS

My heartfelt thanks to those who encouraged and supported me in writing this book. Both times.

Many kind people spent time reading and commenting on drafts for the first edition and for this second edition: Kilroy Hughes, Ralph LaBarge, Dana Parker, Geoff Tully, Jerry Pierce, Steve Taylor, Robert Lundemo Aas, Chad Fogg, Roger Dressler, Tristan Savatier, Van Ling, Julia Taylor, Andy Parsons, and Leo Backman. This book is much richer because of them.

Thanks also Bob Stuart, Tom Holman, Charles Poynton, Mike Schmit, Dave Schnuelle, the guys at Microsoft, and many others who took time to explain things. I'm also grateful to the DVD List members for sharing their knowledge, as well as to the denizens of the rec.video.dvd Internet newsgroups, the Home Theater Forum, and other Internet watering holes of DVD news and information.

Extremely special thanks to Ari Zagnit and Mark Johnson for authoring the sample disc and to Samantha Cheng at TPS for making it happen. Thanks to Chuck Crawford and his wife for contributing their house to the project. I'm indebted to all the generous people who made the sample disc possible: Willie Chu for music; Laurie Smith for art; Thomas Bennett, ace Easter Frog hunter, Jamie Pickell for help with audio; Mark Waldrep at AIX Media Group; Ralph LaBarge at AlphaDVD; Richard Fortenberry at Atomboy; Scott Epstein at Broadcast DVD; Hideo Nagashima of Cinema Craft; Gene Radzik, Bill Barnes, and Roger Dressler at Dolby; Lorr Kramer, Patrick Watson, and Blake Welcher at DTS; David Goodman at DVD International; Gabe Murano at DVant; Bruce Nazarian at Gnome Digital; Henninger Media Services; Rey Umali and Brian Quandt at Heuris; Joe Kane at Joe Kane Productions and Van Ling at Lightstorm; Chris Brown, Cindy Halstead, Lenny Sharp, Todd Collart, and others at InterActual; Chinn Chin and Joe Monasterio at InterVideo; Michele Serra of Library DVD; Microsoft Studios and the Digital Video Services team, as well as Microsoft Windows Hardware Quality Lab (WHQL) and Microsoft Digital Media Division (specific thanks to Andrew Rosen, Craig Cleaver, Kenneth Smith, Randy James, and Eric Anderson); Fred Grossberg at Mill Reef Entertainment; Trai Forrester at New Constellation Technologies; Guy Kuo and Ovation Software; Garrett Smith at Paramount; Sandra Benedetto at Pioneer; Henry Steingieser, Gary Randles, and Randy Berg at Rainmaker New Media; Randy Glenn and Bob Michaels at Slingshot Entertainment; François Abbe at Snell & Wilcox; Paul Lefebvre and Mark Ely at Sonic Solutions; Tony Knight at SpinWare; Greg Wallace, Rainer Broderson, and Gary Hall at Spruce

Technologies; Garrett Maki at Sunset Post; Charles Busslinger, Rick Deane, and Annie Chang at THX; Brad Collar at CVC; Shaun Taylor at Videodiscovery; Jim Babinski at WAMO; Gary Reber at Widescreen Review and Steve Michelson at Steve Michelson Productions; and Blaine Graboyes at Zuma Digital.

Thanks to Steve Chapman at McGraw-Hill for backing me up and bailing me out, and to Beth Brown at MacPS for turning it all into something printable. And special thanks to Fleischman and Arthur, who have been solidly supportive.

JIM TAYLOR

PREFACE

To my surprise, the first edition of *DVD Demystified* quickly became the bible of the industry. I'm profoundly gratified that it has helped so many people understand DVD technology, and I appreciate everyone who has taken the time to send e-mail or tell me in person that they enjoyed the book.

In spite of the publication of the DVD bible in late 1997, DVD technology refused to stand still. Unfinished variations such as writable DVD and DVD-Audio became less unfinished, and new applications such as WebDVD and Divx reshaped markets and perceptions. So, in a brief moment of insanity, having forgotten what an enormous task it was to write this book in the first place, I agreed to do a second edition. If this new edition helps people better understand DVD and use its unique features to achieve their creative visions then my time will have been spent well.

What Is DVD?

Introduction

DVD is the future.[1] No matter who you are or what you do, DVD technology will be in your future, supplying entertainment, information, and enlightenment in the form of video, audio, and computer data. DVD embodies the grand unification theory of entertainment and business media: If it fulfills the hopes of its creators, DVD will replace audio *compact discs* (CDs), videotapes, laserdiscs, CD-ROMs, video game cartridges, and even certain printed publications. By its third birthday, DVD had already become the most successful consumer electronics entertainment product of all time, with over 6 million players sold in the United States, over 10 million players worldwide, and over 30 million DVD-ROM computers.

DVD is a bridge. According to the DVD Entertainment Group, DVD is "the medium of the new millennium." Although undoubtedly there will be more important media in the next thousand years, this is an accurate description for the first decade or so, since DVD is the ideal convergent medium for a converging world. We are witnessing watershed transitions from analog TV to digital TV (DTV), from interlaced video to progressive video, from standard TV to widescreen TV, and from passive entertainment to interactive entertainment. In every case DVD works on both sides, bridging from the "old way" to the "new way."

DVD is excellence. In a world where the prevailing trend is to squeeze in more channels and longer playing time at the sacrifice of quality, DVD is the standout contrarian. As broadcasters convert to DTV, they are more likely to use the extra space provided by digital compression to hold more channels of low quality rather than a few channels of high quality. Digital satellite providers already have taken this approach. Anyone hawking streaming video across the Internet has thrown aesthetics out the window. In contrast, most DVDs are created by people who care passionately about the video experience—people who spend months cleaning up video frame by frame, restoring and remixing audio, reassembling director's cut versions, recording commentaries, researching outtakes and extras, and providing a richness and depth of content seldom seen in other media. DVD is not the ultimate in video quality, but it is the standard bearer for consumer entertainment.

---

[1] As Criswell solemnly intones at the beginning of the *Plan 9 from Outer Space* DVD, "We are all interested in the future, for that is where you and I are going to spend the rest of our lives. And remember, my friends, future events such as these will affect you in the future. You are interested in the unknown, the mysterious, the unexplainable. That is why you are here."

DVD is just DVD.   In the early days of DVD's development, the letters stood for digital video disc. Later, like a stepsister trying to squish her ugly foot into a glass slipper, a few companies tried to retrofit the acronym to "digital versatile disc" in a harebrained attempt to express the versatility of DVD. But just as everyone knows what a VCR and a VHS tape are without worrying what the letters stand for,[2] DVD stands on it own.

But what is DVD? Put simply, DVD is the next generation of CD technology. Improvements in optical technology have made the tightly packed microscopic pits that store data on an optical disc even more microscopic and even more tightly packed. A DVD is the same size as the familiar CD—12 centimeters wide (about 4.7 inches)—but it stores up to 25 times more and is more than nine times faster.

And yet, DVD is much more than CD on steroids. Its increased storage capacity and speed allow it to accommodate high-quality digital video in MPEG-2 format. The result is a small, shiny disc that holds better-than-TV video and better-than-CD audio. A basic DVD can contain a movie over two hours long. A double-sided, dual-layer DVD can hold about eight hours of near-cinema-quality video or more than 30 hours of VHS-quality video. If only still pictures are used, DVD becomes an audio book that can play continuously for weeks.

DVD has many tricks to woo both the weary couch potato and the multimedia junkie alike, such as a widescreen picture, multichannel surround sound, multilingual audio tracks, selectable subtitles, multiple camera angles, karaoke features, seamless branching for multiple storylines, navigation menus, instant fast forward/rewind, and more.

Just as audio CD has its computer counterpart in CD-ROM, DVD has DVD-ROM, which goes far beyond CD-ROM. DVD-ROM holds from 4.4 to 16 gigabytes of data—25 times as much as a 650-megabyte CD-ROM—and sends it to the computer faster than a comparable CD-ROM drive.

DVD is inexpensive. The first few generations of DVD-ROM drives were more expensive than CD-ROM drives, but as the technology has improved and production quantities have increased, the price gap between them has continued to narrow. Once the price gap is insignificant, manufacturers will stop making CD-ROM drives. During the first few years, DVD-Video players were as expensive as high-end VCRs, but mass production of DVD-ROM drives and plummeting costs of audio/video decoder chips are driving the price of consumer DVD players down to the same level as VCRs and CD

---

[2]Some people claim VHS stands for video home system, whereas others insist it stands for video helical scan or vertical helical scan. Just as with DVD, it is better to ignore the fuzzy etymology.

players. DVD discs are produced with much of the same equipment used for CDs, and because they are stamped instead of recorded, they can be produced cheaper and faster than tapes.

DVD is at the crest of a wave bringing significant change to the world of video entertainment and multimedia. It is the first high-quality interactive medium to be affordable to the mass market. Until now, the high-impact visuals of movies, television, and videotape have been linear and unchanging, whereas the dynamic and responsive environment of computer multimedia has suffered from unimpressively tiny video windows with fuzzy, jerky motion. Many artists with the vision to do extraordinary things with an interactive environment have shunned CD-ROM and computers because their creative standards would be compromised. As a result, they have been constricted to the straight and narrow of traditional linear video presentation designed to appeal to the lowest common interests.[3] This does not mean that DVD closes the door on beginning-to-end storytelling, only that it opens new doors for different approaches. DVD is a fresh digital canvas onto which artists can expand their abilities and sketch their nonlinear visions in time and space to be recreated on television screens and computer screens alike as a different experience for each participatory viewer.

DVD-Video versus DVD-ROM

NOTE: *Just as CD audio is not the same as CD-ROM, DVD-Video is not the same as DVD-ROM.*

This book will tell you many things about DVD, but if you take away only a few tidbits, one of them should be that DVD-Video is not the same as DVD-ROM. Just as CD-ROM and CD audio are different applications of the same

---

[3]George Gilder pulls no punches when making this point in his book *Life After Television:* "TV defies the most obvious fact about its customers: their prodigal and efflorescent diversity. TV ignores the reality that people are not inherently couch potatoes; given a chance, they talk back and interact. People have little in common except their prurient interests and morbid fears and anxieties. Necessarily aiming its fare at this lowest-common-denominator target, television gets worse and worse every year."

technology, DVD is two separate things: (1) a computer data storage medium and (2) an audio/video storage medium. You use a DVD-ROM drive —like a CD-ROM drive—to read computer data from a DVD-ROM disc. You use a DVD player—like a VCR or a CD player—to play back video and audio from a DVD (more correctly called *DVD-Video*). As computers become true multimedia systems, this distinction is beginning to disappear, but it is still important to understand the difference.

Technically, DVD is made up of a set of physical media specifications (read-only DVD-ROM, recordable DVD-R, rewritable DVD-RAM, etc.) and a set of application specifications (DVD-Video, DVD-Audio, and other possible specialized formats for applications such as videogame consoles). Unfortunately, the DVD family is plagued with sibling rivalry and competing formats. The compatibilities and incompatibilities of the physical and logical formats are a bewildering muddle.

What Does DVD Portend?

Motion video is one of the most deeply affecting creations of humankind. It combines the ethereal effects of sound and music, the realistic impact of photography, the engrossing drama of theatrical plays, and the variety of visual arts and weaves them all together with the ageless appeal of the storyteller. We are endlessly attempting to improve the richness of this medium with which we replay and reshape our impressions and visions of the world. DVD is one small step in this quest—but a very critical one. It is a milestone in the ascendancy of things digital.

DVD is one of the final nails in the coffin of analog technology. Our depiction of the world is changing from analog forms such as vinyl records, film, and magnetic tape to digital forms such CDs, computer graphics, DVDs, and the Internet. The advantage of information in digital form is that it can be manipulated and processed easily by computer; it can be compressed to achieve cheaper and faster storage or transmission; it can be stored and duplicated without generational loss; its circuitry does not drift with heat or age; it can be kept separate from noise and distortion in the signal; and it can be transmitted over long distances without degrading. As the Internet becomes the dominant paradigm for the way we receive, send, and work with information, DVD will play a vital role. It will take many, many years before the Internet is able to easily carry the immense amounts of data needed for television and movies, interactive multimedia, and even virtual worlds. Until then, DVD is positioned to be the primary vehicle for delivering these information streams. And there will always be a need to archive information while keeping it quickly accessible. DVD and future generations of DVD fit the bill.

Some people believe that DVD heralds the complete convergence of computers and entertainment media. They feel that a technology such as DVD that works as well in the family room as in the office is another reason that the TV and the computer will merge into one. Perhaps. Or perhaps the computer will always be a separate entity with a separate purpose. How many people want to write in a word processor or work on a spreadsheet while sitting on the couch in front of the TV with a cordless keyboard on their lap— especially if the kids want to watch *Animaniacs* or play Ultra Mario instead? Whatever the case, it is inevitable that consumer electronics will gain more of the capabilities of computers, and computers will assimilate more of the features of televisions and stereo systems. The line is blurring, and DVD is rubbing a very big eraser across it.[4]

Who Needs to Know About DVD?

The number of people who are feeling or will feel the effect of DVD is truly astonishing. A large part of this astonished multitude requires a working knowledge of DVD, including its capabilities, strengths, and limitations. This book provides most of this knowledge.

DVD will affect a remarkably diverse range of fields. A few are illustrated in the following pages.

Movies

DVD has significantly raised the quality and enjoyability of home entertainment. According to the *Consumer Electronics Association* (CEA), there were over 19 million American homes with a home theater[5] system by the end of 1999. As DVD player prices drop, the players are becoming increasingly attractive as the central component of home theater systems and as

---

[4]In this age of digital art, it is probably more appropriate to say that DVD is rubbing a very large *smudge tool* across the line.

[5]CEA defines a home theater system as a 25-inch or larger TV, hi-fi VCR or laserdisc player or DVD player, and a surround sound system with at least four speakers.

replacements for aging CD players. DVD video recorders, as they steadily drop in price, will challenge VCRs for the spot next to the television.

Music and Audio

Because audio CD is well established and satisfies the needs of most music listeners, DVD will have less of an effect in this area. The introduction of DVD-Audio trailed DVD-Video by almost 4 years, and the DVD-Video format already incorporated higher-than-CD-quality audio and multichannel surround sound. DVD players also play audio CDs. All this leaves the DVD-Audio format with a tough sales job. However, the unprecedented success of DVD-Video ensures that DVD-Audio will come along for the ride, especially as new players support both formats and the distinctions fade away.

DVD-Audio has a special appeal to music labels because it includes copy protection features that CD never had. Whether those protections will last long and whether they continue to be relevant in the face of Internet music distribution remains to be seen.

DVD is a boon for audio books and other spoken-word programs. Dozens of hours of stereo audio can be stored on one side of a single disc—a disc that is cheaper to produce and more convenient to use than cassette tapes or multi-CD sets.

Music Performance Video

Despite the success of MTV and the virtual prerequisite that a music group cut a video in order to be heard, music performance video has not done as well as expected in the videotape or laserdisc markets. Perhaps it is because you cannot pop a video in a player and continue to read or work around the house. The added versatility of DVD, however, may be the golden key that unlocks the gates for music performance video. With high-quality, long-playing video and multichannel surround sound, DVD music video appeals to a range of fans from opera to ballet to New Age to acid rock. Music albums on DVD can improve their fan appeal by adding such tidbits as live performances, interviews, backstage footage, outtakes, video liner notes, musician biographies, and documentaries. Sales of DVD-Video music titles have already exceeded many expectations.

One of the targets for DVD-Video player sales is the replacement CD player market. CD players inevitably break or are upgraded, and shoppers looking for a new player may be persuaded to spend a little more to get a device that also can play movies. The natural convergence of the audio and video player will bolster the success of combined music and video titles.

Training and Productivity

Until it is eventually replaced by streaming video on broadband Internet, DVD will become the medium of choice for video training. The low cost of hardware and discs, the widespread use of players, the availability of authoring systems, and a profusion of knowledgeable DVD developers and producers will make DVD—in both DVD-Video and multimedia DVD-ROM form—ideal for industrial training, teacher training, sales presentations, home education, and any other application where full video and audio are needed for effective instruction. Especially popular are DVDs that connect to the Internet, making up for the fact that streaming video is too small, too slow, too fuzzy, and too unreliable to be of any use to the average learner.

Videos for teaching skills from accounting to TV repair to dental hygiene, from *tai chi* to guitar playing to flower arranging become vastly more effective when they take advantage of the on-screen menus, detailed images, multiple soundtracks, selectable subtitles, and other advanced interactive features of DVD. Consider an exercise video that randomly selects different routines each day or lets you choose the mood, the tempo, and the muscle groups on which to focus. Or a first-aid training course that slowly increases the difficulty level of the lessons and the complexity of the practice sessions. Or an auto-mechanic training video that allows you to view a procedure from different angles at the touch of a remote control (preferably one with a grease-proof cover). Or a cookbook that helps you select recipes via menus and indexes and then demonstrates with a skilled chef leading you through every step of the preparation. All this on a small disc that never wears out and never has to be rewound or fast-forwarded.

DVD is cheaper and easier to produce, store, and distribute than video-tape. Other products such as laserdisc, CD-i, and even Video CD have done well in training applications, but they require expensive or specialized players. As DVD becomes more established, almost every home and office will have either a standalone player or a DVD computer.

Education

Filmstrips, 16-mm films, VHS tapes, laserdiscs, CD-i discs, Video CDs, CD-ROMs, and the Internet have all had roles in providing images and sound to supplement textbooks and teachers. Most newer technologies lack the picture quality and clarity that is so important for classroom presentations. The exceptional image detail, high storage capacity, and low cost of DVD make it an excellent candidate for use in classrooms, especially since it integrates well with computers. Even though DVD-Video players may not be widely adopted in education, DVD computers will be. CD-ROM has infiltrated all levels of schooling from home to kindergarten to college and will soon pass the baton to DVD as new computers with built-in DVD-ROM drives are purchased. Educational publishers will discover how to make the most of DVD, creating truly interactive applications with the sensory impact and realism needed to stimulate and inspire inquisitive minds.

Computer Software

CD-ROM has become the computer software distribution medium of choice. To reduce manufacturing costs, most software companies use CD-ROMs rather than expensive and unwieldy piles of floppy disks. Yet some applications are too large even for the multimegabyte capacity of CD-ROM. These include large application suites, clip art collections, software libraries containing dozens of programs that can be unlocked by paying a fee and receiving a special code, specialized databases with hundreds of millions of entries, and massive software products such as network operating systems and document collections. Phone books that used to fill six or more CD-ROMs now fit onto a single DVD-ROM. Companies that distribute monthly updates of CD-ROM sets can ship free DVD-ROM drives to their customers and pay for them within a year with the savings on production costs alone.

Computer Multimedia

DVD-ROM products have been slow to appear, largely because the multimedia CD-ROM market is so tough. CD-ROM publishers struggling to produce a product within budget, convince distributors to carry it, and get it onto limited shelf space are reluctant to complicate their lives with a new format.

Still, many multimedia producers are stifled by the narrow confines of CD-ROM and yearn for the wide open spaces and liberating speed of DVD-ROM. Microsoft's Encarta encyclopedia has overflowed onto more than one CD-ROM but can expand for years to come without filling up its single DVD-ROM. The National Geographic collection that is spread across 31 CD-ROMs is much more usable on 4 DVD-ROMs.

In addition to space for more data, DVD brings along high-quality audio and video. Many new computers have hardware or software decoders that can be used to play DVD movies. These DVD-enabled computers will be even more effective for realistic simulations, games, education, and "edutainment." DVD eventually will make blocky, quarter-screen computer video a distant, dismal memory.

Video Games

The capacity to add high-quality, real-life video and full surround sound to three-dimensional game graphics is attractive to video game manufacturers. The major game console manufacturers, Nintendo, Sega, and Sony—and Microsoft with its Xbox—are all using DVD in the newest versions of their systems. A combination video game/CD/DVD player is very appealing.

Many past attempts to combine video footage with interactive games have been met with yawns, but the technology will improve until it finally clicks. Video games that make extensive use of full-screen video—even multimedia games traditionally available only for computers—are appearing in DVD-Video editions that play on any home DVD player and on DVD computers. For example, the venerable Dragon's Lair, a groundbreaking arcade video game that used a laserdisc for movie-quality animation, is now available on DVD-Video.

Information Publishing

The Internet is a wonderfully effective and efficient medium for information publishing, but it lacks the bandwidth needed to do justice to large amounts of data rich with graphics, audio, and motion video. DVD, with storage capacity far surpassing CD-ROM plus standardized formats for audio and video, is perfect for publishing and distributing information in our ever more knowledge-intensive and information-hungry world.

Organizations can use DVD-Video and DVD-ROM to quickly and easily disseminate reports, training material, manuals, detailed reference handbooks, document archives, and much more. Portable document formats such as Adobe Acrobat and HTML are perfectly suited to publishing text and pictures on DVD-ROM. Recordable DVD soon will be available and affordable for custom publishing of discs created on the desktop.

Marketing and Communications

DVD-Video and DVD-ROM are well suited to carry information from businesses to their customers and from businesses to businesses. A DVD can hold an exhaustive catalog able to elaborate on each product with full-color illustrations, video clips, demonstrations, customer testimonials, and more at a fraction of the cost of printed catalogs. Bulky, inconvenient videotapes of product information can be replaced with thin discs containing on-screen menus that guide the viewer on how to use the product, with easy and instant access to any section of the disc.

The storage capacity of DVD-ROM can be exploited to put entire software product lines on a single disc that can be sent out to thousands or millions of prospective customers in inexpensive mailings. The disc can include demo versions of each product, with protected versions of the full product that can be unlocked by placing an order over the phone or the Internet.

And More . . .

- **Picture archives.** Photo collections, clip art, and clip media have long since exceeded the capacity of CD-ROM. DVD-ROM allows more content or higher quality or both. As recordable DVD becomes affordable and easy to use, it also will enable personal media publishing. Anyone with a digital camera, some video editing software, and a recordable DVD drive can put pictures and home movies on a disc to send to friends and relatives who have a DVD player or a DVD computer.

- **Set-top boxes, digital receivers, and personal video recorders.** Savvy designers are already at work combining DVD players with the boxes used for interactive TV, digital satellite, digital cable, hard-disk–based video recorders, and other digital video applications. All these systems are based on the same underlying digital compression technology and can benefit from shared components.

- **WebDVD.** Many announcements have been made of products combining the multimedia capabilities of DVD-ROM and DVD-Video with the timeliness and interactivity of the Internet. Data-intensive media such as audio, video, and even large databases simply do not travel well over the Internet. Since broadband Internet will not arrive anytime soon, DVD is a perfect candidate to deliver the lion's share of the content.

■ **Home productivity and "edutainment."** DVD covers the gamut from PlayStation to WebTV to home PC. DVD-Video can be used for reference products such as visual encyclopedias, fact books, and travelogues; training material such as music tutorials, arts and crafts lessons, and home improvement series; and education products such as documentaries, historical recreations, nature films, and more, all with accompanying text, photos, sidebars, quizzes, and so on.

About This Book

DVD Demystified is an introduction and reference for anyone who wants to understand DVD. It is not a production guide, nor is it a detailed technical handbook, but it provides an extensive technical grounding for anyone interested in DVD. This second edition is completely revised and expanded from the first edition.

Chapter 2, "The World Before and After DVD," provides historical context and background. Any top analyst or business leader will tell you that extrapolating from prior technologies is the best way to predict technology trends. This chapter takes a historical stroll through the developments leading up to DVD and the first few years after its birth.

Chapter 3, "DVD Technology Primer," explains concepts such as aspect ratios, digital compression, and progressive scan. This chapter is a gentle technical introduction for nontechnical readers, but it should be useful for technical readers as well.

Chapter 4, "DVD Overview," covers the basic features of DVD, particularly DVD-Video, as well as topics such as compatibility, regional management, copy protection, and licensing. It also includes a section on myths and misunderstood characteristics of DVD.

Chapter 5, "Disc and Data Details," delves deeper into the viscera of DVD, examining physical structures, disc production, file formats, and other low-level details. This chapter is not essential to understanding or using DVD, but should be instructive for anyone desiring a strong grasp of the underpinnings.

Chapter 6, "Application Details—DVD-Video and DVD-Audio," reveals the particulars of the video and audio specifications for DVD. It lays out the data structures, stream composition, navigation information, and other elements of the DVD application formats.

Chapter 7, "What's Wrong with DVD," explores the shortcomings of DVD and how these might be overcome as the technology develops.

Chapter 8, "DVD Comparison," examines the relationships between DVD and other consumer electronics technologies and computer storage media, including advantages and disadvantages of each, plus discussions of compatibility and interoperability.

Chapter 9, "DVD at Home," helps you decide if DVD is for you, gives advice on selecting a player, and describes how a DVD player connects into a home theater or stereo system.

Chapter 10, "DVD in Business and Education," discusses how DVD applies to these areas, how it can be used best, and what effect it may have.

Chapter 11, "DVD on Computers," clarifies the variety of ways DVD can be used with computers. It also explains how to create DVD-Video discs that are enhanced for use on computers.

Chapter 12, "Essentials of DVD Production," entirely new for this second edition, dips into many of the mysteries of producing content for DVD-Video and DVD-ROM. This chapter provides a thorough grounding for anyone interested in creating DVDs or simply learning more about how DVDs are created.

The last chapter, "The Future of DVD," is a quick peek into the crystal ball to see what possibilities lie ahead for DVD and the world of digital video.

Units and Notation

DVD is a casualty of an unfortunate collision between the conventions of computer storage measurement and the norms of data communications measurement. The SI[6] abbreviations of k (thousands), M (millions), and G (billions) usually take on slightly different meanings when applied to bytes, in which case they are based on powers of 2 instead of powers of 10 (see Table 1.1).

NOTE: *In the world of DVD, a gigabyte is not always a gigabyte.*

---

[6]Système International d'Unités—the international standard of measurement notations such as millimeters and kilograms.

TABLE 1.1

Meanings of
Prefixes

Abbreviation	SI Prefix	IEC Prefix	IEC Abbr.	Common Use	Computer Use	Difference
k or K	kilo	kibi	Ki	1000 (10^3) [k]	1024 (2^{10}) [K]	2.4%
M	mega	mebi	Mi	1,000,000 (10^6)	1,048,576 (2^{20})	4.9%
G	giga	gibi	Gi	1,000,000,000 (10^9)	1,073,741,824 (2^{30})	7.4%
T	tera	tebi	Ti	1,000,000,000,000 (10^{12})	1,099,511,627,776 (2^{40})	10%

The problem is that there are no universal standards for unambiguous use of these prefixes. One person's 4.7 GB is another person's 4.38 GB, and one person's 1.321 MB/s is another's 1.385 MB/s. Can you tell which is which?[7]

The laziness of many engineers who mix notations such as KB/s, kb/s, and kbps with no clear distinction and no definition compounds the problem. And since divisions of 1000 look bigger than divisions of 1024, marketing mavens are much happier telling you that a dual-layer DVD holds 8.5 gigabytes rather than a mere 7.9 gigabytes. It may seem trivial, but at larger denominations the difference between the two usages—and the resulting potential error—becomes significant. There is almost a 5 percent difference at the mega- level and over a 7 percent difference at the giga- level. If you are planning to produce a DVD and you take pains to make sure your data takes, up just under 4.7 gigabytes (as reported by the operating system), you will be surprised and annoyed to discover that only 4.37 gigabytes fit on the disc. Things will get worse down the road with a 10 percent difference at the tera- level.

Since computer memory and data storage, including that of DVD-ROM and CD-ROM, usually are measured in megabytes and gigabytes (as opposed to millions of bytes and billions of bytes), this book uses 1024 as the basis for measurements of data size and data capacity, with abbreviations of KB, MB, and GB. However, since these abbreviations have become so

---

[7]The first (4.7 GB) is the typical data capacity given for DVD: 4.7 billion bytes, measured in magnitudes of 1000. The second (4.38 GB) is the "true" data capacity of DVD: 4.38 gigabytes, measured using the conventional computing method in magnitudes of 1024. The third (1.321 MB/s) is the reference data transfer rate of DVD-ROM measured in computer units of 1024 bytes per second. The fourth (1.385 MB/s) is DVD-ROM data transfer rate in thousands of bytes per second. If you do not know what any of this means, don't worry—by Chapter 4 it should all make sense.

ambiguous, the term is spelled out when practical. In cases where it is necessary to be consistent with published numbers based on the alternative usage, the words *thousand, million,* and *billion* are used, or the abbreviations k bytes, M bytes, and G bytes are used (note the small k and the spaces).

To distinguish kilobytes (1024 bytes) from other units such as kilometers (1000 meters), common practice is to use a large K for binary multiples. Unfortunately, other abbreviations such as M (mega) and m (micro) are already differentiated by case, so the convention cannot be applied uniformly to binary data storage. And in any case, too few people pay attention to these nuances.

In 1999, the *International Electrotechnical Commission* (IEC) produced new prefixes for binary multiples[8] (see Table 1.1). Although the new prefixes may never catch on, or they may cause even more confusion, they are a valiant effort to solve the problem. The main strike against them is that they sound a bit silly. For example, the prefix for the new term *gigabinary* is *gibi-,* so a DVD can be said to hold 4.37 gibibytes, or GiB. The prefix for *kilobinary* is *kibi-,* and the prefix for *terabinary* is *tebi-,* yielding kibibytes and tebibytes. Jokes about "kibbles and bits" and "teletebis" are inevitable.

As if all this were not complicated enough, data transfer rates, when measured in bits per second, are almost always multiples of 1000, but when measured in bytes per second, are sometimes multiples of 1000 and sometimes multiples of 1024. For example, a 1X DVD drive transfers data at 11.08 million bits per second (Mbps), which might be listed as 1.385 million bytes per second or might be listed as 1.321 megabytes per second. The 150 KB/s 1X data rate commonly listed for CD-ROM drives is "true" kilobytes per second, equivalent to 153.6 thousand bytes per second.

This book uses 1024 as the basis for measurements of byte rates (computer data being transferred from a storage device such as a hard drive or DVD-ROM drive into computer memory), with notations of KB/s and MB/s. For generic data transmission, generally measured in thousands and millions of bits per second, this book uses 1000 as the basis for bit rates, with notations of kbps and Mbps (note the small k). See Table 1.2 for a listing of notations.

Keep in mind that when translating from bits to bytes, there is a factor of 8 and when converting from bit rates to data capacities in bytes, there is an additional factor of 1000/1024.

---

[8]The new binary prefixes are detailed in *IEC 60027-2-am2 (1999-01): Letter symbols to be used in electrical technology. Part 2: Telecommunications and electronics, Amendment 2.*

TABLE 1.2

Notations Used in
This Book

Notation	Meaning	Magnitude	Variations	Example
b	bit	(1)		
kbps	thousand bits per second	10^3	Kbps, kb/s, Kb/s	56 kbps modem
Mbps	million bits per second	10^6	mbps, mb/s, Mb/s	11.08 Mbps DVD data rate
B	byte	(8 bits)		
KB	kilobytes	2^{10}	Kbytes, KiB	2 KB per DVD sector
KB/s	kilobytes per second	2^{10}	KiB/s	150 KB/s CD-ROM data rate
MB	megabytes	2^{20}	Mbytes, MiB	650 MB in CD-ROM
M bytes	million bytes	10^6	Mbytes, MB	682 M bytes in CD-ROM
MB/s	megabytes per second	2^{20}	MiB/s	1.32 MB/s DVD data rate
GB	gigabytes	2^{30}	Gbytes, GiB	4.37 GB in a DVD
G bytes	billion bytes	10^9	Gbytes, GB	4.7 G bytes in a DVD

Other Conventions

Spelling   The word *disc,* in reference to DVD or CD, should be spelled with a c, not a k. The generally accepted rule is that optical discs are spelled with a c, whereas magnetic disks are spelled with a k. For magneto-optical discs, which are a combination of both formats, the word is spelled with c because the discs are read with a laser. *The New York Times,* after years of head-in-the-sand usage of k for all forms of data storage, revised its manual in 1999 to conform to industry practice. Standards bodies such as ECMA and the *International Standards Organization* (ISO) persist in spelling it wrong, but what can you expect from bureaucracies? Anyone writing about DVD who spells it as disk instead of disc immediately puts his or her readers on notice that he or she does not understand what he or she is talking about.

PC The term *PC* means "personal computer." If a distinction is needed as to the specific type of hardware or operating system, brand names such as Windows, Mac OS, and Linux are added.

Aspect Ratios This book usually normalizes aspect ratios to a denominator of 1, with the 1 omitted most of the time. For example, the aspect ratio 16:9, which is equivalent to 1.78:1, is represented simply as 1.78. Normalized ratios such as 1.33, 1.78, and 1.85 are easier to compare than unreduced ratios such as 4:3 and 16:9. Note also that the ratio symbol (:) is used to indicate a relationship between width and height rather than the dimension symbol ($\times$), which implies a size.

Widescreen When the term *widescreen* is used in this book, it generally means an aspect ratio of 1.78 (16:9). The term *widescreen* as traditionally applied to movies has meant anything wider than the standard 1.33 (4:3) television aspect ratio, from 1.5 to 2.7. But since the 1.78 ratio has been chosen for DVD, digital television, and widescreen TVs, it has become the commonly implied ratio of the term *widescreen*.

DVD Format Names The term *DVD* is often applied both to the DVD family as a whole and specifically to the DVD-Video format. This book follows the same convention for simplicity and readability but only when unambiguous. When a clear distinction is needed, the book uses the terms *DVD-Video* (or DVD-V), *DVD-Audio* (or DVD-A), and *writable DVD* for the various record-once (DVD-R) and rewritable formats (DVD-RAM, DVD-RW, DVD+RW). The abbreviations *DVD-R(G)* and *DVD-R(A)* are used to distinguish "DVD-R for General" from "DVD-R for Authoring."

Television Systems There are basically two mutually incompatible television recording systems in common use around the world. Each is supported by corresponding digital encoding formats used by DVD. One system uses 525 lines scanned at 60 fields per second with NTSC color encoding and is used primarily in Japan and North America. The other system uses 625 lines scanned at 50 fields per second with PAL or SECAM color encoding and is used in most of the rest of the world. This book generally uses the technically correct terms of 525/60 (simplified from 525/59.94) and 625/50 but also uses the terms *NTSC* and *PAL* in the generic sense.

Colorspaces DVD uses the standard ITU-R BT.601 (formerly CCIR 601) component digital video colorspace of 4:2:0 $Y'C_bC_r$ nonlinear luma and chroma signals. Some DVD video playback systems include analog component video output in $Y'P_bP_r$ format, which is also correctly called Y'/B'-Y'/R'-Y' but incorrectly called $Y'C_bC_r$ (it is analog, not digital). When a technical distinction is not critical or is clear from the context, this book uses the general term *YUV* to refer to the component video signals in nonlinear color-difference format. This book also uses the general term *RGB* to refer to nonlinear R'G'B' video.

Most of these terms and concepts are explained in further detail in Chapter 3.

CHAPTER 2

The World Before DVD

A Brief History of Audio Technology

In 1877, Thomas Edison recorded and played back the words "Mary had a little lamb" on a strip of tinfoil, presaging a profound change in the way we record events. Instead of relying on written histories and oral retellings, we began to capture audible information in a way that enabled us to reproduce the events later. Since then, we have worked continuously to improve the verisimilitude of recording. See Figure 2.1 for a timeline of audio recording technology.

By the 1890s, 12-inch shellac gramophone disks that could play up to 41/2 minutes of sound at 78 rpm had become popular. The Victrola appeared in 1901. Radio followed soon after, with the first commercial broadcast in 1920. Acoustical recording (where sound vibrations were converted directly to wavy grooves on a wax disc) was replaced in the early 1920s by electrical recording (where sound vibrations were first converted to electrical impulses that could be amplified and mixed before being used to drive an electromechanical cutting head to cut grooves). Performers no longer had to cluster around a large horn that gave too much emphasis to the most powerful instruments and the loudest voices. Electrical technology also was applied to sound reproduction, resulting in the birth of the loudspeaker.

Recording technology took a major leap in 1948 when the *long-playing record* (LP) was introduced by Columbia Records. New microgroove technology allowed 25 to 30 minutes of sound to fit on a 12-inch disk turning at the slow speed of 33 1/3 rpm. Columbia's LP was developed under the direction of Peter Goldman, who later went on to develop the first commercial closed-circuit color television system. A year after the LP appeared, RCA Victor introduced a similar 7-inch disk that turned at 45 rpm and played for about 8 minutes. These two new record types quickly replaced the unwieldy 5-minute 78-rpm records.

Magnetic recording appeared in the laboratory in the 1890s but was not integrated into an actual product until about 1940. The first systems recorded onto a thin wire, which later was replaced by polyester tape with a thin coating of magnetic particles. A major advantage of storing electrical sound impulses using the alignment of magnetic particles is that the recording process is as easy as the reproduction process, and the same head can be used for recording and playback.

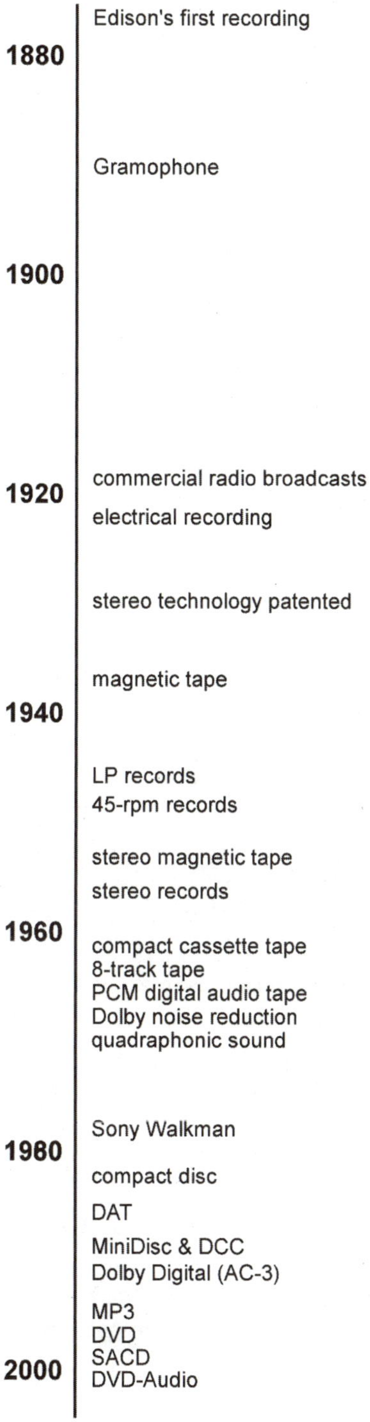

Figure 2.1
Timeline of audio
technology

1880	Edison's first recording
	Gramophone
1900	
1920	commercial radio broadcasts
	electrical recording
	stereo technology patented
	magnetic tape
1940	
	LP records
	45-rpm records
	stereo magnetic tape
	stereo records
1960	compact cassette tape
	8-track tape
	PCM digital audio tape
	Dolby noise reduction
	quadraphonic sound
	Sony Walkman
1980	compact disc
	DAT
	MiniDisc & DCC
	Dolby Digital (AC-3)
	MP3
	DVD
	SACD
2000	DVD-Audio

Up to this point, sound recording had been monophonic, meaning it was recorded as a single-point source. The first patent on stereo sound was issued in 1931, but commercial stereophonic tape systems were not developed until 1956, with stereo phonographs following in 1958. These systems added a spatial component to the sound by storing a second channel, giving the reproduction a much greater sense of realism. Stereo technology debuted to an apathetic reception; many people felt that the slight improvement was not worth the added complexity, and some claimed that stereo recording would never be practical for mainstream applications. This was soon proven shortsighted as engineers became accustomed to the new technology and artists began to take advantage of it.

The natural step beyond stereo was four-channel audio. Quadraphonic systems appeared in early 1970, but technical problems were compounded by three incompatible standards that confused and divided the marketplace and doomed the entire movement to quick extinction.

Audiocassettes using 1/8-inch tape were introduced by Philips in 1963, originally as a business dictation device that subsequently became popular for music. Prerecorded tapes were released in 1966 and steadily encroached on the sales of LPs. Eight-track (and less-successful four-track) audiotapes emerged in 1965 to enjoy a brief spurt of success mainly in automobile sound systems. Dolby noise reduction appeared in 1969 and did much to reduce the problem of tape hiss. The theories and techniques behind noise reduction were refined steadily and today are the basis of the Dolby Digital surround-sound system used by DVD. Sony attempted to launch Elcaset, a high-fidelity 1/4-inch audio cassette, in 1977, but quality was not improved sufficiently to displace the established cassette format, which became firmly entrenched after the introduction of the Sony Walkman in 1979.

By 1982, music sales on cassette surpassed vinyl. An attempt was again made to create a four-channel system; this time for cassette tapes. However, no one was able to agree on a standard, and the benefits did not seem to outweigh the technical complexity of extending the existing cassette format.

Philips again led the development of consumer electronics technology when it partnered with Sony to introduce Compact Disc Digital Audio (CD) in 1982. Digital audio recording is based on sampling, that is, converting electrical signal levels measured over 44,000 times per second into discrete numbers represented by binary digits of zero and one that are stored as pulses. This *pulse-code modulation* (PCM) technology was invented in 1937 and was applied to communications theory by C. E. Shannon in 1949. The first digital PCM audiotape recorder was developed 1967 at the NHK Technical Research Institute in Japan. PCM is used by most digital audio systems, including CD, *digital audiotape* (DAT), laserdisc, and DVD. Denon Records created the first digitally mastered audio recording for commercial release in 1974. The advantages of recording and storing sound digitally include exceptional frequency response (the rendition of frequencies covering the entire range of human hearing from 20 to 20,000 Hz), excellent dynamic range (the reproduction of very soft to very loud sounds with little or no extraneous noise), and no generational loss from copying. The improvement in quality over LP records (arguments of audio purists notwithstanding) plus other advantages, such as longer playing time, smaller size, near-instant track access, no wear from being played, and better resistance to dust and scratches, led CD sales to pass LP sales around 1988.[1]

Of course, a major drawback to audio CD is that it does not record. All attempts to produce a recordable successor to CD have failed to become significant consumer successes. DAT, Philips's *digital compact cassette tape* (DCC), Sony's MiniDisc, and other technologies have languished or died for many reasons, including high cost, incompatibility with existing standards, limited manufacturer support, and politics—which come to the fore when the ability to make perfect digital copies increases the importance of copyright protection.

Sony and Philips continued their work on the CD format, mostly for computer and multimedia applications. That part of the story is covered later in this chapter.

Many other companies are busy developing or promoting new audio storage techniques, each in the hope that its technology will become the successor to tape or CD. These include magnetic cards, solid-state electronic cards, superdense miniature discs, rewritable CDs (CD-RW), and DVD-audio. In addition, there is the MP3 audio format, which focuses less on quality and more on availability. Although DVD does little to advance the art of digital audio other than providing slightly higher quality and multi-channel surround sound, many people have high hopes for the recordable

---

[1]It may seem reasonable that a similar comparison can be made to support the prediction that DVD will replace VHS videotape. However, the analogy must be tempered because VHS does not share the many deficiences of LP records, such as breakable disks, fragile tone arms, no protection from abuse, crackles and pops from dust and scratches, a tendency to skip or repeat, and very high cost for high-fidelity sound.

version. Once available, with sufficient support, DVD audio recorders finally may bring consumer audio recording completely into the digital realm. Of course, copy protection and all its associated baggage will come along for the ride.

A Brief History of Video Technology

It is unknown when the principle of persistence of vision was first discovered. Since it can be illustrated by waving a burning stick in the darkness, it undoubtedly has been known since prehistoric times. The phenomenon, caused by the brain holding an image for a fraction of a second after the optical stimulus is removed, led to some of the first optical illusions, such as tracing letters in the air with a bright light. Much later it was discovered that a quickly flickering light appears to be continuous. At low light levels the flicker threshold—or *fusion frequency*—is around 50 times a second.[2] Even later, in the 1830s, it was learned that a series of

---

[2]Critical fusion frequency depends greatly on the ambient light level. In a dark movie theater, a flicker rate of 48 times per second is sufficient (each frame is flashed twice). European PAL television refreshes at 50 times a second (50 Hz), which is just at the limit for home viewing and is noticeable by some people, especially in well-lit environments or with bright video images. American NTSC television, at 60 Hz, is generally adequate for most environments. Many computer monitors flicker at 72 Hz or higher in order to abate eyestrain and mental fatigue in brightly lit offices.

[3]Most texts on video and motion pictures explain that persistence of vision is what makes it possible. Although there are conflicting opinions, this is apparently inaccurate—a myth that has passed through generations since Peter Mark Roget's definition in 1824. (Roget's term was misappropriated, since he was not describing perception of motion.) It is true that there is a physiologic mechanism by which the eye sees afterimages (some positive, some negative) after a light source ceases, but the scientific consensus is that it has little or nothing to do with flicker fusion or apparent motion. This has been understood since at least the beginning of the twentieth century. That humans perceive a series of changing still images as motion, without registering periods of blackness in between, is a psychologic characteristic of the brain (sometimes called the phi phenomenon), mostly unrelated to persistence of images on retinal photoreceptors or within the optic nerve. Research in high-speed filming and computer video games has shown that we perceive motion quality improvements at frame rates of 60 and 70 Hz and higher—far faster than the 24 Hz of film used for DVDs. If motion perception were based primarily on a buildup of afterimages, we would see blurry streaks or multiple images, especially when moving our eyes or head. For example, waving your hand in front of a strobe light demonstrates physiological persistence of vision: You see more than five fingers. However, a waving hand on a film or television screen produces no distortion and the hand looks normal. This is so partly because the still images of the hand are blurred by motion captured during shutter exposure time, but it is mostly due to the psychological illusion of apparent motion.

Figure 2.2
Timeline of video
technology

1820	revolving picture toys
	photographs
1840	
	color photographs
1860	
	Muybridge's photographs
1880	Nipkow's scanning disc; Eastman's celluloid film
	Kinetograph
1900	mechanical TVs
1920	talking pictures; Farnsworth's electronic TV
	Baird's video disc, Technicolor
	commercial TV broadcasts in Germany and England
1940	cable TV, professional videotape
	color added to TV
1960	Ampex helical scan, color videotape
	Cinemascope
	PAL and SECAM adopted
	Sony Betamax, JVC VHS
	Dolby Stereo added to movies, video stores
1980	laserdisc
	HiFi VCRs, stereo TV broadcasts
	S-VHS, 8 mm
	HiVision, VideoCD
	digital satellite (DBS)
	digital videotape (DV), DVD
2000	HDTV (DTV), personal video recorders (PVRs)

Figure 2.3
Revolving picture toy

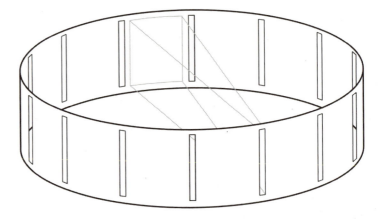

still images shown in rapid succession produces the illusion of motion (see Figure 2.2 for an outline of the history of video technology). Revolving-picture toys with strange names such as *zoetrope, phenakistiscope, thaumatrope, praxinoscope,* and *fantascope* (see Figure 2.3) began to capitalize on this particular feature of human perception, which is what makes modern motion-picture and video technology possible.[3] Early experimenters discovered that a continuous stream of images merely causes blurring as the eye attempts to track the motion. They learned that each image has to be motionless long enough for the brain to acquire it. Most of the moving-image toys used slits through which the image was momentarily viewable. Happily, the brain ignores the absence of the image. This technique of using a slit or shutter to briefly show a still picture as it moves is still used today in movie projectors.

Many years later it was discovered that the eye employs high-frequency tremors to create snapshots as part of its image-acquisition mechanism. By rapidly jerking back and forth, the eye creates multiple still images that it then passes on to the brain for processing.[4]

---

[4]Involuntary saccadic motion occurs about 10 to 25 times a second, typically over about 10 degrees of arc. Voluntary saccadic motion (such as when reading or during intense visual search) happens about 3 to 5 times a second.

Captured Light

Early motion-picture toys used drawings, which were intriguing but fell short of the real thing. The first attempts at capturing a direct visual representation of reality began about the same time, in the 1820s, with Niépce and Daguerre's photographic plates. Black and white (or sepia and white) and hand coloring were the only options for about 40 years, until Scottish physicist James Clerk Maxwell studied vision and determined that all colors could be represented with combinations of red, green, and blue. In 1861, he produced the first color photograph from a three-color process. Photography steadily improved but remained motionless for almost 60 years after its invention. In 1877, the same year Edison invented the phonograph, photographer Eadweard Muybridge captured images of a moving horse as it tripped strings attached to 12 cameras in sequence.[5] He realized that the motion could be recreated by placing the photographs on a rotating wheel and projecting light through them. This led to another string of oddly named "magic lantern" gadgets such as the *zoopraxiscope, phantasmagoria, chronophotographe,* and *zoogyroscope.* The early recording process was very cumbersome, requiring that dozens of cameras be painstakingly set up. In 1882, Étienne-Jules Marey was inspired by Muybridge's work to create a single camera, patterned after a rifle, that exposed 12 images in 1 second on a rotating glass plate. This was a great improvement, but it made for a very short viewing time and did not bode well for popcorn sales. It was not until 7 years later that the celluloid roll film developed by George Eastman—the founder of Kodak—was used by Thomas Edison's engineers to create the Kinetograph camera and the Kinetoscope viewing box for movies lasting up to 15 seconds. Lessons from the past were apparently missed by these and other pioneers, who tried to use continuous film motion only to rediscover that repeated still images were required. Edison's lead engineer, William Dickson, shot the first film in the United States, *Fred Ott's Sneeze.*

The Lumière brothers were the first to project moving photographic pictures to a paying audience in 1895. Their first film, no less inspired than Dickson's, was the spine-tingling *Workers Leaving the Lumière Factory.* The motion picture industry was born and began to steadily improve the technology and the content. Projection speeds varied from 15 to 24 frames per

---

[5]The story goes that Muybridge was hired by a former governor of California in an attempt to win a bet that all four of a horse's hooves left the ground during a gallop. Muybridge's success left the governor $25,000 richer, although he may have paid Muybridge more than that to develop the experiments.

second,[6] with 16 frames per second being the typical speed, requiring a three-bladed shutter to flash the picture 48 times per second to sufficiently reduce flicker. Films at first were silent, although phonographs were used with film projectors even before the 1900s. Many methods of adding sound were tried, some less dismal than others, and by 1927, Warner Bros. (one of the modern-day contributors to the development of DVD) and Fox had developed a practical synchronized sound technology. Within two short years, most films had soundtracks. By this time the industry had settled on a film speed of 24 frames per second, with a two-bladed shutter to flash each frame twice. After the limited success of two-color film in the 1920s, three-color film was introduced by Technicolor in 1932, although it took another 25 years before most films were shot in color. At about the same time, in an effort to improve the movie theater experience and battle the threat of television, movies were made almost twice as wide. Cinemascope was introduced in 1953 with the biblical epic *The Robe*. Some widescreen systems used more than one projector, but the most successful early widescreen systems were based on anamorphic lenses that had been invented by Dr. Henri Chretien for tank periscopes in World War I. These lenses squeezed the image sideways to fit in a standard film frame and then unsqueezed it during projection. The same technique is still used occasionally for movies and is now used by DVD for widescreen video; the squeezing and unsqueezing are done by computers rather than glass lenses.

Less successful improvements to movies were attempted, such as Smell-O-Vision in the 1960s and many variations of three-dimensional images. Another failure, Sensurround, reappeared in a new form as a niche home theater product: motion transducers that attach to the frame of your couch to add extra kick to explosions and low-frequency sounds.

More recent advances in motion picture film technology include stereo audio and 70-mm film—twice the width of standard 35-mm film. Prestige formats such as IMAX and OmniMax use 70-mm film with large square frames to achieve presentations of breathtaking impact.

In 1976, Dolby Laboratories made an affordable surround-sound system using standard film soundtracks, and in 1992, it released multichannel Dolby Digital (AC-3) for the movie *Batman Returns*. In 1993, *Jurassic Park* debuted the sonic realism of *Digital Theater Systems* (DTS) Digital Sound,

---

[6]Since there were no firm standards for film speed in cameras or projectors, films came with a cue sheet telling the projectionist the speed, in feet per second, at which to run the film. Unscrupulous theater owners would speed the films up to squeeze in more showings, even to the point of cutting the running time in half!

which uses an optical synchronization track on the film coupled with multi-channel digital audio from a CD. The primary multichannel audio system of DVD-Video is Dolby Digital. DTS optionally can be placed on the disc in addition to the primary Dolby Digital or PCM audio track. In 1999, Dolby and Lucasfilm THX added a new rear center channel option to Dolby Digital. The new Surround EX format first appeared on *Star Wars: Episode 1— The Phantom Menace.* Not to be outdone, DTS came out with an *Extended Surround* (DTS-ES) rear center channel decoder that could decode Surround EX. DTS also released a new discrete 6.1-channel format, which provided a separate center surround channel instead of relying on matrix encoding.

Movie production has now entered the era where entire scenes or entire movies are generated inside a computer, never being touched by light to trace the images in silver halide crystals on a film negative. Pixar's *A Bug's Life* was the first motion picture feature to be transferred directly to DVD without being printed to film. The resulting picture quality was astounding, especially on a progressive-scan digital playback system such as a computer. *A Bug's Life* also pioneered the concept of reframing scenes and repositioning characters and objects to create a full-screen 4:3 version of the film. Digital movie projection, or d-cinema, has been surprisingly successful in early tests, many of which used DVD-ROM to distribute the electronic movie files to theaters around the world. As the waves of digital convergence lap higher onto the shores of Hollywood, the digital nature of DVD will become increasingly important for best-quality distribution of movies.

Dancing Electrons

About the same time as Muybridge was stretching strings across racetracks, scientists around the world were envisioning television. Once again, the new ideas were christened with fanciful names derived from familiar technology such as *radioscope, phonoscope, cinematophone, radiovision, telephone eye,* and the jaw-breaking *chronophotographoscope.* The word *television* is credited to science writer Hugo Gernsback,[7] who published popular science journals beginning in 1908.

Many early television ideas were developed almost simultaneously around the world and involved mechanical systems that transmitted each

---

[7]Hugo Gernsback may be familiar to science fiction fans as the eponym of the Hugo Award. His name also may be familiar to readers of Popular Electronics magazine, produced by Gernsback Publications.

individual picture element on a separate connection. These were dropped in favor of rapidly scanning the image one line at a time. In 1884, Paul Nipkow applied for a patent in Germany on his image-scanning design that used a rotating disc with holes arranged in a spiral, essentially an updated version of the phenakistoscope. Television technology improved slowly, still based on clumsy mechanical designs, until the first experimental broadcasts were made in the 1920s. Meanwhile, a 14-year-old Utah farm boy named Philo Farnsworth was looking at plowed furrows and dreaming of deflecting beams of electrons in similar rows. Six years later, on January 7, 1927, he submitted the patent applications that established him as the inventor of the all-electronic television. The honor of creating the first working electronic television system also belongs to Kenjito Takayanagi of Tokyo, who transmitted an image of Japanese writing to a cathode-ray tube on Christmas Day, 1926.

The first nonexperimental television broadcasts began in 1935 in Germany and included coverage of the 1936 Olympics in Berlin. The German system had only 180 lines. In Britain, John Logie Baird steadily improved a mechanically scanned television system from 50 lines to 240, and the *British Broadcasting Company* (BBC) used it for experimental broadcasts. Improved "high definition" commercial broadcasts using a 405-line system were begun by the BBC in Britain in 1936, although they were shut down 3 years later by World War II (with a sign-off broadcast of a Mickey Mouse cartoon) and did not return until 1946 (with a repeat of the same cartoon). In the United States, the *National Broadcasting Company* (NBC) demonstrated television to entranced crowds at the 1939 World's Fair, and RCA's David Sarnoff declared that 10,000 sets would be sold before the end of the year. However, at a price equivalent to more than $2500 and with sparse programming from scattered transmitters, only a few thousand sets were sold. NBC began commercial broadcasts on July 1, 1941. *Columbia Broadcasting System* (CBS) also switched from experimental to commercial broadcasts at about the same time.

About 10,000 televisions were scattered around the United States in 1946, and when World War II production restrictions were lifted, another 10,000 were soon sold. Three years later, there were over 1 million sets, and after only 2 more years, in 1951, there were over 10 million. In 1950, the *Federal Communications Commission* (FCC) selected a new "field sequential" color system from CBS that used a three-color wheel spinning inside each TV. This was chosen over RCA's fully electronic version, which didn't have sufficient quality at the time. Luckily circumstances and the Korean War delayed implementation long enough that CBS abandoned its mechanical system as RCA improved its electronic version. The official *National*

Television Standards Committee (NTSC) color television system was implemented in 1954, and widespread purchase of color television sets began in 1964. The NTSC system piggybacked color information onto the black-and-white signal so that the millions of existing sets would still work with new broadcasts. NTSC was adopted by Japan in 1960.

Meanwhile, television in Great Britain after World War II experienced the same rapid growth as in the United States. In 1967, when it came time to implement color television, Britain adopted the PAL system that had been developed in Germany. PAL was technologically superior to NTSC, but it was incompatible with existing British television sets and had to be phased in gradually over many years. In 1967, France and the Soviet Union settled on the SECAM system. In 1968, the number of televisions passed 78 million in the United States and 200 million worldwide. By 1972, half of all American homes had a television set. At the end of the 1970s, almost every country had adopted one of the three standards, as television began to saturate the planet. NTSC is still the most widely used system, PAL is common in most of Europe, and SECAM holds a distant third.

As television was gaining a foothold, *Community Antenna TV* (CATV, the first version of cable television) was founded in rural Pennsylvania by a shopkeeper who built an antenna on a hill to improve reception in his store and then began connecting up the receivers of his customers and neighbors. From its beginnings in 1949, cable TV's promise of more channels and better reception has attracted millions of customers, and in just under 40 years, over half the television homes in the United States were connected to cable.

The first television signal was beamed via satellite across the Atlantic Ocean in 1962, and in 1975, satellites were put in use to distribute programming to cable TV centers. Satellite dishes eventually were made available to consumers and ironically became a competitor to cable TV, especially with the advent of *direct broadcast satellite* (DBS) television, which uses the same digital video compression system as DVD.

Metal Tape and Plastic Discs

Visual information is hundreds of times more dense than audio information. Sound waves are simple vibrations that the human ear can detect from about 20 per second to over 20,000 per second (20 Hz to 20 kHz). Visible images, on the other hand, are formed from complex light waves that the eye can detect within the range of 430 to 750 trillion Hz. Clearly, recording a video signal is a much bigger challenge than recording audio.

For magnetic tape, raising the speed at which the tape moves past the head increases the amount of information that can be recorded in a given period of time, but the amount of tape required quickly becomes preposterous. Nevertheless, in 1951, Bing Crosby Enterprises demonstrated a commercial videotape recorder that used tape traveling at 100 inches per second (almost 6 mph) across 12 heads. A few similarly cumbersome and very expensive systems were developed for television studios, but their high price tags barred them from wider use. The "eureka factor" did not arrive for another 10 years until engineers at Ampex hit on the idea of moving the head past the tape in addition to moving the tape past the head. A revolving head can record nearly vertical stripes on a slow-moving tape, greatly increasing the recording efficiency (see Figure 2.4). The first helical-scan videotape recorder appeared in 1961, and color versions followed within a few years.

Videotape recording was first used only by professional television studios. A new era for home video was ushered in when Philips and Sony produced black-and-white, reel-to-reel videotape recorders in 1965, but at $3000, they did not grace many living rooms. Sony's professional 1/4-inch U-matic videocassette tape appeared in 1972, the same year as the video game Pong. More affordable color video recording reached home consumers in 1975 when Sony introduced the Betamax videotape recorder. The following year JVC introduced VHS, which was slightly inferior to Betamax but won the battle of the VCRs because of extensive licensing agreements with equipment manufacturers and video distributors. The first Betamax VCR cost $2300 (over $6700 adjusted for inflation), and a 1-hour blank tape cost $16 (over $46 in today's dollars). The first VHS deck was cheaper at $885 (equivalent to more than $2500 today), and Sony quickly introduced a new Betamax model for $1300, but both systems were out of the financial reach of most consumers for the next few years.

Figure 2.4
Tape head comparison

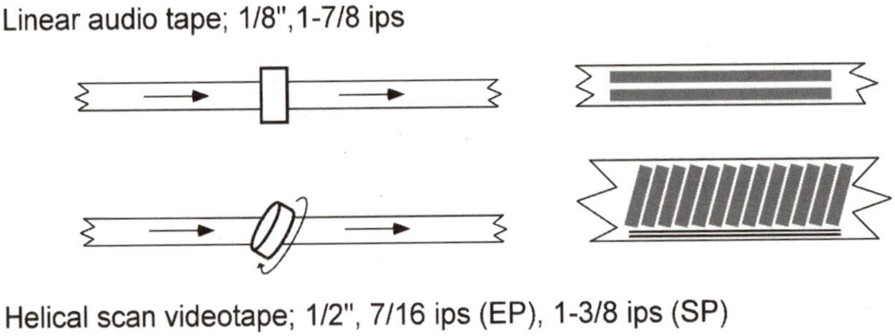

Linear audio tape; 1/8",1-7/8 ips

Helical scan videotape; 1/2", 7/16 ips (EP), 1-3/8 ips (SP)

MCA/Universal and Disney studios sued Sony in 1976 in an attempt to prevent home copying with VCRs. Eight years later, the courts upheld consumer recording rights by declaring Sony the winner. Studios are fighting the same battle today with DVD but have switched to trying to change the technology and the law.

In 1978, Philips and Pioneer introduced videodiscs, which actually had been demonstrated in rudimentary form in 1928 by John Logie Baird. The technology had improved in 50 years, replacing wax discs with polymer discs and delivering an exceptional analog color picture by using a laser to read information from the disc. The technology got a big boost from General Motors, which bought 11,000 players in 1979 to use for demonstrating cars. Videodisc systems became available to the home market in 1979 when MCA joined the laserdisc camp with its DiscoVision brand. A second videodisc technology, called *capacitance electronic disc* (CED), introduced just over 2 years later by RCA, used a diamond stylus that came in direct contact with the disc—as with a vinyl record. CED went by the brand name SelectaVision and was more successful initially, but eventually failed because of its technical flaws. CED was abandoned in 1984 after less than 750,000 units were sold, leaving laserdisc to overcome the resulting stigma. JVC and Matsushita developed a similar technology called *video high density* or *video home disc* (VHD), which used a grooveless disc read by a floating stylus. VHD limped along for years in Japan and was marketed briefly by Thorn EMI in Great Britain, but it never achieved significant success.

For years, laserdisc was the Mark Twain of video technology, with many exaggerated reports of its demise as customers and the media confused it with the defunct CED. Adding to the confusion was the addition of digital audio to laserdisc, which in countries using the PAL television system required the relaunch of a new, incompatible version. Laserdisc persevered, but because of its lack of recording, high price of discs and players, and inability to show a movie without breaks (a laserdisc cannot hold more than 1 hour per side), it was never more than a niche success catering to videophiles and penetrating less than 2 percent of the consumer market in most countries. Laserdisc systems did achieve modest success in education and training, especially after Pioneer's bar-code system became a popular standard in 1987 and enabled printed material to be correlated to random-access visuals. In some Asian countries, laserdisc achieved as much as 50 percent penetration almost entirely because of its karaoke features.

Despite the exceptional picture quality of both laserdisc and CED, they were quickly overwhelmed by the eruption of VHS VCRs in the late 1970s, which started out at twice the price of laserdisc players but quickly dropped below them. The first video rental store in the United States opened in

1977, and the number grew rapidly to 27,000 in the late 1990s. Direct sales to customers, known as *sellthrough,* began in 1980. Today, the home video market is a $16 billion business in the United States alone. Over 87 percent of households have at least one VCR, creating a market of over a hundred million customers who, in 1996, rented about 3 billion tapes and bought over 580 million tapes. There are now over 25,000 VHS titles available in the United States, compared with 9,000 on laserdisc.

In the late 1980s, a new video recording format based on 8-mm tape with metal particles was introduced. The reduced size and improved quality were not sufficient to displace the well-established VHS format, but 8-mm and Hi-8 tapes were quite successful in the camcorder market, where smaller size is more significant.

Around this time, minor improvements were made to television, with stereo sound added in the United States in 1985 and, later, closed captions.

In 1987, JVC introduced an improved Super VHS system called *S-VHS.* Despite being compatible with VHS and almost doubling the picture quality, S-VHS was never much of a success because there were too many barriers to customer acceptance. Players and tapes were much more expensive, VHS tapes worked in S-VHS players but not vice versa. A special cable and an expensive TV with an s-video connector were required to take advantage of the improved picture, and S-VHS was not a step toward *high-definition television* (HDTV), which was receiving lavish attention in the late 1980s and was expected to appear shortly. It is interesting to note that DVD has many of the same strikes against it, even the specter of HDTV again. The major contribution of S-VHS to the industry was the popularization of the s-video connector, which carried better-quality signals than the standard composite format. S-video connectors are still mistakenly referred to as S-VHS connectors.

The Digital Face-Lift

As television began to show its age, new treatments appeared in an attempt to remove the wrinkles. Major European broadcasters rolled out PALplus in 1994 as a stab at *enhanced-definition television* (EDTV) that maintains compatibility with existing receivers and transmitters. PALplus achieves a widescreen picture by using a letterboxed[8] image for display on conven-

---

[8]Letterboxing is the technique of preserving a wide picture on a less wide display by covering the gap at the top and bottom with black bars. Aspect ratios and letterboxing are covered in Chapter 3.

tional TVs and hiding helper lines in the black bars so that a widescreen PALplus TV can display the full picture with extended vertical resolution.

Other ways of giving television a face-lift include *improved-definition televisions* (IDTV) that double the picture display rate or use digital signal processing to remove noise and to improve picture clarity.

None of these measures are more than stopgaps while we wait for the old workhorse to be replaced by *high-definition television* (HDTV). HDTV was first demonstrated in the United States in 1981, and the process of revamping the "boob tube" reached critical mass in 1987 when 58 broadcasting organizations and companies filed a petition with the FCC to explore advanced television technologies. The Advanced Television Systems Committee (ATSC) was formed and began oozing toward consensus as 25 proposed systems were evaluated extensively and either combined or eliminated.

In Japan, a similar process was underway but was unfettered by red tape. An HDTV system called *HiVision,* based on MUSE compression, was developed quickly and was put into use in 1991.[9] Ironically, the quick deployment of the Japanese system was its downfall. The MUSE system was based on the affordable analog transmission technology of its day, but soon the cost of digital technology plummeted as its capability skyrocketed. Back on the other side of the Pacific, the lethargic U.S. HDTV standards-making process was still in motion while video technology graduated from analog to digital. In 1993, the ATSC recommended that the new television system be digital. The Japanese government and the Japanese consumer electronics companies—which would be making most of the high-definition television sets for use in the United States—decided to wait for the American digital television standard and adopt it for use in Japan as well. A happy by-product of HiVision is that the technology-loving early Japanese adopters served as guinea pigs, supporting the development of high-resolution widescreen technology and paving the way for HDTV elsewhere. A European HDTV system, HD-MAC, was even more short-lived; it was demonstrated at the Albertville Olympics but was abandoned shortly thereafter.

In 1992, the ATSC outlined a set of proposed industry actions for documenting a standard. In 1993, the three groups that had developed the four final digital systems—AT&T and Zenith; General Instrument and MIT; and

---

[9]The HiVision system uses 1125 video scan lines. In the charmingly quirky style of Japanese marketing, HiVision was introduced on November 25 so that the date (11/25) corresponded with the number of scan lines.

Philips, Thomson, and the David Sarnoff Research Center—formed a "grand alliance" and agreed to merge their best features into a final standard. The ATSC made its proposal for a *digital television* (DTV) standard to the FCC in November 1995. The motion-picture industry and the computer industry had been aware of the proceedings but apparently had not bothered to become sufficiently involved. At the eleventh hour they both began to complain loudly that they were not happy with the choice of widescreen aspect ratio or the computer-unfriendly parts of the proposal. In December of 1996, with this opposition in mind, the FCC approved all but the video format constraints (aspect ratios, resolutions, and frame rates) of the ATSC DTV proposal, putting these aside as a voluntary standard. This "grand compromise" freed broadcasters to begin implementation but wisely left the standard open for additional video formats. HDTV, originally promised for the early 1990s, finally concluded its long gestation period in December 1998, only to begin an even longer battle for ascendancy.

In the meantime, numerous new formats have been developed to store video digitally and convert it to a standard analog television signal for display. As with early videotape, the elephantine size of video is a problem. Uncompressed standard-definition digital television video requires at least 124 *million bits per second* (Mbps).[10] Compare this with hard drives, which run at around 25 to 100 Mbps; high-speed T-1 digital telephone lines, which run at 1.5 Mbps, and audio CD, which runs at a meager 1.4 Mbps. Obviously, some form of compression is needed. Many proprietary and incompatible systems have been developed, including Intel's DVI in 1988 (which had been developed earlier by the Sarnoff Institute). That same year, the *Moving Picture Experts Group* (MPEG) committee was created by Leonardo Chairiglione and Hiroshi Yasuda with the intent to standardize video and audio for CDs. In 1992, the International Organization for Standardization (ISO) and the *International Electrotechnical Commission* (IEC) adopted the standard known as MPEG-1. Audio and video encoded by this method could be squeezed down to fit the limited data rate of the single-speed CD format. CD-i first used MPEG-1 video playback in 1992 to achieve near-VHS quality. The CD-i MPEG-1 Digital Video format was used as the basis for Karaoke CD, which became Video CD, a precursor to DVD. Video CD has done quite well in markets where VCRs were not already established. At

---

[10]124 Mbps is the data rate of active video at ITU-R BT.601 4:2:0 sampling with 8 bits, which provides an average of 12 bits per pixel ($720 \times 480 \times 30 \times 12$ or $720 \times 576 \times 25 \times 12$). The commonly seen 270-Mbps figure comes from studio-format video that uses 4:2:2 sampling at 10 bits and includes blanking intervals. At higher 4:4:4 10-bit sampling, the data rate is 405 Mbps.

the beginning of 2000, there were about 40 million Video CD players in Asia, but Video CD has not fared as well in Europe, and it may qualify as an endangered species in the United States. MPEG-1 is also commonly used for video and audio on *personal computers* (PCs) and over the Internet. The notorious MP3 format is a nickname for MPEG-1 Layer III audio.

The MPEG committee extended and improved its system to handle high-quality audio/video at higher data rates. MPEG-2 was adopted as an international standard in 1994 and is used by many new digital video systems, including the ATSC's DTV. *Direct broadcast satellite* (DBS), with convenient 18-inch dishes and digital video, also appeared in 1994. With sales of over 3 million units by the end of 1996, DBS was the most successful home entertainment product until DVD came along. DBS was introduced just as MPEG-2 was being finalized. Consequently, most early DBS systems used MPEG-1 but have since converted to MPEG-2. Digital videocassette tape (DV) appeared in 1996. With near-studio quality, it is aimed at the professional and "prosumer" market and is priced accordingly. New competitors to cable and satellite TV are also based on MPEG-2; these include video dialtone (video delivered over phone lines by the phone company), digital cable, and wireless cable (terrestrial microwave video transmission). An updated version of Video CD, called Super Video CD, uses MPEG-2 for better quality. MPEG-2 is also the basis of DVD-Video, augmented with the Dolby Digital (AC-3) multichannel audio system—developed as part of the original work of the ATSC. DVD-Video is intended for the home video market, where it provides the highest resolution yet from a consumer format.

A Brief History of Data Storage Technology

Rewind almost 200 years. In 1801, Joseph-Marie Jacquard devised an ingenious method for weaving complex patterns using a loom controlled by punched metal cards. The same idea was borrowed over 30 years later by Charles Babbage as the storage device for his mechanical computer, the Analytical Engine.[11] Ninety years after Jacquard, Herman Hollerith used a

---

[11]Unlike Jacquard, whose system enjoyed widespread success, Babbage seemed incapable of finishing anything he started. He never completed any of his mechanical calculating devices, although his designs were later proven correct when they were turned into functioning models by other builders.

similar system of punched cards for tabulating the U.S. Census after getting the idea from watching a train conductor punch tickets. Fifty years later, the first electronic computers of the 1940s also employed punched cards and punched tape—using the difference between a surface and a hole to store information. The same concept is used in modern optical storage technology. Viewed through an electron microscope, the pits and lands of a CD or DVD would be immediately recognizable to Jacquard or Hollerith (see Figure 2.5). However, the immensity and variety of information stored on these miniature pockmarked landscapes would truly amaze them. Hollerith's cards held only 80 characters of information and were read at a glacial few per second.[12]

The other primary method of data storage—magnetic media—was developed for the UNIVAC I in 1951. Magnetic tape improved on the storage density of cards and paper tape by a factor of about 50 and could be read significantly faster. Magnetic disks and drums appeared a few years later and improved on magnetic tape, but cards and punched tape were still much cheaper and remained the primary form of data input until the late 1970s. In the 1970s, flexible floppy disks appeared—first unwieldy 8-inch behemoths, then smaller 5.25-inch disks, and finally, in 1983, 3.5-inch diskettes small enough to fit in a shirt pocket. Each successive version held two or three times more data than its predecessor despite being smaller.

Optical media languished during the heyday of magnetic disks, never achieving commercial success other than for analog video storage (i.e., laser videodiscs). It was not until the development of compact disc digital audio in the 1980s that optical media again proved its worth in the world of bits and bytes, setting the stage for DVD (see Figure 2.6).

Other innovations of the 1980s included removable hard disk cartridges, high-density floppy disks from IOmega based on the Bernoulli principle, and erasable optical media based on *magneto-optical* (MO) technology.

Figure 2.5
Punched card and optical disc

---

[12]Modern card readers of the 1970s still used only 72 to 80 characters per card but could read over 1000 cards per second.

Magneto-optical discs use a laser to heat a polyphase crystalline material that can then be aligned by a magnetic field. Features of MO technology were later adapted for DVD-RAM.

Innovations of CD

Sony and Philips reinvigorated optical storage technology when they introduced Compact Disc Digital Audio in 1982. This was known as the "Red Book standard" because of the red covers on the book of technical specifications. The first CD players cost around $1000 (over $1700 in 2000 dollars).

Three years later, a variation for storing digital computer data—CD-ROM—was introduced in a book with yellow covers. CD-ROM did not take hold immediately, especially since the first drives cost over $1000. But many people understood the seeds of technological revolution carried in CD-ROM. The first Microsoft CD-ROM conference in March of 1986 had a sell-out crowd of a thousand people, and by 1992, the conference had metamorphosed into the Intermedia trade show and was attended by hundreds of thousands. As CD-ROM entered the mainstream, original limitations such as slow data rates and glacial access times were overcome with higher spin rates, bigger buffers, and improved hardware. CD-ROM became the

Figure 2.6

Capacities and costs of data storage media

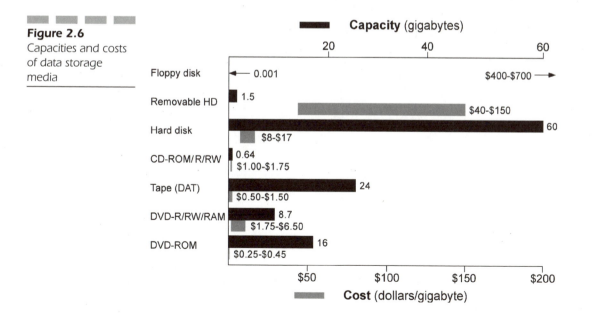

preeminent instrument of multimedia and the standard for delivery of software and data in general (see Figure 2.7).

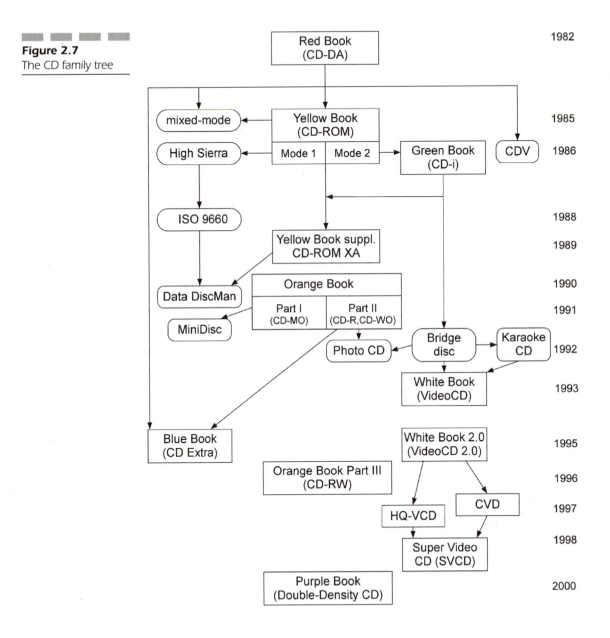

Figure 2.7
The CD family tree

A clever variation on audio CDs called *CD+G* was developed around 1986. A CD+G disc uses six previously unused bits in each audio sector (subcode channels R through W) to hold graphic bitmaps. The player collects snippets of a graphic from each sector as it plays the audio, and after a few thousand sectors, it has accumulated enough to display a picture. CD+G was never widely supported and is available mostly on karaoke CD players and CD-i players.

CD-Video (CDV) was developed in 1986 as a hybrid of laserdisc and CD. Part of the disc contained 5 to 6 minutes of standard audio tracks and could be played on a CD player, whereas the other part contained about 20 minutes of analog video and digital audio in laserdisc format. The standard was presented in a blue book but is not considered part of the canon of colors. The CDV format has mostly disappeared.

After introducing CD-ROM, Sony and Philips continued to refine and expand the CD family. In 1986, they produced *Compact Disc Interactive* (CD-i, the "Green Book standard"), intended to become the standard for interactive home entertainment. CD-i incorporated specialized file formats and custom hardware with the OS-9 operating system running on the Motorola 68000 microprocessor. Unfortunately, CD-i was obsolete before it was finished, and the few supporting companies dropped out early, leaving Philips to stubbornly champion it alone. After consuming over a billion dollars in development and scattershot marketing, CD-i is still limping along as a niche product. Philips, not one to give up easily, announced it would develop a DVD player that could accept a CD-i add-in card, but such a beast has yet to appear.

Some early CD-ROM producers developed their own file systems to organize the data on the disc, requiring a specific device driver. This forced pioneering users to either reboot and load a different driver or use multiple CD-ROM drives, one for each title. In the face of proliferating proprietary CD-ROM file formats, a group of industry representatives met in 1986 at the High Sierra Hotel and Casino near Lake Tahoe, Nevada, and proposed a common structure that became known as the "High Sierra format." In 1988, ISO adopted this format, with a few modifications, as the ISO 9660 CD-ROM file interchange standard. Unfortunately, it was a lowest-common-denominator solution designed to support MS-DOS, so it did not properly support other systems such as the Apple Macintosh *Hierarchical File System* (HFS) and many UNIX systems. These operating systems could read ISO 9660 CDs but had to add their own incompatible extensions to make full use of the advantages of their native file systems. When Microsoft moved beyond DOS, it also rolled its own "Joliet" extensions to ISO 9660 for long filenames with multilingual characters.

In 1989, the CD-ROM "Yellow Book standard" was augmented with an updated format called *CD-ROM XA* (eXtended Architecture). XA was based on ideas developed for CD-i, including the interleaving of data, graphics, and ADPCM compressed audio. CD-ROM XA required newer hardware to read the mixed sector types on the disc and to take advantage of the interleaved streams. Most CD-ROM drives are now XA-compatible, but the interleaving features are almost never used. CD-ROM XA also introduced the so-called bridge disc, which is a type of disc with extra information that can be used in a CD-i player or a computer with a CD-ROM XA drive.

The "Orange Book standard" was developed in 1990 to support *magneto-optical* (MO) and *write-once* (WO) technology. MO has the advantage of being rewritable but is incompatible with standard CD drives and has remained expensive. CD-WO has been largely superseded by "Orange Book Part II," *recordable CD* (CD-R). CD-R can only be written to once, which is sufficient for many uses. The Orange Book also added multisession capabilities to allow CDs to be written to in chunks across different recording sessions. Recordable technology revolutionized CD-ROM production by enabling developers to create fully functional CDs for testing and for submission to disc replicators. As prices have dropped, CD-R also has become popular for business and even personal archiving. Although tape backup systems are cheaper, the familiarity of CD and the widespread availability of CD-ROM drives have made CD-R very appealing. Tape also lacks the quick random access of CD-R.

Around 1990, music CD singles became popular in Japan, and the format subsequently was adopted for use with Sony's Data Discman. The 8-centimeter discs had a capacity of 200 megabytes and commonly held modified CD-ROM XA applications. The Data Discman and its cousin the Bookman had piddling success outside Japan. Sony also developed the portable MMCD player based on an Intel 286 processor and an LCD display, which it concentrated on to the exclusion of CD-i. However, as with CD-i, heavy marketing failed to make the MMCD a success, and Sony finally gave up around 1994. The name was later reused for Sony's proto-DVD format.

Kodak and Philips developed the Photo CD format in 1992, based on the CD-ROM XA and Orange Book standards. Photo CDs are bridge discs that hold up to one hundred 35-mm photographs written in one or more sessions. Special Photo CD players and CD-i players can show the pictures on a television, but the concept never appealed much to home photographers. Nevertheless, since multisession CD-ROM XA drives can read Photo CDs, the format is popular for professional photo production work and for stock photography libraries.

In 1991, Sony introduced the 8-cm MiniDisc, a portable music format based on the rewritable Orange Book's MO standard and using Sony's

ATRAC compression. The poor audio quality of MiniDisc earned it a bad reputation, but it has since been improved considerably and is now in over 10 percent of Japanese households, accounting for 30 percent of prerecorded music sales in Japan. MiniDisc had less success in other countries, and even the recent resurgence of support may be too late to save it if a recordable 8-cm DVD-Audio standard is developed.

MPEG-1 video from a CD was demonstrated by Philips and Sony on their CD-i system in 1991. About 50 movies were released in CD-i Digital Video format before a standardized version appeared as Video CD (VCD) in 1993, based on proposals by JVC, Sony, and Philips and documented as the "White Book specification." Not to be confused with *CD-Video* (CDV), Video CD uses MPEG-1 compression to store 74 minutes of near-VHS-quality audio and video on a CD-ROM XA bridge disc. Philips developed a $200 Video CD add-on for its CD-i players. Hardware and software were made available for computers to play Video CDs. Video CD 2.0 was finalized in 1994. The format has done very well in countries where VHS was not well established. Over 6 million Video CD players were sold in China alone in 1996.

By 1994, drive speeds—and resulting data transfer rates—had risen to quadruple speed. Once again, the price for the new drives was around $1000. Two years later, 8X CD-ROM drives appeared, but the introductory price was down to $400. Later the same year, 12X drives appeared for around $250. Tests began to show that in many cases data transfer rates were not much better than those of 6X drives because the high speeds increased errors and required repeated reads.

After CD-ROM–based multimedia reached mass-market potential, music artists began to look for ways to add multimedia to their music. The idea was to create a CD that would play in a music player but also include software for use on a computer. The various techniques are grouped under the moniker of *Enhanced CD*. The first attempts—in the late 1980s—used mixed-mode CDs that put computer data on the first track and music on the remaining tracks. The downside was that some CD players attempted to play the computer data as if it were music, producing noise that could damage speakers. Stickers on CDs warning of potential equipment damage were not felt to be a good way to increase sales, so more ingenious methods were pursued. A trick of placing the computer data at index zero (in the pregap area before the beginning of the music) worked well with most players, which begin playing music at index one. It was still possible to use the rewind feature on some players to back up into the data, so clever implementations included an audio message warning the listener to stop the player before reaching the noise. Since CD-ROM drives begin at index zero, they are able to read the computer data. That is, they were until Microsoft released a new SCSI CD-ROM driver that began reading at index 1, thus skipping over the data intended for the computer.

By this time, Sony and Philips, with the help of Apple and Microsoft, were working on a more foolproof solution. The answer was stamped multisession, which uses a feature of the Orange Book format—originally designed for recordable discs—to put one session of Red Book audio data on the CD followed by a session of Yellow Book computer data. Combining red, yellow, and orange normally produces deep orange, but in this case blue covers were chosen for the book that documented the new CD Plus standard, released in 1995, which was almost immediately renamed to CD Extra because of threatened lawsuits by the owner of the CD Plus trademark. The CD Extra format is completely compatible with audio CD players but does not work on all computers. Many CD-ROM drives are multisession-capable, but only about 50 percent of multimedia PCs have the requisite multisession software drivers. Fortunately, computers are easier to upgrade than CD players, and sales of Enhanced CDs are expected to increase tenfold to over $750 million by the year 2001. Since the DVD-Audio specification is built on top of the DVD-ROM standard, DVDs with both audio and computer data will be easy to produce and will be compatible with all DVD-Audio players and DVD computers.[13]

Erasable CD, part III of the "Orange Book standard," did not appear until 1997. The standard was finalized in late 1996, and the official name of *Compact Disc Rewritable* (CD-RW) was chosen in hopes of avoiding the disturbing connotations of "erasable." Unfortunately, CD-RW requires new drive hardware to read it. None of more than 700 million existing CD-ROM drives and CD players are able to read CD-RW discs, although it has succeeded well in the prolonged absence of recordable DVD.

An erasable CD format called *THOR* actually was announced by Tandy in 1988, eight years before CD-RW, but this ambitious technology—with its familiar-sounding promise of storing music, video, and computer data—was plagued by technical problems and never made it out of the laboratory. Tandy later joined with Microsoft to create the CD-based *Video Information System* (VIS), which also was mercifully short-lived.

Another notable failure was 1991's *Commodore Dynamic Total Vision* (CDTV), a consumer multimedia console along the lines of CD-i. CDTV employed a CD-ROM–equipped Amiga computer that was hooked up to a TV. Proprietary standards and the decline of Commodore's business doomed CDTV to museum shelves.

Around 1999, the prices of CD-R/RW drives and blank CD-R discs became cheap enough that they began being widely used for recording cus-

---

[13]That's the theory, at least. In practice, sloppy engineering and cheap players result in problems with discs that contain anything other than video or audio. See the compatibility sections of Chapters 4 and 7 for the sordid details.

tom music CDs. This, plus the proliferation of MP3 music files across the Internet, drove a worldwide demand for 1.3 billion CD-Rs in 1999, with an estimated 30 to 40 percent being used for music.

One of the latest variations on the CD theme is from DataPlay, which in 2001 will introduce miniature optical discs and players intended to compete with flash memory devices.

The Long Gestation of DVD

Shortly after the introduction of CDs and CD-ROMs, prototypes of their eventual replacement began to be developed. Systems using blue lasers achieved four times the storage capacity, but these were based on large, expensive gas lasers, since cheap semiconductor lasers at that wavelength were not available. In 1993, ten years after the worldwide introduction of CDs, the prototypes began to approach reality. Nimbus first demonstrated a double-density CD proof of concept in January of 1993. Optical Disc Corporation followed suit in October. Philips and Sony announced that they had a similar project underway, and later Toshiba claimed it had been working on something similar since 1993. Some of these first attempts simply used CD technology with smaller pits to create discs that could hold twice as much data. Although this was far beyond original CD tolerances, the optics of most drives were good enough to read the discs, and it would even be possible to connect the digital output of a regular CD player to an MPEG video decoder.[14] However, it was recognized that CD technology tended to falter when pushed too far. Philips reportedly put its corporate foot down and said it would not allow CD patent licensees to market the technology because it could not guarantee compatibility and a good user experience. When all things were considered, it seemed better to develop a new system (see Figure 2.8).

Hollywood Weighs In

The stage was set in September 1994 by seven international entertainment and content providers, Columbia Pictures (Sony), Disney, MCA/Universal

---

[14]Pioneer commercialized its Alpha system for karaoke in Japan using higher-density CDs. Nimbus created the Presto prototype video player using MPEG-2 video on CDs.

Figure 2.8
Timeline of DVD
development

	Hollywood proposal
1995	Sony/Philps MMCD proposal
	SD Alliance proposal
	computer industry objectives announced
	UDF recommended
	reconciliation announced
1996	combined DVD standard announced
	product announced for fall 1996
	Digital Recording Act attempted
	copy protection agreement (Macrovision) announced
	DVD-ROM and DVD-Video specification version 1.0 published
1997	copy protection agreement (CSS) announced
	players appear in Japan
	drives appear in Japan
	players appear in U.S.
	movie titles appear in U.S.
1998	Warner takes U.S. movie title distribution nationwide
	DVD-R 1.0 drives appear
	DVD consortium changes to DVD Forum, opens to all companies
	first hack for copying DVD on PCs
	Dolby Digital included on 625/50 (PAL) format; *DVD Demystified* published
	first Macintosh DVD upgrade kit
	soft launch of DVD in Europe
	Divx trial program begins; DVD-Audio 0.9 spec published
	DVD-RAM 1.0 drives appear
	Divx goes nationwide; DMCA becomes law
1999	first DTS DVDs ship
	DVI final draft released
	SACD players released in Japan
	Divx shuts down; DVDA formed
	DVD-R 1.9 drives appear
	DVD-RAM 2.0 drives appear
	progressive-scan players
	DVD wins an Emmy; DeCSS released; official DVD launch in Europe
	DVD becomes best-selling CE product; first DVD-18 ships
	DVD-Audio players and DVD-RW recorders appear in Japan
2000	Sony Playstation 2 released in Japan
	DVD-Audio players appear in U.S.
	MPAA wins DeCSS trial in New York
	10 millionth DVD player sold in U.S.
	Sony Playstation 2 released in U.S.

(Matsushita), MGM/UA, Paramount, Viacom, and Warner Bros. (Time Warner), who called for a single worldwide standard for the new generation of digital video on optical media. These studios formed the Hollywood Digital Video Disc Advisory Group and requested the following:

- Room for a full-length feature film, about 135 minutes, on one side of a single disc
- Picture quality superior to high-end consumer video systems such as laserdisc
- Compatibility with matrixed surround and other high-quality audio systems
- Ability to accommodate three to five languages on one disc
- Copy protection
- Multiple aspect ratios for wide-screen support
- Multiple versions of a program on one disc, with parental lockout

Preparations and proposals began, but two incompatible camps soon formed. Like antagonists in some strange mechanistic mating ritual, each side boasted of its prowess and attempted to line up the most backers. At stake was the billion-dollar home video industry as well as millions of dollars in patent licensing revenue.

Dissension in the Ranks

On December 16, 1994, Sony and its partner Philips independently announced their own standard: a single-sided, 3.7-billion-byte *Multimedia CD* (MMCD, renamed from HDCD, which it turned out was already taken). The remaining cast of characters jointly proposed a different standard just over a month later on January 24, 1995. Their *Super Disc* (SD) standard was based on a double-sided design holding 5 billion bytes per side.

The SD Alliance was led by 7 companies—Hitachi, Matsushita (Panasonic), Mitsubishi, Victor (JVC), Pioneer, Thomson (RCA/GE), and Toshiba (business partner of Time Warner)—and attracted about 10 other supporting companies, mostly home electronics manufacturers and movie studios.[15] Philips and Sony assembled a rival gang of about 14 companies, mostly

---

[15]Other companies supporting SD included MCA (then owned by Matsushita), MGM/UA (owned by Turner), Nippon Columbia, Samsung, SKC, Turner Home Entertainment (independent at the time, now merged with Time Warner), WEA (Time Warner's giant CD-ROM manufacturing arm), and Zenith.

peripheral hardware manufacturers.[16] Neither group had support from major computer companies. At this stage, the emphasis was on video entertainment, with computer data storage as a sideline goal.

Two days after the SD announcement, Sony head Norio Ohga told the press that Sony "may make concessions but . . . will not join" and indicated that Sony would hold out for a third standard incorporating more of its own specifications, even to the point of asking the Ministry of International Trade and Industry to arbitrate the unification. Sony and Philips held the lion's share of CD technology patents and hoped to include as many of them as possible in the new format. The companies in the SD Alliance, including Sony's arch rival Matsushita, planned to use their own newly patented technology and slow the flow of patent revenue to the competition.

Sony and Philips played up the advantages of MMCD's single-layer technology, such as lower manufacturing costs and CD compatibility without the need for a dual-focus laser, but the SD Alliance was winning the crucial support of Hollywood with its dual-layer system's longer playing time. On February 23, Sony played catch-up by announcing a two-layer, one-side design licensed from 3M that would hold 7.4 billion bytes.

The Referee Shows Up

The scuffling continued, and the increasing emphasis on data storage by consumer electronics companies began to worry the computer industry. At the end of April 1995, five computer companies, Apple, Compaq, HP, IBM, and Microsoft, formed a technical working group that met with each faction and urged them to compromise. The computer companies flatly stated that they did "not plan to choose between these proposed new formats" and provided a list of nine objectives for a single standard:

- One format for both computers and video entertainment
- A common file system for computers and video entertainment
- Backward compatibility with existing CDs and CD-ROMs
- Forward compatibility with future writable and rewritable discs
- Cost similar to current CD media and CD-ROM drives
- No mandatory caddy or cartridge
- Data reliability equal to or better than CD-ROM

---

[16]Others backing the MMCD format included Acer, Aiwa, Alps, Bang & Olufson, Grundig, Marantz, Mitsumi, Nokia, Ricoh, TEAC, and Wearnes.

- High data capacity, extensible to future capacity enhancements
- High performance for video (sequential files) and computer data

Sony refused to budge and a month later said there would be "no adjustment in its DVD standards." Norio Ohga said, " . . . a split on the standard is unavoidable because we are in a world of democracy." He rejected the possibility of a third standard and defended his decision on the grounds of "liberalism and democracy." Both sides invited the other to give in, Toshiba inviting "Sony/Philips to engage in serious discussion to resolve this issue," and Sony pronouncing that "we would, of course, be happy to discuss our proposal with all interested parties." Meanwhile, the SD group announced on May 11 that Matsushita had developed a transparent bonding technology that allowed both substrates to be read from a single side. It then tried to stack the deck by announcing the development of recordable SD technology.

On August 14, 1995, the computer industry group, now up to seven members with the addition of Fujitsu and Sun, concluded that the most recent versions of the two formats essentially met all their requirements except the first one: that there be a single, unified standard. In order to best support the requirements for a cross-platform file system and read-write support, they recommended adoption of the *Universal Disk Format* (UDF) developed by the *Optical Storage Technology Association* (OSTA). OSTA had already agreed to refine the UDF standard for interchange compatibility between read-only and nonsequential read-write applications for both television products and computers and had held the first of a set of technical meetings on July 25. IBM reportedly told Sony and Philips that it intended to settle on the higher-capacity SD format and gave them a few weeks to come up with a compromise. Faced with the hazard of no support from the computer industry or the possibly worse prospect of a standards war reminiscent of Betamax versus VHS, the two DVD camps announced at the Berlin IFA show that they would discuss the possibility of a combined standard. The companies officially entered into negotiations on August 24. The computer companies expressed their preference that the MMCD data storage method and dual-layer technology be combined with SD's bonded substrates and better error-correction method.

Reconciliation

At the end of August 1995, Philips and Sony proposed a new MCD combined format to the SD Alliance. Conflicting reports appeared in the press, some saying a compromise was imminent and others claiming that officials from Sony and Toshiba had denied both the compromise news and the rumors that MMCD would be standardized for computer data applications while SD would be used for video. On September 15, 1995, the SD Alliance announced that "considering the computer companies' requests to enhance reliability," it was willing to switch to the Philips/Sony method of bit storage despite a capacity reduction from 5 billion to 4.7 billion bytes. It complained that circuit designs would have to be changed and that "reverification of disc manufacturability" would be required but conceded that it could be done. It did not concede the naming war, however, proposing that the SD name be retained. Sony and Philips made a similar conciliatory announcement, and thus, almost a year after they began, the hostilities officially ended.

Henk Bodt, executive vice president of Philips, said that the next step was to publish in October a comprehensive specification. Mr. Bodt made some notable predictions, stating that the new players would be "substantially more expensive" than VHS players, at around $800. He also thought that issues of data compression probably would make recordability realistic only in the professional field. "Certainly I don't think that these players will replace the videocassette recorder."

On September 25, 1995, OSTA announced establishment of the UDF file system interchange standard, a vast improvement over the old ISO 9660 format, finally implementing full support for modern operating systems along with recording and erasing. Work on support for application of the UDF format to DVD was still underway.

Seeing an opportunity to make its own recommendations, the *Interactive Multimedia Association* (IMA) and the *Laser Disc Association* (LDA, which later changed its name to the *Optical Video Disc Association—OVDA*) held a joint conference on October 19 to determine requirements for "innovative video programming" based on years of experience with laserdisc and CD-ROM.[17] The general consensus was that better video and audio were insufficient and that DVD movie players required interactivity to be of more interest to the worldwide mass market. The group recommended that baseline interactivity be required in all DVD-Video players and also recommended that the design allow optional addition of proven features such as player control using printed bar codes, on-the-fly seamless branching under program control, and external control. Some of the group's recommendations, such as random access to individual frames, graphic overlay, and mul-

---

[17]At this meeting your humble yet foresighted author predicted that DVD would not appear in the United States in meaningful numbers before 1997, despite manufacturer claims that it would be ready in early 1996. It turned out that DVD did not appear in the United States in any numbers at all before 1997.

tiple audio tracks, were mostly supported by the tentative DVD standards, but the remaining recommendations were largely ignored.

On October 30, 1995, OSTA announced completion of an appendix to the UDF file system specification that described the restrictions and requirements for DVD media formatted with UDF. This simplified version, which became known commonly as MicroUDF, allowed DVD-Video players to implement low-cost circuitry for locating and reading movie files.

Undeterred by the prospect of retooling its standard, Toshiba announced on November 7 a prototype SD-ROM drive and claimed that its data transfer rate of nine times CD-ROM speed had not yet been achieved with CD-ROM technology.

The two DVD groups continued to hammer out a consensus that finally was announced on December 12, 1995. The new format covered the basic DVD-ROM format and video standards, taking into account the recommendations made by movie studios and the computer industry.[18] A new alliance was formed—the DVD Consortium—consisting of Philips and Sony, the big seven from the SD camp, and Time Warner. When all was said and done, Matsushita held 25 percent of the approximately 4000 patents; Pioneer and Sony each had 20 percent; Philips, Hitachi, and Toshiba were left with 10 percent of the pie; Thomson had 5 percent; and the remaining members—Mitsubishi, JVC, and Time Warner—held negligible slivers (see Figure 2.9). On top of DVD-specific patents, the MPEG LA controversially claimed 44 essential patents from 12 companies: Columbia University, Fujitsu, General Instruments, Kokusai Denshin Denwa, Matsushita, Mitsubishi, Philips, Samsung, Scientific Atlanta, Sony, Toshiba, and Victor; Dolby, of course, had a finger in the pie with Dolby Digital (AC-3) patents. Fraunhofer, Thomson, and others held MPEG audio patents. Additional fundamental optical disc technology patents are held by Pioneer, Discovision, and others. All this led to a complex dance of cross-licensing that worked reasonably well for the major contributors but left other companies with no recourse but to pay licensing royalties.

---

[18]The SD physical format of two 0.6-mm bonded substrates was selected, with both Matsushita's dual-layer system (one on each substrate) from the SD format and 3M's dual-layer technology (a second photopolymer layer on a single substrate) from the MMCD format. Toshiba's 8/15 modulation was replaced by MMCD's more reliable EFMPlus 8/16 modulation. SD's more robust Reed-Solomon Product Code error correction was chosen with 32-kilobyte blocks instead of 23-kilobyte blocks. The maximum pit length was reduced from 2.13 to 1.87 microns for higher data density, and the scanning velocity was raised from 3.27 to 3.49 meters per second. These last two changes resulted in the channel bit rate improving from SD's 25.54 Mbps to 26.16 Mbps and helped compensate for the increased modulation (see Table A.8 for more details).

Figure 2.9
The DVD patent pie

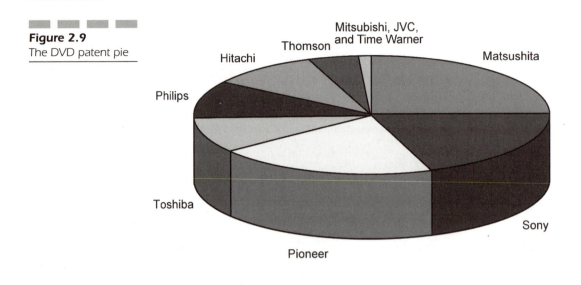

Product Plans

In January 1996, companies began announcing their DVD plans for the new year. Philips targeted late 1996 for its hardware release. Most other companies pegged a slightly earlier fall release. Thomson stunned everyone by announcing that its player would be available by summer for $499—$100 less and months sooner than the rest.[19]

The road ahead looked smooth and clear until the engineers in their rose-colored glasses, riding their forecast-fueled marketing machines, crashed headlong into the protectionist paranoia of Hollywood.

As the prospect of DVD solidified, the movie studios began to obsess about what would happen when they released their family jewels in pristine digital format with the possibility that people could make high-quality videotape copies or even perfect digital dubs. Rumors began to surface that DVD would be delayed because of copyright worries, but Toshiba and others confidently stated that everything was on track for the planned fall release. On March 29, the *Consumer Electronics Manufacturers Association* (CEMA) and the *Motion Picture Association of America* (MPAA) announced that they had agreed to seek legislation that would protect intellectual property and consumers' rights concerning digital video recorders. They hoped their proposal would be included in the Digital Recording Act of 1996

---

[19]Almost exactly a year later, Thomson reannounced its RCA-brand players at $599 and $699 for limited spring 1997 release, with full availability delayed until fall. The Thomson-designed players were built by Matsushita.

that was about to be introduced in the U.S. Congress.

Their recommendation was intended to:

- allow consumers to make home video recordings from broadcast or basic cable television.
- allow analog or digital copies of subscription programming, with the qualification that digital copies of the copy could be prevented.
- allow copyright owners to prohibit copying from pay-per-view, video-on-demand, and prerecorded material.

The two groups hailed their agreement as a landmark compromise between industries that often had been at odds over copyright issues. They added that they welcomed input from the computer industry, and input they got! A week later a very upset industry group fired off a "list of critiques" of the technical specifications that had been proposed by Hollywood and the consumer electronics companies. The *Information Technology Industry* (ITI) Council, a group of 30 computer and communications companies including Apple, IBM, Intel, Motorola, and Xerox was less than thrilled with the MPAA and CEMA attempting to unilaterally dictate hardware and software systems to keep movies from being copied onto personal computers. "No way will we simply accept it as is," said IBM's Dr. Alan Bell, chair of the Technical Working Group, in reference to the proposal. The computer industry said it preferred voluntary standards for copy protection and objected to being told exactly how to implement things. "Any mandatory standard that was legislated and then administered by a government body is anathema to the computer industry," said an ITI spokesperson, who also pointed out that legal protection for copyrights "should cover all digital media including records, movies, images, and texts" rather than focusing on motion pictures as MPAA and CEMA had proposed. ITI announced that it would have a counterproposal ready for an April 29 meeting, recognizing that it was "an important issue to Hollywood, and we don't want to take money out of the studios' pockets." Hollywood countered that the computer industry had been invited to participate early on but did not, either from laziness or arrogance.

Turbulence

The news media was filled with reports of DVD being "stalled," "embattled," and "derailed." A few weeks later, on May 20, Toshiba's executive vice president, Taizo Nishimuro, was reported to have said that a copy protection deal between the computer industry's Technical Working Group and the

DVD Consortium would be signed at a June 3 meeting. Apparently, however, the agreement had only been that all three parties would refrain from introducing copy protection legislation before the June 3 meeting, and Toshiba officials later denied the reports of any sort of settlement. Despite all this, executives at Thomson said that they were confident the issues would be resolved and that they were ready for launch as early as summer.

As DVD progress was being attacked by the popular press, standardization efforts were having their own problems. Sony threw a monkey wrench into the process by proposing a completely new DVD-Audio format, *Direct Stream Digital* (DSD). Sony claimed that its single-bit format was not tied to specific sampling frequencies and sizes, thus eliminating downsampling and oversampling and resulting in less noise. The ARA did not agree and continued to push for PCM. Then, in the May 1996 issue of *CD-ROM Professional,* guest columnist Hugh Bennet publicized a serious flaw that had apparently gone unrecognized or was being ignored by DVD engineers. The dye used in recordable CD media (CD-R) did not reflect the smaller-wavelength laser light used in DVD, thus rendering the discs invisible. He pointed out that more than 2 million CD recorders were expected to be in use by the end of 1996 and that the recorders were expected to have written between 75 and 100 million CD-R discs by then. A new type II CD-R that would work with CD and DVD had been proposed but would take years to supersede the current format, leaving millions of CD-R users out in the cold when it came time to switch to DVD readers.

Glib Promises

At a June 1996 DVD conference, participants glossed over the technical difficulties and downplayed the copy protection schism, expecting a solution to be announced before the end of the month. Toshiba executive Toshio Yajima said, "It was a misunderstanding between industries." Sixty executives from the three sides were committed to meeting once a week to resolve the copyright protection disputes. At the same conference, one of DVD's most ardent cheerleaders, Warner Home Video President Warren Lieberfarb, told attendees in his keynote speech that 250 movie titles would be ready for a fall launch. Lieberfarb also gave projections that player sales would be from 2.8 million to 3.7 million units in the first year. Other first-day speakers were equally optimistic, and everyone was assured that the fall launch was on schedule, that the 10 companies of the DVD Consortium had agreed to establish a one-stop agency for licensing, and that the preliminary DVD 0.9 specification book was immediately available for a paltry $5000. A few days into the conference, Thomson admitted that prospects for a fall rollout of its player were only "50/50," and Toshiba let it be known that it would delay its product launch until October.

Attendees at the conference were even less sanguine. Many thought December would be a more realistic target date, even though it would be too late for the Christmas buying spree. An industry executive was quoted as saying, "What the DVD industry needs is a healthy dose of realism. Somebody ought to just stand up and announce that it's a '97 product, and all this speculation would come to an end." Over the next few months, however, realism did not make much of an appearance, and speculation refused to leave the stage.

In anticipation of the imminent release of DVD, and with copy protection details unresolved, the MPAA and CEMA announced draft legislation, called the *Video Home Recording Act* (VHRA), designed to legally uphold copy protection. The two groups had been working on the legislation since 1994, without directly consulting with the computer industry. Consequently, ITI and the Technical Working Group reportedly filed a letter of protest against the proposed legislation, requesting that an encryption and watermarking scheme be used. The original legislation covered the insertion of a 2-bit identifier in the video stream to mark copy-prohibited content. The copy protection technology eventually became the much more complex CSS, and the legislation likewise mutated into the rather ambiguous *Digital Millennium Copyright Act* (DMCA).

Sliding Deadlines and Empty Announcements

On June 25, Toshiba announced that the 10 companies in the consortium had agreed to integrate standardized copy protection circuits in their players, including a regional management system to control the distribution and release of movies in different parts of the world. Many people took this to mean that the copy protection issue had been laid to rest, not understanding that this copy protection agreement dealt only with analog copying by using the Macrovision signal-modification technology to prevent recording on a VCR. The manufacturers, still clutching their dreams of DVD players under Christmas trees and hoping to keep enthusiasm high, conveniently failed to mention that Hollywood was holding out for digital copy protection as well.

On July 11, Matsushita announced that it would begin shipping DVD players in Japan in September, with machines reaching the United States in October. Matsushita blamed copyright protection and licensing issues for the delay. Again, more pragmatic viewpoints appeared, such as that of Jerry Pierce, director of MCA's Digital Video Compression Center, who flatly stated that "the DVD launch will be in 1997." By this time Philips—the most conservative of the bunch—had moved its release date to the spring of 1997. Matsushita also said its DVD-ROM drives for PCs would not be introduced until early 1997 but that within 2 to 3 years every PC would have a DVD drive.

At the end of July, over a month after a copy protection settlement was to have been made, the DVD Consortium—the hardware side of the triangle—agreed to support a copy protection method proposed by Matsushita. The Copy Protection Technical Working Group—representing all three sides of the triangle—agreed to look into the proposal, which used content encryption to prevent movie data from being copied directly using a DVD-ROM drive.

Not unexpectedly, licensing discussion had by now fallen apart. On August 2, Philips made a surprise announcement that it had been authorized by Sony to begin a licensing program for their joint DVD technology. "In an effort to avoid further undesirable delays, Philips and Sony have decided to move forward in the best interest of the DVD system and its future licensees." They called on other companies to join in pooling patents, but Thomson, for one, declined. Many less-than-responsible journalists had a field day, reporting that the DVD standard had been sabotaged or that Philips and Sony were trying to steal patent revenue from the other companies.

By the end of August, there was still no copy protection agreement. At the gigantic CeBIT Home exhibition in Hanover, Germany, Jan Oosterveld of Philips explained that key issues such as copy protection, regional coding, and software availability were unresolved. "The orchestra is assembled, the musicians have their positions, but they have not decided which tune they will play and how much time they need to rehearse and tune their instruments," he said. He elaborated that copy protection was complicated by export and import restrictions and that key technology was still not available to non-Japanese companies. The regional control issue—on the table for almost a year—still "created confusion on how the world should be divided." The software supply for DVD-ROM drives also looked bleak, with only 12 companies working on DVD-ROM titles. Oosterveld reiterated that the music industry was expected to take the lead with DVD-Audio and that there was no decision on whether to use phase-change or magneto-optical technology for DVD-RAM. Despite this, Philips was optimistic that by the year 2000 around 10 percent of all optical drives, or 25 million out of 250 million, would be based on DVD. Compared with other predictions, this was rather pessimistic, but given Philips's slightly better track record at forecasting, it may have been the most realistic.

At the same time, Sakon Nagasaki of Matsushita told the press that completion of the DVD specification and resolution of the encryption problem were two separate issues, saying that encryption "is a problem in the United States, not in Japan." Matsushita announced that it could not wait any longer for a copy protection agreement, that its two movie player models would be available in Japan on November 1 for 98,000 and 79,800 yen, and that if a copy protection agreement were reached, the players would be introduced a few weeks later in the United States for $599 and $699.[20] Although Sony announced that it was delaying the release of its players until spring, Hitachi, Pioneer, and Toshiba followed Matsushita's lead and promised players before Christmas.

A few weeks later, on September 12, 1996, Toshiba announced its first home PCs and reaffirmed its commitment—despite ongoing copyright protection issues—to bring out DVD-Video and DVD-ROM players, stating that PC makers "should have the first units with our implementation of the copyright protection toward the end of September, and we think we can start shipping those units in volume by mid-November."

The following week was the IMA Expo, with a special focus on DVD. The glowing news from Toshiba was that every DVD-ROM PC would have hardware or software to play DVD movies, DVD-RAM would be ready within a year, and DVD-ROM would be such a success that Toshiba would no longer be making CD-ROM drives in the year 2000. Estimates from International Data Corporation (IDC) placed DVD-ROM drive sales at 10 million in 1997 and 70 million in 2000. During the Expo, Pioneer announced November 22 as the release date in Japan of a combination laserdisc and DVD player at 133,000 yen and a DVD-only player at 83,000 yen, plus two other DVD karaoke players. U.S. prices were expected to be $1200 and $750. By this time, the DVD Consortium had managed to fit most of the world into six geographic regions for release-control purposes, but Mexico, Australia, and New Zealand were still bouncing from region to region. According to Warner Advanced Media, there were supposedly 30 DVD-ROM software titles in development for early 1997. Matsushita representatives at the Expo were privately admitting that their players would not be out in the United States until February.

[20]When the Matsushita (Panasonic) players were finally released in March of 1997, the price of the DVD-A100 was unchanged, but that of the DVD-A300 had risen by $50 to $749.95.

At about this time, the supposedly final 1.0 version of the $5000 DVD-ROM and DVD-Video technical specifications appeared. The books listed a publication date of August, but it took a while for them to be shipped. A few features such as NTSC Closed-Caption support had been changed, and a complex country-specific ratings system had reverted to a simpler version from the earlier spec. Copy protection details had been left for a later appendix, which when it finally appeared only covered the superficial basics, since the decision had been made to keep copy protection separate from the technical specifications.

Pioneer announced on September 27 that beginning in the latter half of October it would sell a mix-and-match system of DVD players and receivers, bringing its total to five models placed on sale in 1996. European introduction of the mix-and-match models was planned for spring 1997, with the United States to follow in the summer.

On October 5, 1996, Samsung announced that November 1 would be the debut date of its DVD player in Korea and cited predictions of a global market for DVD players at 400,000 before the end of 1996. The affiliated Samsung Entertainment Group expected to release at least 10 DVD movie titles by the end of the year, with more than 100 in 1997 and over 500 in the year 2000. Samsung also expected to commercialize its DVD-ROM drive by the end of 1996.

Hopes for a DVD-Audio specification before the end of 1996 dwindled. Philips announced that it would team up with Sony to develop bitstream-based *Direct Stream Digital* (DSD) technology for DVD.

The Birth of DVD

The big news finally arrived on October 29, 1996: the Copy Protection Technical Working Group announced that a tentative agreement had been reached. The modified copy protection technology developed by Matsushita and Toshiba, called *Content Scramble System* (CSS), would be licensed through a nonprofit entity. On the same day, Pioneer confessed that it would delay U.S. shipment of DVD players until January, pushing back the December release date it had announced just the day before. At this point only Toshiba and Matsushita were still promising to make DVD hardware available in December, although most of the press and even more of the public were confused, thinking release dates in Japan applied to the United States.

Matsushita and Toshiba delivered DVD players in Japan as promised. News reports from November 1 described a dismal rainy day with lackluster player sales and a handful of discs, mostly music videos. A student was quoted as saying, "Prerecorded discs are not particularly appealing, but it would be great if you can record your favorite films on the disc as many times as you want," which undoubtedly did not help convince the studios to rush their movies onto DVD.

The PR machines did not rest, with Fujitsu claiming on November 6 that it was the first company to market a DVD-ROM–equipped computer, Toshiba stating on November 7 that it expected rewritable DVD-RAM to be available within a year or soon thereafter, and Matsushita announcing on November 11 the development of the world's first DVD-Internet linkage system for playing video from an Internet Web page through the use of a DVD-ROM drive. Toshiba's news on November 18 was more down to earth: postponement of its DVD player release in the United States until late January or early February.

Most other announcements were put on hold for Comdex, the huge computer technology exhibition in Las Vegas. At Comdex, there were dozens of announcements of hardware and software, including the expected release of Toshiba's DVD-ROM drive worldwide in January 1997. Intel CEO Andy Grove's keynote speech included a DVD demo of *Space Jam* using Compcore's software-only playback system.

November came and went with no sightings of DVD players for sale in North America. On December 2, Akai announced its own DVD player to be launched in Japan at the end of January and worldwide in April.

By December 13, 1996, Toshiba's Precia PC, which was to have been rolled out in Japan in November, had been pushed back to January. Toshiba blamed it on bureaucratic problems with copy protection chips. "We're hoping it's resolved any day now, but we've been hoping for that for two or three weeks," went the now-too-familiar refrain. Other sources claimed the real reason for the delays was that the Japanese government had barred Toshiba from releasing hardware to the United States because of concerns over the export of encryption technology. The U.S. delivery date for Toshiba's DVD-equipped PCs slipped to March.

Things began to look up on December 20 when Warner Home Video began sales in Japan of four major movie titles—*The Assassin, Blade Runner, Eraser,* and *The Fugitive*—with another four announced for January release.

By this time, the delay in the United States was being blamed on lack of titles rather than on copy protection issues, but in the first week of January 1997, at the Consumer Electronics Show in Las Vegas, DVD finally began to look like a real product. At least six studios announced the release of over 60 DVD-Video titles for the March-April time frame: New Line Home Video, Warner Home Video, Sony's Columbia TriStar Home Video, Sony Music

Entertainment/Sony Wonder, MGM Home Entertainment, and Philips' Polygram. Perennial favorites *Casablanca* and *Singin' in the Rain* were announced by two studios each.

Also at the Consumer Electronics Show, DVD players were demonstrated or announced by Akai, Denon, Faroudja, Fisher, JVC, Meridian, Mitsubishi, Panasonic, Philips/Magnavox, Pioneer, RCA, Samsung, Sony, Toshiba, Yamaha, and Zenith, most with a U.S. release date between March and summer. This brought the lineup of player manufacturers to at least 20, including others who had previously announced DVD movie plans: Goldstar, Hitachi, Hyundai, and Sharp. Compared with typical industry support of new launches—even now-ubiquitous products such as telephones, televisions, VCRs, and CDs—this was an astonishing endorsement that flew in the face of the predictions of gloom and doom for DVD as a consumer video product.

And on the computer side, where most people expected DVD to be a slam dunk, there were now over 45 companies developing hardware or playback software. As it turned out, many of them abandoned their efforts within a year.

Players from Panasonic and Pioneer finally began to appear in the United States in February as more movies and music performance videos were announced from Lumivision, Warner Bros. Records, and others. Eager customers bought the players, only to discover that no titles other than those from Lumivision were scheduled to appear before the end of March. Warner, the primary supplier, limited its release to seven test cities. Pioneer announced that its $11,500 DVD-R recorder would be available, complete with 40 to 50 blank discs, around June of 1997. Computer makers geared up for both hardware and software DVD-Video playback, but once again, the copy protection issue settled on the scene like a wet blanket: Decryption/descrambling licenses were available for hardware but not yet allowed for software. The studios were unconvinced that DVD players implemented in software would completely protect their assets. They were right, of course, but that did not mollify the computer makers and software developers who were counting on cheap software playback.

DVD had finally embarked, late and lacking some of its early luster, on the rocky road to acceptance.

Disillusionment

The two final competing DVD-RAM proposals were merged, and an official format was announced on April 14, 1997. The approval procedure was allegedly changed so that members of the original SD camp could steamroll their combined proposal over the objections and abstentions of Sony and

Philips, who were promoting a format more closely related to their own CD-RW. The fresh veneer of cooperation was cracking, revealing the possibility of a competing recordable DVD format. Toshiba, which apparently announced the DVD-RAM arrangement before it was approved, made almost comical claims that drives would be available before the end of 1997 for only $350. More sober predictions targeted 1998 and prices of $800 or more. At about the same time, Pioneer revealed that its DVD-R drive would cost $6000 more than originally expected, raising the price to about $17,000. The good news was that DVD-R was expected to be compatible with most players and drives. The bad news was that DVD-RAM would not be compatible with anything. The worse news, surfacing about a month later, was that Sony and Philips were apparently striking off on their own with a renegade recordable format, along with partner Hewlett Packard. A Philips spokeswoman said, "We feel the DVD-RAM spec 1.0 agreed upon by the DVD Consortium is not an ideal format. Our format will bring far better compatibility with the CD products that are already on the market." NEC also announced its own competing recordable format. The DVD family was losing unity. To top it off, various technology announcements began appearing from DVD companies, announcing research on high-density DVD. To DVD manufacturers, these were forward-looking announcements, necessary in the jostling for position that would precede work on the next generation of DVD. To new customers just learning about DVD, especially those used to the slow turnover of traditional technologies such as TVs and record players, news hinting at the imminent obsolescence of the new DVD format was disconcerting.

Rockley Miller, editor of the *Multimedia Monitor* newsletter, characterized the acceptance cycle of new technologies. He posited that every new technology hits an initial peak of euphoria, often before it is released to market, where wildly exaggerated forecasts are piled on claims of vast improvements. Of course, the new devices never live up to the impossible expectations, leading to Miller's second cycle: disillusionment. Sales are lower than expected, equipment does not work perfectly, and development happens more slowly than predicted. The press always jumps in at this point, sometimes even proclaiming the failure of the format. From the trough of disillusionment, however, comes the gradual rise to the third phase of the cycle: real product and real work. Bugs are worked out, better products are released, and people begin to figure out how to use the new technology for what it does best.

In the middle of 1997, DVD was in the trough of disillusionment. On the home video front, Time Warner and player manufacturers had not expanded beyond the six trial cities, although discs and players usually could be had via mail order. Many of the major studios were keeping their wait-and-see stance. Some sources claimed that the studios were pressuring the hardware makers for a kickback. Meanwhile, rumors spread of a pay-per-view version of DVD called *ZoomTV,* supposedly under development for use by the holdout studios such as Disney and DreamWorks. Home consumers who heard the rumors were less than thrilled at the prospect of having their movie-viewing habits held hostage to the vagaries of telephone systems and remote authentication servers. Amid the litany of boredom and frustration, one of the few bright spots was the initial success of disc sales. By the end of April, less than one month after the release of the first 40 or so DVD titles in the United States, over 50,000 copies had been sold, exceeding most expectations. The title wave of DVD had begun to swell. Over 30,000 players had been bought in the first 15 weeks of U.S. sales, and over 150,000 players had been sold in Japan since DVD's introduction six months earlier. Initial bad reviews of in-store demos were supplanted by glowing reports from early purchasers who made side-by-side comparisons of DVD and laserdisc and proclaimed DVD the clear winner.

On the computer side, however, little was happening. Creative Labs' DVD-ROM upgrade kit with hardware DVD-Video playback appeared in April but showed signs of a rushed release. Other DVD-ROM products failed to make their already delayed debuts, and both Apple and Microsoft pushed back release dates for DVD support in their operating systems. There were still no movie decryption licenses allowed for software playback, and industry observers began to doubt that DVD computers would be available in more than miniscule numbers before 1998. This was increasingly irksome to those who simply wanted a DVD-ROM data drive and had no interest in using their computers to play movies. Many in this group began hoping for a quick and early death for DVD-Video so that they could get on with the real business of DVD-ROM.

The audio-only initiative met yet another setback in June when Philips and Sony announced they were working on their own Super Audio CD format based on their DSD audio coding process. Undaunted, the DVD Forum's *Working Group 4* (WG-4) was moving ahead, based on the recommendations of the *International Steering Committee* (ISC) formed by the *Recording Industry Association of America* (RIAA), Europe's *International Federation of Phonographic Industries* (IFPI), and the *Recording Industry Association of Japan* (RIAJ). (See Table 2.1.)

In June, Hitachi demonstrated a home DVD video recorder containing a DVD-RAM drive, a hard disk drive acting as a buffer, two MPEG-1 encoders, and one MPEG-2 decoder. No production date was mentioned.

In July of 1997, the MPEG LA (licensing authority) opened for business in Denver, Colorado, as a one-stop shop for MPEG-2 patents, since they rep-

TABLE 2.1

Recommendations
of the ISC for
DVD-audio

Very high quality multichannel audio with downmixing

Studio mastering and archiving with no loss of quality

Playing time of at least 74 minutes

Additional data such as video, lyrics, and still pictures

On-screen navigation menus

Simple navigation, similar to audio CD players

Copyright protection and anti-piracy measures

Conditional access

Compatibility with DVD-Video format wherever possible

Compatibility with existing audio CD players

No caddy, no proscribed package size

Durability equal to or better than CD

Single-sided, 12-cm discs preferred

resented about 90 percent of the MPEG-2 patent holders. Because of the delay in establishing a patent royalty program, and because of the precedent set when MPEG-1 patents were not enforced, some people felt that the MPEG-2 patent holders had lost the right to make money from their patents. Most of the consumer electronic companies appeared on the list of MPEG LA's licensees, but computer hardware and software companies remained conspicuously absent.

There were flickers of good news here and there, as Universal Home Video announced in July that they would enter the DVD market, and finally, in August, Warner broadened distribution of discs from the original seven test cities to countrywide. Image Entertainment, longtime supporter of laserdiscs, realized that the only way to recover from a precipitous drop in laserdisc sales was to embrace the DVD format with more alacrity. Panasonic announced the first DVD notebook PC.

Divx: A Tale of Two DVDs

In September of 1997, ZoomTV surfaced, sporting a new name and heavy-duty financial backing. Circuit City and the Hollywood law firm of Ziffren, Brittenham, Branca & Fischer announced the formation of Digital Video

Express, a company that planned to release Divx, a "DVD rental" format, in the summer of 1998. Furious controversy burst out, with backers hailing it as an innovative approach to video rental with cheap discs you could get almost anywhere and keep for later viewings and detractors spurning it as an insidious evil scheme for greedy studios to control what you see in your own living room (Table 2.2). Of course, most of the detractors already owned DVD players and were annoyed that they could not play Divx discs on them.

Contrary to many claims, Divx was not a competing format—it was a pay-per-viewing-period variation of DVD. The discs were designed to be purchased from any retail outlet at a price close to rental fees. The $4.50 purchase price covered the first viewing. Once inserted into a "Divx-enhanced" DVD player, the disc would play normally for the next 48 hours, allowing the viewer to pause, rewind, and even put in another disc before finishing the first disc. Once the 48 hours were up, the "owner" of the disc could pay $3.25 to unlock it for another 48 hours. Divx DVD players, which initially cost about $100 more than a regular player but later dropped to a premium of only $50, included a built-in modem so that they could call a toll-free number during the night to upload billing information. Since the player only called once a month or so, it did not have to be hooked permanently to a phone line. There was a DivxSilver program that allowed most discs to be converted to unlimited play for an additional fee of $20 or so. The idea was to let buyers purchase a disc at low cost, try it out, and then pay full price if they liked it enough to watch repeatedly. Unlimited-playback DivxGold discs were announced but never shipped.

Divx players were able to play regular DVD discs, but Divx discs were designed not to play in standard DVD players. Each Divx disc was serialized with a bar code in the standard burst cutting area so that it could be uniquely identified and tracked by the player for billing and encryption purposes. In addition to normal CSS copy protection, Divx discs used modified channel modulation (making normal drives physically incapable of reading protected data from the disc), triple DES encryption (three 56-bit keys), and watermarking of the video. Ironically, the company had nothing to fear from pirates who made bit-for-bit copies, since copied discs would still require authorization and billing. A pirate outfit would simply be providing free Divx replication services. Actual cracking of the encryption for copying video to standard DVD discs was a potential threat, but Divx technicians claim that no one ever got through even the first layer of protection.

Limited trials of Divx players began June 8, 1998, in San Francisco and Richmond. The only available player was from Zenith (which at the time was in Chapter 11 bankruptcy), and the promised 150 movies had dwindled to 14. The limited nationwide rollout (with one Zenith player model and 150 movies in 190 stores) began on September 25, 1998. By the end of 1998,

TABLE 2.2	Advantages of Divx	Disadvantages of Divx
Pros and Cons of Divx	▪ Viewing could be delayed, unlike rentals.	▪ Higher player cost.
	▪ Discs need not be returned. No late fees.	▪ Although discs did not have to be returned, the viewer still had to go to the effort of purchasing the disc. Cable/satellite pay per view is more convenient.
	▪ Movie could be watched again for a small fee. Initial cost of "owning" a disc was reduced.	
	▪ Discs could be unlocked for unlimited viewing (Divx Silver), an inexpensive way to preview before deciding to purchase.	▪ Higher cost than for regular DVD rental.
		▪ Casual, quick viewing (looking for a name in the credits, playing a favorite scene) required paying a fee.
	▪ Unlike rental discs, the discs were new and undamaged.	
	▪ The "rental" market was opened up to other retailers, including mail order.	▪ Most Divx titles did not come in widescreen format and did not include extras such as foreign language tracks, supplemental video, and commentaries.
	▪ Studios got more control over the use of their content.	▪ The player had to be hooked to a phone line (but only once a month or after about 10 viewings).
	▪ Special offers received from studios in your Divx mailbox.	
		▪ The Divx central computer ostensibly collected information about viewing habits. Of course, cable/satellite pay-per-view services and rental chains do the same.
		▪ Divx players included a "mailbox" for companies to send you unsolicited offers.
		▪ Those who did not password-protect their Divx player could receive unexpected bills when children or visitors played Divx discs.
		▪ Divx discs would not play in regular DVD players or DVD computers.
		▪ Unlocked (DivxSilver) discs only worked in players on the same account. Playback in a friend's Divx player would incur a charge.
		▪ You could not give Divx discs as gifts unless you were sure that the recipient owned a Divx player.
		▪ There was little market for used Divx discs.
		▪ Divx players were never available outside the United States and Canada.

about 87,000 Divx players (from four models available) and 535,000 Divx discs were sold (from about 300 titles available). These numbers were quite impressive, representing 10 percent of all player sales. By March 1999, 420 Divx titles were available, compared with over 3500 open DVD titles. Despite the small numbers, it was again an impressive feat to produce over 400 titles in only 6 months. Support from studios and player manufacturers increased, motivated by incentives in the form of guaranteed licensing payments totaling over $110 million. Initial backers Disney (Buena Vista), DreamWorks SKG, Paramount, and Universal were joined by MGM and Twentieth Century Fox, although Time Warner and Sony's Columbia/Tristar actively combated Divx by creating rental programs. Harman/Kardon, JVC, Kenwood, and Pioneer announced their intention to join Matsushita (Panasonic), Thomson (Proscan/RCA), and Zenith in making DVD players. Thomson even demonstrated a high-definition version of Divx, which appealed to studios that were even more fearful of releasing movies on high-definition DVD than on standard DVD. Retailer support was weak, partly because chains were reluctant to sell a format that would pad the pockets of competitor Circuit City.

Meanwhile, the vitriolic reaction to Divx showed no signs of abating. Dozens of anti-Divx Web sites peppered the Internet. The big psychological hurdle was that many people were uncomfortable buying a disc that they did not control. Despite a promising beginning, the company could not shake its bad press. In March, shares of Circuit City rose more than 12 percent from speculation about the future of Divx and rumors of investment by Blockbuster, among others. The investments fell through, however, allegedly scuttled by studios that backed Divx but did not want to be beholden to Blockbuster. Richard Sharp, the CEO of Circuit City who originally brokered the Divx deal, finally gave in to pressure from stockholders and others. On June 16, 1999, less than a year after initial product trials, Digital Video Express announced that it was shutting down operations. E-cheers rang out across the Internet, with the anti-Divx sites claiming victory in torpedoing the foe. A popular DVD news site, *The Digital Bits,* received 628 reader e-mails about the death of Divx by 11 A.M. on the day of the announcement. The swell of resentment on the Internet undoubtedly contributed to the demise of Divx. In an auto-obituarial press release, Richard Sharp grumbled that "unfortunately, we have been unable to obtain adequate support from studios and other retailers." Industry analyst Ted Pine was much closer to the mark when he said, "This is the first technology to run up against the *vox populi* of the Internet." Circuit City took a $114 million after-tax loss. *Variety* magazine estimated the total loss to be $337 million. Digital Video Express provided $100 rebates, making

Divx players a great deal, especially since they generally had excellent features and quality. The Divx computer shuts down on June 30, 2001, after which no Divx disc will be playable. The players will continue to play standard DVDs.

When all was said and done, Divx did not confuse or delay development of the DVD market nearly as much as many people predicted.[21] In fact, it probably helped by stimulating Internet rental companies to provide better services and prices, by encouraging manufacturers to offer more free discs with player purchases, and by motivating studios to develop rental programs. The world will never know if Divx could have been a success, but if Circuit City had managed to hold on for a few more years, to the point where Divx discs were available as impulse purchases at supermarket checkout counters and corner convenience stores, it probably would have become the primary DVD format.

Ups and Downs

Meanwhile, the roller coaster ride continued in fall of 1997. Disney announced that it would finally enter the market with DVD discs, but the planned titles did not include any of the coveted animated features. Four days later, Disney announced that it would also release discs in Divx format.

In October, the 10-member DVD Consortium changed its name to the DVD Forum and opened up membership to all interested companies. By the time the first DVD Forum general meeting was held in December, the organization had grown to 120 members.

3.95G DVD-R drives finally appeared. In spite of the $17,000 price tag, DVD developers desperate for an easy way to test their titles snapped them up. After all, compared to the $150,000 price of the first CD-R recorders, DVD-R recorders were a bargain.

In November 1997, the first public breach of DVD copy protection occurred. A program called softDVDcrack was posted to the Internet, allowing digital copies of movies to be made on computers. The press reported that CSS encryption had been cracked, but in actuality the program hacked

---

[21]Truthful disclosure time: I complained about Divx from time to time, usually from the perspective that it would confuse customers and hold back DVD. My point was that it should have been included in the DVD format from the very beginning so that all DVD owners would have the option to purchase Divx discs. The Divx developers did propose Divx to the DVD Consortium while DVD was still under development, but they were turned down.

the Zoran software DVD player to get to the decrypted, decompressed video. Zoran plugged the leak, but pirated copies of the earlier Zoran player were available on the Internet. Still, the process was so complicated that only the most dedicated hackers were interested in spending half a day of their time copying a two-hour movie.

The continuing difficulty of getting equipment to encode multichannel MPEG audio finally became such a problem that on December 5, 1997, the DVD Forum Steering Committee voted to amend the DVD specification to add Dolby Digital as one of the mandatory audio format options on 625/50 (PAL/SECAM) discs. Previously, PAL/SECAM discs had to have either PCM or MPEG audio tracks. The change allowed disc producers to exclusively use Dolby Digital soundtracks. Philips, the primary supporter of MPEG audio, and its partner Sony voted against the change, but they were overruled by the other eight companies. There was scattered cheering among European DVD enthusiasts who preferred the much wider selection of Dolby Digital-equipped audio gear. There was grumbling from Philips, which said, "The decision made is based on incorrect information. Philips will continue to support MPEG-2 multichannel audio for PAL/SECAM countries."

In December, Microsoft released DirectShow 5.2 (also called DirectShow 2.0), an extension to the Windows operating system that supported DVD playback. Up to this point, vendors of DVD hardware and software for Windows PCs had used the old MCI system, each implementing things slightly differently. DirectShow promised to provide a robust and standard interface for developers of DVD software, along with support for the full DVD-Video feature set. Unfortunately, as it turned out, most DVD software decoder vendors were more interested in signing OEM deals to get their software bundled into new PCs than in supporting Microsoft's standard. The shift from proprietary decoders to DirectShow-compatible decoders, which should have happened in less than a year, barely reached the three-quarters point three years later. The resulting lack of a single DVD development target for Windows contributed to the retarded growth of multimedia DVD software.

At the end of December, 340,000 players had been sold in the U.S. This was below "expectations" of 1.2 million. Dozens of reports appeared, lamenting or disparaging DVD's performance. Few bothered to mention that most of the expectations had been unreasonably optimistic. Given that DVD had been launched only nine months previously, with the first five months limited to seven test cities, player sales were actually quite impressive. Rockley Miller's predicted cycles of euphoria leading to disillusionment were holding true.

The Second Year

1998 began with many predictions that it would be "the year of DVD." These predictions were about a year early, even though things began to look up for the fledgling format. The DVD specification was treated to a minor freshening: version 1.01 of DVD-ROM and version 1.1 of DVD-Video. A draft version of DVD-Audio was announced. The first dual-layer discs, DVD-9s, began to be produced, even though yields—the number of usable discs in a replication batch—were only at 30 percent. Warner Home Video noted that it made over $50 million in revenue from DVD sales. E4 finally shipped the first DVD upgrade kit for Macintosh computers. Panasonic announced the DVD-L10, the first portable player, which was a big hit with travelers and those doing DVD presentations. Pioneer announced that its LD-V7200 industrial player would be available in the spring. With features such as barcode control, external control, mouse and keyboard input, video blackboard, genlock, and high reliability, the player met the needs of specialized applications such as museum installations, video walls, training centers, and kiosks. Sonic Solutions announced DVDit[22], a simple, low-cost production tool to convert PowerPoint presentations, HyperCard stacks, and Premiere projects into DVD-Video titles. Initial excitement faded as the product underwent numerous metamorphoses and rebirths before materializing over two years later.

Announcements of DVD titles steadily rolled in, many from smaller studios. A few of the major Hollywood studios had yet to take off their hat, let alone throw it in the ring, although Fox tipped its bowler slightly and announced it would release Divx discs.

Since DVD still had not been officially launched outside of Japan and the U.S., a market for imported discs began to grow in other countries, fed by fans who bought region 1 players from the U.S. In February, the UK *Federation Against Copyright Theft* (FACT), raided a High Street record outlet and seized imported region 1 discs. A pair of British entrepreneurs tried to get around restrictions by setting up a region 1 DVD import business in France, but they were sentenced and fined. However, since many European countries had weak or no restrictions on imported movies, and since most countries, even the UK, allow individuals to legally purchase titles from

---

[22]Sonic puts an exclamation mark after the name of the product. Such abuses of punctuation by out-of-control marketing departments must be rigorously opposed whenever possible.

outside the country, international Internet DVD mail order business flourished. Plans for a large-scale launch of DVD in Western Europe fizzled, resulting in a "soft launch" in May 1998, which amounted to a few manufacturers and a few studios releasing a few players and a few titles to join the existing trickle.

In March 1998, in spite of half-hearted opposition from the DVD-RAM camp, the DVD Forum officially adopted the DVD-RW format, developed primarily by Pioneer as a rewritable variation of DVD-R. The recordable side of the DVD family began to get more confusing. In April, the DVD Forum publicly requested that HP, Philips, and Sony stop using the letters DVD in their product name. Philips responded that no one owned the letters. A few months later, on June 16th, HP, Mitsubishi, Philips, Sony, and Yamaha announced formation of the DVD+RW Compatibility Alliance to promote the DVD+RW format. Sony announced that its DVD+RW drive would be available at the end of 1998. Also in June, the DVD Forum released a preliminary version 0.9 of the DVD-Audio specification, with expectations that players would be available by the end of the year. By this time, jaded DVD technology watchers knew to add 6 months to any announced timetable. But for DVD+RW and DVD-Audio, adding 24 months would still not have been a sufficient reality adjustment.

Philips and Sony continued to work half in and half out of the Forum, announcing that SACD would be licensed to existing CD licensees at no additional charge. They promoted features such as watermarking and a hybrid disc that would also play in existing CD players.

In April, Paramount, one of the last holdout studios, announced that it would begin releasing movies on DVD.

New 350-MHz and 400-MHz Intel Pentium II processors had reached the point where the CPU could handle most of the work of DVD decoding and processing. Low-cost, good quality, software-only DVD playback began to be feasible. In June, DVD-RAM drives appeared, only 6 months late. Prices were reasonable, at around $800, with a surprise entry from Creative Labs at $500. 2X and 3X DVD-ROM drives also began to appear, which at speeds equivalent to18X and 24X CD-ROM drives, and with prices under $175, began to be slightly more competitive. Sony and others had already announced that 5X drives would be available soon. Still, the DVD-ROM market was not taking off as most had expected, causing many software developers to abandon DVD-ROM products and plans. Some ventures that had focused heavily on DVD-ROM, such as Divion, went out of business completely.

The Second Wind

In June, Divx was launched. Time Warner and others had already rolled out new rental programs in anticipation. Sears announced that it was canceling plans to carry Divx players, apparently because of negative publicity. Thomson (maker of RCA and ProScan brands) joined the Divx program, helping dilute image problems caused by Zenith's Chapter 11 bankruptcy. At the July Video Software Dealers Association conference, retailers were down on Divx but were moving quickly to build up DVD sales and rental programs based on reports of record player sales. At the show, Warner Home Video, DVD's biggest supporter, announced that it had generated over $110 million in revenue from DVD sales in the first 6 months of 1998. NetFlix, the first Internet-based DVD rental store, had rented over 20,000 copies since opening its e-doors in April. JVC's announcement of its D-VHS digital videotape system, a potential competitor to DVD, did little to dampen DVD enthusiasm, especially since D-VHS did not include any provision for pre-recorded tapes.

In August, Fox announced it would produce "open DVDs" along with Divx versions. This left DreamWorks SKG as the only major studio not committed to open DVD. A month later DreamWorks bowed to the inevitable and announced non-Divx titles. Still, blockbuster movies from Steven Spielberg (the S in SKG) were nowhere to be seen. A DVD release of *Back to the Future* was announced, only to be withdrawn a week later, not to reappear for over two years. Digital Theater Systems (DTS) announced that titles supporting the optional DTS audio format would finally appear by year's end. As it turned out, the direct-to-video Mulan disc was the only one to appear on schedule.

At the September 1998 DVD Forum conference in San Francisco, Warren Lieberfarb, defender of the DVD faith, predicted that DVD would be "one of the hottest consumer electronics products to be launched in the last few decades." Dan Russel of Intel predicted that more than 40 million DVD-ROM drives would be available by the end of 1999. Warren's crystal ball was crystal clear, if not quite optimistic enough, while Dan's crystal ball turned out to be running a bit fast.

The new Toshiba SD-7108 progressive-scan DVD players, able to display almost twice the resolution as standard players from the same DVD, didn't make it out of the warehouse. Concerns about lack of copy protection on progressive-scan output kept them there for another year. A few weeks later, Genesis Microchips announced a new de-interlacing chip that would eventually find its way into most of the progressive-scan DVD players released in 1999 and 2000.

In September, the Chinese government announced the new *Super Video CD* (SVCD) format, similar to DVD with MPEG-2 video, onscreen menus, and multilanguage subtitles, but using standard CDs, partly to avoid the high royalties demanded by DVD patent holders.

Hitachi announced that its DVD-RAM camcorder would be out by the end of 1999. The reality distortion field of this announcement was set to about 1.2 years.

As a publicity stunt, NetFlix made DVDs of Clinton's grand jury testimony in the Monica Lewinsky case available for 2 cents within a week of the broadcast. NetFlix received unexpected extra publicity when some customers were surprised to discover an X-rated movie in the package. In an unrelated story, Project X, the rumored game cum DVD player, was given the official moniker of Nuon. The company had wisely recognized that it would do much better by positioning its technology as an enhanced DVD player rather than yet another game console. Nuon-based players were promised for 1999. Blockbuster, well-known for its refusal to rent any movie with a rating above R, began DVD rentals in 500 stores in September. Philips and Blockbuster partnered on a rental promotion program, as did Sony with NetFlix, and Warner Bros. and Columbia TriStar with West Coast Video and Hollywood Entertainment. The flurry of rental programs just happened to coincide with the nationwide expansion of Divx. By this time NetFlix, the online rental service, had become the largest DVD rental "store" with over 2,000 titles. Online sales of DVDs by companies such as DVDExpress, Reel.com, Buy.com, and Amazon.com had become fiercely competitive, with many titles being sold at a 50-percent discount.

The UK's FACT made headlines again with a raid on Laser Enterprises in Essex, England to seize imported U.S. DVDs. A month later, in October, the official launch of DVD in Western Europe occurred; sort of. At least there were more players and more discs. This was followed by another less-than-spectacular official launch, that of HDTV in the U.S. on November 1, 1998. Approximately 40 TV stations began broadcasting bits and pieces of HDTV programming to a few thousand HDTV sets around the country. The next non-event was the release of the DVD-Audio spec, with promises of players in 1999 once the copy protection issues were resolved. As the format that cried wolf, DVD-Audio was losing credibility among DVD technology watchers. The release date of new 4.7G DVD-R drives was bumped to the second quarter of 1999, to the dismay of developers desperate to use the larger-capacity format for testing works in progress.

In November, as Comdex rolled around, DVD-ROM drive speeds were notched up to 6X. Various combinations of player features and drive speeds were touted as "2nd generation" and "3rd generation," or even "DVD II" and "DVD III," leading to much confusion among customers, many of whom thought that a generation or two of DVD had already become obsolete. Microsoft released *Encarta* on DVD, the one-millionth DVD player was sold to dealers in the U.S., and *Lost In Space*, the first major PC-enhanced movie, sold an incredible 200,000 copies in the first week. At the time there were less than 500,000 players in homes, so the obvious conclusion—given that such a lame movie couldn't possibly appeal to one out of every two player owners—was that DVD PC owners were buying movies in much higher numbers than anyone expected. InterActual, the company that developed the PCFriendly software used on *Lost In Space*, began clinching deals with most of the major Hollywood studios for PC-enhanced titles.

Demolishing opinions that dual-layer, double-sided discs (DVD-18s) would never be achieved, WAMO announced a new process to make them. This was immediately followed by rumors that the 195-minute Titanic would be the first DVD-18 release, since director James Cameron was insisting that the DVD contain both fullscreen and widescreen versions. DVD fans had been miffed when the videotape version went on sale September 1 sans the DVD version. They had a long time to wait for the DVD, although it still came out too soon for DVD-18.

DVD had a very merry Christmas in 1998. Musicland sold $5 million worth of DVDs in the last week before Christmas, compared to sales of $50 million for all of 1998. Over 5 million DVD titles were sold in the last 5 weeks of the Christmas buying season. By the end of the year, 1.4 million players had been sold to dealers in the U.S. with about 1 million installed in homes. Europe trailed with about 125,000 players. Warner Home Video, perennial barometer of DVD, made $170 million in U.S. DVD sales; 17 percent of all sell-through video rental. Shortages of standard DVD players apparently led to brisk sales of premium-priced Divx players. However, DVD-ROM was not doing quite so well. Microsoft's predictions of 15 million DVD-ROM drives were off by 6 to 7 million.[23] Over 100,000 DVD-RAM drives had shipped since the middle of the year, which was nothing to sneer at, but rather anticlimactic for a technology with the potential to take over from CD-RW.

---

[23]For those of you keeping track, yes, by this time I was contracting to Microsoft as a DVD Evangelist. I was quietly predicting 10 million DVD-ROM drives instead of the 15 million number thrown out by others at Microsoft.

The Year of DVD

1999, by most reckonings, was the year of DVD. It came off a highly successful Christmas season and continued to beat many expectations. Internet sales bloomed—Image Entertainment reported that over 23 percent of its revenue came from Internet retailers. DVD fulfilled its role as the standard-bearer of quality video: nearly 75 percent of movie titles were enhanced for 16:9 widescreen, 10 percent were collector's editions, and 20 percent took up dual-layer DVD-9 discs. The winter CES show was filled with demonstrations of writable DVD products, both for computer use and for home video recording, although no products were expected before late 1999 or 2000. The bombshell at CES was Thomson's demonstration of a prototype high-definition Divx player. Thomson was counting on cautious movie studios being happier to release high-definition movies on a secure platform such as Divx. What Thomson wasn't counting on was that Divx would be defunct six months later.

Pioneer's LD-V7200 industrial player finally appeared, about 8 months late. eMachines set a new entry level for DVD PCs with the announcement of a 333-MHz Celeron PC with a 5X DVD-ROM drive, without monitor, for $600. The first few DTS DVDs trickled out. Also trickling out were rumors that Sony was developing a new DVD-based version of its PlayStation game console.

In February, Philips, Sony, and Pioneer officially began their patent licensing program for DVD players and discs. Things were busy on the copyright front as well. Hitachi, IBM, NEC, Pioneer, and Sony (dubbed the Galaxy group) announced they had agreed on a digital watermarking technology to enhance DVD copyright protection. In the same month, apparently as a move to combat bootleggers, *Titanic* was released in China as a 4-disc Video CD set. An estimated 3.5 million pirated copies had already been sold by the time the official version hit the streets. And the *Digital Display Working Group* (DDWG) announced on February 23 the completion of the *Digital Visual Interface* (DVI) specification for final draft review. Studios were looking forward to this replacement for computer VGA output, since DVI included a copy protection mechanism. On March 3, IBM, Intel, Matsushita, and Toshiba (dubbed the "4C"), announced a content protection framework for DVD-Audio, the fruit of over 12 months of work with music labels BMG, EMI, Sony Music Entertainment, Universal Music Group, and Warner Music Group. This would later become CPSA (content protection system architecture) and cover both DVD-Video and DVD-Audio.

On March 1, 1999, Sony Computer Entertainment officially announced the PlayStation 2. "We have not come to a decision at this time whether we will put in the ability to play DVD [video] or not," said Kaz Hirai, executive vice president and chief operating officer.

DVD Gets Connected

On March 14, director John Frankenheimer hosted a live event for owners of the *Ronin* DVD. As he answered questions over the Internet, he played back hidden content from the disc to illustrate behind-the-scenes events and to explain details of producing the movie. Since it was impossible to stream DVD video over the Internet, the trick was to use InterActual's PCFriendly software to remotely control the DVD-ROM drives of the tens of thousands of chat participants. WebDVD had reached a new milestone. Research indicated that more than half of DVD-Video player owners also owned DVD-ROM PCs. Time Warner Chairman and CEO Gerald Levin called WebDVD the "key to consumer entertainment and information in the next decade." The second European DVD Summit, held in Dublin, also saw growing interest in Web-connected and enhanced DVDs. Nuon, the forerunner in creating an enhanced DVD player for the consumer electronics market, announced at the DVD Summit that the first Nuon players would be available in spring 2000, later than originally projected

On April 20, another DVD milestone was set with the release of *A Bug's Life*, the first feature film to be created and distributed from beginning to end using all-digital technology. The visual quality of the disc was stunning, especially when played back in high resolution on a computer. The producers of the film remarked that the DVD rendition, unlike the film release, finally achieved the look they had created on their CG workstations.

DVD began to pick up speed in Europe, with many new titles, and DVD-Video players selling for under £250 in UK stores such as Woolworth.

At the NAB conference in April, 1999, Sonic Solutions re-announced DVDit, now reborn as a $500 authoring package, for July. Apple released a public beta of QuickTime 4.0, without the hoped-for support for MPEG-2 or DVD. Daikin announced that it was partnering with InterActual to produce an *Enhanced DVD Kit* (EDK) to integrate WebDVD production with its Scenarist authoring system. Matsushita revealed that it planned to ship two DVD-Audio players under the Panasonic and Technics brands, priced at $800 and $1700, with delivery date tentatively set for fall. Sony released its SACD player in Japan at the tear-inducing price of $4,500. Pioneer shipped the first samples of the 4.7G DVD-R drive, based on version 1.9 of the specification. Not surprisingly, the final 2.0 version was held up by work on copy protection features. HP announced that its first DVD+RW drive would be out in June. The *Consumer Electronics Manufacturers Association* (CEMA) announced that digital television (DTV) sales over the previous seven months came to a total of 25,694.

At the E3 conference in May, Nintendo said that its upcoming DVD-based game console, code-named Dolphin, would play DVD movies. Later it was revealed that only the Matsushita (Panasonic) versions of the box would play video, since Nintendo was concentrating on a bare-bones, lowest-cost version. At the same conference Nuon admitted that players using its chip would not be out in 1999. Nuon, which had once been a unique avant-garde technology, was losing its avantness and gaining competitors.

Patents and Protections

In June 1999, the other patent pool, comprising Hitachi, Matsushita, Mitsubishi, Time Warner, Toshiba, and JVC commenced worldwide joint licensing of patents essential for DVD-Video players, DVD-ROM drives, DVD decoders, and DVD-Video and DVD-ROM discs. Sony announced that it had developed a single-chip laser with dual wavelengths; 650 nm for reading DVDs, and 780 nm for reading CDs. Oddly, Sony said that it intended to use this breakthrough technology only in the PlayStation 2.

The 4C group said that at its June 11 meeting it expected to pick a watermarking technology for its content-protection framework. The competing proposals were from the Galaxy group (Hitachi, IBM, NEC, Pioneer, and Sony) and the Millennium group (Macrovision, Digimarc, and Philips). June came and went with no decision. 1999 came and went with no decision. 2000 came and looked to go with no decision. Related efforts by the *Secure Digital Music Initiative* (SDMI) fared better, when at the June 23-25 meeting, 100 companies from the music, consumer electronics, and information technology industries adopted a specification for portable devices for digital music.

On June 16, Divx announced its own obituary.

At the PC Expo, on June 22, Philips announced that its 3.0G DVD+RW drive would be out in September 1999 for $700. At the same conference a year earlier, Sony had made a similar announcement of DVD+RW availability in 1998.

About this time, 4.7G DVD-R drives finally began shipping to anxious developers, many of whom would have gladly paid the old $17,000 price, rather than the new $5,000 price, if it would have gotten the drives to them any sooner. The ability to quickly test full DVD-5 volumes on DVD-R was a critical step in the development of the DVD industry, where year-to-date sales of players had just passed 1 million.

Also in June, the *DVD Association* (DVDA) was formed. Many in the industry recognized the need for an association to support non-Hollywood DVD developers. The *Interactive Digital Media Association* (IDMA), long-time home to CD-i developers, stepped forward to host the new association.

On July 6, 1999, another obituary of sorts was reported by Pioneer Entertainment, which announced completion of its transition out of the laserdisc business. After years of being the leading laserdisc supplier, Pioneer Entertainment had shifted to DVD and VHS only, which by then accounted for more than 90 percent of the company's title business. Although laserdisc players were still made by Pioneer New Media Technologies for education and industrial applications, the decision by Pioneer Corporation's entertainment group to abandon laserdiscs marked the end of an era.

Panasonic, hoping to create a new era of digital tape for consumers, announced again that it would begin to sell its HD-capable D-VHS VCR. NEC announced another potential competitor to DVD, its GigaStation video recorder based on *Multimedia Video Disc* (MVDisc), the latest incarnation of it the *Multimedia Video File* (MMVF) technology it had announced more than a year before. The recorders were expected to appear in Japan in September, followed by worldwide release in 2000. Panasonic reiterated that its DVD-Audio players would ship in the U.S. in October. As it turned out, all three of these long-delayed technologies had yet more delays in store.

Hitachi, Sega, and Nippon Columbia announced that they had developed a conditional-access hybrid DVD that incorporated a programmable ROM chip to store information about what parts of the disc had been purchased. Sonopress, a leading disc manufacturer, announced a different kind of hybrid, christened DVD Plus, that combined DVD and CD layers in a single disc. American Airlines announced that in September it would become the first airline to offer DVD movies on scheduled flights.

Blockbusters and Logjams

At the end of August 1999, James Cameron's *Titanic* was finally released on DVD-9, in widescreen letterbox format. The long delay and non-anamorphic version dampened enthusiasm. The disc didn't do nearly as well as *The Matrix*, which came out three weeks later and quickly became the first DVD to sell a million copies.[24] *The Matrix* gained notoriety as buyers reported problems playing it on dozens of different player models. While there were a couple of errors on the disc itself, it turned out that many players had not been properly engineered to handle a disc that aggressively exercised DVD features and included extra content for use on PCs. The success of *The Matrix* amplified the apparent magnitude of the problem. Under pressure from disgruntled customers, manufacturers released firmware upgrades to correct the flaws in their players. In November, the first major Steven Spielberg movie, *Saving Private Ryan*, was released on DVD. Customers reported video problems on the disc, only to be told that the flaring, distorted video in the beach scene was an intentional effect in the film.

Toshiba shipped its progressive-scan DVD player, which had been languishing in warehouses for a year. Panasonic also released a progressive-scan player. The true potential of DVD video quality was finally being unlocked in places other than computers.

Digital Theater Systems moved more towards the mainstream when the company released DTS encoders in October. Previously, all DTS encoding had been done by the company. DVD took a giant step in mainstream consciousness when, on October 12, 1999, it received an Emmy award from the National Academy of Television Arts & Sciences. The award was presented to Dolby, Matsushita, Philips, Sony, Time Warner, and Toshiba.

DVD technology on PCs continued to improve, with 10X drives. Another important landmark was the release of Stephen King's *The Stand*, the first commercial release on DVD-18. And on November 23, DVD officially became the hottest-selling home entertainment product in history. HDTV was not doing so well, with almost half the nation's 1,600 television stations supporting the Sinclair Broadcast Group's campaign to revise the broadcast standard. Still, factory-to-dealer sales of digital televisions continued to grow, reaching a year-to-date total of over 97,000.

At the last minute, Sony, Philips, and HP cancelled their planned DVD+RW launch. The DVD+RW camp had decided to retrench, abandoning the 3.0G format, which would have been incompatible with every existing DVD-ROM drive and player, and focusing on the improved 4.7G version that promised backward compatibility with most DVD readers.

Pioneer announced that it would release a DVD-Audio player in Japan without copy protection, since that was the only part that was unresolved.

---

[24]It took two years for DVD to reach the first million-copy point. Audio CD took four years with George Michael's *Faith*, and it took eleven years before a million VHS copies of *Top Gun* were shipped.

The player would only be able to play unprotected DVD-Audio discs until it was updated with final copy protection support. This turned out to be a reasonable compromise, since final decisions on DVD-Audio encryption and watermarking were almost a year away.

Crackers

In November 1999, the SDMI approved selection of Verance's watermarking technology for the portable music device standard. Former competing companies Aris Technologies and Solana Technology Development had merged earlier to form Verance. The same watermarking technology would be later chosen for DVD-Audio.

In November and December, fallout began to hit from a Windows software program called DeCSS that had spread across the Internet in late October. The program was designed to remove CSS encryption from discs and to copy the video files to a hard drive. DeCSS was written by 16-year-old Jon Johansen, of Norway, based on code created by a German programmer, member of an anonymous group called *Masters of Reverse Engineering* (MoRE). The MoRE programmers reverse engineered the CSS algorithm and discovered that the Xing software DVD player had not encrypted the key it used to unlock protected DVDs. Given the general weakness of the CSS design, additional keys were quickly generated by computer programs that guessed at values and tried them until they worked. Johansen claimed his intent in turning the MoRE code into an application was to be able to play movies using the Linux operating system, which had no licensed CSS implementation.

Anyone familiar with CSS was surprised that it had taken so long for the system to be cracked. After all, the first edition of this book had predicted three years earlier that CSS would be compromised. In spite of this, frenetic press reports portrayed a "shocked" movie industry, taken aback by the failure of the system that was supposed to protect its assets. DVD-Audio player manufacturers announced a six-month delay, presumably to counter the threat of DeCSS by reworking the planned CSS2 copy protection system. Those familiar with the lack of DVD-Audio titles and the incompleteness of CSS2 saw the announcement as a convenient excuse to delay products that would not have been ready in any case.

Other DVD "ripping" software had been available for years, but DeCSS was different because it directly decrypted CSS rather than intercepting video after it was decrypted and decompressed by a legitimate software DVD player. This direct implementation of the CSS protection method could be considered illegal circumvention under the DMCA and the WIPO treaties.

There were rumors that the key used by the Xing player had been revoked by removing it from the set of keys hidden on new discs, but the rumors turned out to be false. Causing a legal player to stop working, even though it had not properly protected the CSS secrets, would not have been a good move, and by this time the entire set of 400 player keys had been guessed at, so revoking one key would have done little good.

On December 27, the *DVD Copy Control Association* (DVD CCA), the corporate entity responsible for licensing CSS, sued 21 individuals and 72 Web sites, along with hundreds of "Does"—as in John Doe—to be filled in later. They were accused of posting or linking to DeCSS software as a misappropriated trade secret—a rather shaky argument. The *Electronic Frontier Foundation* (EFF) responded on behalf of the defendants with a shaky argument of their own, tied to free speech. "It is EFF's opinion that this lawsuit is an attempt to architect law to favor a particular business model at the expense of free expression. It is an affront to the First Amendment (and UN human rights accords) because the information the programmers posted is legal. EFF also objects to the DVD CCA's attempt to blur the distinction between posting material on one's own Web site and merely linking to it (i.e., providing directions to it) elsewhere." On December 29, the California court enjoined the posting of DeCSS, but denied the injunction request against linking to DeCSS, as the court rightly feared the side effects of banning the mere act of linking to information on the Internet, citing such an order as "overbroad and extremely burdensome."

Thus ended the year of DVD. Over 4 million new players had been sold in the U.S., bringing the installed base to around 5.5 million. Player sales had generated over $300 million. About 2.5 million players existed in European homes, and the worldwide base of DVD PCs was estimated at somewhere around 40 million. Fifty million discs shipped in the U.S. during the Christmas season alone. Total U.S. interactive entertainment sales for the year topped $7 billion.

The Medium of the New Millennium

The DeCSS saga continued. On January 14, 2000, the seven top U.S. movie studios—Disney, MGM, Paramount, Sony (Columbia/TriStar), Time Warner, Twentieth Century Fox, and Universal—backed by the MPAA, filed lawsuits in Connecticut and New York in a further attempt to stop the distribution of DeCSS on Web sites in those states. These suits were based on the DMCA and alleged circumvention of DeCSS. On January 24, Jon Johansen, the Norwegian programmer who first distributed DeCSS, was questioned by local police, who raided his house and confiscated his computer equipment and cell phone.

This strengthened the viewpoint of many observers that Hollywood was twisting the law to bully and intimidate the opposition in what was a losing battle. By this time the DeCSS source code was widely available on hundreds of Web sites, on thousands of T-shirts, and had even been made publicly available by the DVD CCA itself in court records.

Around this time a new wrinkle appeared, under a confusingly familiar name. A major drawback of attempting to copy DVDs was that they quickly filled even huge hard drives and took literally a week to download using a 56K modem. Copies could be made on DVD-Rs, but blank discs cost more than the original DVD. As a result, a new DVD redistribution technology called DivX ;-) appeared. (Yes, the smiley face is part of the name.) DivX ;-) was a simple hack of Microsoft's MPEG-4 video codec, coupled with MP3 audio, allowing DeCSSed video to be re-encoded at a lower data rate (and lower quality) so that it could be downloaded more easily and be played using the Windows Media Player.

In another ironic twist, Sigma Designs, the leading producer of DVD playback add-in cards, announced at the beginning of February that it would add Linux support to its NetStream 2000 DVD playback card.

At the CES show in January, most DVD player manufacturers either showed or talked about progressive-scan players, even though the number of potential customers with the needed progressive-scan displays was miniscule. DVD-Audio players were also on display, but few company representatives were willing to guess as to when they might finally be for sale.

On February 14, Jack Valenti, head of the MPAA, speaking of DeCSS and other threats to Hollywood's intellectual property, referred to the Internet, "where some obscure person sitting in a basement can throw up on the Internet a brand new motion picture, and with the click of a button have it go with the speed of light to 6 billion people around the world, instantaneously." This led people to wonder where they could get these new ISL modems[25] that 6 billion other people already had. This Chicken Little attitude, shared by so many in the motion picture industry, painted digital video as a dangerous new technology that opened the floodgates of piracy.

---

[25]ISL = instantaneous speed of light, of course.

What they seemed unable or unwilling to recognize was that any analog source such as laserdisc or even VHS tape can be digitized. Once digitized, the file can be copied and distributed the same as any other digital video file. And given the loss of image quality caused by compression methods such as Divx ;-), there is no discernable difference.

Region coding, part of the CSS license, came under attack in London when British supermarket group Tesco wrote to Warner Home Video demanding an end to the policy.

That's No Moon, That's a PlayStation!

George Lucas, under pressure from an increasingly vocal campaign to have Star Wars released on DVD, wrote a letter to fans on February 20 explaining why Star Wars would not be available on DVD any time soon. Essentially his excuse was that it had to be done right, which meant he had to do it, but that he was too busy at the moment.

In March, things improved slightly on the compatibility front. Toshiba announced a new combination CD-RW/DVD-ROM drive, and the *Optical Storage Technology Association* (OSTA) heralded the release of MultiRead 2, a specification requiring DVD units to read all CD formats as well as DVD-RAM discs.

Sony's PlayStation 2, which shipped more than 1 million units in its first 12 days, embarrassed the company with a flaw that allowed users to get around DVD region coding restrictions. Since a memory card glitch had also been discovered, Sony said, "We are asking buyers to return memory cards or consoles for checks and repairs while at the same time investigating the reasons for the glitches." Most owners declined to send their cards in to have the region code trick disabled. A week later brought the discovery of yet another loophole using the game console's analog RGB output to get around Macrovision protection in order to copy DVDs to videotape. But the glitches and loopholes were insignificant compared to the overall success of PlayStation 2, which in less than a month doubled the number of DVD players in Japan. DVD industry analysts worldwide quickly updated their forecasts to account for the expected impact of this new kind of DVD player.

Throughout March, the MPAA kept busy, sending cease and desist letters to any Web sites it found posting or linking to DeCSS.

In April 2000, Sonic Solution's DVDit finally made its debut as a $500 retail product for Windows. Two years late, and much different from the original conception, it nevertheless heralded a new generation of DVD authoring tools that were affordable to just about anyone interested in doing professional DVD production.

JVC announced a consumer version of D-VHS. The digital tape format, originally designed only to record data, had been reworked with a standard way of recording and playing back digital video, along with now requisite copy protection. JVC had high hopes that movies would be released on pre-recorded tapes. In a surprise move, Apple bought Astarte, an up-and-coming developer of DVD authoring software for Macintosh computers. Sonic, the main developer of Macintosh DVD authoring systems, announced hDVD, an unofficial variation of DVD-Video that incorporated high-definition video. The target market was high-end computers with fast DVD-ROM drives and powerful processors to decode the HD video. Ravisent, Sonic's partner in the initiative, was providing the software decoder.

In May, Circuit City recalled the Apex DVD player, one of its best-selling items. The recall and the high sales had the same cause—loopholes in the player that allowed it to be easily modified to play discs from any region and to disable Macrovision copy protection. The recall occurred primarily because of pressure from Macrovision.

Also in May, Sony announced plans for a third release of SACD player models, dropping the price from over $3,000 to just over $700, and shifting focus from audiophiles to mainstream audio buyers. Philips moved in the opposite direction, releasing a $7,500 SACD player. Less than a hundred SACD titles were available by this time. Pioneer announced that its DVD video recorder had sold 25,000 units in seven months, since its release in Japan.

Macrovision was back in the news in June with the announcement that it had implemented copy protection technology for 525-line progressive-scan output. Constellation 3D, the company that had been busy for months issuing press releases about its *fluorescent multilayer disc* (FMD) technology, announced that it had adapted the design to be readable with standard red laser technology. This raised the interesting possibility that DVD players with minor modifications would be able to read 25 gigabyte FMDs. On the other hand, C3D, a relative newcomer to the field, did not explain how discs with six or ten or more layers could be reliably mass-produced, given how hard it had been just to get two-layer DVD-9s to work. Nevertheless, C3D confidently predicted that the new players would be available by summer of 2001.

On June 12, Hollywood Entertainment announced that it was closing the e-commerce portion of its Web site, Reel.com. The company had experienced losses in 1999, and Reel.com was blamed for much of the continuing losses in 2000. Given the cutthroat discounting on Internet DVD sites, this was no surprise.

Good news on compatibility came on June 27, 2000, when the DVD Forum announced a plan to deal with conflicts among recordable formats. A new specification, called DVD Multi, defined requirements of physical compatibility between all the Forum-sanctioned formats. Any player with a DVD Multi logo would read all discs, including DVD-RAM and DVD-RW, and any recorder with the DVD Multi logo would write to all formats.

In July 2000, over three years after the introduction of DVD-Video and DVD-ROM in the U.S., DVD-Audio players shipped. It was not what anyone would call a grand coming out. There was no marketing or PR hoopla and the players were in short supply and hard to find. Verance watermarking technology, the same system chosen by the SDMI, was included. Meanwhile, the *Watermarking Review Panel* (WaRP)—the successor to the *Data-Hiding Sub-Group* (DHSG)—of the *Copy Protection Technical Working Group* (CPTWG) had still not chosen between Millennium and Galaxy watermarking proposals for DVD-Video. DVD-Audio production equipment—and watermarking equipment—was still so scarce that music studios did not expect to be able to release more than a few titles for the Christmas 2000 season. Sonic Solutions, the only company working on a commercial DVD-Audio authoring system, had only pre-release versions of the software available for use.

Tony Faulkner, an audiophile's audiophile, had been leading a grassroots effort calling for open listening tests to prove that watermarking would not destroy the high-quality audio that the format had been so carefully designed to preserve. Invoking the specter of copycode, a disastrous earlier attempt to hide copyright data in audio streams, he pointed out that results of private tests on the new watermarking methods had not been released, and that almost no testing had been done by members of the audio production community. Verance itself admitted that it had done no testing on 192 KHz content. Soon after, watermarking "tests" for DVD-Audio were conducted in London, but since the decision had already been made to use Verance technology, it was more of a demo effort to woo the audio community. Participants complained that the samples were poorly chosen and the listening environment was terrible. Faulkner, who participated in the tests, claimed that he had been able to correctly identify the watermarked streams in six out of eight cases. He worried that if he was able to detect watermarked samples with only 2-bit copy management payload, what would happen if a full 72-bit payload were used? The poor selection of "sub-Walkman standard" test pieces—analog recordings with tape hiss and "dreadfully recorded pop"—and the use of loudly whirring computer hard drives to play the music in the listing, was alarming to many, who wondered if any of the engineers working on watermarking techniques truly understood high-fidelity audio. A few music industry executives were quick to point out that watermarking was optional.

On July 5, Sony clouded the optical disc waters further with the announcement of a new format that was halfway between CD and DVD. *Double-density CD* (DDCD) versions of CD, CD-R, and CD-RW reduced track pitch and pit length to increase capacity from 650 million bytes to 1.3 billion bytes. The CIRC error correction and addressing information (ATIP) were tweaked slightly to accommodate the higher density of data, and a copy protection scheme was added. The new specification, "Purple Book," was supposed to be finalized by September 2000. New DDCD discs were not compatible with existing players.

The convergence process took another step forward when EchoStar Communications demonstrated the first satellite television receiver combined with a DVD player. The entire system, complete with satellite dish, and receiver/DVD player was priced at only $400.

Pioneer released software to upgrade version 1.9 DVD-R drives to version 2.0. New version 2.0 discs, with CPRM copy-protection features, could only be written in 2.0 drives. Older 1.9 and 1.0 discs could still be written in 2.0 drives, although 1.9 media was no longer being made.

On August 1, 2000, Warner Home Video announced that *The Matrix* DVD was the first disc to sell over three million copies in the U.S.

On August 17, Judge Kaplan, presiding over the DeCSS suit in New York, granted the requested injunction against the Web site maintained by 2600: The Hacker Quarterly. 2600 had long since removed the DeCSS code, but it maintained that it had a right to link to Web sites containing DeCSS or information about DeCSS—thus the suit. The district court granted a permanent injunction against (1) posting on any Internet site, or in any other way manufacturing, importing or offering to the public, providing, or otherwise trafficking in DeCSS or any other technology primarily designed to circumvent CSS, and (2) linking any Internet web site, either directly or through a series of links, to any other Internet web site containing DeCSS. In response to the injunction, 2600 revised its list of some 450 Web sites so that the URLs were listed without embedded links. Users interested in going to any of the sites had to copy and paste the URLs into their Web browser. This was the sole tangible outcome of the MPAA's suit. In an inspired bit of irony, the 2600 Web site also directed visitors to the Web search engine at Disney's Go.com, which provided hundreds of links to sites such as the DeCSS Distribution Center. In theory, all Internet search engines were in violation of DMCA as long as a single copy of DeCSS existed, even in countries where it was not illegal. Following the favorable outcome of the suit, the MPAA sent new rounds of threatening letters to Internet sites posting or linking to DeCSS. "If we have to file a thousand lawsuits a day, we'll do it," said Valenti. "It's less expensive than losing control of all your creative works."

DVD Turns Four

Fall 2000. Four years since the introduction of DVD in Japan. *Jurassic Park* is finally coming out on DVD, joined by dozens of other high-profile movies such as *Braveheart*, *American Beauty*, *Toy Story*, *Toy Story 2*, and *The Rocky Horror Picture Show*. George Lucas has committed to releasing Star Wars on DVD...some day. Nuon-based DVD players are...right around the corner. A variety of home DVD video recorders, at least one of each flavor of writable DVD, are about to become the newest toys for those who can afford the $2,000 to $4,000 price tags. Hitachi is poised to release a $2,300 camcorder using 8-cm DVD-RAM discs. Students at seven dental schools will carry DVDs in place of two million pages of textbooks and manuals—400 pounds of material that would normally be used in the course of four years of dental school.

Sales of players are on target to hit 10 percent penetration of U.S. households—over 10 million—before the end of the year. Over 8,500 movies are available on DVD, to be played in over 15 million players and over 60 million DVD computers worldwide. DVD-ROM titles have yet to take off, and DVD-ROM drive placement in new computers is still low, but the transition to DVD-ROM technology in PCs is still inevitable.

Looking back, it seems silly to have ever doubted that DVD would be anything but a rousing success.

DVD
Technology
Primer

Introduction

This chapter explains some of the basic technology that is part of DVD technology, such as audio and video encoding, aspect ratios, and video scanning formats.

Gauges and Grids: Understanding Digital and Analog

We live in an analog world. Our perceptions are stimulated by information in smooth, unbroken form, such as sound waves that apply varying pressure on our eardrums, a mercury thermometer showing infinitely measurable detail, or a speedometer dial that moves continuously across its range. Digital information, on the other hand, is a series of snapshots of analog values coded as numbers, like a digital thermometer that reads 71.5 degrees or a digital speedometer that reads 69 mph.[1]

The first recording techniques all used analog methods—changes in physical material such as wavy grooves in plastic disks, silver halide crystals on film, or magnetic oxides on tape. After transistors and computers came on the scene, it was discovered that information signals could be isolated from their carriers if they were stored in digital form. One of the big advantages of digital information is that it is infinitely malleable. It can be processed, transformed, and copied without losing a single bit of information. Analog recordings always contain noise (such as tape hiss) and random perturbations, so each successive generation of recoding or

---

[1]There are endless debates about whether the "true" nature of our world is analog or digital. Consider again the thermometer. At a minute enough level of detail, the readings cannot be more accurate than a molecule of mercury. Physicists explain that the sound waves and photons that excite receptors in our ears and eyes can be treated as waves or as particles. Waves are analog, but particles are digital. Research shows that we perceive sound and video in discrete steps so that our internal perception is actually a digital representation of the analog world around us. There are a finite number of cones in the retina, similar to the limited number of photoreceptors in the CCD of a digital camera. At the quantum level, all of reality is determined by discrete quantum energy states that can be thought of as digital values. However, for the purposes of this discussion, referring to gross human perception, it is sufficiently accurate to say that sound and light, and our sensation of them, are analog.

transmission is of lower quality. Digital information can pass through multiple generations, such as from a digital video master, through a studio network, over the Internet, into a computer bus, out to a recordable DVD, back into the computer, through the computer graphics chips, out over a FireWire connection, and into a digital monitor, all with no loss of quality. Digital signals representing audio and video also can be processed numerically. Digital signal processing is what allows AV receivers to simulate concert halls, surround sound headphones to simulate multiple speakers, and studio equipment to enhance video or even correct colors.

When storing analog information in digital form, the trick is to produce a representation that is very close to the original. If the numbers are exact enough (like a thermometer reading of 71.4329 degrees) and repeated often enough, they closely represent the original analog information.[2]

Digital audio is a series of numbers representing the intensity, or amplitude, of a sound wave at a given point. In the case of DVD, these numbers are "sampled" over 48,000 times a second (as high as 192,000 times a second for super-high–fidelity audio), providing a much more accurate recording than is possible with the rough analogues (pun intended) of vinyl records or magnetic tape. When a digital audio recording is played back, the stream of numerical values is converted into a series of voltage levels, creating an undulating electrical signal that drives a speaker.

Digital video is a sheet of dots, called *pixels,* each holding a color value. It is similar to drawing a picture by coloring in a grid, where each square of the grid can only be filled in with a single color. If the squares are small enough and a sufficient range of colors is available, the drawing becomes a reasonable facsimile of reality. For DVD, each grid of 720 squares across by 480 or 576 squares down represents a still image, called a *frame.* Thirty frames are shown each second to convey motion. (For PAL DVDs, 25 frames are shown each second.)

---

[2]Ironically, digital data is stored on analog media. The pits and lands on a DVD are not of a uniform depth and length, and they do not directly represent ones and zeros. (They produce a waveform of reflected laser light that represents coded runs of zeros and transition points.) Digital tape recordings use the same magnetic recording medium as analog tapes. Digital connections between AV components (digital audio cables, IEEE-1394/Firewire, etc.) encode data as square waves at analog voltage levels. However, in all cases, the digital signal threshold is kept far above the noise level of the analog medium so that variations do not cause errors when the data is retrieved.

Birds Over the Phone:
Understanding Video Compression

After compact discs appeared in 1982, digital audio became a commodity. It took many years before the same transformation could begin to work its magic on video. The step up from digital audio to digital video is a doozy, for in any segment of television there is about 250 times as much information as in the same-length segment of CD audio. Despite its larger capacity, however, DVD is not even close to 250 times more spacious than CD-ROM. The trick is to reduce the amount of video information without significantly reducing the quality of the picture. The solution is digital compression.

In a sense, you employ compression in daily conversations. Picture yourself talking on the phone to a friend. You are describing the antics of a particularly striking bird outside your window. You might begin by depicting the scene and then mentioning the size, shape, and color of the bird. But when you begin to describe the bird's actions, you naturally do not repeat your description of the background scene or the bird. You take it for granted that your friend remembers this information, so you only describe the action —the part that changes. If you had to continually refresh your friend's memory of every detail, you would have very high phone bills. The problem with TV is that it has no memory—the picture has to be refreshed continually, literally. It is as if the TV were saying, "There's a patch of grass and a small tree with a 4-inch green and black bird with a yellow beak sitting on a branch. Now there's a patch of grass and a small tree with a 4-inch green and black bird with a yellow beak hanging upside down on a branch. Now there's a patch of grass and a small tree with a 4-inch green and black bird with a yellow beak hanging upside down on a branch trying to eat some fruit," and so on, only in much more detail, redescribing the entire scene 30 times a second. In addition, a TV individually describes each piece of the picture even when they are all the same. It would be as if you had to say, "The bird has a black breast and a green head and a green back and green wing feathers and green tail feathers and . . ." (again, in much more meticulous detail) rather than simply saying, "The bird has a black breast, and the rest is green." This kind of conversational compression is second nature to us, but for computers to do the same thing requires complex algorithms. Coding only the changes in a scene is called *conditional replenishment*.

The simplest form of digital video compression takes advantage of spatial redundancy—areas of a single picture that are the same. Computer pictures are made up of a grid of dots, each one a specified color. But many of

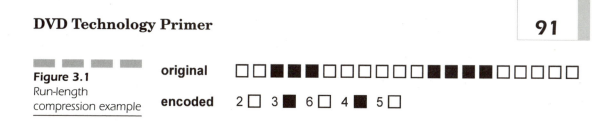

Figure 3.1
Run-length
compression example

the dots are the same color. Therefore, rather than storing, say, a hundred red dots, you store one red dot and a count of 100. This reduces the amount of information from 100 pieces to 3 pieces (a marker indicating a run of similar colored dots, the color, and the count) or even 2 pieces (if all information is stored as pairs of color and count) (Figure 3.1). This is called *run-length compression*. It is a form of *lossless* compression, meaning that the original picture can be reconstructed perfectly with no missing detail. Run-length compression is great for simple pictures and computer data but does not reduce a large, detailed picture enough for most purposes.

DVD-Video uses run-length compression for subpictures, which contain captions and simple graphic overlays. The legibility of subtitles is critical, so it is important that no detail be lost. DVD limits subpictures to four colors at a time, so there are lots of repeating runs of colors, making them perfect candidates for run-length compression. Compressed subpicture data makes up less than one-half of 1 percent of a typical DVD-Video program.

In order to reduce picture information even more, *lossy* compression is required. In this case, information is removed permanently. The trick is to remove detail that will not be noticed. Many such compression techniques, known as *psychovisual* encoding systems, take advantage of a number of aspects of the human visual system.

1. The eye is more sensitive to changes in brightness than in color.

2. The eye is unable to perceive brightness levels above or below certain thresholds.

3. The eye cannot distinguish minor changes in brightness or color. This perception is not linear. In other words, certain ranges of brightness or color are more important visually than others. For example, variegated shades of green such as leaves and plants in a forest are more easily discriminated than various shades of dark blue such as in the depths of a swimming pool.

4. Gentle gradations of brightness or color (such as a sunset blending gradually into a blue sky) are more important to the eye and more readily perceived than abrupt changes (such as pinstriped suits or confetti).

The human retina has three types of color photoreceptor cells, called *cones*.[3] Each is sensitive to different wavelengths of light that roughly correspond to the colors red, green, and blue. Because the eye perceives color as a combination of these three stimuli, any color can be described as a combination of these primary colors.[4] Televisions work by using three electron beams to cause different phosphors on the face of the television tube to emit red, green, or blue light, abbreviated to RGB. Television cameras record images in RGB format, and computers generally store images in RGB format.

RGB values are a combination of brightness and color. Each triplet of numbers represents the intensity of each primary color. As just noted, however, the eye is more sensitive to brightness than to color. Therefore, if the RGB values are separated into a brightness component and a color component, the color information can be more heavily compressed. The brightness information is called *luminance* and is often denoted as Y.[5] Luminance is essentially what you see when you watch a black-and-white TV. Luminance is the range of intensity from 0 percent (black) through 50 percent (gray) to 100 percent (white). A logical assumption is that each RGB value would contribute one-third of the intensity information, but the eye is most sensitive to green, less sensitive to red, and least sensitive to blue, so a uniform average would yield a yellowish green image instead of a gray image.[6] Consequently, it is necessary to use a weighted sum corresponding to the spec-

---

[3]*Rods,* another type of photoreceptor cell, are only useful in low-light environments to provide what is commonly called *night vision.*

[4]You may have learned that the primary "colors" are red, yellow, and blue. Technically, these are magenta, yellow, and cyan and usually refer to pigments rather than colors. A magenta ink absorbs green light, thus controlling the amount of green color perceived by the eye. Since white light is composed of equal amounts of all three colors, removing green leaves red and blue, which together form magenta. Likewise, yellow ink absorbs blue light, and cyan ink absorbs red light. Reflected light, such as that from a painting, is formed from the character of the illuminating light and the absorption of the pigments. Projected light, such as that from a television, is formed from the intensities of the three primary colors. Since video is projected, it deals with red, green, and blue colors.

[5]The use of Y for luminance comes from the XYZ color system defined by the *Commission Internationale de L'Eclairage* (CIE). The system uses three-dimensional space to represent colors, where the Y axis is luminance and X and Z axes represent color information.

[6]Luminance from RGB can be a difficult concept to grasp. It may help to think of colored filters. If you look through a red filter, you will see a monochromatic image composed of shades of red. The image would look the same through the red filter if it were changed to a different color, such as gray. Since the red filter only passes red light, anything that is pure blue or pure green will not be visible. To get a balanced image, you would use three filters, change the image from each one to gray, and average them together.

Figure 3.2
Color and luminance
sensitivity of the eye

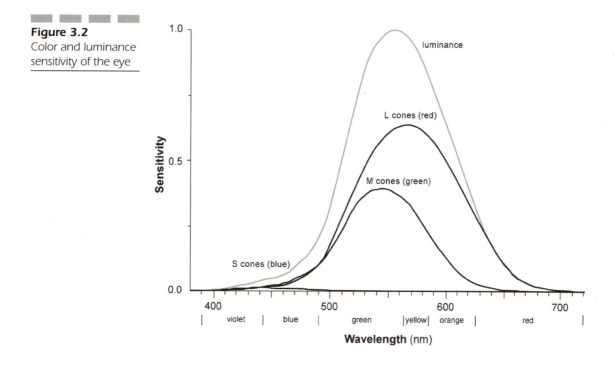

tral sensitivity of the eye, which is about 70 percent green, 20 percent red, and 10 percent blue (Figure 3.2).

The remaining color information is called *chrominance* (denoted as *C*), which is made up of *hue* (the proportion of color: the redness, orangeness, greenness, etc.), and *saturation* (the purity of the color, from pastel to vivid). For the purposes of compression and converting from RGB, however, it is easier to use *color difference* information rather than hue and saturation. In other words, the color information is what is left after the luminance is removed. By subtracting the luminance value from each RGB value, three color difference signals are created—R-Y, G-Y, and B-Y. Only three stimulus values are needed, so only two color difference signals need be included with the luminance signal. Since green is the largest component of luminance, it has the smallest difference signal (G makes up the largest part of Y, so G-Y results in the smallest values). The smaller the signal, the more it is subject to errors from noise, so B-Y and R-Y are the best choice. The green color information can be recreated by subtracting the two difference signals from the Y signal (roughly speaking). Different weightings are used to derive Y and color differences from RGB, such as YUV, YIQ, and YC_bC_r.

DVD uses YC_bC_r as its native storage format. Details of the variations are beyond the scope of this book.

As mentioned earlier, the sensitivity of the eye is not linear, and neither is the response of the phosphors used in television tubes. Therefore, video is usually represented with corresponding nonlinear values, and the terms *luma* and *chroma* are used. These are denoted with the prime symbol as Y' and C', as is the corresponding R'G'B'. Details of nonlinear functions are also beyond the scope of this book.

Compressing Single Pictures

An understanding of the nuances of human perception led to the development of compression techniques that take advantage of certain characteristics. Just such a development is *JPEG compression*, which was produced by the Joint Photographic Experts Group and is now a worldwide standard. JPEG separately compresses Y, B-Y, and R-Y information, with more compression done on the latter two, to which the eye is less sensitive.

To take advantage of another human vision characteristic—less sensitivity to complex detail—JPEG divides the image into small blocks and applies a *discrete cosine transform* (DCT), a mathematical function that changes spatial intensity values to spatial frequency values. This describes the block in terms of how much the detail changes and roughly arranges the values from lowest frequency (represented by large numbers) to highest frequency (represented by small numbers). For areas of smooth colors or low detail (low spatial frequency), the numbers will be large. For areas with varying colors and detail (high spatial frequency), most of the values will be close to zero. A DCT is an essentially *lossless transform,* meaning that an inverse DCT function can be performed on the resulting set of values to restore the original values. In practice, integer math and approximations are used, causing some loss at the DCT stage. Ironically, the numbers are bigger after the DCT transform. The solution is to *quantize* the DCT values so that they become smaller and repetitive.

Quantizing is a way of reducing information by grouping it into chunks. For example, if you had a set of numbers between 1 and 100, you could quantize them by 10. That is, you could divide them by 10 and round to the nearest integer. The numbers from 5 to 14 would all become 1s, the numbers from 15 to 24 would become 2s, and so on, with 1 representing 10, 2 representing 20, and so forth. Instead of individual numbers such as 8, 11, 12, 20, and 23, you end up with "3 numbers near 10" and "2 numbers near 20." Obviously, quantizing results in a loss of detail.

Quantizing the DCT values means that the result of the inverse DCT will not exactly reproduce the original intensity values, but the result is close and can be adjusted by varying the quantizing scale to make it finer or coarser. More important, since the DCT function includes a progressive weighting that puts bigger numbers near the top left corner and smaller numbers near the lower right corner, quantization and a special zigzag ordering result in runs of the same number, especially zero. This may sound familiar. Sure enough, the next step is to use run-length encoding to reduce the number of values that need to be stored. A variation of run-length coding is used that stores a count of the number of zero values followed by the next nonzero value. The resulting numbers are used to look up symbols from a table. The symbol table was developed using Huffman coding to create shorter symbols for the most commonly appearing numbers. This is called *variable-length coding* (VLC). See Figures 3.3 and 3.5 for examples of DCT, quantization, and VLC.

The result of these transformation and manipulation steps is that the information that is thrown away is least perceptible. Since the eye is less sensitive to color than to brightness, transforming RGB values to luminance and chrominance values means that more chrominance data can be selectively thrown away. And since the eye is less sensitive to high-frequency color or brightness changes, the DCT and quantization process removes mostly the high-frequency information. JPEG compression can reduce a picture to about one-fifth the original size with almost no discernible difference and to about one-tenth the original size with only slight degradation.

Compressing Moving Pictures

Motion video adds a *temporal* dimension to the spatial dimension of single pictures. Another worldwide compression standard from the *Moving Picture Experts Group* (MPEG), was designed with this in mind. MPEG is similar to JPEG but also reduces redundancy between successive pictures of a moving sequence.

Just as your friend's memory allows you to describe things once and then only talk about what's changing, digital memory allows video to be compressed in a similar manner by first storing a single picture and then only storing the changes. For example, if the bird moves to another tree, you can tell your friend that the bird has moved without needing to describe the bird over again.

MPEG compression uses a similar technique called *motion estimation* or *motion-compensated prediction*. Since motion video is a sequence of still

Figure 3.3
Block transforms and quantization

Encoding

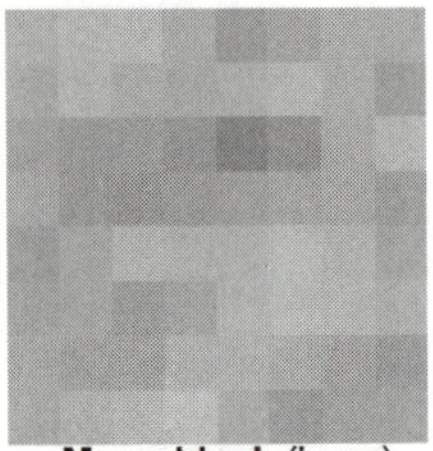

Macroblock (luma)

Pixel values

134	142	145	131	114	122	131	130
129	143	134	130	135	144	134	118
123	117	118	111	97	109	130	143
129	116	112	116	120	126	130	118
118	127	141	138	138	148	141	125
125	129	119	127	143	149	145	136
131	126	128	142	141	135	126	116
131	140	146	154	133	118	124	124

DCT coefficients

+1037	-1	-6	+1	-12	+8	-4	-4
-16	+1	+28	-6	-14	0	+4	0
+19	+32	-7	-19	+2	-1	-4	-3
+29	-9	-14	+13	-10	-6	+1	0
+4	+14	-6	-13	-2	+7	+1	+2
-26	+2	+16	+2	+11	+6	+1	+1
-10	-11	+27	-18	+4	+1	0	0
-2	+1	+1	-19	-1	+6	+6	0

Quantized coefficients
$$Q\mathring{A}(DCT*16)/(QM*8)$$

+130	0	0	0	-1	0	0	0
-2	0	+3	0	-1	0	0	0
+2	+3	0	-1	0	0	0	0
+3	-1	-1	+1	0	0	0	0
0	+1	0	-1	0	0	0	0
-2	0	+2	0	0	0	0	0
-1	-1	+2	-1	0	0	0	0
0	0	0	-1	0	0	0	0

Decoding

Reconstructed coefficients (dequantize)

+1040	0	0	0	-9	0	0	0
-12	0	+24	0	-10	0	0	0
+14	+24	0	-10	0	0	0	0
+24	-8	-9	+10	0	0	0	0
0	+9	0	-10	0	0	0	0
-19	0	+10	0	0	0	0	0
-9	-10	+21	-12	0	0	0	0
0	0	0	-14	0	0	0	0

Reconstructed values (IDCT)

136	141	138	125	119	125	132	134
136	137	133	130	134	139	133	121
121	125	122	112	107	117	130	138
125	123	119	117	121	128	127	122
123	129	135	139	140	139	136	131
129	125	124	130	139	144	141	136
129	130	134	139	140	135	127	120
132	138	144	141	131	122	122	127

Difference

-2	+1	+7	+6	-5	-3	-1	-4
-7	+6	+1	+0	+1	+5	+1	-3
+2	-8	-4	-1	-10	-8	0	+5
+4	-7	-7	-1	-1	-2	+3	-4
-5	-2	+6	-1	-2	+9	+5	-6
-4	+4	-5	-3	+4	+5	+4	0
+2	-4	-6	+3	+1	0	-1	-4
-1	+2	+2	+13	+2	-4	+1	-3

Quantization matrix

8	16	19	22	26	27	29	34
16	16	22	24	27	29	34	37
19	22	26	27	29	34	34	38
22	22	26	27	29	34	37	40
22	26	27	29	32	35	40	48
26	27	29	32	35	40	48	58
26	27	29	34	38	46	56	69
27	29	35	38	46	56	69	83

pictures, many of which are very similar, each picture can be compared with the pictures near it. The MPEG encoding process breaks each picture into blocks, called *macroblocks,* and then hunts around in neighboring pictures for similar blocks. If a match is found, instead of storing the entire block, the system stores a much smaller *vector* describing how far the block moved (or did not move) between pictures. Vectors can be encoded in as little as 1 bit, so backgrounds and other elements that do not change over time are compressed extremely efficiently. Large groups of blocks that move together, such as large objects or the entire picture panning sideways, are also compressed efficiently.

MPEG uses three kinds of picture storage methods. *Intra* pictures are like JPEG pictures, in which the entire picture is compressed and stored with DCT quantization. This creates a reference, or *information,* frame from which successive pictures are built. These *I frames* also allow random access into a stream of video and in practice occur about twice a second. *Predicted* pictures, or *P frames,* contain motion vectors describing the difference from the closest previous I frame or P frame. If the block has changed slightly in intensity or color (remember, frames are separated into three channels and compressed separately), then the difference (*error*) is also encoded. If something entirely new appears that does not match any previous blocks, such as a person walking into the scene, then a new block is stored in the same way as in an I frame. If the entire scene changes, as in a cut, the encoding system is usually smart enough to make a new I frame. The third storage method is a *bidirectional* picture, or *B frame.* The system looks both forward and backward to match blocks. In this way, if something new appears in a B frame, it can be matched to a block in the next I frame or P frame. Thus P and B frames are much smaller than I frames.

Experience has shown that two B frames between each I or P frame work well. A typical second of MPEG video at 30 frames per second looks like I B B P B B P B B P B B I B B P B B P B B P B B P B B (Figure 3.4). Obviously, B frames are much more complex to create than P frames, requiring time-consuming searches in both the previous and subsequent I or P frame. For this reason, some real-time or low-cost MPEG encoders only create I and P frames. Likewise, I frames are easier to create than P frames, which require searches in the subsequent I or P frame. Therefore, the simplest

Figure 3.4

Typical MPEG picture sequence

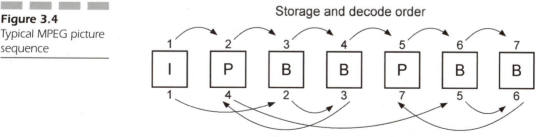

encoders only create I frames. This is less efficient but may be necessary for very inexpensive real-time encoders that must process 30 or more frames a second.

MPEG-2 encoding can be done in real time (where the video stream enters and leaves the encoder at display speeds), but it is difficult to produce quality results, especially with *variable bit rate* (VBR). VBR allows varying numbers of bits to be allocated for each frame depending on the complexity. Less data is needed for simple scenes, whereas more data can be allocated for complex scenes. This results in a lower average data rate and longer playing times but provides room for data peaks to maintain quality. DVD encoding frequently is done with VBR and is usually not done in real time, so the encoder has plenty of time for macroblock matching, resulting in much better quality at lower data rates. Good encoders make one pass to analyze the video and determine the complexity of each frame, forcing I frames at scene changes and creating a compression profile for each frame. They then make a second pass to do the actual compression, varying quantization parameters to match the profiles. The human operator often "tweaks" minor details between the two passes. Many low-cost MPEG encoding hardware or software for personal computers uses only I frames, especially when capturing video in real time. This results in a simpler and cheaper encoder, since P and B frames require more computation and more memory to encode. Some of these systems can later reprocess the I frames to create P and B frames. MPEG also can encode still images as I frames. Still menus on a DVD, for example, are I frames.

The result of the encoding process is a set of data and instructions (Figure 3.5). These are used by the decoder to recreate the video. The amount of compression (how coarse the quantizing steps are, how large a motion estimation error is allowed) determines how closely the reconstructed video resembles the original. MPEG decoding is *deterministic*—a given set of input data always should produce the same output data. Decoders that properly implement the complete MPEG decoding process will produce the same numerical picture even if they are built by different manufacturers.[7] This does not mean that all DVD players will produce the same video picture. Far from it, since many other factors are involved, such as conversion from digital to analog, connection type, cable quality, and display quality. Advanced decoders may include extra processing steps such as block filter-

---

[7]Technically, the *inverse discrete cosine transform* (IDCT) stage of the decoding process is not strictly prescribed, and is allowed to introduce small statistical variances. This should never account for more than an occasional least significant bit of discrepancy between decoders.

Figure 3.5
MPEG video
compression example

DC	Symbols
130	01 0

Run codes (zeros, value)	Symbols (from lookup table)
1,-2	0011 01
0,+2	1100
3,+3	0000 0001 1100 0
0,+3	0111 0
0,+3	0111 0
1,-1	0101
2,-1	0010 11
1,-1	0101
0,-1	101
0,-1	101
0,+1	100
0,-2	1101
0,-1	101
2,+1	0010 10
7,-1	0000 1001
0,+1	100
0,-1	101
2,+2	0000 1110
9,-1	1111 0001
1,-1	0101
EOB	0110

Bytestream (14 bytes: 22 symbols)

46 E0 0E 1C E5 2D 6D 9B 4A 09
94 3B C5 58

Compression
4.6:1 (64:14)

134	142	145	131	114	122	131	130
129	143	134	130	135	144	134	118
123	117	118	111	97	109	130	143
129	116	112	116	120	126	130	118
118	127	141	138	138	148	141	125
125	129	119	127	143	149	145	136
131	126	128	142	141	135	126	116
131	140	146	154	133	118	124	124

Pixel values

+130	0	0	0	-1	0	0	0
-2	0	+3	0	-1	0	0	0
+2	+3	0	-1	0	0	0	0
+3	-1	-1	+1	0	0	0	0
0	+1	0	-1	0	0	0	0
-2	0	+2	0	0	0	0	0
-1	-1	+2	-1	0	0	0	0
0	0	0	-1	0	0	0	0

Quantized coefficients

	1	5	6	14	15	27	28
2	4	7	13	16	26	29	42
3	8	12	17	25	30	41	43
9	11	18	24	31	40	44	53
10	19	23	32	39	45	52	54
20	22	33	38	45	51	55	60
21	34	37	47	50	56	59	61
35	36	48	49	57	58	62	63

Zig-zag scan sequence

ing and edge enhancement. Also, many software MPEG decoders take shortcuts to achieve sufficient performance. Software decoders may skip frames and use mathematical approximations rather than the complete but time-consuming transformations. This results in lower-quality video than from a fully compliant decoder.

Encoders, on the other hand, can and do vary widely. The encoding process has the greatest effect on the final video quality. The MPEG standard prescribes a *syntax* defining what instructions can be included with the encoded data and how they are applied. This syntax is quite flexible, and leaves much room for variation. The quality of the decoded video depends very much on how thoroughly the encoder examines the video and how clever it is about applying the functions of MPEG to compress it. In a sense, MPEG is still in its infancy, and much remains to be learned about efficient encoding. DVD video quality steadily improves as encoding techniques and equipment get better. The decoder chip in the player will not change—it doesn't need to be changed—but the improvements in the encoded data will provide a better result. This can be likened to reading aloud from a book. The letters of the alphabet are like data organized according to the syntax of language. The person reading aloud from the book is similar to the decoder—the reader knows every letter and is familiar with the rules of pronunciation. The author is similar to the encoder—the writer applies the rules of spelling and usage to encode thoughts as written language. The better the author, the better the results. A poorly written book will come out sounding bad no matter who reads it, but a well-written book will produce eloquent spoken language.[8]

It should be recognized that random artifacts in video playback (aberrations that appear in different places or at different times when the same video is played over again) are not MPEG encoding artifacts. They may indicate a faulty decoder, errors in the signal, or something else independent of the MPEG encode-decode process. It is impossible for fully compliant, properly functioning MPEG decoders to produce visually different results from the same encoded data stream.

MPEG (and most other compression techniques) are *asymmetric,* meaning that the encoding process does not take the same amount of time as the decoding process. It is more effective and efficient to use a complex and time-consuming encoding process because video generally is encoded only once before being decoded hundreds or millions of times. High-quality MPEG encoding systems can cost hundreds of thousands dollars, but since most of the work is done during encoding, decoder chips cost less than $20, and decoding can even be done in software.

Some analyses indicate that a typical video signal contains over 95 percent redundant information. By encoding the changes between frames, rather than reencoding each frame, MPEG achieves amazing compression ratios. The difference from the original generally is imperceptible even when compressed by a factor of 10 to 15. DVD-Video data typically is compressed to approximately one-thirtieth of the original size (Table 3.1).

---

[8]Obviously, it would sound better if read by James Earl Jones than by Ross Perot. But the analogy holds if you consider the vocal characteristics to be independent of the translation of words to sound. The brain of the reader is the decoder, the diction of the reader is the post-MPEG video processing, and the voice of the reader is the television.

TABLE 3.1

Compression Ratios

Native data	Native rate (kbps)	Compression	Compressed Rate (kbps)*	Ratio	Percent
720 × 480 × 12 bits × 24 fps	99,533	MPEG-2	3,500	28:1	96
720 × 480 × 12 bits × 24 fps	99,533	MPEG-2	6,000	17:1	94
720 × 576 × 12 bits × 24 fps	119,439	MPEG-2	3,500	34:1	97
720 × 576 × 12 bits × 24 fps	119,439	MPEG-2	6,000	20:1	95
720 × 480 × 12 bits × 30 fps	124,416	MPEG-2	3,500	36:1	97
720 × 480 × 12 bits × 30 fps	124,416	MPEG-2	6,000	21:1	95
352 × 240 × 12 fps × 24 bits	24,330	MPEG-1	1,150	21:1	95
352 × 288 × 12 fps × 24 bits	29,196	MPEG-1	1,150	25:1	96
352 × 240 × 12 fps × 30 bits	30,413	MPEG-1	1,150	26:1	96
2 ch × 48 kHz × 16 bits	1,536	Dolby Digital 2.0	192	8:1	87
6 ch × 48 kHz × 16 bits	4,608	Dolby Digital 5.1	384	12:1	92
6 ch × 48 kHz × 16 bits	4,608	Dolby Digital 5.1	448	10:1	90
6 ch × 48 kHz × 16 bits	4,608	DTS 5.1	768	6:1	83
6 ch × 48 kHz × 16 bits	4,608	DTS 5.1	1,536	3:1	67
6 ch × 96 kHz × 20 bits	11,520	MLP	5,400	2:1	53
6 ch × 96 kHz × 24 bits	13,824	MLP	7,600	2:1	45

*MPEG-2 and MLP compressed data rates are an average of a typical variable bit rate

Birds Revisited: Understanding Audio Compression

Audio takes up much less space than video, but uncompressed audio coupled with compressed video uses up a large percentage of the available bandwidth. Compressing the audio can result in a small loss of quality, but if the resulting space is used instead for video, it may improve the video quality significantly. In essence, reducing both the audio and the video is more effective. Usually video is compressed more than audio, since the ear is more sensitive to detail loss than the eye.

Just as MPEG compression takes advantage of characteristics of the human eye, modern audio compression relies on detailed understanding of the human ear. This is called *psychoacoustic* or *perceptual* coding.

Picture again your telephone conversation with a friend. Imagine that your friend lives near an airport, so that when a plane takes off, your friend cannot hear you over the sound of the airplane. In a situation like this, you quickly learn to stop talking when a plane is taking off, since your friend will not hear you. The airplane has *masked* the sound of your voice. At the opposite end of the loudness spectrum from airplane noise is background noise, such as a ticking clock. While you are speaking, your friend cannot hear the clock, but if you stop, then the background noise is no longer masked.

The hairs in your inner ear are sensitive to sound pressure at different frequencies (pitches). When stimulated by a loud sound, they are incapable of sensing softer sounds at the same pitch. Because the hairs for similar frequencies are near each other, a stimulated audio receptor nerve will interfere with nearby receptors and cause them to be less sensitive. This is called *frequency masking*.

Human hearing ranges roughly from low frequencies of 20 Hz to high frequencies of 20,000 Hz (20 kHz). The ear is most sensitive to the frequency range from about 2 to 5 kHz, which corresponds to the range of the human voice. Because aural sensitivity varies in a nonlinear fashion, sounds at some frequencies mask more neighboring sounds than at other frequencies. Experiments have established certain *critical bands* of varying size that correspond to the masking function of human hearing (Figure 3.6).

Another characteristic of the human audio sensory system is that sounds cannot be sensed when they fall below a certain loudness (or amplitude). This sensitivity threshold, as with everything else, is not linear. In other words, the threshold is at louder or softer points at different frequencies. The overall threshold varies a little from person to person—some people have better hearing than others. The threshold of hearing is adaptive; the ear can adjust its sensitivity in order to pick up soft sounds when not over-loaded by loud sounds. This characteristic causes the effect of *temporal masking,* in which you are unable to hear soft sounds for up to 200 milliseconds after a loud sound and for 2 or 3 milliseconds before a loud sound.[9]

Perceptual Coding

DVD uses three audio data reduction systems: Dolby Digital (AC-3) coding, MPEG audio coding, and DTS (Coherent Acoustics) coding. All use mathematical models of human hearing based on sensitivity thresholds, frequency masking, and temporal masking to remove sounds that you cannot hear. The resulting information is compressed to about one-third to one-twelfth the original size with little to no perceptible loss in quality (see Table 3.1).

Digital audio is *sampled* by taking snapshots of an analog signal thousands of times a second. Each sample is a number that represents the amplitude (strength) of the waveform at that instance in time. Perceptual

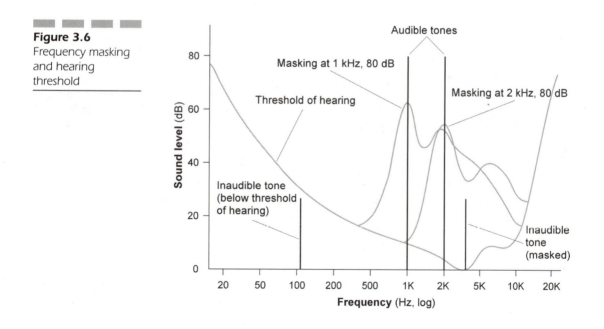

Figure 3.6
Frequency masking and hearing threshold

---

[9]How can masking work backward in time? The signal presented by the ear to the brain is a composite built up from stimuli received over a period of about 200 milliseconds. A loud noise effectively overrides a small portion of the earlier stimuli before it can be accumulated and sent to the brain.

audio compression takes a block of samples and divides them into frequency bands of equal or varying widths. Bands of different widths are designed to match the sensitivity ranges of the human ear. The intensity of sound in each band is analyzed to determine two things: (1) how much masking it causes in nearby frequencies and (2) how much noise the sound can mask within the band. Analyzing the masking of nearby bands means that the signal in bands that are completely masked can be ignored. Calculating how much noise can be masked in each band determines how much compression can be applied to the signal within the band. Compression uses quantization, which involves dividing and rounding, and this can create errors known as *quantization noise.* For example, the number 32 quantized by 10 gives 3.2 rounded to 3. When reexpanded, the number is reconstructed as 30, creating an error of 2. These errors can manifest themselves as audible noise. After masked sounds are ignored, remaining sounds are quantized as coarsely as possible so that quantization noise is either masked or is below the threshold of hearing. The technique of noise masking is related to *noise shaping* and is sometimes called *frequency-domain error confinement.*

Another technique of audio compression is to compare each block of samples with the preceding and following blocks to see if any can be ignored on account of temporal masking—soft sounds near loud sounds—and how much quantization noise will be temporally masked. This is sometimes called *temporal-domain error confinement.*

Digital audio compression also can take advantage of the redundancies and relationships between channels, especially when there are six or eight channels. A strong sound in one channel can mask weak sounds in other channels, information that is the same in more than one channel need only be stored once, and extra bandwidth can be temporarily allocated to deal with a complex signal in one channel by slightly sacrificing the sound of other channels.

MPEG-1 Audio Coding

MPEG-1 digital audio compression carries either monophonic or stereophonic audio. It divides the signal into frequency bands (typically 32) of equal widths. This is easier to implement than the slightly more accurate variable widths.

MPEG-1 has three *layers,* or compression techniques, each more efficient but more complicated than the last. Layer II is the most common and is the only one allowed by DVD. Layer II compression typically uses a sample block size of 23 milliseconds (1152 samples). Layer III, commonly called MP3, is a popular format for compressed music on the Internet, but it is not directly supported by the DVD-Video or DVD-Audio formats. There are some DVD players that can play MP3 audio files from CD-ROM or DVD-ROM discs.

MPEG-2 Audio Coding

MPEG-2 digital audio compression adds multiple channels and provides a mode for backward compatibility with MPEG-1 decoders. This backward-compatible mode is required for DVD. Five channels of audio are *matrixed* into the standard left/right stereo signal, which is encoded in the normal MPEG-1 Layer II format. *Phase matrix encoding* is the process of mixing multiple audio channels into two channels according to a defined mathematical relationship relying on different audio signal phases. This relationship can then be used to later reconstruct a total of four channels (left, right, center, and surround).[10] The advantage of matrixing is that the two-channel audio can still be played on standard stereo audio systems. MPEG provides predetermined matrixing formulas depending on the intended audience for the stereo audio signal. One is a conventional stereo signal, and another is designed to deliver a signal that is compatible with Dolby Surround decoding to recreate a center channel and left/right surround channels. Additional discrete channel separation information, plus the low-frequency channel, is put in an *extension stream* so that an MPEG-2 audio decoder can recreate six separate signals. For eight channels, an additional layer is provided in another extension stream. The extension streams are compressed in a way that reduces redundant information shared by more than one channel. Because of the need to be backward compatible with MPEG-1 decoders, the center and surround channels that are matrixed into the two MPEG-1 channels are duplicated in the extension stream. This allows the MPEG-2 decoder to subtract these signals from the matrixed signals, leaving the original left and right channels. This duplication of information adds an overhead of approximately 32 kbps, making the backward-compatible mode somewhat inefficient.

---

[10]A signal containing two main channels with additional channels matrixed onto them is often referred to as L_t/R_t, with the t standing for "total." A pure stereo signal that does not carry phase-shifted audio intended for a decoder, is sometimes identified as L_o/R_o, with the o standing for "only."

The hierarchical structure of MPEG-1 plus extension streams means that a two-channel MPEG decoder need only decode a two-channel data stream, and a six-channel MPEG decoder need only process a six-channel data stream. The eight-channel MPEG decoder is the only one that must decode the entire contents of an eight-channel stream. Therefore, an MPEG-1 decoder "sees" only the MPEG-1–compatible data to produce stereo audio, whereas an MPEG-2 decoder combines it with the first layer of extension data to produce 5.1-channel audio or, with both layers of extension data, to produce 7.1-channel audio. This clever technique provides an advantage to MPEG-2 over other encoding schemes, since cheap MPEG-1 decoders can be used when necessary. However, the cost of MPEG-2 decoders dropped in a few years to about the same level as MPEG-1. Nevertheless, there are several drawbacks to the backward-compatibility scheme, in addition to the inefficiency of matrixed channel duplication. The original two-channel MPEG-1 encoding process was not developed with matrixed audio in mind and therefore may remove surround-sound detail. The matrix-canceling process tends to expose coding artifacts; that is, when the signal from a matrixed channel is removed, the remaining signal may no longer mask the same neighboring frequencies or noise as the original signal, thus unmasking the noise introduced by formerly appropriate levels of quantization. This problem can be mitigated in part with special processing by the encoder, but this makes the encoding process less efficient with a resulting loss in quality at a given bit rate.

MPEG-2 includes a non-backward-compatible process, originally labeled NBC but now known as *advanced audio coding* (AAC). This coding method deals with all channels simultaneously and is thereby more efficient, but it was not developed in time to be supported by DVD.

MPEG-2 allows a variable bit rate in order to handle momentary increases in signal complexity. Unfortunately, this turns out to be difficult to deal with in practice because audio and video are usually processed separately, and with two variable rates there is a danger of simultaneous peaks pushing the combined rate past the limit. This can be controlled by limiting the peak rate, but this is done at the expense of possibly reducing audio quality in difficult passages.

Dolby Digital Audio Coding

Dolby Digital audio compression (known as AC-3 in standards documents) provides for up to 5.1 channels of discrete audio. One of the advantages of Dolby Digital is that it analyzes the audio signal to differentiate short, transient signals from long, continuous signals. Short sample blocks are then used for short sounds, and long sample blocks are used for longer sounds. This results in smoother encoding without transient suppression and block boundary effects that can occur with fixed block sizes. For compatibility with existing audio/video systems, Dolby Digital decoders can downmix multichannel programs to ensure that all the channels are present in their proper proportions for mono, stereo, or Dolby Pro Logic reproduction.

Dolby Digital uses a frequency transform—somewhat like the DCT transform of JPEG and MPEG—and groups the resulting values into frequency bands of varying widths to match the critical bands of human hearing. Each transformed block is converted to a floating-point frequency representation that is allocated a varying number of bits from a common pool, according to the importance of the frequency band. The result is a *constant–bit-rate* (CBR) data stream.

Dolby Digital also includes dynamic range information so that different listening environments can be compensated for. Original audio mixes, such as movie sound tracks, which are designed for the wide dynamic range of a theater, can be encoded to maintain the clarity of the dialogue and to enable emphasis of soft passages when played at low volume in the home.

Dolby Digital was developed from the ground up as a multichannel coder designed to meet the diverse and often contradictory needs of consumer delivery. It also has a significant lead over other multichannel systems in both marketing and standards adoption. By 2000, over 57,000 Dolby Digital decoders were entrenched in living rooms, compared with essentially no MPEG-2 audio decoders, no SDDS decoders, and relatively few DTS decoders. Dolby Digital has been chosen for the U.S. DTV standard and is being used for digital satellite systems and most other digital television systems.

DTS Audio Coding

Digital Theater Systems (DTS) Digital Surround uses the Coherent Acoustics differential subband perceptual audio transform coder, which is similar to Dolby Digital and MPEG audio coders. It uses polyphase filters to break the audio into subbands (usually 32) of varying bandwidths and then uses ADPCM to compress each subband. The ADPCM step is a linear predictive coding that "guesses" at the next value in the sequence and then encodes only the difference. The prediction coefficients are quantized based on psychoacoustic and transient analysis and then are variable-length coded using entropy tables.

DTS decoders include downmixing features using either preset downmix coefficients or custom coefficients embedded in the stream. Dynamic range control and other user data can be included in the stream to be used in post-decoder processes.

The DTS Coherent Acoustics format used on DVDs is different from the one used in theaters, which is Audio Processing Technology's apt-X, a straight ADPCM coder with no psychoacoustic modeling.

MLP Audio Encoding

In addition to the lossy perceptual coding formats from DVD-Video, the DVD-Audio format includes a mathematical encoding technique called *Meridian lossless packing* (MLP). MLP compresses audio data bit for bit, removing redundancy without discarding any data. In addition to storing more data in the same amount of space, thus giving longer playing times, MLP reduces the maximum peak data rate. Since six channels of 96-kHz 24-bit audio have an uncompressed data rate of 13.824 Mbps, reducing the peak data rate to fit into the DVD-Audio maximum of 9.6 Mbps is important. MLP produces variable data rates, providing longer playing times than a fixed-rate scheme.

MLP achieves a typical compression ratio of 2:1 by using a combination of three techniques: lossless matrixing to compress interchannel redundancy, lossless waveform prediction to compress intersample correlation, and entropy coding of the remaining values. MLP also incorporates stream buffering to help deal with transients and hard-to-compress segments so as to limit the peak data rate. Unlike lossy encoding methods, MLP cannot throw away more data to stay within a specified data rate, so in some case it will be unable to sufficiently reduce the data rate. In such cases, the encoder operator must use other options such as reducing the bit size of one or more channels or filtering out high-frequency data.

MLP can encode a 2-channel downmix along with a multichannel mix. It also can carry additional data such as dynamic range control profiles, copy control information, time codes, and descriptive text. The format also has built-in error detection to recover from transmission errors.

Effects of Audio Encoding

Aside from lossless MLP encoding, which has no effect on the audio signal, lossy perceptual audio compression can affect the quality of the audio signal. Audio compression techniques result in a set of data that is processed in a specific way by the decoder. The form of the data is flexible, so improvements in the encoder can result in improved quality or efficiency without changing the decoder. As understanding of psychoacoustic models improves, perceptual encoding systems can be made better.

At minimal levels of compression, around 7:1 or less for today's encoders and even less for future encoders, perceptual encoding removes only the imperceptible information and provides decoded audio that is virtually indistinguishable from the original. At higher levels of compression, and depending on the nature of the audio, *bit starvation* may produce identifiable effects of compression. These include a slightly harsh or gritty sound, poor reproduction of transients, loss of detail, and less pronounced separation and spaciousness.

Both Dolby Digital and MPEG-2 audio on DVD are usually compressed at a factor of about 10:1. Most tests place the quality of Dolby Digital and MPEG-2 audio neck and neck, just short of sounding as good as the original uncompressed source. These tests have shown that the audio quality is completely acceptable to average listeners, many of whom are unable to tell the difference between Dolby Digital and the original uncompressed PCM source.

DTS on DVD is usually compressed at a factor of 6:1 or 3:1. Many listeners claim that DTS audio quality is better than Dolby Digital, but such claims are rarely based on accurate comparisons. DTS tracks are usually encoded at a reference volume level that is 4 decibels higher, and most DTS soundtracks are mixed differently than their Dolby Digital counterparts, including different volume levels in the surround and LFE tracks. This makes it nigh impossible to compare them objectively, even using a disc that contains a soundtrack in both formats. The limited semiscientific comparisons that have been done indicate that there is little perceptible difference between the two and that any difference is noticeable only on high-end audio systems.

A Few Timely Words about Jitter

Apart from aspect ratios and anamorphic conversions, *jitter* is one of the most confusing aspects of DVD.[11] Luckily the average DVD owner does not really need to worry about jitter, which is good thing because even the experts disagree about its effects. Part of the problem is that jitter means

---

[11]When I wrote the first edition of this book, I thought the hardest part would be covering all the technical details of the format. I soon discovered that the hardest part was explaining DVD's aspect-ratio features in a way that was easy to understand without taking up an entire chapter. Jitter also could easily take an entire chapter.

many things, most of them quite technical. Modern episodes of *Star Trek* come closest to providing a comprehensive definition. Since it is not very dramatic to say, "Captain, we have detected jitter!" crew members instead say, "We have encountered a temporal anomaly!" In general terms, jitter is inconsistency over time.

When most people speak of jitter, they mean *time jitter,* also called *phase noise,* which is a time-base error in a clock signal—deviation from the perfectly spaced intervals of a reference signal. Figure 3.7 compares the simplified square wave of a perfect digital signal to the same signal after being affected by things such as poor-quality components or poorly designed components, mismatched impedance in cables, logic-level mismatches between *integrated circuits* (ICs), interference and fluctuations in power supply voltage, *radiofrequency* (RF) interference, and reflections in the signal path. The resulting signal contains aberrations such as phase shift, high-frequency noise, triangle waves, clipping, rounding, slow rise/fall, and ringing. The binary values of the signal are encoded in the transition from positive voltage to negative voltage, and vice versa. In the distorted signal, the transitions no longer occur at regularly spaced intervals.

However, looking closely at Figure 3.7 reveals something interesting. Even though the second signal is misshapen to the point of displacing the transition points, the sequence of ones and zeros is still reconstructed correctly, since each transition is within the interval timing window. In other words, there is no data loss, and there is no error. The timing information is distorted, but it can be fixed. This is the key to understanding the difference between correctable and uncorrectable jitter. In the digital domain, jitter is almost always inconsequential. Minor phase errors are easily corrected by resynchronizing the data. Of course, large amounts of jitter can cause data errors, but most systems specify jitter tolerances at levels far below the error threshold.[12] Jitter is an interface phenomenon—it only becomes a problem when moving from the analog to the digital world or from the digital world to the analog world. For example, jitter in the sampling clock of an analog-to-digital converter causes uneven spacing of the samples, which

---

[12]The AES/EBU standard for serial digital audio specifies a 163-nanosecond clock rate with ±20 nanoseconds of jitter. The full 40-nanosecond range is 24 percent of the unit interval. Testing has shown that correct data values are received with bandwidths as low as 400 kHz. Jitter in the recovered clock is reduced with wider bandwidths up to 5 MHz. The CD Orange Book specifies a maximum of 35 nanoseconds of jitter but also recommends that total jitter in the readout system be less then 10 percent of the unit interval (that is, 23 out of 230 nanoseconds). The DVD-ROM specification states that jitter must be less than 8 percent of the channel bit clock period (8 percent of 38 nanoseconds comes to approximately 3 nanoseconds of jitter).

Figure 3.7
Effects of interface
jitter

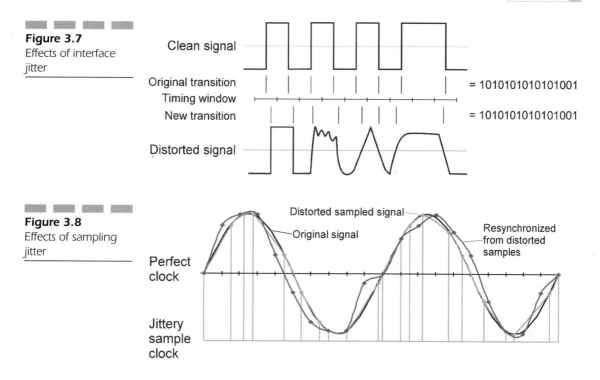

Figure 3.8
Effects of sampling
jitter

results in a distorted measurement of the waveform (Figure 3.8). On the other end of the chain, jitter in a digital-to-analog converter causes voltage levels to be generated at incorrect moments in time, resulting in audio waveform distortions such as spurious tones and added noise, causing what is often described as a "harsh sound." Jitter that passes into an analog speaker signal degrades spatial image, ambience, and dynamic range. Actual data errors produce clicks or pops or periods of silence.

NOTE: There are two kinds of jitter: harmful and harmless. Even harmful jitter can usually be corrected.

In some cases there is nothing you can do about jitter (other than buy better equipment). In other cases, power conditioners and good-quality cables with good shielding reduce certain kinds of jitter. Before you do anything, however, it is important to understand the various types of jitter and which ones are worth worrying about. Many a shrewd marketer has capitalized on the fears of consumers worried about jitter and sonic quality, bestowing on the world such products as colored ink that supposedly reduces reflections from the edge of the disc, disc stabilizer rings that claim to reduce rotational variations, foil stickers alleged to produce "morphic resonance" to rebalance human perception, highly damped rubber feet or hardwood stabilizer cones for players, cryogenic treatments, disc polarizing devices, and other technological nostrums that are intimate descendents of Dr. Feelgood's Amazing Curative Elixir.

Basically, there are five types of jitter that are relevant to DVD.[13]

■ *Oscillator jitter*   Oscillating quartz crystals are used to generate clock signals for digital circuitry. The quality of the crystal and the purity of the voltage driving it determine the stability of the clock signal. Oscillator jitter is a factor in other types of jitter, since all clocks are "fuzzy" to some degree.

■ *Sampling jitter (recording jitter)*   This is the most critical type of jitter. When the analog signal is being digitized, instability in the clock results in the wrong samples being taken at the wrong time (see Figure 3.8). Reclocking at a later point can fix the time errors but not the amplitude sampling errors. There is nothing the consumer can do about jitter that happens at recording time or during production, because it becomes a permanent part of the recording. Sampling jitter also occurs when the analog signal from a DVD player is sent to a digital processor (such as an AV receiver with DSP features or a video line multiplier). The quality of the DAC in the receiving equipment determines the amount of sampling jitter. Using a digital connection instead of an analog connection avoids the problem altogether.

■ *Media jitter (pit jitter)*   This type of jitter is not critical. During disc replication, a laser beam is used to cut the pattern of pits in the glass master. Any jitter in the clock or physical vibration in the mechanism used to drive the laser will be transmitted to the master and thus to every disc that is molded from it. Variations in the physical replication process also can contribute to pits being longer or shorter than they should be. These variations are usually never large enough to cause

---

[13]One particular phenomenon is incorrectly referred to as jitter. When DVD-ROM drives and CD-ROM drives perform *digital audio extraction* (DAE) from audio CDs, they can run into problems if the destination drive cannot keep up with the data flow. Most drives do not have block-accurate seeking, so they may miss or duplicate a small amount of data after a pause. These data errors cause clicks when the audio is played back. This is colloquially referred to as "jitter," and there are software packages that perform "jitter correction" by comparing successive read passes during DAE, but technically this is not jitter. It is a data error, not a phase error.

data errors, and each disc is tested for data integrity at the end of the production line. Media jitter can be worse with recorded discs because they are subject to surface contamination, dust, and vibration during recording. Strange as it may seem, however, recorded media usually have cleaner and more accurate pit geometry than pressed media. However, with both pressed and recorded discs, the minor effects of jitter have no effect on the actual data.

▪ *Readout jitter (transport jitter)* This type of jitter has little or no effect on the final signal. As the disc spins, phase-locked circuits monitor the modulations of the laser beam to maintain proper tracking, focus, and disc velocity. As these parameters are adjusted, the timing of the incoming signal fluctuates. Media jitter adds additional perturbations. Despite readout jitter, error correction circuitry verifies that the data is read correctly. Actual data errors are extremely rare. The data is buffered into RAM, where it is clocked out by an internal crystal. In theory, the rest of the system should be unaffected by readout jitter because an entirely new clock is used to regenerate the signal.

▪ *Interface jitter (data-link jitter)* This type of jitter may be critical or it may be harmless, depending on the destination component. When data is transmitted to another device, it must be modulated onto an electrical or optical carrier signal. Many factors in the transmission path (such as cable quality) can induce random timing deviations in the interface signal (see Figure 3.7). There is also *signal-correlated jitter,* where the characteristics of the signal itself cause distortions.[14] As a result of interface jitter, values are still correct (as long as the jitter is not severe enough to cause data errors), but they are received at the wrong time. If the receiving component is a digital recorder that simply stores the data, interface jitter has no effect. If the receiving component uses the signal directly to generate audio and does not sufficiently attenuate the jitter, it will cause audible distortion. A partial solution is to use shorter cables or cables with more bandwidth and to properly match impedance.

To restate the key point, when the component that receives a signal is designed only to transfer or store the data, it need only recover the data, not the clock, so jitter below the error threshold has no effect. This is why digital

---

[14]For example, a string of ones or a string of zeros may travel faster or slower than a varying sequence because the transmission characteristics of the cable are not uniform across the signal frequency range.

copies can be made with no error. However, when the component receiving a signal must reconstruct the analog waveform, then it must recover the clock as well as the data. In this case, the equipment should reduce jitter as much as possible before regenerating the signal. The problem is that there is a tradeoff between data accuracy and jitter reduction. Receiver circuitry designed to minimize data errors sacrifices jitter attenuation.[15] The ultimate solution is to decouple the data from the clock. Some manufacturers have approached this goal by putting the master clock in the DAC (which is probably the best place for it) and having it drive the servo mechanism and readout speed of the drive. RAM-buffered time-base correction in the receiver is another option. This *reclocks* incoming bits by letting them pile up in a line behind a little digital gate that opens and closes to let them out in a retimed sequence. This technique removes all incoming jitter but introduces a delay in the signal. The accuracy of the gate determines how much new jitter is created.

The quality of digital interconnect cables makes a difference, but only up to a point. The more bandwidth in the cable, the less jitter there is. Note that a "digital audio" cable is actually transmitting an analog electrical or optical signal. There is a digital-analog conversion step at the transmitter and an analog-digital conversion step at the receiver. This is the reason interface jitter can be a problem. However, the problem is less serious than when an analog interface cable is used because no resampling of analog signal values occurs.

Much ado is made about high-quality transports—disc readers that minimize jitter to improve audio and video quality. High-end systems often separate the transport unit from other units, which ironically introduces a new source of jitter in the interface cable. Jitter from the transport is a function of the oscillator, the internal circuitry, and the signal output transmitter. In theory, media jitter and transport jitter should be irrelevant, but in reality, the oscillator circuitry is often integrated into a larger chip, so leakage can occur between circuits. It is also possible for the servo motors to cause fluctuations in the power supply that affect the crystal oscillator, especially if they are working extra hard to read a suboptimal disc. Other factors such as instability in the oscillator crystal, temperature, and physical vibration may introduce jitter. A jitter-free receiver changes every-

---

[15]The jitter tolerance characteristic of a PLL circuit is inversely proportional to its jitter attenuation characteristic. That is, the more "slack" the circuit allows in signal transition timing, the more jitter gets through. This situation can be improved by using two PLLs to create a two-stage clock recovery circuit.

thing. If the receiver reclocks the signal, you can use the world's cheapest transport and get better quality than with an outrageously expensive, vacuum-sealed, hydraulically cushioned transport with a titanium-lead chassis. The reason better transports produce perceptibly better results is that most receivers do not reclock or otherwise sufficiently attenuate interface jitter. Even digital receivers with DSP circuitry usually operate directly on the bit stream without reclocking it.

Interface jitter affects all digital signals coming from a DVD player: PCM audio, Dolby Digital, DTS, MPEG-2 audio, and so on. In order to stay in sync with the video, the receiver must lock the decoder to the clock in the incoming signal. Since the receiver depends on the timing information recovered from the incoming digital audio signal, it is susceptible to timing jitter.

There is an ongoing tug of war between engineers and critical listeners. The engineers claim to have produced a jitterless system, but golden ears hear a difference. After enough tests, the engineers discover that jitter is getting through somewhere or being added somewhere, and they go back to the drawing board. Eventually, the engineers will win the game. Until then, it is important to recognize that most sources of jitter have little or no perceptible effect on the audio or video.

Pegs and Holes: Understanding Aspect Ratios

The standard television picture is restricted to a specific shape: a third again wider than it is high. This aspect ratio is designated as 4:3, or 4 units wide by 3 units high, also expressed as 1.33.[16] This rectangular shape is a fundamental part of the NTSC and PAL television systems—it cannot be changed without redefining the standards.[17]

---

[16]There is no special meaning to the numbers 4 and 3. They are simply the smallest whole numbers that can be used to represent the ratio of width to height. An aspect ratio of 12:9 is the same as 4:3. This also can be normalized to a height of 1, but the width becomes the repeating fraction 1.33333 . . . , which is why the 4:3 notation is generally used. For comparison purposes, it is useful to use the normalized format of 1.33:1 or 1.33 for short.

[17]The next generation of television—known as HDTV, ATV, DTV, etc.—has a 1.78 (16:9) picture that is much wider than current television. However, the new digital format is incompatible with existing standard recording and display equipment.

The problem is that movies are wider than television screens. Most movies are 1.85 (about 5.5:3). Extrawide movies in Panavision or Cinemascope format are around 2.35 (about 7:3). Thus, the trick is to somehow fit a wide movie shape into a not-so-wide television shape (Figure 3.9).

Fitting a movie into television is like the old conundrum of putting a square peg in a round hole, but in this case it is a rectangular peg. Consider a peg that is twice as wide as a square hole (Figure 3.10). There are essentially three ways to get the peg in the hole:

1. Shrink the peg to half its original size or make the hole twice as big (Figure 3.11).

2. Slice off part of the peg (Figure 3.12).

3. Squeeze the sides of the peg until it is the same shape as the hole (Figure 3.13).

Now, think of the peg as a movie and the hole as a TV. The first two peg-and-hole solutions are used commonly to show movies on television. Quite often you will see horizontal black bars at the top and bottom of the picture.

Figure 3.9
TV shape versus movie shape

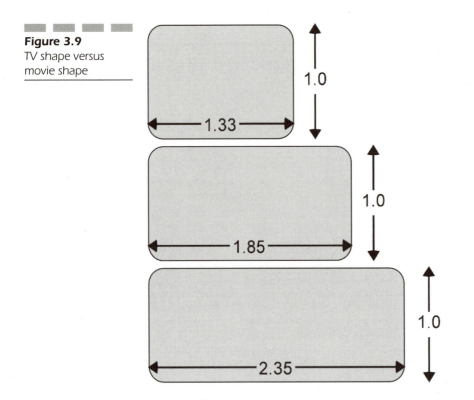

1.0

1.33

1.0

1.85

1.0

2.35

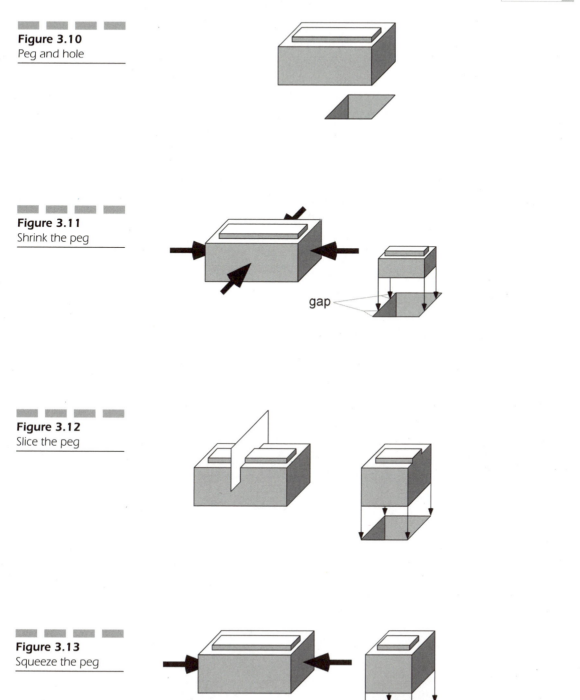

Figure 3.10
Peg and hole

Figure 3.11
Shrink the peg

gap

Figure 3.12
Slice the peg

Figure 3.13
Squeeze the peg

This means that the width of the movie shape has been matched to the width of the TV shape, leaving a gap at the top and the bottom. This is called *letterboxing.* It does not refer to postal pugilism but rather to the process of putting the movie in a black box with a hole the shape of a standard paper envelope. The black bars are called *mattes.*

At other times you might see the words "This presentation has been formatted for television" at the beginning of a movie. This indicates that a *pan and scan* process has been used, where a TV-shaped window over the film image is moved from side to side or up and down or is zoomed in and out (Figures 3.14 and 3.15). This process is more complicated than just chopping off a little from each side; sometimes the important part of the picture is all on one side or mostly on the other side, and sometimes there is more picture on the film above or below what is shown in the theater, so the artist who transfers the movie to video must determine for every scene how much of each side should be chopped off or how much additional picture from above or below should be included in order to preserve the action and story line. For the past 20 years or so, most films have been shot *flat,* sometimes called *soft matte.* The cinematographer has two rectangles in the viewfinder, one for 1.85 (or wider) and one for 4:3 (see Figure 3.13). He or she composes the shots to look good in the 1.85 rectangle while making sure that no crew, equipment, or raw set edges are visible above or below in the 4:3 area. For presentation in the theater, a *theatrical matte* is used to mask off the top and bottom either when the film is printed or with an *aperture*

Figure 3.14
Soft matte filming

1.33

(1.78)

1.85

Figure 3.15
Pan and scan transfer

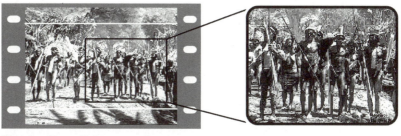

Flat (soft matte): zoom in

Flat (soft matte): full frame

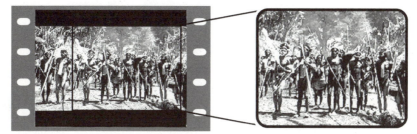

Scope (hard matte): extract

plate on the projector. When the movie is transferred to video for 4:3 presentation, the full frame is available for the pan and scan (and zoom) process.[18] In many cases, the director of photography or even the director approves the transfer to ensure that the intention and integrity of the

---

[18]Contrast this to *hard matte* filming, where the top and bottom are physically—and permanently—blacked out to create a wide aspect ratio. Movies filmed with anamorphic lenses also have a permanently wide aspect ratio, with no extra picture at the top or bottom.

original filming are maintained. Full control over how the picture is reframed is very important. For example, when the mattes are removed, close-up shots become medium shots, and the frame may need to be zoomed in to recreate the intimacy of the original shot. In a sense, the film is being composed anew for the new aspect. The pan and scan process has the disadvantage of losing some of the original picture but is able to make the most of the 4:3 television screen and is able to enlarge the picture to compensate for the smaller size and lower resolution as compared with a theater screen.

The third peg-and-hole solution has been used for years to fit widescreen movies onto standard 35-mm film. As filmmakers tried to enhance the theater experience with ever wider screens, they needed some way to get the image on the film without requiring new wider film and new projectors in every theater. They came up with the *anamorphic* process, where the camera is fitted with an anamorphic lens that squeezes the picture horizontally, changing its shape so that it fits in a standard film frame. The projector is fitted with a lens that unsqueezes the image back to its original width when it is projected (Figure 3.16). It is as if the peg were accordion-shaped so that it can be squeezed into the square hole and then pop back into shape after it is removed. You may have seen this distortion effect at the end of a Western movie where John Wayne suddenly becomes tall and skinny so that the credits will fit between the edges of the screen.

How It Is Done with DVD

DVD mixes and matches all the preceding techniques. Three standard methods are targeted for 4:3 displays, while a newer format is intended for widescreen displays. The four options provided by DVD are as follows (Figure 3.17):

1. *Full frame ("the peg fits the hole")*   Most material shot for television is already in 4:3 format. Older movies, such as *The Wizard of Oz,* were filmed in 4:3 aspect ratio.

2. *Pan and scan ("chop off the sides")*   This is the traditional "fill the frame" way of showing video on a standard TV. When the film is converted to video, the transfer artist (also called *colorist* or *telecine artist*) uses a variety of techniques to make the picture fill the screen and best follow the story, including zooming in and out and scanning up, down, left, and right. The zoom technique is often used with soft matte movies to preserve the nuances of close-ups.

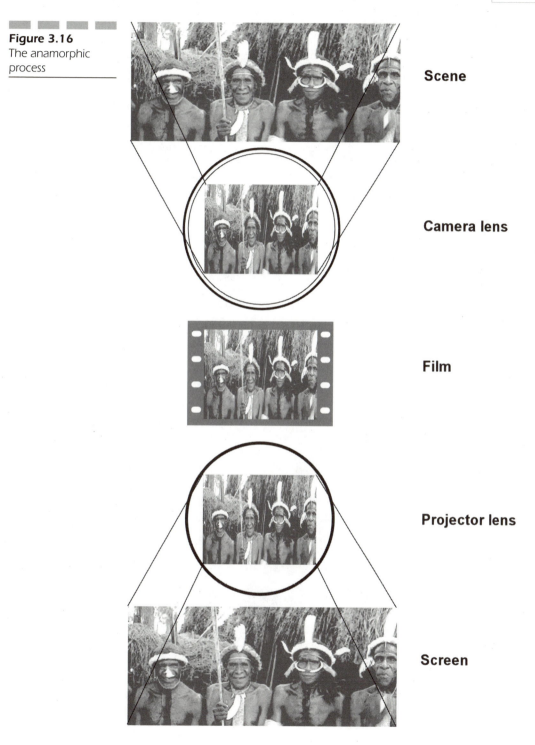

Figure 3.16
The anamorphic
process

Scene

Camera lens

Film

Projector lens

Screen

Figure 3.17a-o
Aspect ratios, conversions, and displays

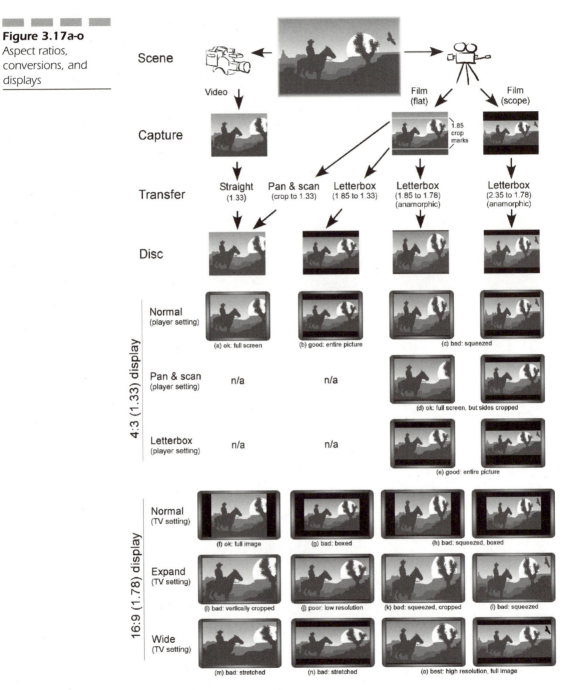

Scene

Capture

Transfer

Disc

Video

Film (flat)

Film (scope)

1.85 crop marks

Straight (1.33)

Pan & scan (crop to 1.33)

Letterbox (1.85 to 1.33)

Letterbox (1.85 to 1.78) (anamorphic)

Letterbox (2.35 to 1.78) (anamorphic)

4:3 (1.33) display

Normal (player setting)

(a) ok: full screen

(b) good: entire picture

(c) bad: squeezed

Pan & scan (player setting)

n/a

n/a

(d) ok: full screen, but sides cropped

Letterbox (player setting)

n/a

n/a

(e) good: entire picture

16:9 (1.78) display

Normal (TV setting)

(f) ok: full image

(g) bad: boxed

(h) bad: squeezed, boxed

Expand (TV setting)

(i) bad: vertically cropped

(j) poor: low resolution

(k) bad: squeezed, cropped

(l) bad: squeezed

Wide (TV setting)

(m) bad: stretched

(n) bad: stretched

(o) best: high resolution, full image

3. *Letterbox ("shrink and matte")*   This is the alternate way of showing widescreen video on a standard TV, preferred by videophiles and popularized by laserdiscs. The original theatrical image is boxed into the 4:3 frame by adding black mattes to the top and bottom of the picture.

4. *Widescreen ("accordion squeeze")*   One of the advantages of DVD-Video for home theater systems is widescreen support. DVD supports wide images by using a 16:9 (1.78) aspect ratio that is anamorphically squeezed into a 4:3 TV shape before being stored on the disc.[19] (The 4:3 ratio can be expressed as 12:9, so in order to get from 16:9 to 12:9, the width needs to be reduced by 25 percent, from 16 down to 12.) Widescreen televisions have a wider scanning pattern to display the full 16:9 shape. DVD players also can display widescreen video on a standard 4:3 TV.[20] There are three ways to do this, including two options similar to those performed during video transfer, but in this case they are performed by the player.

- *Automatic letterbox*   All DVD players can add letterbox mattes when displaying widescreen video on a 4:3 display. The player actually squeezes the image vertically by 25 percent (the same amount it was squeezed horizontally in the anamorphic process) so that its proper proportions are restored.

- *Automatic pan and scan*   Center-of-interest information can be included with the widescreen video to tell the player which part to extract. The player chops off the indicated amount from each side and then unsqueezes the remaining picture to create a 4:3 image for the TV.

- *Lie to the player*   A DVD player has no way to know what kind of TV you have, so you can tell it to send a 16:9 picture to a 4:3 TV. You will

---

[19]Some people prefer to think of anamorphic video as being stretched vertically rather than squeezed horizontally. The difference is largely a matter of semantics. If anamorphic video were stretched vertically from a letterboxed source after being transferred from film to video, it would lose resolution, but in practice, it happens during the transfer process, so it is largely a matter of perspective whether you think of the video as being squeezed horizontally or stretched vertically. Matching the source to the height of the 4:3 shape and squeezing horizontally comes out the same as matching the width of the 4:3 shape and stretching vertically. A widescreen TV increases horizontal sweep to stretch anamorphic video from a standard NTSC signal, whereas some 4:3 TVs decrease vertical scan pitch to achieve a 16:9 aspect ratio, so either point of view is valid.

[20]Anamorphic video is not unique to DVD. There are a few anamorphic laserdiscs and even rare anamorphic videotapes. The problem is that they can only be viewed properly on a widescreen TV. Unlike DVD players, standard laserdisc players and VCRs are unable to adapt anamorphic video for standard TVs.

see the unchanged anamorphic picture, making Hardy look like Laurel. You are not supposed to do this, but just as there are no "mattress tag police," there are no "aspect ratio police" to come and take your player away.

How It Wasn't Done with DVD. There are other possible solutions for dealing with widescreen video that could have been used. DVD could have stored the full-width, undistorted image, but this would have used up more storage space[21] and would have required reducing the amount of video on a disc or reducing the video quality.

Another option would have been to always letterbox the video before storing it on DVD. The problems with this approach are that vertical information would have been lost, and storage space that could have held picture information would have been used to hold black mattes instead. A few widescreen films are put on DVD this way, usually because a letterboxed transfer from film to video is available and either the original elements are no longer available or the studio does not want to pay for a new anamorphic transfer.

Variable anamorphic squeeze also could have been used. The wider the video, the more it would be squeezed. The advantage to this approach is that every pixel of video storage space would be used to hold video, no matter its shape. The problem is that more expensive circuitry would be required to handle multiple squeeze ratios.

The designers of DVD chose the reasonable compromise of using the anamorphic technique to fit the most amount of information into the standard television image space, and they settled on the standard 16:9 wide aspect ratio (see "Why 16:9?" following). Having the DVD player shrink the anamorphic picture vertically for letterbox display on a 4:3 TV gives the same result (and the same information loss) as shrinking and letterboxing the picture before storing it on the disc, yet preserves more picture information for widescreen TVs.

Unfortunately, there is no standardized package labeling for anamorphic DVDs. The following terms are all used to mean the same thing: *enhanced for widescreen TVs, enhanced for 16:9 TVs, 16:9, 16:9 fullscreen version, widescreen 16 × 9, anamorphic video, anamorphic widescreen, 1.78 edge-to-edge,* and *widescreen.* In general, look for the word *enhanced* or *anamorphic.*

---

[21]Thirty-three percent more, to be exact, since 1.78 is 33 percent larger than 1.33. Even more data would be needed to store movies in their original aspect ratio. Most movies have an aspect ratio of 1.85, which would require 39 percent more data. Panavision and Cinemascope movies with a 2.35 ratio would require 76 percent more data.

Some people complain that the term *anamorphic* should apply only to the optical process used in films, or even that the term as applied to DVD is incorrect. However, the term has a very clear meaning for DVD, independent of the aspect ratio of the film or the TV, and is the clearest and most unambiguous way of specifying the form of the video on the disc. The recommended label for DVD packaging is *anamorphic widescreen.*

Widescreen TVs

Widescreen displays are quite flexible in the way they deal with different input formats. They can display 4:3 video with black bars on the sides—a kind of sideways letterbox that is sometimes called a *windowbox*—and they also have display modes that enlarge the video to fill the entire screen.

Wide mode stretches the picture horizontally (Figure 3.18). This is sometimes called *full* mode. This is the proper mode to use with anamorphic video, but it makes everything look short and fat when applied to 4:3 video. Some widescreen TVs have a *parabolic* or *panorama* version of wide mode, which uses nonlinear distortion to stretch the sides more and the center less, thus minimizing the apparent distortion. This mode should not be used with anamorphic DVD output or very strange fun-house-mirror effects will occur.

Expand mode proportionally enlarges the picture to fill the width of the screen, thus losing the top and bottom (Figure 3.19). This is sometimes called *theater* mode. Expand mode is for use with letterboxed video because it effectively removes the mattes. If used with standard 4:3 picture, this mode causes a Henry VIII "off with their heads" effect.

Most widescreen TVs also have other display modes that are variations of the three basic modes. Some standard 4:3 TVs, especially in Europe, can adjust vertical scan size to create a letterboxed display from anamorphic pictures. The advantage is that the picture is full resolution, since no pixels are lost by having the player do the letterboxing.

Figure 3.18
Wide (full) mode on a widescreen TV

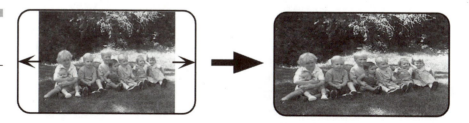

Figure 3.19
Expand (theater)
mode on widescreen
TV

NOTE: *Expand mode should only be used for video that is already letterboxed into a 4:3 picture, as on a laserdisc, a letterbox-only DVD, or other nonanamorphic source. Letting the DVD player autoletterbox anamorphic video causes a loss of resolution, which is amplified when the TV expands the picture (see Figure 3.17j).*

When DVD contains widescreen video, the different output modes of the player can be combined with different widescreen TV display modes to create a confusing array of options. Figure 3.17 shows how the different DVD output modes look on a regular TV and on a widescreen TV. Note that there is one "good" way to view widescreen video on a regular TV (see Figure 3.17e) but that a very large TV or a widescreen TV is required to do it justice. Also note that there is only one good way to view widescreen video on a widescreen TV, and this is with widescreen (anamorphic) output to wide mode (see Figure 3.17o).

Clearly, it is very easy to display the wrong picture in the wrong way. In some cases, the equipment is smart enough to help out. The player can send a special signal embedded in the video blanking area or via the s-video connector to the widescreen TV, but everything must be set up properly:

1. Connect the DVD player to the widescreen TV with an s-video cable.

2. Set the widescreen TV to s-video in (using the remote control or the front-panel input selector).

3. Set the TV to *automatic* or *normal* mode.

4. Set the DVD player to 16:9 output mode (using the on-screen setup feature with the remote control or with a switch on the back of the player).

If everything is working right and the TV is equipped to recognize widescreen signaling, it will automatically switch modes to match the format of the video.

Aspect Ratios Revisited

To review, video comes out of a DVD-Video player in basically four ways:

1. Full frame (4:3 original)
2. Pan and scan (widescreen original)
3. Letterbox (widescreen original)
4. Anamorphic (widescreen original)

All four can be displayed on any TV, but the fourth is specifically intended for widescreen TVs. This may seem straightforward, but it gets much more complicated. The problem is that very few movies are in 16:9 (1.78) format. The preceding discussions dealt with 16:9 widescreen in a general case. Until 16:9 cameras become more widespread, however, there will be very little video created in 16:9.

Most movies are usually 1.85 or wider, although European movies are often 1.66. DVD only supports aspect ratios of 1.33 (4:3) and 1.78 (16:9) because they are the two most common television shapes. Movies that are a different shape must be made to fit, which brings us back to pegs and holes. In this case, the hole is DVD's 16:9 shape (which is either shown in full on a widescreen TV or formatted to letterbox or pan and scan for a regular TV). There are essentially four ways to fit a 1.85 or wider movie peg into a 1.78 hole:

1. *Letterbox to 16:9*   When the movie is transferred from film, black mattes are added to box it into the 16:9 shape. These mattes become a permanent part of the picture. The position and thickness of the mattes depend on the shape of the original.

   a. For a 1.85 movie, the mattes are very small. On a widescreen TV or in automatic pan and scan on a regular TV (where the player is extracting a vertical slice from the letterboxed picture), the mattes are hidden in the *overscan* area.[22] On a standard TV in automatic letterbox mode (where the player is letterboxing an already letterboxed picture), the thin permanent mattes merge imperceptibly with the thick player-generated mattes.

   b. For a 2.35 movie, the permanent mattes are much thicker. In this case, the picture has visible mattes no matter how it is displayed.

---

[22]*Overscan* refers to covering the edges of the picture with a mask around the screen. Overscan was implemented originally to hide distortion at the edges. Television technology has improved to the point where overscan is not usually necessary, but it is still used. Most televisions have an overscan of about 4 to 5 percent. Anyone producing video intended for television display must be mindful of overscan, making sure that nothing important is at the edge of the picture. It should be noted that when DVD-Video is displayed on a computer, there is no overscan, and the entire picture is visible.

When the picture is letterboxed by the player, the permanent mattes merge with the player-generated mattes to form extrathick mattes on the television. These mattes are the same size as if the movie had been letterboxed originally for 4:3 display (as with a laserdisc).

    c. For a 1.66 movie, thin mattes are placed on the sides instead of at the top and bottom and generally will be hidden in the overscan area.

2. *Crop to 16:9* For 1.85 movies, slicing 2 percent from each side is sufficient and probably will not be noticeable. The same applies to 1.66 movies, except that about 3 percent is sliced off the top and off the bottom. However, to fit a 2.35 movie requires slicing 12 percent from each side. This procrustean approach throws away a quarter of the picture and is not a likely to be a popular solution.

3. *Pan and scan to 16:9* The standard pan and scan technique can be used with a 16:9 window (as opposed to a 4:3 window) when transferring from film to DVD. For 1.85 movies, the result is essentially the same as cropping and is hardly worth the extra work. For wider movies, pan and scan is more useful, but the original aspect ratio is lost, which goes against the spirit of a widescreen format. When going to the trouble of supporting DVD's widescreen format, it seems silly to pan and scan inside it, but if the option is there, someone is bound to use it.

4. *Open the soft matte to 16:9* When going from 1.85 to 16:9, a small amount of picture from the top and bottom of the full-frame film area can be included. This stays close to the original aspect ratio without requiring a matte, and the extra picture will be hidden in the overscan area on a widescreen TV or when panned and scanned by the DVD player. Even wider movies are usually shot full frame, so the soft matte area can be included in the transfer (Figure 3.20).

Most movies are converted to anamorphic DVD using methods 1 and 4. Many directors are opposed for artistic reasons to the pan and scan process, especially if it is done mechanically by the player. They may choose to make

Figure 3.20
Opening the frame
from 1.85 to 1.78

1.85

1.78

their movies on DVD viewable only in widescreen or letterbox format. Or they may choose to do a full-frame transfer in 4:3 format. This brings up the issue of different transfers, which will be discussed after a brief digression into why 16:9 is the widescreen aspect ratio of choice.

Why 16:9?

The 16:9 ratio has become the standard for widescreen. Most widescreen televisions are this shape, it is the aspect ratio used by almost all high-definition television standards, and it is the widescreen aspect ratio used by DVD. You may be wondering why this ratio was chosen, since it does not match television, movies, computers, or any other format. But this is exactly the problem: There is no standard aspect ratio (Figure 3.21).

Current display technology is limited to fixed physical sizes. A television picture tube must be built in a certain shape. A flat-panel *liquid-crystal display* (LCD) screen must be made with a certain number of pixels. Even a video projection system is limited by electronics and optics to project a certain shape. These constraints will remain with us for decades until we progress to new technologies such as scanning lasers or amorphous holographic projectors. Until then, a single aspect ratio must be chosen for a given display. The cost of a television tube is based roughly on diagonal measurement (taking into account the glass bulb, the display surface, and the electron beam deflection circuitry), but the wider a tube is, the harder it is to maintain *uniformity* (consistent intensity across the display) and *convergence* (straight

Figure 3.21
Common aspect ratios

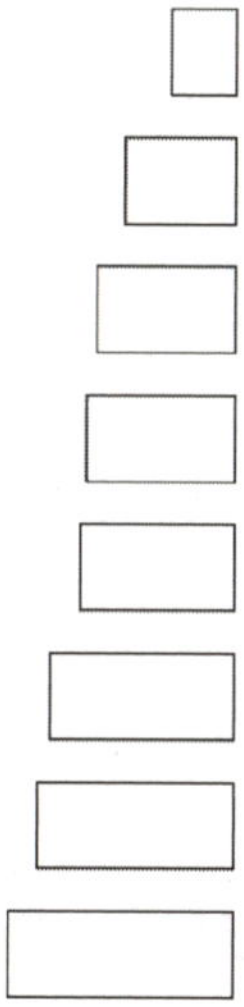

	7:9	2.3:3	0.77:1	Paper (8.5 x 11)
	12:9	**4:3**	1.33:1	Television
	15:9	5:3	**1.66:1**	Photographs, many European movies
	16:9	5.3:3	1.78:1	DVD, HDTV, most widescreen TV
	16.7:9	5.6:3	**1.85:1**	Most movies
	19.9:9	6.6:3	**2.21:1**	Wide movies (70 mm)
	21.2:9	7.1:3	**2.35:1**	Wide movies (Panavision, Cinemascope)
	24.3:9	8.1:3	**2.7:1**	Extra-wide movies (Ultra Panavision)

horizontal and vertical lines). Therefore, too wide a tube is not desirable.

The 16:9 aspect ratio was chosen in part because it is an exact multiple of 4:3. That is, $4/3 \times 4/3 = 16/9$. The clean mathematical relationship between 4:3 and 16:9 makes it easy to convert between the two. Going from 4:3 to 16:9 merely entails adding one horizontal pixel for every three ($3{\rightarrow}4$), and going from 16:9 to 4:3 requires simply removing one pixel from every four ($4{\rightarrow}3$).[23] This makes the scaling circuitry for letterbox and pan and scan functions much simpler and cheaper. It also makes the resulting picture cleaner.

The 16:9 aspect ratio is also a reasonable compromise between television and movies. It is very close to 1.85, and it is close to the mean of 1.33 and 2.35. That is, $4/3 \times 4/3 \times 4/3 \approx 2.35$. Choosing a wider display aspect ratio, such as 2:1, would have made 2.35 movies look wonderful but would have required huge mattes on the side when showing 4:3 video (as in Figure 3.17f, but even wider).

Admittedly, the extra space could be used for *picture outside picture* (POP, the converse of PIP), but it would be very expensive extra space. To make a 2:1 display the same height as a 35-inch television (21 inches) requires a width of 42 inches, giving a diagonal measure of 47 inches. In other words, to keep the equivalent 4:3 image size of 35-inch television, you must get a 47-inch 2:1 television. Figure 3.22 shows additional widescreen display sizes required to maintain the same height of common television sizes.

Figure 3.22
Display sizes at equal heights

Diagonal size (width x height)

1.33 (4:3)	1.78 (16:9)	2.0
27" (22 x 16)	33" (29 x 16)	36" (32 x 16)
32" (26 x 19)	39" (34 x 19)	43" (38 x 19)
34" (27 x 20)	42" (36 x 20)	46" (41 x 20)
36" (29 x 22)	44" (38 x 22)	48" (43 x 22)
42" (34 x 25)	51" (45 x 25)	56" (50 x 25)
46" (37 x 28)	56" (49 x 28)	62" (55 x 28)
50" (40 x 30)	61" (53 x 30)	67" (60 x 30)
53" (42 x 32)	65" (57 x 32)	71" (64 x 32)
60" (48 x 36)	73" (64 x 36)	80" (72 x 36)
65" (52 x 39)	80" (69 x 39)	87" (78 x 39)
84" (67 x 50)	103" (90 x 50)	113" (101 x 50)
100" (80 x 60)	122" (107 x 60)	134" (120 x 60)
120" (96 x 72)	147" (128 x 72)	161" (144 x 72)

[23]In each case, a weighted scaling function generally is used. For example, when going from 4 to 3 pixels, 3/4 of the first pixel is combined with 1/4 of the second to make the new first, 1/2 of the second is combined with 1/2 of the third to make the new second, and 1/4 of the third is combined with 3/4 of the fourth to make the new third (see Figure 6.25). This kind of scaling causes the picture to become slightly softer but is generally preferable to the cheap alternative of simply throwing away every fourth line. Similar weighted averages can be used when going from 3 to 4 (see Figure 6.27).

Figure 3.23 demonstrates the area of the display used when different image shapes are letterboxed to fit it (that is, the dimensions are equalized in the largest direction to make the smaller box fit inside the larger box).[24] The 1.33:1 row makes it clear how much smaller a letterboxed 2.35:1 movie is: Only 57 percent of the screen is used for the picture. On the other hand, the 2:1 row makes it clear how much expensive screen space goes unused by a 4:3 video program or even a 1.85:1 movie. The two middle rows are quite similar, so the mathematical relationship of 16:9 to 4:3 gives it the edge.

In summary, the only way to support multiple aspect ratios without mattes would be to use a display that can physically change shape—a "mighty morphin' television." Since this is currently impossible (or outrageously expensive), 16:9 is the most reasonable compromise. This said, the designers of DVD could have improved things slightly by allowing more than one anamorphic distortion ratio. If a 2.35 movie were stored using a 2.35 anamorphic squeeze, then 24 percent of the internal picture would not be wasted on the black mattes, and the player could automatically generate the mattes for either 4:3 or 16:9 displays. This was not done, probably because of the extra cost and complexity it would add to the player. The limited set of aspect ratios presently supported by the MPEG-2 standard (4:3, 16:9, and 2.21:1) also may have had something to do with it.

Figure 3.23
Relative display sizes
for letterbox display

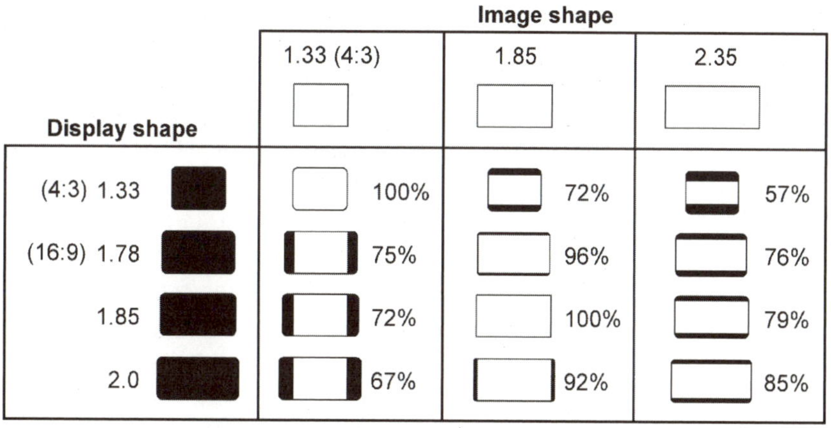

Display shape	Image shape		
	1.33 (4:3)	1.85	2.35
(4:3) 1.33	100%	72%	57%
(16:9) 1.78	75%	96%	76%
1.85	72%	100%	79%
2.0	67%	92%	85%

[24]If you wanted to get the most for your money when selecting a display aspect ratio, you would need to equalize the diagonal measurement of each display because the cost is roughly proportional to the diagonal size. This approach is sometimes used when comparing display aspect ratios and letterboxed images, and it has the effect of emphasizing the differences. The problem is that wider displays end up being shorter (for example, a 2:1 display normalized to the same diagonal as a 4:3 display would be 4.47:2.25, which is 12 percent wider but 25 percent shorter). In reality, no one would be happy with a new widescreen TV that was shorter than their existing TV. Therefore, it is expected that a widescreen TV will have a larger diagonal measurement and will cost more.

The Transfer Tango

Of course, the option still remains to transfer the movie to DVD's 4:3 aspect ratio instead of 16:9. At first glance, there may seem to be no advantage in doing this because 1.85 movies are so close to 16:9 (1.78). It seems simpler to do a 16:9 transfer and let the player create a letterbox or pan and scan version. But there are disadvantages to having the player automatically format a widescreen movie for 4:3 display: The vertical resolution suffers by 25 percent, the letterbox mattes are visible on movies wider than 1.85, and the player is limited to lateral motion. In addition, many movie people are averse to what they consider as surrendering creative control to the player. Therefore, almost every pan and scan DVD is done in the studio and not enabled in the player. During the transfer from film to video, the engineer has the freedom to use the full frame or zoom in for closer shots, which is especially handy when a microphone or a piece of the set is visible at the edge of the shot.

Many directors are violently opposed to pan and scan disfigurement of their films. Director Sydney Pollack sued a Danish television station for airing a pan and scan version of his *Three Days of the Condor,* which was filmed in 2.35 Cinemascope. Pollack feels strongly that the pan and scan version infringed his artistic copyright. He believes that "The director's job is to tell the film story, and the basis for doing this is to choose what the audience is supposed to see, and not just generally but exactly what they are to see." Years later, when talking about DVD, Pollack said, "DVD hasn't changed my approach to filmmaking, but what it has changed radically is my emotional reaction to the afterlife of the films that I do. I was always terribly disturbed by the fact that the overwhelming majority of people who see the work that you do as a filmmaker do not ever see it in its original form. . . . More and more people were watching videos, which were more and more often being altered by panning and scanning. Those pictures were getting butchered on video. . . . You have superb quality with DVD. Plus, you can see the movies in their original widescreen format, framed as they were intended."[25] Some directors, such as Stanley Kubrick, accept only the original aspect ratio. Others, such as James Cameron, who closely supervise the transfer process from full-frame film, feel that the director is responsible for making the pan and scan transfer a viable option by recomposing the movie to make the most of the 4:3 TV shape.

---

[25]Interview in *Widescreen Review,* Issue 37, 2000.

About two-thirds of widescreen movies are filmed at 1.85 (flat) aspect ratio. When a 1.85 film is transferred directly to full-frame 4:3 by including the extra picture at the top and bottom, the actual size of the images on the TV are the same as for a letterbox version. In other words, letterboxing only covers over the part of the picture that also was covered in the theater.

A pan and scan transfer to 4:3 makes the "I didn't pay good money for my 30-inch TV just to watch black bars" crowd happy. But a letterbox transfer to anamorphic 16:9 is still needed to appease the videophiles who demand the theatrical aspect ratio and to keep the "I paid good money for my widescreen TV" crowd from revolting. Ordinarily, this would mean two separate products, but not with DVD. The producer of the disc can put the 4:3 version on one side (or one layer) and the letterboxed 16:9 version on the other. This more or less doubles the premastering cost and slightly increases the mastering and replication costs, but with production runs of 100,000 copies or more, this adds less than a dollar to the unit cost. On the other hand, the widescreen letterbox transfer is sometimes reserved for a special edition and sold as a separate product at a higher price.

As widescreen TVs and HDTVs slowly replace traditional TVs, 4:3 transfers will become less common, and even letterboxed 4:3 transfers will become more appreciated. In Japan and Europe, where widescreen TVs already outsell standard TVs, letterboxed video is more popular.

Summary

Putting everything together (ignoring the option of cropping during transfer) gives the following variations to the four basic output formats:

1. 4:3 full frame (4:3 original)
   a. Direct transfer
2. 4:3 pan and scan (wide original)
   a. Pan and scan transfer
   b. Automatic pan and scan done by player
      (1) On direct transfer (16:9 original)
      (2) On pan and scan transfer (not 16:9 original)
      (3) On letterbox transfer (not 16:9 original)
3. 4:3 letterbox (wide original)
   a. Letterbox transfer
   b. Automatic letterbox done by player
      (1) On direct transfer (16:9 original)

    (2) On pan and scan transfer (not 16:9 original)

    (3) On letterbox transfer (not 16:9 original)

4. 4:3 anamorphic (wide original)

    a. Direct transfer (16:9 original)

    b. Pan and scan transfer (not 16:9 original)

    c. Letterbox transfer (not 16:9 original)

This may be clearer in the form of Table 3.2.

    All these variations are possible, but only a few of them are regularly used, such as 1a, 2a, 3a, and 4c. The automatic pan and scan feature of DVD

TABLE 3.2

Combinations of Output Formats and Video Transfers[a]

	4:3 Original	16:9 Original		Non-16:9 Original	
	Stored in 4:3	Stored in 4:3	Stored in 16:9	Stored in 4:3	Stored in 16:9
Full frame (1)	Direct transfer (a)	n/a	n/a	n/a	n/a
Pan and scan (2)	P&S transfer from full frame (a)	P&S transfer (a)	Auto P&S by player (b1)	P&S transfer (a)	Auto P&S by player on P&S transfer (b2) or on LB transfer (b3)
Letterbox (3)	LB transfer from inside soft matte (a)	LB transfer (a)	Auto LB by player (b1)	LB transfer (a)	Auto LB by player on P&S transfer (b2) or on LB transfer (b3)
Anamorphic (4)	Anamorphic transfer from 16:9 soft matte (a)	n/a	Anamorphic transfer (a)	n/a	P&S transfer to anamorphic (b) or LB transfer to anamorphic (c)

[a]Labels in parentheses refer to the outline in text.

players is rarely used; in many cases, both a 4:3 pan and scan version and a widescreen letterbox version will be included on a single disc. In other cases, where a new video transfer is deemed too expensive or the original film is no longer available, whatever existing transfer is available will be used, such as 4:3 pan and scan or 4:3 letterbox.

The Pin-Striped TV: Interlaced versus Progressive Scanning

One of the biggest problems facing early television designers was displaying images fast enough to achieve a smooth illusion of motion. Early video hardware was simply not fast enough to provide the required flicker fusion frequency of around 50 or 60 frames per second. The ingenious expedient solution was to cut the amount of information in half by alternately transmitting every other line of the picture (Figure 3.24). The engineers counted on the persistence of the phosphors in the television tube to make the two pictures blur into one.[26] For a 525-line signal, first the 262 (and a half) odd lines are sent and displayed, followed by the 262 (and a half) even lines. This is called *interlaced scanning*. Each half of a frame is called a *field*. For the NTSC system, 60 fields are displayed per second, resulting in a rate of 30 frames per second. There are 480 active lines of video, meaning that only 240 lines are visible at a time. For the PAL and SECAM systems, there are 50 fields per second, resulting in a rate of 25 frames per second. There are 576 active lines out of a total of 625, giving 288 lines per field.

The alternative approach, *progressive scanning,* displays every line of a complete frame in one sweep. Progressive scanning requires twice the frequency in order to achieve the same refresh rate. Progressive scan monitors are more expensive and generally are used for computers. *High-definition television* (HDTV) also includes progressive scanning.

Progressive scan provides a superior picture, overcoming many disadvantages of interlaced scanning. In interlaced scanning, small details, especially thin horizontal lines, appear only in every other field. This causes a

---

[26]Many texts refer to "persistence of vision" as the phenomenon that allows interlaced video and motion pictures in general to create a seemingly continuous moving image. This is largely incorrect (see Chapter 2).

Figure 3.24

Interlaced scan and
progressive scan

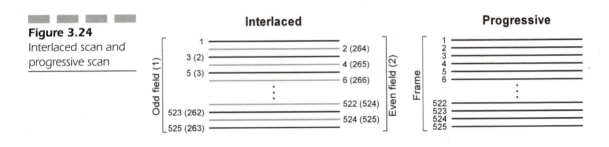

disturbing flicker effect, which you can see when someone on TV is wearing stripes. A common practice in video production is to filter the video to eliminate vertical detail smaller than two scan lines. This improves the stability of the picture but cuts the already poor resolution in half. NTSC video frames must be reduced to 200 lines of detail before interline flicker disappears. The flicker problem is especially noticeable when computer video signals are converted and displayed on a standard TV. The alternating black and white horizontal lines in Macintosh window titles were especially problematic. In addition to flicker, line crawl occurs when vertical motion matches the scanning rate. Interlaced scanning also causes problems when the picture is paused. If objects in the video are moving, they end up in a different position in each field. When two fields are shown together in a freeze-frame, the picture appears to shake. One solution to this problem is to show only one field, but this cuts the picture resolution in half. You may have seen this effect on a VCR: When the tape is paused, much of the detail disappears.

Since film runs at 24 progressive frames per second, displaying it at NTSC rates of 60 video fields per second requires a process called *2–3 pulldown,* where one film frame is shown as two fields, and the following film frame is shown as three fields. This pattern results in pairs of 24-per-second film frames converted to 60-per-second TV fields [(2 + 3) × 12 = 60] (Figure 3.25). Unfortunately, this causes side effects. One is that film frame display times alternate between 2/60 of a second and 3/60 of a second, causing a *motion judder* artifact—a jerkiness that is especially visible when the camera pans slowly. Another side effect is that two of every five television frames contain fields derived from two different film frames, which does not cause problems during normal playback but can cause problems when pausing or playing in slow motion. A minor problem is that NTSC video

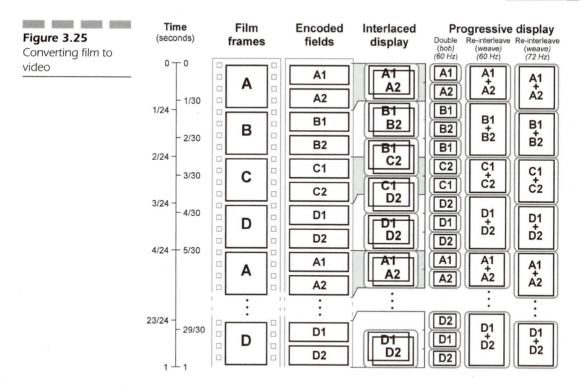

Figure 3.25
Converting film to video

actually plays at 59.94 Hz, so the film runs 0.1 percent slow and the audio must be adjusted to match. Displaying film at PAL rates of 50 video fields per second is simpler and usually is achieved by showing each film frame as two fields and playing it 4 percent faster.[27] This is sometimes called *2–2 pulldown*.

Most video is encoded from videotape. The videotape is created by a *telecine* machine, which performs 2–3 pulldown when making an NTSC tape. Since it would be inefficient to encode the extra fields, they are not duplicated in the MPEG-2 stream. A good encoder recognizes and removes the duplicate fields. This is called *inverse telecine* (no, it is *not*

[27]Since the video is sped up 4 percent when played, the audio must be adjusted before it is encoded. In many cases the audio speedup causes a semitone pitch shift that the average viewer will not notice. A better solution is to digitally shift the pitch back to the proper level during the speedup process.

called 3–2 pushup). There are flags in the MPEG-2 stream that indicate which fields to show when and which fields to repeat when. The encoder sets the repeat_first_field flag on every fifth field, which instructs the decoder to repeat the field, thus recreating the 2–3 pulldown sequence needed to display the video on an interlaced TV. In other words, the decoder in the player performs 2–3 pulldown, but only by following the instructions in the MPEG-2 stream. Therefore, a film on DVD must be encoded for the intended display rate—either NTSC or PAL, but not both. Some players can convert PAL to NTSC or NTSC to PAL; this is covered in Chapter 6.

Technically, DVD-Video can only be stored in interlaced format.[28] Signals from standard video cameras are already in interlaced format. Film, which is inherently progressive, is encoded into MPEG-2 as paired fields. Even though the encoding is field-based, the frame-based nature of the source can be preserved. This allows progressive-scan players to put Humpty Dumpty back together again.

Progressive DVD Players

When DVD was developed, there were about 1 billion interlaced TV sets in the world and less than 100,000 progressive TVs (not counting computer monitors). Not surprisingly, DVD is biased toward encoding and displaying in interlaced format. This does not mean that DVDs cannot be displayed in progressive-scan format, but it does mean that it is not a trivial process. Nevertheless, it is worth doing. A major advantage of DVD is that computers and progressive players can deinterlace the MPEG-2 video and display it progressively with considerably better quality than on standard interlaced displays. Progressive players work with all standard DVD titles but look best with video encoded from a progressive source such as film. The result is a significant increase in perceivable vertical resolution for a more detailed and filmlike picture.

---

[28]In MPEG-2 encoding, the decision between progressive and interlaced format can be made all the way down at the macroblock level. The DVD-Video specification limits MPEG-2 video to non-progressive sequences, which can include both progressive and interlaced frames. Progressive frames are still encoded for display as two fields, but they are identified as progressive. Interlaced frames can further include both progressive and interlaced macroblocks. Since progressive macroblocks are more efficient (using one motion vector instead of two), even interlaced source is often encoded with more than 50 percent progressive macroblocks. However, each frame is represented as two fields of 720 × 240 pixels each for NTSC or 720 × 288 pixels each for PAL/SECAM.

A progressive-scan DVD player converts the interlaced (480i) video from DVD into progressive (480p) format for connection to a progressive display at 31.5 kHz or higher. It is also possible to buy an external *line multiplier* to convert the output of a standard DVD player to progressive scanning. All DVD computers are progressive players because the video is displayed on a progressive monitor. However, quality of deinterlacing and video playback varies wildly from computer to computer.

Converting interlaced DVD video to progressive video involves much more than putting film frames back together. There are essentially four methods of converting from interlaced video to progressive video:

1. *Reinterleave* (also called *weave;* Figure 3.26). If the original video is from a progressive source, the two fields can be recombined into a single frame.

2. *Line-double* or *line-multiply* (also called *bob;* Figure 3.27). If the original video is from an interlaced source, simply combining two fields will cause motion artifacts (the effect is reminiscent of a zipper), so instead, each line of a single field is repeated twice to form a frame. Better line doublers use *interpolation* to produce new lines that are a combination of the lines above and below. The term *line doubler* is vague and misleading because cheap line doublers only bob, whereas expensive line doublers (those which contain digital signal processors)

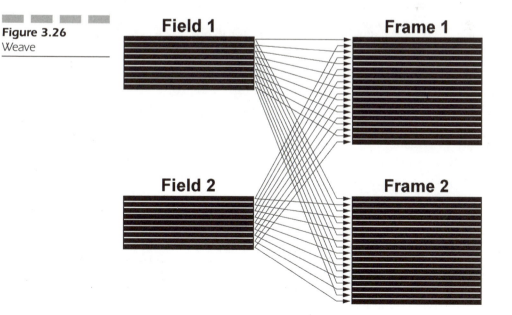

Figure 3.26
Weave

Figure 3.27
Bob

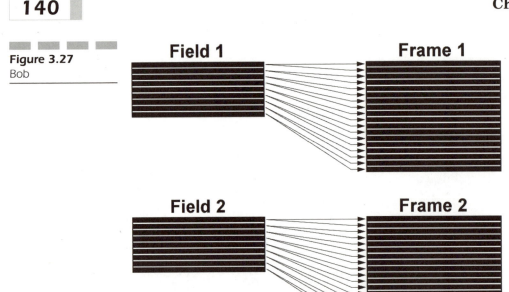

also can weave, and in many cases, the number of lines is more than doubled.

3. *Field-adaptive deinterlacing.* This method examines individual pixels across three or more fields and selectively weaves or bobs regions of the picture as appropriate. Regions of pixels that do not change across frames can be reinterleaved without creating motion artifacts. Sections where there is motion can be bobbed or can be averaged across fields to create a motion blur effect. The cost of field-adaptive deinterlacing is $10,000 and up, so it will be a while before we see it in consumer DVD players.

4. *Motion-predictive deinterlacing.*[29] This method uses massive image processing to identify moving objects in order to selectively weave or bob regions of the picture as appropriate. These systems decompose the picture into two-dimensional or three-dimensional representations that are used to generate a progressive version of the picture. For example, converting interlaced video of a basketball game to progressive format would require that a model be generated for the frame of reference, for the ball, and for each player (each of which

---

[29]The term *motion-adaptive* originally applied only to systems that perform object motion analysis, but it now tends to be used for both field-adaptive and motion-predictive deinterlacing.

might be moving in simple motion across the screen or might be moving toward or away from the camera, growing and shrinking in apparent size). Some systems use MPEG-2 motion vectors as clues to guide the analysis process, but since motion vectors also can be used to replicate similar areas that appear anywhere in the frame, they cannot be relied on exclusively for motion detection. High-quality motion-predictive systems generally cost $50,000 and up.

The three common categories of deinterlacing systems are:

1. *Integrated.* This is usually best, where the deinterlacer is integrated with the MPEG-2 decoder so that it can read MPEG-2 flags and analyze the encoded video to determine when to bob and when to weave. Most DVD computers use this method.

2. *Internal.* The decoded digital video is passed from the MPEG-2 decoder to a separate deinterlacing chip. A potential disadvantage is that flags, motion vectors, and other information in the MPEG-2 stream may no longer be available to help the deinterlacer determine the original format and cadence of the progressive source.

3. *External.* Analog video from the DVD player is passed to a separate line multiplier or to a display with a built-in line doubler. In this case, the video quality is slightly degraded as it is converted to analog, back to digital, and often back again to analog. In addition, ancillary data from the MPEG-2 stream are not available. For high-end projection systems, a separate line multiplier (which bobs, weaves, and interpolates to a variety of scanning rates) may achieve the best results.

As deinterlacers become better, using field-adaptive and motion-predictive techniques, the differences in quality between the three categories, or at least the first two, will mostly disappear.

For better quality, any deinterlacing process other than simple bobbing also must undo the 2–3 pulldown process on film source. Since the MPEG-2 encoder already did this, the deinterlacer simply must ignore the repeat field flags. This could be called "synthetic inverse telecine." An external deinterlacer has a harder time because it receives the signal after the player has repeated the fields.

A progressive DVD player has to determine whether the video should be line-doubled or reinterleaved. When reinterleaving film-source video, the player also has to deal with the difference between film frame rate (24 Hz) and TV frame rate (30 Hz). Since the 2–3 pulldown trick cannot be used to spread film frames across progressive video frames, there are worse motion artifacts than with interleaved video. Progressive video is commonly displayed at 60 Hz, twice the normal rate, so frames are repeated in a 2–3 sequence, which means that the smoothing effect of interlaced fields is lost. However, the increase in resolution more than makes up for it. Advanced progressive players and DVD computers can get around the problem by displaying at multiples of 24 Hz, such as 72 and 96 Hz, etc.

A progressive player also has to deal with problems such as video that does not have clean cadence (such as when it is edited after being converted to interlaced video, when bad fields are removed during encoding, or when the video is sped up or slowed down to match the audio track). Figure 3.25 shows how film frames cross video frames (frames B and D). If the video is cut on any of these frames, the frame coherency is lost, and the deinterlacer no longer has a clean sequence of paired fields to weave back together. The MPEG encoder is also affected because it becomes harder to do inverse telecine.

Another problem is that many DVDs are encoded with incorrect MPEG-2 flags, so a reinterleaver that uses these flags has to recognize and deal with pathological cases. In some instances it is practically impossible to determine if a sequence is 30-frame interlaced video or 30-frame progressive video.

A related problem is that many TVs with progressive input do not allow the aspect ratio to be changed. When a nonanamorphic signal is sent to these TVs, they stretch it horizontally instead of properly windowboxing or proportionally enlarging it.

Just as early DVD computers did a poor job of progressive-scan display of DVDs, the first generation of progressive consumer players also were a bit disappointing. As techniques improve, however, as DVD producers become more aware of the steps they must take to ensure that their content looks good on progressive displays, and as more progressive displays appear in homes, the experience will undoubtedly improve, bringing home theaters closer to real theaters.

CHAPTER 4

DVD Overview

Introduction

This chapter deals with the family, fundamental formats, and features of DVD technology, with a focus on DVD-Video. It also explores, explains, and explodes some of the myths and misconceptions that have grown up around DVD.

The DVD Family

The DVD family started off as promisingly as the *Brady Bunch*. Mr. Laserdisc and Mrs. CD-ROM produced rotund twins that everyone loved: DVD-ROM and DVD-Video. Little brother DVD-R came next and got along well enough despite a case of split personality when he was 2 years old. Like the sibling squabbles of the Bradys, however, rivalries and frictions soon popped up. DVD-RAM followed DVD-R but refused to play with the others. DVD-Audio, after an interminable gestation, was so different from DVD-Video that at first the two could not play together. Cousin DVD+RW was ostracized by the rest of the family, even though she tried hard to fit in. And the young triplets, DVD-Video Recording, DVD-Audio Recording, and DVD-Stream Recording, were odd enough that it would take the rest of the family a while to be able to handle them. The happy ending, where differences are resolved and everyone gets along again, which always came at the end of every *Brady Bunch* episode, is much longer in coming for the DVD family.

Figure 4.1 shows the overall relationships between the various members of the DVD family. DVD-ROM is the base format that underlies everything. The writable formats are all variations of DVD-ROM, each with its particular good and bad points. DVD-R can record data once, whereas DVD-RAM, DVD-RW, and DVD+RW can be rewritten thousands of times. When first released, DVD-R and DVD-RAM were available for computers only, but by 2001 all recordable formats also were being used in home video recorders. DVD-R, based on organic dye media similar to CD-R, is generally compatible with other DVD drives and players. The related DVD-RW format, based on phase-change technology similar to CD-RW, is also generally compatible with other drives and players. Just as with CD-R, problems reading DVD-R and DVD-RW will soon disappear and be forgotten. On the other hand, DVD-RAM, a concoction of magneto-optical and phase-change technologies, was not compatible with anything when it was released. It took more than a year before DVD-ROM drives that could read DVD-RAM discs were released. It

Figure 4.1
The DVD family

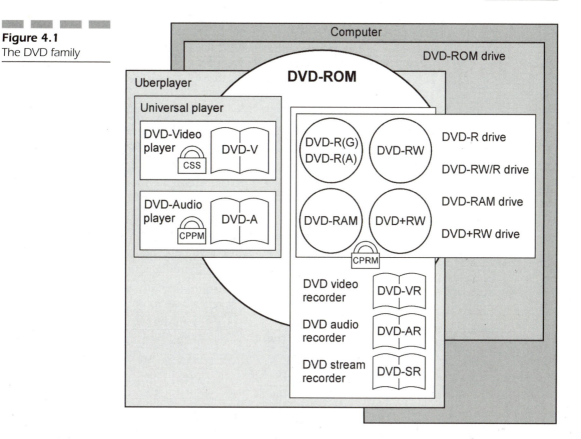

took more than 2 years before compatible DVD-Video players began to slowly trickle out. DVD+RW, championed by Philips, Sony, and Hewlett Packard, is not an official member of the DVD family. Similar to DVD-RW, it has about the same level of compatibility, although competing manufacturers may be hesitant to modify their drives and players to accommodate DVD+RW.

Manufacturers of DVD-RAM, DVD-RW, and DVD+RW claim that specific features make their format better or more suited for particular uses, but the reality is that the technical distinctions make little difference, especially as drives get faster and buffers get bigger. They all record data on a rewritable disc.

Beyond the various physical formats of DVD, there are logical formats, or application formats, that define how data are organized on the disc for a specific purpose. DVD-Video was the first application format, designed for video and audio. DVD-Audio, with specific features aimed at extra-high-fidelity audio, was launched as a separate format, but after a few years, it will have mostly merged with DVD-Video. In 2000 and 2001, additional application formats for recording video (DVD-VR), audio (DVD-AR), and streaming data such as from a camcorder or digital satellite receiver (DVD-SR) were rolled out. These application formats are designed to handle real-time recording and custom playlists.

Just as it is important to understand the difference between DVD-ROM and DVD-Video (or DVD-Audio), it is crucial to understand the difference between DVD data recorders and DVD video recorders (or audio recorders). Data recorders were released for computers beginning in 1997, while *audio/visual* (A/V) recorders were not available for another 3 years or so. Of course, video and audio are just another kind of data; so data recorders connected to computers can be used to write discs with audio and video on them, but it has to be processed in the computer. A/V recorders, like VCRs, have TV tuners and external inputs for analog audio and video, as well as built-in real-time encoders for MPEG-2 audio and Dolby Digital video. Because of the incompatibility of the new DVD-VR and DVD-AR recording formats with existing drives and players, some recorders provide the option to write the standard DVD-Video or DVD-Audio formats even though they were never optimized for real-time recording.

Details of the physical formats are presented in Chapter 5. Details of the application formats are presented in Chapter 6.

The DVD Format Specification

The DVD Forum is the voluntary association of manufacturers, content developers, and other interested companies that establishes the DVD formats and promotes their acceptance. The official specification for DVD is documented in a series of books published by the DVD Forum. The books are divided into various parts for physical specification, file system specification, and application specifications (Tables 4.1 and 4.2). Forum members participate in *working groups* (WGs) assigned to develop and maintain various areas of the specification (Table 4.3). The books are available under NDA and license from the DVD *Format and Logo Licensing Corporation* (FLLC). The books do not include information about copy protection schemes, which are licensed and documented separately (see "Licensing," a later section). The DVD formats incorporate various standards defined in other documents (see Appendix B).

The DVD format specification defines only the discs and their content. It does not define players. It provides guidelines on player features and basic design, but manufacturers generally are free to implement players as they

TABLE 4.1

DVD Specification
Books

Book	Letter[a]	Physical	File System[③]	Application	First Published
DVD-ROM	A	Part 1[②]	Part 2		August 1996
DVD-Video	B	Part 1[②]	Part 2	Part 3[①]	August 1996
DVD-VR (video recording)	B			Part 3[①]	October 1999
DVD-SR (stream recording)	B			Part 3[①]	Mid 2001
DVD-Audio	C	Part 1[②]	Part 2	Part 4[④]	March 1999
DVD-AR (audio recording)	C			Part 4[④]	End 2000
DVD-R 1.0 (recordable)	D	Part 1[⑥]	Part 2		July 1997
DVD-R(G) 2.0 (recordable)	D	Part 1[⑥]	Part 2		March 2000
DVD-R(A) 2.0 (recordable)	D	Part 1[⑥]	Part 2		July 2000
DVD-RW (rerecordable)	D	Part 1[⑥]	Part 2		November 1999
DVD-RAM 1.0 (rewritable)	E	Part 1[⑤]	Part 2		July 1997
DVD-RAM 2.0 (rewritable)	E	Part 1[⑤]	Part 2		September 1999

[a]Archaic. Books are now identified by name rather than letter.
① through ⑥ indicate responsibility of DVD Forum working groups (Table 4.3).

desire. This encourages innovation but also leads to inconsistencies and incompatibilities among players.

Compatibility

Attempting to understand DVD compatibility could drive a granite plinth insane. It does not help that compatibility has many facets and meanings. No specification exists for DVD players, only for physical discs, file systems,

TABLE 4.2

DVD Specification
Timeline

Book	Version	Published	Product Available
DVD-ROM	0.9	April 1996	
DVD-Video	0.9	April 1996	
DVD-ROM	1.0	August 1996	Q1 1996 (Q3 1996 in Japan)
DVD-Video	1.0	August 1996	Q1 1996 (Q3 1996 in Japan)
DVD-R	0.9	April 1997	
DVD-RAM	0.9	April 1997	
DVD-R	1.0	July 1997	Q4 1997
DVD-RAM	1.0	July 1997	Q2 1998
DVD-ROM	1.01	December 1997	
DVD-Video	1.1	December 1997	
DVD-Audio	0.9	May 1998	
DVD-RAM	1.9	October 1998	
DVD-R	1.9	November 1998	Q2 1999
DVD-RW	0.9	October 1999	
DVD-VR	0.9	January 1999	
DVD-Audio	1.0	March 1999	Q3 2000
DVD-R	1.01	April 1999	
DVD-Video	1.11	May 1999	
DVD-Audio	1.1	May 1999	
DVD-RAM	2.0	September 1999	Q3 2000
DVD-ROM	1.02	September 1999	
DVD-VR	1.0	October 1999	Q4 2000 (Q4 1999 in Japan)
DVD-RW	1.0	November 1999	Q4 2000 (Q4 1999 in Japan)
DVD-RAM	2.1	February 2000	
DVD-R(A)	2.0	July 2000	Q3 2000 (firmware upgrade)
DVD-VR	1.1	March 2000?	
DVD-Audio	1.2	May 1999	
DVD-AR	0.9	Mid 2000	
DVD-R(G)	2.0	October 2000	Q4 2000
DVD-RW	1.1	Ocober 2000	
DVD-AR	1.0	Early 2001	
DVD-SR	0.9	Mid 2000	
DVD-SR	1.0	Mid 2001	

Note: File system specifications are omitted for simplicity.

TABLE 4.3

DVD Forum
Working Groups

WG-1	DVD-Video applications
WG-2	Physical specifications for DVD-ROM
WG-3	File system specifications for all DVD variations
WG-4	DVD-Audio applications
WG-5	Physical specifications for DVD-RAM
WG-6	Physical specifications for DVD-R and DVD-RW
WG-9	Copy protection liaison
WG-10	Professional applications of DVD

and applications. Player makers are free to implement these specifications as they see fit. Thus, physical compatibility, file system compatibility, application compatibility, and implementation compatibility all exist. Unfortunately, failures of compatibility occur at all these levels (Table 4.4). The upshot is that consumers cannot assume that any disc with "DVD" in its name will work in any player or computer with "DVD" in its name. If someone records video onto a DVD to send to someone else, he or she cannot assume that it will play on the recipient's DVD hardware.

NOTE: DVD compatibility breakdowns happen at the physical level, logical level, and implementation level.

Physical Compatibility

Physical compatibility is not an issue with DVD-ROM. Every player, drive, and recorder is physically able to read data from DVD-ROM discs. They may not know what to do with it or may not even attempt to read certain sections of the disc, but they are capable of reading the bits. Physical compatibility only becomes a problem with the writable formats, as illustrated in Table 4.4. Physical compatibility also applies to other formats such as CD and CD-R, where it is up to the manufacturer to decide whether or not to support a particular physical medium.

TABLE 4.4	Problem	Incompatibility	Explanation
Examples of Compatibility Problems	A DVD player cannot play a DVD-Audio disc.	Application	Unless the player was designed to read the DVD-Audio data format, it will not recognize the contents of the disc. Luckily, many DVD-Audio discs include audio in DVD-Video format so that they will play in DVD-Video players and DVD computers.
	A DVD player cannot play a recorded DVD.	Physical	The disc may be a type that the player cannot physically read, such as DVD-RAM.
		Application	The disc may be recorded using an application format that the player does not recognize, such as DVD-VR or DVD-SR.
	A DVD player cannot play a PC-enhanced disc.	Application	The set-top DVD player does not recognize the computer applications and HTML pages on the disc. It can play the contents of the DVD-Video zone, and perhaps, the DVD-Audio zone, but not the other zones.
		Implementation	Some players were not designed to properly deal with extra files and extra directories on the disc. Even though this is allowed by the DVD specification, it confuses certain players to the point that they cannot even play the DVD-Video content.
	A player cannot play a CD-R but plays a CD-RW.	Physical	The player does not have a second laser at the wavelength needed to read a CR-R. CD-RW reflectively is different, so the laser for reading DVDs can read a CD-RW (and a CD).
	A player cannot play a CD.	Physical	If the disc is recorded on CD-R media, the player may not be able to physically read it.
		Application	Every DVD player has been designed to read CDs, but the CD may have computer data on it, not CD-Audio.

continues

**TABLE 4.4
Cont.**

Examples of
Compatibility
Problems

Problem	Incompatibility	Explanation
A DVD player cannot play a Video CD or a Super VCD.	Physical	If the disc is recorded on CD-R media, the player may not be able to physically read it.
	Application	Even if the player can read the disc, the manufacturer may have chosen not to add firmware and circuitry needed to play other video formats.
A DVD computer cannot play a Sony PlayStation DVD.	Application	Although the computer can read the data, it does not know how to execute the application or interpret the proprietary file formats.
A computer cannot play or copy some movies.	Application	Most movies are encrypted with CSS. This prevents direct copying of files without an authentication process between the decoder and the drive. If the computer does not have a software player and decoder that implements CSS, it is not able to decrypt and play the files.
	Implementation	Bugs may be present in the computer driver or player software that prevent proper playback, or the data may have been incorrectly formatted on the disc.
A DVD player cannot play certain movies.	Application	A DVD-Audio–only player cannot play movies. It was not designed to read the DVD-Video files.
	Implementation	The disc may have been authored or encoded in a way that is not compliant with the DVD specification, or the information on the disc may be compliant with the specification but not handled properly by the player.

continues

	PROBLEM	INCOMPATIBILITY	EXPLANATION
TABLE 4.4 Cont. Examples of Compatibility Problems	A DVD player cannot read a disc recorded in a computer.	Physical	The disc may be a type that is not physically readable in the computer drive.
		File system	DVD players use the UDF file system to find and read the files containing audio and video. The computer may not have used UDF. DVD players also expect the files to be physically contiguous. The computer may not have ordered the video files in contigous order from the beginning of the disc.
		Application	The data may be recorded as a set of MPEG-2 files or MP3 files, which the player is not designed to recognize and process.
	One computer cannot read or play the disc written on another computer.	Physical	The disc may be a type that is not physically readable in the computer drive, such as DVD-RAM or DVD+RW.
		File system	The second computer may not recognize the file system used by the first. For example, newer versions of Windows natively use FAT32, which many other operating systems cannot read.
		Application	The second computer may not have the software needed to play the disc.
		Implementation	Bugs may be present in the drivers or formatting software of the first or second computer.

File System Compatibility

File system compatibility is generally not a problem. The file system determines how the data is organized and accessed. UDF and ISO 9660 are the two common file systems on DVD; in fact, most discs contain both. Because DVDs are simply storage media, other file systems such as Microsoft FAT, NTFS, Macintosh HFS, UNIX, and so on can be used to write data files to the disc. Compatibility problems generally occur only with these specialized file systems, such as when a disc formatted with Windows FAT32 is placed in the drive of a Mac.

Application Compatibility

Application compatibility is the most confusing area. It is not always clear why a disc does not work in a specific player because it may not be obvious which application formats the player supports. For example, a DVD-Video player can read the data and UDF files on a DVD-Audio disc, but if it is not built to read and process data from the DVD-Audio files in the AUDIO_TS directory, it will not play the audio. Both the player and the disc say "DVD" on them, but it may not be clear why they do not work together. The proliferation of DVD Forum application formats such as DVD-VR and DVD-AR, along with other custom formats for computers and game consoles, places a burden on the consumer to understand which discs use which application formats and which players can play them.

Implementation Compatibility

Implementation compatibility has to do with flaws or omissions in players (including computers and any other device that can play DVDs). Some players are poorly designed, whereas others behave in unexpected ways with unanticipated content. Each player implements the DVD specification in a slightly different way, complicated by the fact that the specification is ambiguous and confusing. The result is that discs may play differently or not play at all in different players. Implementation errors also can occur on the other side as well. That is, a bug might exist in the encoder or in the system used to author the disc, or the person who authored the disc may have done something that is not allowed by the DVD specification and thus does not work on some or all players. Implementation problems are discussed in more detail under "Playback Incompatibilities" in Chapter 7.

New Wine in Old Bottles: DVD on CD

Another compatibility problem, which falls between application and implementation, has to do with putting DVD content on media other than DVD. Many people would like to be able to put short DVD-Video programs onto CDs or inexpensive CD-Rs. The idea of a "mini-DVD" is very appealing, particularly for testing titles during development and for viewing short programs such as music video singles, home movies, or corporate marketing clips. However, each would-be clever inventor is disappointed to learn that it does not work on set-top players. Only a few odd player models designed around DVD-ROM drives can play the discs. All other DVD-Video players fail to play DVD-Video content from CD media. A number of reasons for this are:

1. The player does not expect DVD content on CD media. The first thing most players do after a disc is inserted is check focus depth and reflectivity. If nothing can be read at DVD focus, the player switches focus (and usually switches lens and laser as well) to see if it can read data at CD focus depth. If it determines that a CD is in the drive, it goes to CD mode. It checks for audio CD content, and it might check for Video CD content or MP3 files, but it does not look for DVD files.

2. The drive unit is not fast enough. It is simpler and cheaper for players to spin CDs at 1X speed rather than the 9X speed needed to provide the 11 Mbps data rate required by DVD-Video content.

3. Many players cannot read CD-R discs. A player without a laser at CD wavelength cannot read CD-R media.

Hollywood may be concerned about movies being copied easily and cheaply to CD-Rs that would play in DVD players, although the quality would be poor even if the video were spread across more than one CD. Of course, player manufacturers could deal with all three of these obstacles, but they do not believe that the demand justifies the extra expense. Some also may accuse them of not wanting to sustain the CD market when they can make more money (or pay fewer royalties) with DVD.

Computers are more forgiving. Most of them are media agnostic when it comes to DVD content. DVD-Video files from any source with fast enough data rates, including CD-R or CD-RW, or even hard drives or Jaz drives, with or without UDF formatting, will play back on almost any DVD computer as long as the drive can read the media. In fact, many DVD authoring systems include the option to write small volumes to CD-R or CD-RW.

An alternative is to put Video CD or Super Video CD content on CD-R or CD-RW media for playback in a DVD player. About half the set-top player models can play VCDs, although very few can play SVCDs. The limitations of VCD apply (MPEG-1 video and audio, 1.152 Mbps, 74 minutes of playing time[1]). All DVD-ROM computers able to read recordable CD media can play recorded VCD discs if they have the necessary playback software. An MPEG-2 decoder and SVCD player software are needed to play SVCDs.

Compatibility Initiatives

Many efforts have been made to deal with compatibility problems at various levels. In October 1997, the *Optical Storage Technology Association* (OSTA) produced the MultiRead specification, which provides a logo for devices—including DVD drives and players—that read CD audio, CD-ROM, CD-R, and CD-RW. The follow-up MultiRead 2 specification from December 1999 added DVD-ROM and DVD-RAM to the list. The DVD Forum manages a certification program for discs and players (see Appendix C). In 2000, the DVD Forum began finalizing the DVD Multi program to promote physical compatibility and limited application compatibility for reading DVD-ROM, DVD-R, DVD-RW, and DVD-RAM discs and writing DVD-R, DVD-RW, and DVD-RAM discs. A player or drive with the DVD Multi logo is guaranteed to read a certain set of disc and application formats (Table 4.5).

The state of affairs is far from satisfactory, but at least the industry is drifting toward read compatibility across formats, if not write compatibility.

---

[1]These limitations can be stretched with the so-called DVCD and DSVCD formats, which squeeze more tracks on a disc to extend the capacity. In the case of DVCD, playing time is increased from 74 to 100 minutes at the expense of minor physical compatibility problems on some VCD and DVD players.

TABLE 4.5

DVD Multi Compatibility

	DVD Multi Consumer Electronics Device				DVD Multi Computer	
	Player	**Audio-Only Player**	**Video-Only Player**	**Recorder**	**Drive**	**Record**
DVD-ROM disc	Will not play	Will not play	Will not play	Will not play, cannot record	Reads	Reads, cannot write
DVD-Video disc	Plays	Will not play	Plays	Plays, cannot record	Plays[a]	Plays,[a] cannot write
DVD-Audio disc	Plays	Plays	Will not play	Plays, cannot record	Plays[b]	Plays,[b] cannot write
DVD-R disc	Plays	Plays (audio)	Plays (video)	Plays, records	Reads	Reads, writes
DVD-RW disc	Plays	Plays (audio)	Plays (video)	Plays, records	Reads	Reads, writes
DVD-RAM disc	Plays	plays (audio)	plays (video)	Plays, records	Reads	Reads, writes

[a]Can play if computer has DVD-Video decoder.

[b]Can play if computer has DVD-Audio decoder.

Bells and Whistles: DVD-Video and DVD-Audio Features

The creators of DVD realized that in order to succeed, DVD had to be more than just a roomier CD or a more convenient laserdisc. Hollywood started the ball rolling by requesting a digital video consumer standard that would hold a full-length feature film, had better picture quality than existing high-end consumer video with wide-screen aspect ratio support, contained multiple versions of a program and parental control, supported high-quality surround audio with soundtracks for at least three languages, and had built-in copy protection. Then the computer industry added its requirements of a single format for computers and video entertainment with a common cross-platform file system, high performance for both movies and computer data, compatibility with CDs and CD-ROMs, compatible recordable and rewritable versions, no mandatory caddy or cartridge, and high data capacity with reliability equal to or better than CD-ROM. Later on, Hollywood decided that it wanted a copy protection and locking system to control release across different geographic regions of the world.

The designers threw in a few more features, such as multiple camera angles and graphic overlays for subtitling or karaoke, and DVD was born. Unlike CD, where the computer data format was cobbled on top of the digital music format, the digital data storage system of DVD-ROM is the base standard. DVD-Video is built on top of DVD-ROM using a specific set of file types and data types. A DVD-ROM can contain digital data in almost any conceivable format, as long as a computer or other device can make use of it. DVD-Video, on the other hand, requires simple and inexpensive video players, so its capabilities and features are strictly defined.

From the beginning, the plan was to create a separate DVD-Audio format based on input from the music industry. Requirements included copyright identification and copy protection, compatibility with DVD-ROM and DVD-Video, CD playback (including an optional hybrid DVD/CD format that could play in CD players), navigation and random access similar to DVD-Video but also usable on players without an attached video display, slideshow features, and of course, superior sound quality. DVD-Audio supports a subset of DVD-Video features, although eventually every new DVD-Video player also will play DVD-Audio discs. Apart from audio-only players such as small, portable devices, the distinctions between formats and players eventually should disappear.

The following sections present the features of DVD-Video. The term *player* also applies to software players on computers, as well as other devices such as videogame consoles that have the ability to play DVD-Video discs. DVD-Audio features are mentioned when relevant.

Over 2 Hours of High-Quality Digital Video and Audio

Over 95 percent of Hollywood movies are shorter than 2 hours and 15 minutes, so 135 minutes was chosen as the goal for a digital video disc. Uncompressed, this much video could take up 255 gigabytes.[2] DVD uses MPEG-2 compression to fit high-resolution digital video onto a single disc. The MPEG-2 encoding system compresses video in two ways: spatially, by reducing areas of repetitive detail and removing information that is not perceptible, and temporally, by reducing information that does not change over time. Reducing the video information by a factor of almost 50 enables it to fit in less than 5 gigabytes. Unfortunately, compression can cause unwanted effects such as blockiness, fuzziness, and video noise. However, the variable data rate of DVD enables extra data to be allocated for more complex scenes. Carefully encoded video is almost indistinguishable from the original studio master.

As mentioned earlier, the 135-minute length (or the absurdly precise 133-minute length) is a rough guideline based on estimates of average video compression and number of audio tracks. The length of a movie that can fit on a standard DVD depends almost entirely on how many audio tracks are available and how heavily the video is compressed. Other factors come into play, such as the frame rate of the source video (24, 25, or 30 frames per second), the quality of the original (soft video is easier to compress than sharp film grain, and clean video is easier to compress than noisy or dirty video), and the complexity (slow, simple scenes are easier to compress than rapid motion, frequent changes, and intricate detail).

In any case, the average Hollywood movie easily fits on one side of a DVD. This overcomes one of the big objections to laserdisc—that you had to flip the disc over or wait for the player to flip it over after each hour of playing time.

For DVD-Audio the quality was bumped up to the next level with double the PCM sampling rates of DVD-Video and lossless packing to increase playing times without lossy compression. A single-layer DVD-Audio disc can play

---

[2]Digital studio masters generally use 4:2:2 10-bit sampling, which at 270 Mbps eats up over 32 megabytes every second.

74 minutes of super-fidelity multichannel audio or over 7 hours of CD-quality stereo audio.

Widescreen Movies

Television and movies shared the same rectangular shape until the early 1950s when movies began to get much wider. Television has stayed unchanged until recently. Widescreen TVs are appearing slowly, and DVD is bound to cause a huge jump in demand. Movies can be stored on DVD in widescreen format to be shown on widescreen TVs close to the width envisioned by the director. DVD includes techniques to show these widescreen movies on regular televisions and straddles old and new television, since HTDV is a widescreen format. These different aspect ratios are discussed in detail in Chapter 3.

Multiple Surround Audio Tracks

The DVD-Video standard provides for up to eight soundtracks to support multiple languages and supplemental audio. Each of these audio tracks can include surround sound with 5.1 channels of discrete audio.[3] DVD surround-sound audio uses Dolby Digital (AC-3) encoding, DTS Digital Surround, or MPEG-2 audio encoding. The 5.1-channel digital tracks can be downmixed by the player with Dolby Surround encoding for compatibility with regular stereo systems and Dolby Pro Logic audio systems. An option is also available for better-than-CD-quality linear PCM audio. Almost all DVD players include digital audio connections for high-quality output.

The usefulness of multiple audio tracks was discovered when digital audio was added to laserdiscs, leaving the old analog tracks free. Visionary publishers such as Criterion used the analog tracks to include audio commentary from directors and actors, musical sound tracks without lyrics, foreign-language audio dubs, and other fascinating or obscure audio tidbits.

DVD-Audio improves on the audio features of DVD-Video with higher sampling rates for PCM and improved support for multichannel PCM audio tracks and audio downmixing.

---

[3]*Discrete* means that each channel is stored and reproduced separately rather than being mixed together (as in Dolby Surround) or simulated. The .1 refers to a *low-frequency effects* (LFE) channel that connects to a subwoofer. MPEG-2 and SDDS audio allow 7.1 channels, but this feature is unlikely to be used for home products.

Most DVD players allow the owner to select a preferred language so that the appropriate menus, language track, and subtitle track can be selected automatically when available. In many cases, the selection also determines the language used for the player's on-screen display.

Karaoke

DVD keeps karaoke fans singing because it includes special karaoke audio modes to play music without vocals or add vocal backup tracks. More importantly, DVD's subtitle feature breaks the language barrier with up to 32 different sets of lyrics in any language, complete with bouncing balls or word-by-word (or ideogram-by-ideogram) highlighting.

Karaoke support is optional for players. Karaoke players have the ability to mix karaoke audio tracks (guide/vocal tracks and melody tracks) into the base stereo tracks. Provisions are also included for identifying the music and singing, such as male vocalist, female soloist, chorus, and so on.

Subtitles

Video can be supplemented with one of 32 subpicture tracks for subtitles, captions, and more. Unlike existing closed captioning or teletext systems, DVD subpictures are graphics that can fill the screen. The graphics can appear anywhere on the screen and can create text in any alphabet or symbology. Subpictures can be Klingon characters, karaoke song lyrics, *Monday Night Football*-style motion diagrams, pointers and arrows, highlights and overlays, and much more. Subpictures are limited to a few colors at a time, but the graphics and colors can change with every frame, which means subpictures can be used for simple animation and special effects. Some DVDs use subpictures to show silhouettes of the people speaking on a commentary track, in the style of the *Mystery Science Theater 3000* TV show. Being able to see the gestures and pointing of the commentators can enhance the audio commentary.

The transparency effect can be used to dim down areas of the picture and make other areas stand out. The same video can be shown with or without this highlighting effect. This can be used as a great effect for educational video and documentaries. Other options include covering parts of a picture for quizzing, drawing circles and arrows, and even creating overlay graphics to simulate a camcorder, night-vision goggles, or a jet fighter cockpit.

Different Camera Angles

One of the most innovative features of DVD is an option to view scenes from different angles. A movie can be filmed with multiple cameras so that the viewer can switch at will between nine different viewpoints. In essence, camera angles are multiple simultaneous video streams. As you watch, you can select one of the nine video tracks just as you can select one of the eight audio tracks.

This feature presents a paradigm shift that could be as significant as the way sound changed motion pictures. The storytelling opportunities are fascinating to contemplate. Imagine a movie about a love triangle that can be watched from the point of view of each main character; a murder mystery with multiple solutions; or a scene that can be played at different times of the day, different seasons, or different points in time. Music videos can include shots of each performer, enabling viewers to focus on their current favorites or to pick up instrumental techniques. Classic sports videos can be designed so that armchair quarterbacks have complete control over camera angles and instant replay shots. Exercise videos may allow viewers to choose their preferred viewpoints. Instructional videos can provide close-ups, detail shots, and picture insets containing supplemental information. The options are endlessly diverse and merely require new tools and new approaches to filmmaking and video production.

The disadvantage of this feature is that each camera angle requires that additional footage be created and stored on the disc. A program with three camera angles available the entire time can only be one-third as long if it has to fit in the same amount of space.

Multistory Seamless Branching

A major drawback of almost every previous video format, including laserdiscs, Video CD, and even computer-based video such as QuickTime, is that any attempt to switch to another part of the video causes a break in play. DVD-Video finally achieves completely seamless branching. For example, a DVD can contain additional director's cut scenes for a movie but jump right over them without a break to recreate the original theatrical version.

This opens up endless possibilities for mix-and-match variety. At the start of a movie, the viewer could choose to see the extended director's cut, alternate ending number four, and the punk rock club scene rather than the jazz club scene, and the player would jump around the disc, indistinguishably stitching scenes together. It is even possible for a disc to tell the player to randomly select alternate sequences so the experience will be different every time. Of course, this requires significant additional work by the director or producer. Most mass-market releases skip this option, leaving it to small, independent producers with more creative energy.

Parental Lock

DVD includes parental management features to block playback and to provide multiple versions of a movie on a single disc. Players can be set to a specific parental level using password-protected onscreen menus. If a disc with a rating above this level is put in the player, it will not play. In some cases, different programs on the disc have different ratings.

A disc also can be designed so that it plays a different version of the movie depending on the parental level that has been set in the player. By taking advantage of the branching feature of DVD, objectionable scenes can be skipped over automatically or substituted during playback, usually without a visible pause or break. For example, a PG-rated scene can be substituted for an R-rated scene, along with dialog containing less profanity. This requires that the disc be carefully authored with alternate scenes and branch points that do not cause interruptions or discontinuities in the soundtrack.

Unfortunately, not even 1 percent of DVDs use the multirating feature. Hollywood studios are not convinced that the demand is big enough to justify the extra work involved, which includes shooting extra footage, recording extra audio, editing new sequences, creating branch points, synchronizing the soundtrack across jumps, submitting new versions for MPAA rating, dealing with players that do not implement parental branching properly, having video store chains refuse to carry discs with unrated content, and much more. The few discs that have multirated content do not have standard package labeling or other way to be easily identified.

Another option is to use a software player on a computer that can read a "play list" telling it where to skip scenes or mute the audio. Play lists can be used to "retrofit" the thousands of DVD movies that have been produced without parental control features.

Menus

In order to provide access to many advanced features, the DVD-Video standard includes on-screen menus. The video can stop at any point for interaction with the viewer, or selectable hot spots can be on live video. Menus are used to select from multiple programs, choose different versions of program content, navigate through multilevel or interactive programs, activate features of the player or the current disc, and more.

For example, a movie disc may have a main menu from which you can choose to watch the movie, view a trailer, watch a "making of" featurette, or peruse production stills. Another menu may also be available from which you can choose to hear the regular sound track, foreign-language sound track, or director's commentary. Selecting the supplemental information option from the main menu may bring up another menu with options such as production stills, script pages, storyboards, and outtakes.

Interactivity

In addition to menus, DVD can be even more interactive when the creator of the program takes advantage of the rudimentary command language that is built into all DVD players. DVD-Video can be programmed for simple games, quizzes, branching adventures, and so on. DVD brings a new level of personal control to video programs. While it is not apparent just how much control the average couch potato is interested in having, directing the path and form of a presentation is definitely an appealing option. "Choose your own ending" stories have graduated from paper to video. The creative community could embrace an entire new genre of nonlinear cinema.

For example, a music video disc could provide an editing environment where the viewers can choose music, scenes, and so on to create their own custom version. An instructional video can include comprehension check questions. If the wrong answer is chosen, a special remedial segment can be played to further explain the topic.

On-Screen Lyrics and Slideshows

The DVD-Audio format includes features for displaying lyrics on the screen as the audio plays and optionally highlighting parts of the lyrics in time with the audio. This is also possible using the subtitle feature of the DVD-Video format, but it is not as straightforward.

The DVD-Audio format also includes a slideshow feature for showing pictures as the audio plays. The pictures can appear automatically at preselected points in the program, or the viewer can choose to browse through them at will, independent of the audio. The DVD-Video format also supports programmed slideshows but not browsable pictures that don't interrupt the audio.

Customization

As mentioned, DVD players can be customized with a parental lock. Many other options can be set on a DVD player to customize the viewing experience. Most DVD players can be set for the preferred soundtrack language and subtitle language and even menus in the chosen language, when available. Preferred aspect ratio—widescreen, letterbox, or pan and scan—also can be set.

If you were studying French, for example, you could set your preferences to watch movies with French dialog and English subtitles. When these are available on the disc, they will be selected automatically.

Instant Access

Consumer surveys indicate that one of the most appealing features of DVD is that you never have to rewind or fast forward it. It is amazing how important time and convenience can be to consumers, but consider our penchant for microwave ovens, electric pencil sharpeners, and escalators. A DVD player can obligingly jump to any part of a disc—program, chapter, or time position—in less than a second.

Special Effects Playback

In addition to near-instantaneous search, most DVD players include features such as perfect freeze-frame, frame-by-frame advance, slow motion, double-speed play, and high-speed scan. Most DVD players scan backward at high speed. But due to the nature of MPEG-2 video compression, most cannot play at normal speed in reverse or step a frame at a time in reverse. This is only possible on advanced players that have more sophisticated video processors.

Access Restrictions

DVDs include a feature enabling the author of the disc to restrict *user operations* (UOPs) such as fast forward, chapter search, and menu access. Almost every button on the remote control can be blocked at any point on the disc. This is not always a benefit to the viewer (as when you are locked into the FBI warning or advertisements at the beginning of a disc), but it is helpful in complicated discs to keep button-happy viewers from going to the wrong place at the wrong time.

Durability

Unlike tape, DVDs are impervious to magnetic fields. A DVD left on a speaker or placed too close to a motor will be unharmed. Discs are also less sensitive to extremes of heat and cold. Because they are read by a laser that never touches the surface, the discs will never wear out—even your favorite one that you play six times a week or the kids' favorite one that they play six times a day. DVDs are susceptible to scratching, but their sophisticated error-correction technology can recover from minor damage.

Programmability

Some DVD players are viewer-programmable, similar to CD players. Chapters can be selected for playback in specified order. You can rearrange the sequence of tracks in a music video to your own taste. You can even drive your friends crazy by having the catchiest song reappear at strategically annoying points. Multidisc players can be programmed to show a demo of your favorite scenes from different discs to impress visitors. Note that the access restrictions mentioned previously may make it difficult to program jumps into any arbitrary part of a disc.

Availability of Features

Obviously, most DVD features entail extra work by the producer of the disc. Adding additional scenes, multiple language tracks, subtitles, ratings information, menus, branch points, and more demands additional effort and expense. The extent to which movie producers support these features depends largely on how much customers demand them and how much they are willing to pay for them. In the laserdisc market, a thriving special-edition industry emerged, titillating videophiles with restored footage, outtakes, director's commentaries, production photos, and documentaries. These special editions required hundreds of hours of extra work by dedicated or obsessed professionals, and they generally sold for over $100—three times the cost of a regular edition. Special editions of DVDs are even more common and more popular than those on laserdisc, yet the price typically is increased by only $10 or so, if at all.

Beyond DVD-Video and DVD-Audio Features

DVDs can contain much more than the limited selection supported by home DVD players. Computer software such as screen savers, games, and interactive enhancements can be included. Along with standard DVD computers, home video game systems and Internet WebTV boxes can support enhanced features. A single DVD could contain a movie, a video game based on the movie, an annotated screenplay with hotlinks to related scenes and storyboards, the searchable text of the novelization complete with illustrations and hyperlinks. In addition, it could contain links to Internet Web sites with more information, fan discussion forums, related merchandise, and special promotions. See the "WebDVD" section of Chapter 11.

DVD is becoming the most common component of multipurpose set-top boxes, such as a combination digital video recorder, cable TV receiver, and DVD player/recorder, or a video game console that is also a DVD player and a Web browser.

DVD Myths

Numerous myths have evolved concerning DVD. Apparently, some people had nothing better to do while waiting for it to appear than to sit around and misconstrue its characteristics. Some myths quickly met their deserved deaths once DVD proved itself, but many others continue to circulate, like urban legends of microwaved cats and kidney thefts. This section deals with the most common misperceptions of DVD.

Myth: "DVD Is Revolutionary"

DVD is evolutionary, not revolutionary. The printing press was revolutionary. Television was revolutionary. Even CDs can be considered revolutionary because they were a completely new way of storing digital audio and computer data on a compact optical disc. But DVD is not fundamentally more than the evolution of CD and the refinement of Video CD. Other than digital video and some clever features, nothing is radically different between DVD and VHS, between DVD and laserdisc, or between DVD-ROM and CD-ROM.

Myth: "DVD Will Fail"

As DVD was being developed, many pundits predicted that it would be a flop, joining the neglected ranks of other consumer electronic innovations such as quadraphonic sound, the 8-track tape, the Tandy/Microsoft VIS, and the digital compact cassette. In less than 3 years, however, DVD became the most successful consumer electronics product ever. Hundreds of companies supply DVD products and services: all major consumer electronics manufacturers (and many minor ones), all major Hollywood movie studios (and scores of independent filmmakers), many major music labels (as well as indie labels), all major computer hardware manufacturers, countless audio/video production houses, and rapidly growing ranks of corporate A/V departments. On the consumer entertainment side, DVD has begun to fulfill its destiny of replacing VHS tape in a decade or two. On the computer side, DVD-ROM drives and recordable DVD drives are inexorably replacing CD technology, heading to the point where it becomes almost impossible to buy a PC without a DVD drive.

Some people still claim that DVD will never amount to much or that it will be quickly superseded by something newer and better. On close inspection, these arguments do not hold water. The possibility did exist that DVD-Video would never capture more than a niche market, similar to laserdisc, but the success of DVD-ROM was virtually secured from the beginning. The ever-expanding needs of computer data and multimedia require a capacious medium for storage and distribution. CD-ROM was the undisputed king of the realm, but DVD-ROM is the crown prince—the guaranteed successor, since nothing else provides a similarly compatible improvement.

The window of opportunity for new technology grows smaller all the time, as evidenced by such not-quite-failures as S-VHS, DAT, and MiniDisc. But none of these can be compared with DVD, which has a mainstream computer counterpart holding open the door to acceptance. In fact, DAT is arguably the most successful of these other products because it also can be used for computer data backup.

DVD-Video has more backing than any new entertainment product in the history of consumer electronics. The annual sales income of the 10 founding DVD companies alone is over $350 billion, more than the gross domestic product of many countries. Staggering amounts of money were spent to develop DVD, and even more is being spent to produce and market it.

Myth: "DVD Is a Worldwide Standard"

If only this were so. DVD is still closely tied to the NTSC and PAL television formats. All PAL DVD players can play NTSC discs, but very few NTSC DVD players can play PAL discs. Even worse, DVD includes regional codes that can prevent a disc from being played on players sold in other countries. See Chapter 7 for more explanation and Chapter 6 for the minutiae.

Technically, DVD is not a "standard" at all in the formal sense. Just like CD, it is a proprietary but open standard created by a group of companies motivated by mutual interests and anticipated profits. Existing standards such as MPEG video and the UDF file format were adopted for DVD. Some of the fundamental parts of the DVD specification, such as the physical formats for read-only and writable discs, have been submitted and approved by official standards bodies such as ECMA and ISO. However, the important parts of the standard, such as the application formats for video and audio, are proprietary to the DVD Forum. Both the official standards and the proprietary specifications are subject to patent royalties.

Myth: "Region Codes Do Not Apply to Computers"

Regional codes apply to DVD-Video discs played in DVD-ROM drives. Every DVD-ROM drive is either set to a region by the manufacturer or must be set by the user before a region-coded disc can be played. Newer (RPC2) drives allow up to five region changes before the region code is set permanently. Of course, there are ways around regional restrictions, just as with standalone DVD-Video players. Regional codes do not apply to PC software or DVD-Audio discs, only to DVD-Video.

Myth: "A DVD-ROM Drive Makes Any PC a Movie Player"

Most DVD computers can play DVD-Video discs, but some, especially those which have had a DVD-ROM drive installed later, do not have everything that is needed. A computer can play DVD-Video movies only if it has the right stuff. A fast computer, such as a 350-MHz Pentium II with accelerated video hardware or a Mac G4, needs only a DVD-ROM drive and DVD playback software. Slower computers require additional DVD playback hardware and must run faster than 100 MHz to handle the load. When DVD-Audio was finally released in late 2000, neither Microsoft nor Apple had plans to add playback support any time soon.

See Chapter 11 for more about DVD PCs.

Myth: "Competing DVD-Video Formats are Available"

This statement is a myth in the sense that inadequately educated friends and "advisors" sometimes caution others not to buy DVD players yet because supposed competing formats exist. As far as DVD-ROM and DVD-Video go, one and only one format exists. The confusion seems to be partly a carryover from the early competition between DVD's progenitor prototypes, SD and MMCD. On the other hand, in the case of competing recordable formats and DVD-Audio discs that will not play in DVD-Video players, this is anything but a myth, as detailed at the beginning of this chapter.

Myth: "DVD Players Can Play CDs"

Even though the DVD specification makes no mention of CD compatibility, all DVD players can play audio CDs—as long as they are the commercial stamped kind. Many DVD-Video players cannot read CD-R discs, although most can read CD-RW discs. Compatibility with other CD formats varies. Only about 50 percent of DVD players can read Video CDs (assuming they are not on CD-R media), and only a very few players can play Super Video CDs or MP3 CDs. And, of course, DVD-Video players cannot play CD-ROMs.

Many people have come up with the idea of putting DVD-Video content onto a CD-R. Aside from physical compatibility problems, these so-called MiniDVDs only play in DVD computers and in a few odd player models designed around DVD-ROM drives. See "Compatibility," from earlier in the chapter, for more detail.

Some early DVD-ROM drives could not read CD-Rs, but all modern DVD drives—apart from the very cheapest models—can read CD-Rs. Most DVD computers can play Video CDs and MP3 CDs, but most lack the software need to play Super Video CDs. Most DVD computers can play DVD-Video content from a CD-R or other CD media.

Myth: "DVD Is Better Because It Is Digital"

Nothing is inherent to digital formats that magically makes them better than analog formats. The celluloid film used in movie theaters is analog, yet few people would say that DVD-Video is better than film. Japan's HiVision television had much higher video quality than DVD, but it was analog. Conversely, the quality of digital video from CD-ROMs is certainly nothing to write home about.

The way DVD stores audio and video in digital form has advantages, not the least of which is the ability to use compression to extend playing times. The quality and flexibility of DVD stand out when compared with similar analog products. It is a mistake, however, to make the generalization that anything digital must be superior to anything analog.

Myth: "DVD Video Is Poor Because It Is Compressed"

Much ado is made of the "digital artifacts" that supposedly plague DVD-Video. While it is true that digital video can appear blocky or fuzzy, a well-compressed DVD exhibits few discernible artifacts on a properly calibrated display. Many early discs, especially demonstration discs, were created with hardware or software that was partially finished or not fully tested. Compression techniques have much room for improvement; video encoders are steadily improving, producing better pictures within the same compression constraints. The improvements will benefit all existing players—no hardware upgrade is required. Minor glitches and quality problems are disappearing as compression engineers improve their craft.

The term *artifact* refers to anything that was not in the original picture. Artifacts can come from film damage, film-to-video conversion, analog-to-digital conversion, noise reduction, digital enhancement, digital encoding, digital decoding, digital-to-analog conversion, NTSC or PAL video encoding, Macrovision, composite signal crosstalk, connector problems, impedance mismatch, electrical interference, waveform aliasing, signal filters, television picture controls, tube misconvergence, projector misalignment, and much more. Many people blame all kinds of visual deficiencies on the MPEG-2 encoding process. Occasionally, this blame is placed accurately, but usually it is not. Only those with training or experience can tell for certain where a particular artifact came from. If an artifact cannot be duplicated in repeated playings of the same sequence from more than one copy of a disc, then it is clearly not a result of MPEG encoding. Here are a few of the most common artifacts:

- *Blocks* are small squares in the video. These may be especially noticeable in fast-moving, highly detailed sequences or video with high contrast between light and dark. This artifact appears when not enough bits are allocated during MPEG compression for storing block detail.

- *Halos* or *ringing* are mall areas of distortion or dots around moving objects or high-contrast edges. This is called the *Gibbs effect* and is also known as *mosquitoes,* or *mosquito wings.* This is an artifact of MPEG encoding, but it is easy to confuse with edge enhancement.

- *Edge enhancement* is a digital picture-sharpening process that is frequently overdone, causing a "chiseled" look or a ringing effect like halos around streetlights at night. This happens before MPEG encoding.

- *Posterization* or *banding.* Bands of colors or shading in what should be a smooth gradation. This can come from the MPEG encoding process or the digital-to-analog conversion process in the player. It also can happen on a computer when the number of video colors is too low.

- *Aliasing* occurs when angled lines have "stair steps" in them. This artifact is usually caused by lines that are too sharp to be properly represented in video, especially when interlaced.

- *Noise* and *snow* refer to the gray or white spots scattered randomly throughout the picture, or graininess. This may result from film grain or low-quality video.

- *Blurriness* refers to low detail and fuzziness of video. This results from low-quality video or too much filtering of the video before encoding.

- *Worms* or *crawlies* are squirming lines and crawling dots. Usually this results from low-quality video or bad digitizing. This also may result from chroma crawl from composite video (either in the original source or from the connection from the DVD player).

The number one cause of bad video is a poorly adjusted TV. The high fidelity of DVD video demands much more from the display. Turn the sharpness and brightness down. See "How to Get the Best Picture" in Chapter 9 for more information.

Myth: "Compression Does Not Work for Animation"

It is often claimed that animation, especially hand-drawn cell animation such as cartoons and Japanese *anime,* does not compress well with MPEG-2. Other people claim that animation is so simple that it compresses better. Neither is generally true.

Supposedly the *jitter* between frames caused by differences in the drawings or in their alignment causes problems. Modern animation techniques produce very exact alignment, so usually no variation occurs between object positions from frame to frame unless it is an intentional effect. Even when objects change position between frames, the motion estimation feature of MPEG-2 can easily compensate for it.

Because of the way MPEG-2 compresses video, it can have difficulty with the sharp edges common in animation. This loss of high-frequency information can show up as *ringing* or blurry spots along high-contrast edges. However, at the data rates commonly used for DVD, this problem does not occur. The complexity of sharp edges tends to be balanced out by the simplicity of broad areas of single colors.

Myth: "Discs Are Too Fragile to Be Rented"

The Blockbuster Video chain allegedly took a stance early on that it would not rent DVDs unless the format included a protective caddy. The designers of DVD, however, having learned from the bad experience of CD-ROM caddies, and not wishing to more than double the cost of discs by requiring a caddy or protective shell, politely ignored such requests. Within 2 years after DVD was released, essentially every video rental chain and outlet carried the discs.

DVDs are, of course, liable to scratches, cracks, accumulation of dirt, and fingerprints. But these occur at the surface of the disc where they are out of focus to the laser. Damage and imperfections may cause minor channel data errors that are easily corrected. A common misperception is that a scratch will be worse on a DVD than on a CD because of higher area density and because the audio and video are compressed. DVD data density is about seven times that of CD-ROM, so it is true that a scratch will affect more data. But DVD error correction is more than 10 times more effective than CD error correction. This improved reliability more than makes up for the density increase. It is also important to realize that MPEG-2 and Dolby Digital compression are partly based on removal or reduction of imperceptible information, so decompression does not expand the data as much as might be assumed. For example, video may be compressed to one-tenth its original size but may only be decompressed to nine-tenths, with the remaining one-tenth permanently removed. Major scratches on a disc may cause uncorrectable errors that will cause an *input-output* (I/O) error on a computer or show up as a momentary glitch in DVD-Video picture, but many schemes exist for concealing errors in MPEG video.

Laserdiscs, music CDs, and CD-ROMs are likewise subject to scratches, but many video stores and libraries rent them. DVD manufacturers are fond of taking a disc, rubbing it vigorously with sandpaper, and then placing it in a player, where it plays perfectly. Disc cleaning/polishing products can repair minor damage. Commercial polishing machines can restore a disc to pristine condition after an amazing amount of abuse.

Videocassettes have their own share of reliability problems: deterioration from repeated play, susceptibility to heat and magnetic fields, broken tape, and broken parts. On balance, DVD does not perform any worse in a rental environment than tapes.

Myth: "Dolby Digital Means 5.1 Channels"

Do not assume that the "Dolby Digital" label is a guarantee of 5.1 channels. Dolby Digital is an encoding format that can carry anywhere from 1 to 6 discrete channels. A Dolby Digital soundtrack can be mono, stereo, Dolby Surround stereo, Dolby Surround EX, and so on. Most movies produced before 1980 had a monophonic soundtrack only. When these movies are put on DVDs, unless a new soundtrack is mixed, the soundtrack is encoded into a single channel of Dolby Digital. In some cases, more than one Dolby Digital version of a soundtrack is available: a 5.1-channel track and a track specially remixed for two-channel Dolby Surround. It is normal for the DVD player to indicate playback of a Dolby Digital audio track while the receiver indicates Dolby Surround. This means that the disc contains a two-channel Dolby Surround signal encoded in Dolby Digital format. The same applies to DTS, although very few DTS tracks are encoded with fewer than 5.1 channels.

Myth: "The Audio Level from DVD Players Is Too Low"

People complain that the audio level from DVD players is too low. In truth, the audio level is too high on everything else. Movie soundtracks are extremely dynamic, ranging from near silence to intense explosions. In order to support an increased dynamic range and hit peaks (near the 2V RMS limit) without distortion, the average sound volume must be lower. This is why the line level from DVD players is lower than from almost all other sources. The volume level among DVDs varies, but it is more consistent than on CDs and laserdiscs. If the change in volume when switching between DVD and other audio sources is annoying, you may be able to adjust the output signal level on the player or the input signal level on the receiver.

Myth: "Downmixed Audio Is No Good Because the LFE Channel Is Omitted"

The LFE channel is omitted for a good reason when Dolby Digital 5.1-channel soundtracks are mixed down to two channels in the player. The LFE channel is intended only for extra bass boost, since the other 5 channels carry full-range bass. Audio systems without Dolby Digital capabilities generally do not have speakers that can properly reproduce very low frequencies, so the designers of Dolby Digital chose to have the decoders throw out the LFE track to avoid muddying the sound on average home systems. Anyone who truly cares about the LFE channel should invest in a receiver with Dolby Digital, bass management, and a separate subwoofer output.

Myth: "DVD Lets You Watch Movies as They Were Meant to Be Seen"

This refers to DVD's 1.78 anamorphic widescreen feature, which is close to the most common movie aspect ratio (1.85). However, many movies have a wider shape than widescreen TVs. Thus, even though they look much better on a widescreen TV, they still have to be formatted to fit the less oblong shape, usually with black bars at the top and bottom. See "Aspect Ratios" in Chapter 3 for more information.

Myth: "DVD Crops Widescreen Movies"

As mentioned in the preceding paragraph, some movies are wider than DVD's widescreen format. Some people assume that the only way to make them fit is to crop the sides of the picture. However, in almost all cases, Cinemascope and similarly wide movies are *letterboxed* to fit the entire original width within DVD's widescreen picture shape. It is true that for standard TV display, widescreen movies are often cropped. See "Aspect Ratios" in Chapter 3 for more information.

Myth: "DVD Will Replace Your VCR"

When DVD was first released, this was a misleading statement, since DVD players could not record. DVD video recorders were introduced in 2000, and as they drop in price over the first few years of the new millennium, they will become viable replacements for VCRs. However, it will take years for DVD to make even a small a dent in the installed base of VCRs. Plus, other technologies—such as personal video recorders, computers, and digital videotape—are also vying to be the VCR of the future.

The incompatibilities between the various recordable DVD formats could seriously delay their acceptance. In the case of videotape, once the battle between VHS and Betamax was over, you could expect that any VHS tape would work in any other VCR. DVD's version of the battle of the formats, bigger and more brutal, could drag on for a long time until a single format emerges victorious or some sort of truce leads to players that read every format.

Myth: "People Will Not Collect DVDs Like They Do CDs"

A common argument against the success of video is that it is not as collectible as music. Music can be listened to over and over, whereas most movies only bear watching a few times. Music can play in the background without disrupting everyday tasks, but movies require devoted watching. It is true that the average household will own more CDs than DVDs. However, music combined with video is more collectible than music alone—and more playable than movies. Research shows that televisions are often left tuned to music channels such as MTV with no one in the room. Six percent of the top-selling video titles are music videos.

Beyond music, the extra features of DVDs make them much more collectible than videotapes. "Special edition" DVDs packed with audio commentaries, outtakes, interviews, featurettes, and other goodies are often too much to be digested in a rental period, making them more likely to be purchased.

Myth: "DVD Holds 4.7 to 18 Gigabytes"

As mentioned in Chapter 1, the abbreviation GB, when referring to storage capacity, sometimes stands for *gigabytes*, which are measured in powers of 2, and sometimes stands for *billions of bytes*, in powers of 10. A DVD holds 4.4 to 15.9 gigabytes, which is the same as 4.7 to 17 billion bytes. Advertising-oriented folks favor the bigger numbers, and in some cases the marketing maniacs have pushed the boundaries of creative mathematics by rounding 17.01 up to 18 (see "Variations and Capacities," later in this chapter).

Myth: "DVD Holds 133 Minutes of Video"

The oft-quoted length of 133 minutes for DVDs is apocryphal. It is simply a rough estimate based on an average 3.5-Mbps video track and three 384-kbps audio tracks. (Typical subtitles are negligibly small.) If there is only one audio track, the average playing time goes up to 159 minutes. The video rate is highly variable—a single-layer DVD-5 actually can hold over 9 hours of VHS-quality video. There are two constants: disc capacity and maximum data rate (which is 9.8 Mbps for video and 10.08 Mbps combined with audio and subpictures). All the rest is variable. Even the capacity varies depending on the number of sides and layers. A dual-layer DVD-9 holds over 4 hours of high-quality video, and a double-sided, dual-layer DVD-18 holds over 8 hours. Using MPEG-1, a DVD-18 can contain a mind-numbing 33 hours of video, which also would be butt-numbing if you tried to watch all of it in one sitting.

It is said that the figure of 133 minutes was originated by the press. The original SD proposal achieved approximately 142 minutes of playing time, but by adopting 8/16 modulation from the MMCD format, manufacturers sacrificed 6.3 percent of disc capacity. Supposedly, a clever but clueless journalist applied 6.3 percent to 142, and the meaninglessly exact figure of 133 minutes has stuck ever since.

Anyone talking or writing about DVD-Video should make things easier for themselves and their audience by simply stating that a single layer holds over 2 hours of video. If more precision is required, the nice round figure of 2 hours and 15 minutes (135 minutes) is just as accurate.

It should be noted that all this applies only to DVD-Video. A DVD-ROM can hold any sort of digitized video to be played back on an endless variety of computer hardware or software. If someone developed a revolutionary new holographic wavelet compression algorithm, a DVD-ROM might hold 3 hours of film-quality video. It is not likely, but the point is that it is important to differentiate between the deliberate restrictions of DVD-Video and the limited capacity of DVD-Video players compared with the wide-open digital expanse of DVD-ROM and computers.

Myth: "DVD-Video Runs at 4.692 Mbps"

This figure is about as meaningless as 133 minutes. The figure of 4.692 Mbps is supposedly the average data rate for a DVD-Video. Table 4.6 shows how it is calculated. But what if only one subpicture track is available? Then the pristine sum is off by an egregious 0.03 kbps (or so, since 10 kbps is only an average). And if only one audio track and one subpicture track is available, the so-called average data rate goes all the way up to 3.894, an error of more than two-tenths!

Sarcasm aside, what usually happens is that the content is compressed to fit the capacity of the disc. If the movie is 110 minutes long and has two audio tracks, the video bit budget can be set at a much higher 4.9 Mbps to achieve better quality. Or a 2 1/2-hour movie might be compressed slightly more than usual if the disc producer determines that the video quality is acceptable.

	Bit Rate	Count	Total
TABLE 4.6			
	3.5 Mbps average video	1	3.500 Mbps
How to Create a Meaninglessly Exact Number	384 kbps audio	3	1.152 Mbps
	10 kbps average subpicture	4	0.040 Mbps
			4.692 Mbps

The probable genesis for this number was that it was calculated from the required 133 minutes of length (which was calculated from the original 135) by figuring out what video data rate was left over after accounting for the audio. Then the subpicture tracks were thrown in to even things up.

The maximum video data rate of DVD is limited to 9.8 Mbps by the DVD-Video specification. The maximum combined rate of video, audio, and subtitles is limited to 10.08 Mbps. The average data rate is almost always lower, usually between 4 and 6 Mbps. Some people assume that DVD is therefore unable to sustain a continuous rate of 9.8 Mbps or higher. This is not the case. All DVD players and drives can maintain an internal data rate of at least 11.08 Mbps. DVD-Video players have a 1-Mbps overhead for navigation data. A movie compressed to a constant bit rate of 10.08 Mbps would play for 62 minutes.

Single-speed DVD-ROM drives can sustain a transfer rate of 11.08 Mbps, with burst rates as high as 100 Mbps or more, depending on the data buffer and the speed of the drive connection. DVD-ROM drives with higher spin rates are accordingly faster, although most multispeed DVD-ROM drives cannot maintain the maximum quoted speed across the entire surface of the disc—they only achieve the maximum data rate when reading from the outer edge.

Myth: "Some Units Cannot Play Dual-Layer or Double-Sided Discs"

Dual-layer compatibility is required by the DVD specification. Almost every DVD-Video player and DVD-ROM drive, even the first ones sold, can read dual-layer discs. Occasional problems with dual-layer discs are caused by faulty disc production, flawed players (which often can be fixed with a firmware upgrade), or bugs in DVD-ROM driver software (which can be upgraded to fix the problem).

All players and drives also read double-sided discs—as long as you flip the disc over. So far, only DVD jukeboxes can switch automatically to the other side of a disc. This capability eventually may appear on a standard player, but since a single side can hold 4 hours or more of continuous video, demand for it is not very high. Most people appreciate the bathroom break. Some combination laserdisc/DVD players can play both sides of a laserdisc but not both sides of a DVD.

Bits and Bytes and Bears

This section provides a brief overview of selected aspects of DVD physical format and data format. More technical details are presented in Chapter 5.

Pits and Marks and Error Correction

Data is stored on optical discs in the form of microscopic *pits* (Figure 4.2). The space between two pits is called a *land*. On writable discs, pits and lands are often referred to as *marks* and *spaces*. Read-only discs are stamped in a molding machine from a liquid plastic such as polycarbonate or acrylic and then coated with a reflective metallic layer. Writable discs are made of material designed to be physically changed by the heat of a laser, creating marks. As the disc spins, the pits (or marks) pass under a reading laser beam and are detected according to the change they cause in the intensity of the beam. These changes happen very fast (over 300,000 times per second) and create a stream of transitions spaced at varying intervals: an encoded digital signal. Many people assume that the digital ones and zeros that make up the data stored on the disc are encoded directly as pits and lands, but it is much more complicated than this. Pits and lands both represent strings of zeros of varying lengths, whereas each transition between them represents a one. In addition, this signal is not a direct representation of the contents of the disc. Half the information has been used to pad and rearrange (*modulate*) the data in

Figure 4.2
DVD pits

sequences and patterns designed to be accurately readable as a string of pulses. About 13 percent of the remaining digital signal is extra information for correcting errors. Errors can occur for many reasons, such as imperfections on the disc, dust, scratches, a dirty lens, and so on. A human hair is about as wide as 150 pits, so even a speck of dust or a minute air bubble can cover a large number of pits. However, the laser beam focuses past the surface of the disc, so the spot size at the surface is much larger and is barely affected by anything smaller than a few millimeters. This is similar to the way dust on a camera lens is not visible in the photographs because the dust is out of focus. As the data is read from the disc, the error correction information is separated out and checked against the remaining information. If it does not match, the error correction codes are used to try to correct the error.

The error correction process is like a number square, where you add up columns and rows of numbers (Figure 4.3). You could play a game with these squares where a friend randomly changes a number and challenges you to find and correct it. If the friend gives you the sums along with the numbers, you can add up the rows and columns and compare your totals against the originals. If they do not match, then you know that something is wrong—either a number has been changed or the sum has been changed.[4] If a number has been changed, then a corresponding sum in the other direction also will be wrong. The intersection of the incorrect row and incorrect column pinpoints the guilty number, and in fact, by knowing what the sums are supposed to be, the original number can be restored. The error correction scheme used by DVD is a bit more complicated than this but operates on the same general principle.

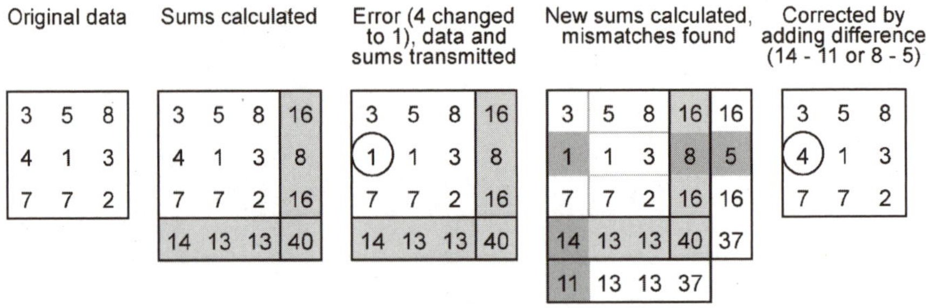

Figure 4.3
Number squares

| Original data | Sums calculated | Error (4 changed to 1), data and sums transmitted | New sums calculated, mismatches found | Corrected by adding difference (14 - 11 or 8 - 5) |

---

[4]It is possible for more than one number to be changed in such a way that the sum still comes out correct. However, the DVD encoding format makes this an extremely rare occurrence.

It is always possible that so much of the data is corrupted that error correction fails. In this case, the player must try reading the section of the disc over again. In the very worst cases, such as an extremely damaged disc, the player will be unable to read the data correctly after multiple attempts. At this point, a movie player will continue on to the next section of the disc, causing a brief glitch in playback. A DVD-ROM drive, on the other hand, cannot do this. Computers will not tolerate missing or incorrect data, so the DVD-ROM drive must signal the computer that an error has occurred so that the computer can request that the drive either try again or give up.

Layers

One of the clever innovations of DVD is to use layers to increase storage capacity. The laser that reads the disc can focus at two different levels so that it can look through the first layer to read the layer beneath. The outside layer is coated with a semireflective material that enables the laser to read through it when focused on the inner layer. When the player reads a disc, it starts at the inside edge and moves toward the outer edge, following a spiral path. If unwound, this path would stretch 11.8 kilometers (7.3 miles), three times around the Indianapolis 500 Speedway. When the laser reaches the end of the first layer, it quickly refocuses onto the second layer and starts reading in the opposite direction—from the outer edge toward the inner. Refocusing happens very quickly, but on most players the video and audio pause for a fraction of a second as the player searches for the resumption point on the second layer. If the player has a large enough buffer, or if the disc is carefully designed to lower the data rate at the layer switch point (so that the buffer will take longer to empty), the laser pickup has time to refocus and retrack without causing a visible break.

The DVD standard does not actually require compatibility with existing CDs. However, manufacturers recognize the vital importance of backward compatibility. If the hardware were unable to read CDs, DVD would not have a snowball's chance in Hollywood of surviving. The difficult part is that the pits on a CD are at a different level than those on a DVD (Figure 4.4). In essence, a DVD player must be able to focus a laser at three different distances. This problem has various solutions, including using lenses that switch in and out, and holographic lenses that are actually focused at more than one distance simultaneously. An additional difficulty is that CD-R discs do not properly reflect the 635- to 650-nanometer wavelength laser required for DVD, so DVD players and DVD-ROM

Figure 4.4
DVD layers

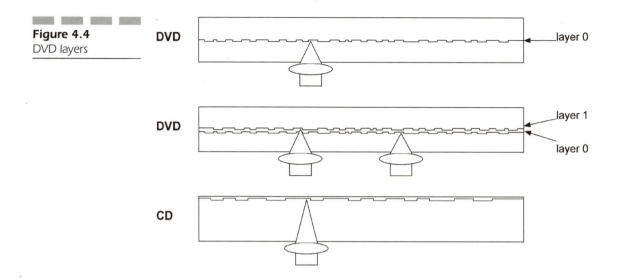

drives intended to read recordable CDs must include a second 780-nanometer laser.

The remaining task to ensure CD compatibility merely requires an extra bit of circuitry and firmware for reading CD-format data. However, the CD family is quite large and includes some odd characters, not all of which fit well with DVD. The prominent members of the CD family are audio CD, Enhanced CD (or CD Plus), CD-ROM, CD-R, CD-RW, CD-i, Photo CD, CDV, and Video CD. It would be technically possible to support all these, but most of them require specialized hardware. Therefore, most manufacturers choose to support only the most common or easy-to-support versions. Some, such as Enhanced CD and Video CD, are easy to support with existing hardware. Others, such as CD-i and Photo CD, require additional hardware and interfaces, so they are not commonly supported. However, since the data on a CD can be read by any DVD system, conceivably any CD format could be supported. DVD-ROM computers support more CD formats than DVD players partly because some are designed for computer applications and partly because specialized CD systems can be simulated with computer software. See Chapter 8 for details of the different CD formats and the compatibility of DVD-Video and DVD-ROM with each.

Variations and Capacities of DVD

Believe it or not, over 144 possible variations of DVD exist. DVDs come in about 24 physical incarnations (Table 4.7) and at least 6 data-format variations (general data, DVD-Video, DVD-Audio, DVD-Video Recording, DVD-Audio Recording, and DVD-Stream Recording). The combinations of these formats make for quite a variety of discs.

Two sizes of discs are available: 12 centimeters (4.7 inches) and 8 centimeters (3.1 inches), both 1.2 millimeters thick. These are the same diameters and thickness as CD, but DVDs are made of two 0.6-millimeter substrates glued together. This makes them more rigid than CDs so that they spin with less wobble and can be tracked more reliably by the laser. The thinner substrate reduces birefringence and improves tilt margins for more accurate data readout. A DVD can be single-sided or double-sided. A single-sided disc is a stamped substrate bonded to a blank, or *dummy,* substrate. A double-sided disc is two stamped substrates bonded back to back. To complicate matters, each side can have one or two layers of data. This is part of what gives DVD its enormous storage capacity. A double-sided, dual-layer disc has data stored on four separate planes. See Chapter 5 for more details on dual-layer construction.

Six configurations of layers and substrates are possible (Figure 4.5 and Table 4.7):

- One single-layer substrate bonded to a blank substrate (one side, one layer; DVD-5)

- Two single-layer substrates bonded together (two sides, one layer each; DVD-10)

- Two single-layer substrates with a transparent bond (one side, two layers; DVD-9)

- One dual-layer substrate bonded to a blank substrate (one side, two layers; DVD-9—uncommon variation)

- One dual-layer substrate bonded to a single-layer substrate (two sides, one and two layers; DVD-14)

- Two dual-layer substrates bonded together (two sides, two layers each; DVD-18)

TABLE 4.7

Physical Format
Variations

Type	Size	Sides and Layers[a]	Billions of Bytes[b]	Giga-bytes[b]	Approx. Playing Time[c]
DVD-ROM (DVD-5)	12 cm	1 side (1 layer)	4.70	4.37	2.25 h
DVD-ROM (DVD-9)	12 cm	1 side (2 layers)	8.54	7.95	4 h
DVD-ROM (DVD-10)	12 cm	2 sides (1 layer each)	9.40	8.75	4.5 h
DVD-ROM (DVD-14)	12 cm	2 sides (1 and 2 layers)	13.24	12.33	6.25 h
DVD-ROM (DVD-18)	12 cm	2 sides (2 layers each)	17.08	15.91	8 h
DVD-ROM	8 cm	1 side (1 layer)	1.46	1.36	0.75 h
DVD-ROM	8 cm	1 side (2 layers)	2.65	2.47	1.25 h
DVD-ROM	8 cm	2 sides (1 layer each)	2.92	2.72	1.5 h
DVD-ROM	8 cm	2 sides (1 and 2 layers)	4.12	3.83	2 h
DVD-ROM	8 cm	2 sides (2 layers each)	5.31	4.95	2.5 h
DVD-R 1.0	12 cm	1 side	3.95	3.67	1.75 h
DVD-R(G) 2.0	12 cm	1 side	4.70	4.37	2.25 h
DVD-R(G) 2.0	12 cm	2 sides	9.40	8.75	4.5 h
DVD-R(A) 2.0	12 cm	1 side	4.70	4.37	2.25 h
DVD-RAM 1.0	12 cm	1 side	2.58	2.40	1.25 h
DVD-RAM 1.0	12 cm	2 sides	5.16	4.80	2.5 h
DVD-RAM 2.0	12 cm	1 side	4.70	4.37	2.25 h
DVD-RAM 2.0	12 cm	2 sides	9.40	8.75	4.5 h
DVD-RAM 2.0	8 cm	1 side	1.46	1.36	0.75 h
DVD-RAM 2.0	8 cm	2 sides	2.92	2.72	1.5 h
DVD-RW 1.0	12 cm	1 side	4.70	4.37	2.25 h
DVD-RW 1.0	12 cm	2 sides	9.40	8.75	4.5 h
DVD+RW 2.0	12 cm	1 side	4.70	4.37	2.25 h
DVD+RW 2.0	12 cm	2 sides	9.40	8.75	4.5 h
CD-ROM	12 cm	1 side	0.68	0.64	0.25 h[d]
DDCD-ROM	12 cm	1 side	1.36	1.28	0.5 h[d]

[a]Writable DVDs have only one layer per side. DVD-14 (and corresponding 8-centimeter size) has one layer on one side and two layers on the other side.

[b]Reference capacities in billions of bytes (10^9) and gigabytes (2^{30}). Actual capacities can be slightly larger if the track pitch is reduced.

[c]Assuming an average aggregate data rate near 4.7 Mbps. Actual playing times can be much longer or shorter.

[d]Assuming that the data from the CD is transferred at typical DVD video data rate, about four times faster than a single-speed CD-ROM drive.

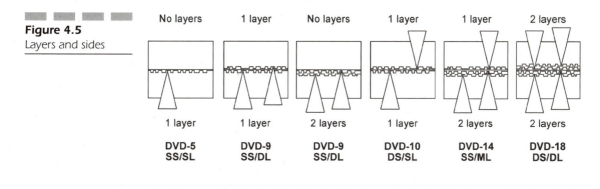

Figure 4.5
Layers and sides

RULE OF THUMB: *It takes about 2 gigabytes to store 1 hour of average video.*

A single-sided, single-layer DVD holds 4.7 billion bytes of data (4.37 gigabytes), 7 times more than a CD-ROM, which holds over 650 megabytes.[5] A double-sided, dual-layer DVD holds just over 17 billion bytes (15.9 gigabytes), which is 25 times what a CD-ROM holds. See "Units and Notation" in Chapter 1 for a discussion of the difference between billions of bytes and gigabytes.

At the end of 1999, approximately 80 percent of DVD video titles were released on DVD-5, with about 10 percent on DVD-9 and 10 percent on DVD-10. The share of DVD-14 and DVD-18 discs was insignificant. In terms of actual disc production, the percentage of DVD-9 discs is much higher.

Hybrids

A *hybrid* disc is one that combines the features of one or more formats. Anyone who speaks in general terms of a hybrid disc is being dangerously vague because as many potential hybrids exist as physical and logical formats. Table 4.8 lists some of the more common DVD hybrids.

Hybrid players are also available: video-capable audio players (DVD-Audio players with video output) and universal players (players that can play both DVD-Video and DVD-Audio). New convergence devices such as digital satellite receivers with DVD players built in are also known as hybrids.

---

[5]The loose tolerance of the CD standard tracks enables to be placed more tightly together, so CDs actually can hold 750 megabytes or more. An example is the so-called 84-minute CD.

TABLE 4.8

DVD Hybrids

Enhanced DVD	A disc that works in both DVD-Video players and DVD-ROM PCs. This is the most common use of the term hybrid.
Cross-platform DVD	A DVD-ROM disc that runs on Windows and Mac OS computers.
WebDVD	A DVD-ROM or DVD-Video disc that also contains HTML content, usually designed to work with a connection to the Internet. Also called a connected DVD.
Universal DVD	A disc that contains both DVD-Video and DVD-Audio content. Also called a DVD-AV.
CD-compatible DVD	A disc with two layers, one that can be read in DVD players and one that can be read in CD players. Also called a legacy disc. Three variations of this hybrid are:
	A CD substrate is bonded to the back of a 0.6-millimeter DVD substrate. The CD substrate is usually thinner than normal, around 0.9 millimeters, which causes problems with some CD readers. The CD substrate can be read by CD players, and the other side of the disc can be read by DVD players. The resulting disc is 0.3 to 0.6 millimeters thicker than a standard CD or DVD, which can cause problems in players with tight tolerances, such as portables. Sonopress, the first company to announce this type of disc, calls it DVDPlus. It is colloquially known as a "fat" disc.
	A 0.6-millimeter CD substrate is bonded to a semitransparent 0.6-millimeter DVD substrate. Both layers are read from the same side, with the CD player being required to read through the semitransparent DVD layer, causing problems with some CD players.
	A 0.6-millimeter CD substrate is given a special refractive coating to create a 1.2-millimeter focal depth. The CD substrate is bonded to the back of a 0.6-millimeter DVD substrate. One side can be read by CD readers, and the other side can be read by DVD readers.
DVD-PROM	A disc with two layers, one containing pressed (DVD-ROM) data and one containing writable (DVD-RAM, DVD-RW, and so on) media for recording. Also called a mixed-media or rewritable sandwich disc. (PROM comes from the computer term *programmable read-only memory*.)
DVD-14	A disc with two DVD layers on one side and one DVD layer on the other.
Chipped DVD	A disc with an embedded memory chip for storing custom usage data and access codes.

Regional Management

Motion picture studios want to control the geographic distribution of DVD-Video titles. This is partly because of home video release timing. A movie may come out on video in the United States when it is just hitting screens in Europe. The other reason for regional management is to preserve exclusive distribution arrangements with local distributors.

The DVD-Video standard includes codes that can be used to prevent playback of certain discs in certain geographic regions. Each disc contains a set of region flags.[6] If a flag is cleared, the disc is allowed to be played in the corresponding region; if the flag is set, the disc is not allowed to be played. Players are branded with the code of the region where they are intended to be sold. The player puts up a message and refuses to play a disc that is not flagged to play in the player's region. This means that discs bought in one country may not play on players bought in another country.

The use of regional codes is entirely optional. Discs with no region locks will play on any player in any country. The codes are not an encryption system; just one bit of information on the disc that the player checks. In many cases, discs are released without any region locks, but in other cases regional control is very important to the business model of movie distribution. Many studios sell exclusive foreign release rights to other distributors. If the foreign distributor can be assured that discs from other distributors will not be competing in its region, then the movie studios can sell the rights for a better price. The foreign distributors are free to focus on their region of expertise, where they may better understand the cultural and commercial environment. Regions apply only to CSS-encrypted DVD-Video titles, not DVD-Audio titles, unencrypted DVD-Video titles, or DVD-ROM titles.

The DVD standard specifies eight regions, also called *locales*.[7] Players and discs are identified by a region number that is usually superimposed on a world globe icon. If a disc plays in more than one region, it will have more than one number on the globe. The regions are broken out as follows (see Table 4.9 for details). Also see Table A.25.

---

[6]Regions apply to disc sides or to PTP layers. Thus it is possible to have a disc that is one region on one side and a different region on the other or has different region settings for each layer.

[7]Since each region is represented by a bit, a single byte can hold 8 region flags. The neighboring byte is reserved, so it would be possible for the DVD Forum to designate a total of 16 regions in the future.

Figure 4.6
Map of DVD regions.

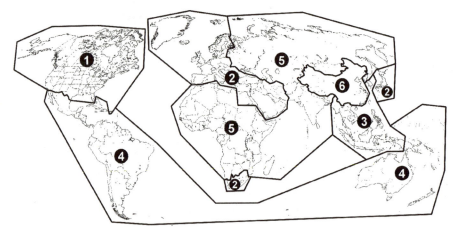

TABLE 4.9

DVD Regions

1 Canada, United States, Puerto Rico, Bermuda, the Virgin Islands, and some islands in the Pacific

2 Japan, western Europe (including Poland, Romania, Bulgaria, and the Balkans), South Africa, Turkey, and the Middle East (including Iran and Egypt)

3 Southeast Asia (including Indonesia, South Korea, Hong Kong, and Macau)

4 Australia, New Zealand, South America, most of Central America, western New Guinea, and most of the South Pacific

5 Most of Africa, Russia (and former Russian states), Mongolia, Afghanistan, Pakistan, India, Bangladesh, Nepal, Bhutan, and North Korea

6 China and Tibet

7 Reserved

8 Special nontheatrical venues (airplanes, cruise ships, hotels)

1. North America

2. Japan and Europe

3. Southeast Asia

4. Australia, New Zealand, and Central/South America

5. Northwest Asia and North Africa

6. China

7. Unassigned

8. Special venues

Whenever a deterrent is artificially imposed, a way around it is inevitably found. For example, many video game systems introduced since 1995 include regional restrictions. Workarounds quickly appeared for buyers who were interested in games from other countries. Not surprisingly, as soon as DVD players were released, numerous ways were found to defeat the regional coding. Some early players could be set to "region 0"[8] with a switch on the circuit board or sequence of keys on the remote control. Movie studios quickly complained, and manufacturers made it more difficult for users to modify the region setting. Nevertheless, *code-free* players can be purchased, even from legitimate manufacturer outlets outside the United States, and after-market "region mod" chips are available for many players. The Internet is replete with details on how certain players can be made region-free. The second salvo from movie studios, particularly Fox, Buena Vista/Touchstone/Miramax, MGM/Universal, and Polygram, was to add program code to some of their discs to check for the proper region in the player. These "smart discs" ("There's Something About Mary" and "Psycho" are examples) query the player for its region code and refuse to work if the player is not set to the single correct region. These discs prevent code-free players from working, so the response from player modification designers was *code-switchable* players, which enable the region code to be changed by using the remote control. *Autoswitching* players also check the region on the disc when it is inserted and then set the player region to match. These players do not always work with smart discs, since a disc can have all its region flags set so that the player does not know which region to switch to.

Some people believe that region codes are an illegal restraint of trade, but no legal cases have occurred to establish this. Conversely, rumors have evolved that the major movie studios in conjunction with the MPAA and consumer electronics companies are pursuing legislation to make region-modification devices illegal in the United States. The only requirement for manufacturers to make region-coded players is the CSS license (see the next section for details). Physically modifying a player will void the warranty but is not illegal.[9]

The average consumer in the United States and Canada need not worry about regions. All the region 1 discs they buy from North American

---

[8]*Region 0* is a common but misleading term. There is no region 0. Region-free players and all-region discs exist, but region 0 players or region 0 discs are nonexistant. A player modified to work in all regions may have all the bits in the region mask set, which means that it is technically a region 65535 or region FFFF player.

[9]At least this is the general consensus. Anyone relying on this book, rather than a lawyer, for legal advice deserves whatever happens to them.

producers will play fine in their region 1 players. Only those who buy imported discs from other regions or move to other countries will run into problems.

Regional codes apply to DVD-ROM systems but may only be used with CSS-encoded DVD-Video discs, not computer software on DVD-ROM discs. Operating systems and DVD decoders check for regional codes before playing movies from a DVD-Video disc. The first few generations of DVD-ROM drives were governed by regional protection control phase 1 (RPC1). Most RPC1 drives had no built-in region code; the operating system or decoder was required to implement regional management. RPC phase 2 (RPC2) took effect January 1, 2000, requiring that every new DVD-ROM drive have built-in region control. See Chapter 11 for more information on regional management in computers.

Content Protection

Before Hollywood would embrace DVD, it had to be assured that DVD would not put Hollywood's bread and butter out on the open market for anyone to make perfect digital copies. Thus, were born various schemes intended to reduce consumer copying of video and audio. See Chapter 7 for more on copy protection.

In some ways, it is a futile exercise, since a completely foolproof protection method also would make it impossible to use the disc—if you can see it or hear it, you can copy it. Alan Bell, chairman of the *Copy Protection Technical Working Group* (CPTWG), points out that "really strong digital encryption is always ultimately defeatable by analog output." He elaborates that watermarking is the best solution, as long as players recognize watermarked analog copies. What many proponents of copy protection apparently fail to recognize is that a digital copy of an analog output is only slightly degraded from the original digital source, and it can be distributed and recopied as easily and endlessly as a digital copy of a digital source. Nevertheless, millions of dollars and hundreds of thousands of person-hours have been spent creating technical measures that make it harder to create digital and analog copies from DVD. The result is that the average DVD buyer cannot simply make a videotape copy of a DVD or copy DVD files to a hard drive or a writable DVD. A determined consumer, on the other hand, will find many ways of getting around copy protection.

Implementation of copy protection incorporates three components: technology (the protection method of encryption in the digital domain and watermarking in the analog domain), licensing (requiring compliant devices from manufacturers), and legislation (enforcement). These must be balanced according to the needs of content owners, manufacturers, and consumers.

Requirements of content owners are:

- No effect on the quality of the content
- Effective against unauthorized consumer use
- Robust and tamper-resistant
- Renewable (to recover from a breach of the system)
- Applicable to all forms of distribution or media
- Suitable for implementation on CE devices and PCs
- Low cost

Requirements of system manufacturers are:

- No effect on normal use of system
- Low additional resource requirement
- Tamper-resistant
- Voluntary
- Low cost

Requirements of consumers are:

- No effect on normal use of system
- No loss of quality
- Fair-use copying
- No additional cost
- No limitations on playback equipment or environment

Since the goals of each group are sometimes in conflict, the resulting protection methods are a compromise. However, each group seems reasonably happy with its ability to use DVD and associated copy protection measures. Manufacturers and studios are busy making products, and consumers are busy buying them.

When DVD was first being developed, content protection was intended to be part of the specification. After numerous lengthy delays, the DVD Forum recognized the need to separate the legal and technical aspects of copy protection from the rest of the DVD specification. Copy protection features are covered under separate licenses, where the makers of DVD playback systems essentially agree to implement content protection features in return for being granted access to the decryption keys and algorithms needed to play back encrypted content. The DVD Forum does not specify copy protection technologies. The industry's CPTWG solicits proposals and makes recommendations. The DVD Forum Working Group 9 is responsible for coordinating with the CPTWG. WG9 reviews copy protection systems and submits them to the DVD Forum for approval. Once a copy protection system is approved, the various working groups of the DVD Forum amend specifications as needed to support the requirements of the copy protection system.

Over time, it was recognized that an overall framework was needed for security and access control across the entire DVD family and beyond. The 4C entity (that is, Intel, IBM, Matsushita, and Toshiba), in cooperation with the CPTWG and the *Secure Digital Music Initiative* (SDMI), developed the *Content Protection System Architecture* (CPSA). CPSA covers encryption, watermarking, playback control, protection of analog and digital outputs, and so on. It is broadly defined to include physical and electronic distribution of analog and digital audio and video in consumer electronics and computer systems.

The CPSA creates a structure that defines the content protection obligations of compliant modules. It defines how *content management information* (CMI) and *copy control information* (CCI) are carried and verified throughout the playback chain. CMI, also known as *usage rules,* specifies the conditions under which the content can be used. It also may contain other information such as triggers telling the player when and how to protect audio and video outputs. CCI, a subset of CMI, constrains how the content can be copied.

Eight forms of content protection apply to DVD. Each is explained below. Technical details, particularly as they apply to computers, are covered in Chapter 11.

Content Scrambling (CSS)   Because of the potential for perfect digital copies, worried movie studios forced a deeper copy protection requirement into the DVD-Video standard. The *Content Scrambling System* (CSS) is a data encryption and authentication scheme intended to prevent copying video files directly from the disc. Occasional sectors containing A/V data (audio, video, or subpicture) are scrambled in such a way that the data cannot be used to recreate a valid signal. Scrambled sectors are encrypted with a combination of a *title key* and a *disc key*. The title key is stored in the sector header, which is normally not readable from a DVD-ROM drive or other DVD reader. Each *video title set* (VTS) on the disc has a separate key. The *disc key* is hidden in the control area of the disc, which is also not directly accessible.

Use of CSS is strictly controlled by licensing. Each CSS licensee is given a *player key* from a master set of 400 keys that are stored on every CSS-encrypted disc. This allows a license to be revoked by removing its key from future discs. The CSS algorithm exchanges player keys with the drive unit to generate an encryption key that is then used to obfuscate the exchange of disc keys and title keys that are needed to decrypt data from the disc.

All standard DVD players have a decryption circuit that decrypts the data before displaying it. The process is similar to scrambled cable channels, except that the average consumer will never see the scrambled video and will have no idea that it has gone through an encryption/decryption process. The process does not degrade the data; it merely shifts the data around and alters it so that the original values are unrecognizable and difficult to decipher. The decryption process completely restores the data. The only case in which someone is likely to see a scrambled video signal is if they attempt to play the disc on a player or computer that does not support CSS or if they attempt to play a copy of the data or the disc. Since the copy does not include the key, the video signal cannot be decrypted and appears garbled or blank.

No unscrambled digital output is allowed until work in progress for secure digital connections is finished. When digital recording devices become available, scrambled video may not be able to be recorded.

On the computer side, DVD decoder hardware and software must include a CSS decryption module. The encrypted data is sent from the drive to the decoder to be decrypted and then MPEG decoded before being displayed. DVD-ROM drives have extra firmware to exchange authentication and decryption keys with the CSS module in the computer so as to protect movies. Beginning in 2000, new DVD-ROM drives are required to support regional management in conjunction with CSS. Makers of equipment used to display DVD-Video (drives, decoder chips, decoder software, display adapters, and so on) must license CSS. CSS is not required of video players or DVD computers; systems without it will not be able to play scrambled movies. See Chapter 11 for more details.

Content Protection for Prerecorded Media (CPPM) CPPM replaces CSS for use with DVD-Audio.[10] It is also known as *4C* for the group of four companies that developed it: IBM, Intel, Matsushita, and Toshiba. CPPM is a method for encrypting and protecting content on prerecorded (read-only) discs. The authentication mechanism is the same as for CSS, so no changes are required to existing drives. A disc with both DVD-Video and

---

[10]A slightly improved system called CSS2, originally intended for use on DVD-Audio discs, was abandoned and replaced by the more sophisticated CPPM when CSS was cracked in 1999.

DVD-Audio content may use both CSS and CPPM. CPPM is planned for use on other prerecorded media such as SD memory cards.

CPPM has some similarities to CSS but is more robust and sophisticated. It has no title keys, and the disc key is replaced by an *album identifier.* The role of the album identifier is to provide a key that cannot be duplicated on recordable media, since it is stored in the control area of the lead-in, which is not accessible on writable discs. Each player has a set of 16 *device keys.* Key sets may be either unique to each device or shared by multiple devices. Device keys are highly confidential. Rather than secretly storing the set of all known device keys on the disc, CPPM stores a *media key block* (MKB) in the DVDAUDIO.MKB file on the disc. The media key block is provided by the 4C entity to disc replicators. The player performs a series of decryptions and mathematical transforms on the media key block with its device key. The resulting *media key* is used with the album identifier to decode the encrypted portions of the disc. If a device key is revoked, the media key block is changed on future discs. Revoked players will then generate an invalid media key that will not decrypt the disc.[11]

As with CSS, only audio, video, subpicture, and still picture sectors are encrypted. Other sectors containing navigation, highlight, and real-time information are not encrypted.

Content Protection for Recordable Media (CPRM)   CPRM is a mechanism that ties a recording to the medium on which it is recorded. It was developed by the same 4C group that created CPPM and shares many features. CPRM is supported by all DVD recorders released after 1999. The goal of CPRM is to enable a recording to be made and played on different devices while ensuring that copies of the recording will not be playable. CPRM is defined for writable DVD formats and for SD memory cards. It is intended to be used for other recordable media such as compact flash cards and microdrives.

Each blank recordable DVD disc has a unique 64-bit *media identifier* (media ID) etched in the burst cutting area (BCA; see Chapter 5). This means that each disc can be uniquely distinguished from all other recordable discs. The media ID does not have to be secret but must be an unalterable value tied to the medium. When protected content is recorded onto the

---

[11]This feature is called *renewability* by the creators of the process, but it does not renew anything. On the contrary, it can make formerly functioning devices cease to work. Perhaps they were thinking of "renew" as used in *Logan's Run*. A problem with revocable devices is that if keys are stolen or the system is cracked so that legitimate keys can be used in unauthorized devices, the keys cannot be revoked without making hundreds or thousands of genuine players stop working.

disc, it is encrypted with a *media unique key* derived from the media ID and the media key. During playback, the media ID is read from the BCA and used to generate a key to decrypt the contents of the disc. If the contents of the disc are copied to another medium, the ID will be absent or wrong, and the data will not be decryptable.

When recording, the recorder generates a title key that is used to encrypt content on the disc. Each disc includes one title key. If something has already been recorded, then the existing title key is used to encrypt new content as well. As with CPRM, each disc contains a media key block that is used to control the validity of device keys. Unlike CPPM, the media key block is pre-recorded in the control area of the disc, where it cannot be changed. The recorder uses its set of 16 device keys to process the media key block and produce a media key. The recorder then uses the media key and the media ID to encrypt the title key before recording it on the disc. The media key and media ID are not used to encrypt recorded content, only to encrypt the title key, which is used (along with CCI) to encrypt the actual A/V sectors.

CPRM improves on the revocation features of CPPM by allowing recorders to update the media key block by writing a *media key block extension* to a file on the disc. This allows new revocation information to be disseminated quickly. Since the media key block is not a secret, it can be sent over the Internet to CPRM devices. Only one media key block extension is included on a disc. It is replaced by newer versions as they become available. Writing a new media key block extension changes the media key, so the title key must be reencrypted with the new key. Media key block extensions can only alter a media key generated by the static media key block (and thus revoke device keys); they cannot generate a media key, so they cannot be hacked to enable new device key sets. Writing media key block extensions is optional for recorders; reading is mandatory.

To play CPRM-encrypted content, a player generates a media key by applying its device key set to the media key block that it reads from the control area of the lead-in. It reads the media ID from the BCA and combines it with the media key to generate a media unique key that it then uses to decrypt the title key after reading it from the video or audio zone on the disc. The decrypted title key can then be used to decrypt the encrypted sectors.

As with CPPM and CSS, only A/V sectors (video, audio subpicture, and so on) are encrypted. Navigation and other sectors, such as the *real-time data information* (RDI) sectors of DVD-VR, are not encrypted.

Analog Protection System (Macrovision)   Copying from DVD to VHS or other analog recording systems is prevented by a Macrovision or similar circuit in the player. The general term is *analog protection system* (APS).

Computer video cards with composite or s-video TV output also must use APS. DVD players or DVD computers can be built without APS, but they will not be licensed to play CSS-protected video.

The Macrovision 7.0 process provides two separate antitaping processes: *automatic gain control* (AGC) and *Colorstripe*. Macrovision AGC technology has been in use since 1985 to protect prerecorded videotapes. It works by adding pulses to the vertical blanking sync signal to confuse the automatic-recording-level circuitry of a VCR, causing it to record a noisy, unstable picture. The Colorstripe technology was developed in 1994 for digital set-top boxes and digital video networks (it cannot be applied to prerecorded tapes). The Colorstripe process produces a rapidly modulated colorburst signal that confuses the chroma processing circuitry in VCRs, resulting in horizontal stripes when the recording is played back.

AGC works on approximately 85 percent of consumer VCRs, and Colorstripe works on approximately 95 percent, but only NTSC models, not PAL or SECAM. Macrovision is intended to affect only VCRs, but unfortunately, it may degrade the picture, especially with old or nonstandard television equipment. Macrovision makes DVD players unusable with some line doublers. Effects of Macrovision may appear as stripes of color, distortion, repeated darkening and brightening, rolling or tearing, and black-and-white picture.

Just as with videotapes, some DVDs are Macrovision-protected and some are not. The discs themselves tell the player whether to enable Macrovision AGC or Colorstripe. The producer of the disc decides what amount of copy protection to allow and pays Macrovision royalties accordingly. Each video object unit of the disc contains "trigger bits" telling the player whether or not to enable Macrovision AGC, with the optional addition of two- or four-line Colorstripe. This finely detailed selective control, occurring about twice a second, enables the disc producer to disable copy protection for scenes that may be adversely affected by the process.

Macrovision protection is provided on the composite and s-video output of all but a few commercial DVD players. Macrovision protection was not present on the interlaced component video outputs of early DVD players but is required by the CSS license for all players (the AGC process only, because no colorburst exists in a component signal). A version of AGC for progressive-scan component output was developed in 1999 to be required in 2001 by the CSS license and incorporated into progressive-scan players (set-top only, not computers).

Macrovision protection can be defeated with inexpensive video processing boxes that clean up the video signal. The Macrovision Corporation has been very aggressive in buying the patents for these technologies in order to take them off the market, but some are always available. Only a few work with the new Colorstripe feature. These devices go under names such as Video Clarifier, Image Stabilizer, and Color Corrector. They are a necessary accessory for people who have combination VCR-TVs, which usually route the video signal through the VCR, thus preventing protected DVDs from being watched on the TV. Many newer digital devices, such as digital camcorders and computer video capture cards, recognize the Macrovision process on incoming video signals and refuse to copy them.

Copy Generation Management (CGMS)   Digital video copying—and some analog video copying—is controlled by information on each disc specifying if the data can be copied. This is a serial *copy generation management system* (CGMS) designed to prevent copies or to prevent copies of copies. The information is embedded in the outgoing analog and digital video signals. Obviously, the equipment making the copy has to recognize the CGMS information and abide by the rules.

The analog (CGMS/A) information is encoded into the XDS service of line 21 of the NTSC television signal. The digital standards (CGMS/D) are incorporated into DTCP and HDCP (see the following). Digital recording devices generally check for CGMS/A information in analog inputs.

The CGMS information indicates whether no copies, one copy, or unlimited copies can be made. If no copies are allowed, the recording device will not make a copy. If one copy is allowed, the recording device will make one copy and change the CGMS information to indicate that no copies can be made from the copy.

As with APS information, CGMS flags are present for each sector on a disc. However, since CGMS information could be tampered with, devices are required to establish the protection status of content by the presence or absence of encryption. If the content is encrypted with CSS, CPPM, or CPRM, the "no copies" status is assumed. If no encryption is present, the "copy freely" status is assumed.

Some DVD devices are designed to check for no-copy flags on recordable media. If a no-copy flag is found, the disc is presumed to be an illegal copy and will not be played.

Obviously, CGMS does not prevent multiple copies from being made from the original. However, it is the most fair and reasonable form of copy protection, in that it allows fair-use copies by consumers for their own personal use.

DVD-Audio extends the concepts of CGMS for recordings on legacy media such as CD-R, Minidisc, and DAT. Unlike digital copies of protected content on writable DVD media, which must be encrypted, unencrypted authorized copies can be made at sound quality no better than audio CD. Specifically, the copies can only have one or two channels of no greater than

48 kHz sampling frequency at no more than 16 bits per sample. The recorder must watermark the copies (see following section) and keep track of what copies have been made so that only one copy of original audio content can be made per recorder unless otherwise authorized by the content owner by setting CCI parameters for number of allowed copies. An ISRC must be included with the content so that the recorder can track the number of copies it makes of any title. The content owner also can specify allowed sound quality for copies (CD-Audio, two-channel full-quality, multichannel full-quality). Aside from analog and CD-quality digital audio (IEC-958) outputs, all other outputs from a copy-protected DVD-Audio disc must be encrypted by a method such as DTCP.

Watermarking   Watermarking is a technical process of embedding information into content in a way that is intended to be transparent to the user of the content, yet which cannot be removed or altered easily. Each digital video frame or segment of digital audio is permanently marked with noise that is supposedly undetectable by human ears or eyes. (As discussed in Chapter 2, there is much debate about how undetectable watermarking is in practice. The amount of watermarking can be varied so that an especially dynamic piece could have a lower level of watermarking to reduce its impact.) The noise carries a digital signature that can be recognized by recording and playback equipment. The signature stays connected to the content regardless of digital or analog transformations. Watermarking does not directly protect the content—it only identifies the status of the content. When used with a content protection system, watermarking usually carries CMI. When the content is played on compliant devices, they recognize the CMI carried in the watermark and abide by its constraints. This only works if a "hook" is present that compels devices to be compliant. Encryption is the carrot to which watermarking is attached. To get the keys and secrets needed to play encrypted content, manufacturers must sign a license, which may require that watermark detection be implemented.

Another use of watermarking is to detect if an analog copy has been made or if the content has been digitally reencoded. A *fragile watermark,* as used by the *Secure Digital Music Initiative* (SDMI), is designed to be destroyed by any processing of the content, thus indicating that it is not the original version.

DVD-Audio uses watermarking technology developed by Verance. All DVD-Audio players licensed to play CPPM or CPRM discs are required to include circuitry to recognize the Verance watermark. Watermarking will be added to DVD-Video at some point, as a requirement for new players only. It will not make existing DVD-Video players obsolete.

DVD-Audio recorders include *remarking* encoders that change the watermark for copy-generation management. The CCI embedded in the watermark from a "copy once" source is changed in the recording to "copy never" or "no more copies." The inclusion of remarking encoders in millions of consumer devices may make the system more vulnerable to being compromised.

Digital Transmission Content Protection (DTCP)   A digital medium such as DVD deserves to be connected digitally to other digital devices such as digital televisions or digital video recorders. Unfortunately, DVD is far ahead of digital interconnect standards, which are perennially held back by copy protection concerns. Thus the pristine digital content from DVD is usually converted to analog by the player and then converted back to digital by the receiving device—kind of like sending a photocopy of a work of art instead of the real thing.

Digital transmission content protection (DTCP) is the leading technology for protecting content sent over digital connections. Often called *5C* for the five companies that developed it (Intel, Sony, Hitachi, Matsushita, and Toshiba), DTCP focuses on IEEE 1394/FireWire but can be applied to other transmission protocols. Under DTCP, devices that are digitally connected, such as a DVD player and a digital TV or a digital VCR, exchange keys and authentication certificates to establish a secure channel. The DVD player encrypts the encoded audio/video signal as it sends it to the receiving device, which must decrypt it. This keeps other connected but unauthenticated devices from hijacking the signal. No encryption is needed for content that is not protected on the disc. Security can be "renewed" by new content (such as new discs or new broadcasts) and by new devices that carry updated key blocks and revocation lists that identify unauthorized or compromised devices. Digital devices that do nothing more than reproduce audio and video are able to receive all data, as long as they can authenticate that they are playback-only devices. Digital recording devices are only able to receive data that is marked as copyable, and the recorders must change the flag to "do not copy" or "no more copies" if the source is marked "copy once." DTCP requires new DVD players with digital connectors (such as those on DV equipment). These new products will not appear until 2001 at the earliest. Since the encryption is done by the player, no changes are needed to existing discs.

High-Definition Output Protection (HDCP for DVI)   In 1998, the *Digital Display Working Group* (DDWG) was formed to create a universal interface standard between computers and displays—a new digital replacement for the venerable analog "VGA" connection standard. Founding group members include Silicon Image, Intel, Compaq, Fujitsu, Hewlett-Packard,

IBM, and NEC. The resulting *Digital Visual Interface* (DVI) specification, released in April 1999, was based on Silicon Image's panelLink technology. DVI quickly gained wide acceptance as the new industry standard for low-cost, high-speed digital links to video displays. DVI supports 1600×1200 (UXGA) resolution, which covers all the HDTV resolutions. Even higher resolution can be supported with dual links. Contemporary IEEE 1394 implementations are limited to 40 Mbps, which is plenty for compressed MPEG video but not in the same league with DVI's 4.95 Gbps. Many new HDTV displays are likely to have both IEEE 1394 and DVI connections.

Intel proposed a security component for DVI: *High-Bandwidth Digital Content Protection* (HDCP). HDCP provides authentication, encryption, and revocation.

Special hardware on the video adapter card and the display monitor encrypts video data before it is sent over the link. HDCP is not mandatory, and early DVI monitors were released before HDCP was ready. When an HDCP-equipped DVI card senses that the connected monitor does not support HDCP, it lowers the image quality of protected content.

HDCP is key to winning Hollywood support for high-definition movies on computers and other digital display systems. CSS allows unprotected VGA output of DVD content, which is worrisome enough to studio executives. The HDCP key exchange process verifies that a receiving device is authorized to display or record video. It uses an array of forty, 56-bit secret device keys and a 40-bit key selection vector—all supplied by the HDCP licensing entity. If the security of a display device is compromised, its key selection vector is placed on the revocation list. The host device has the responsibility of maintaining the revocation list, which is updated by *system renewability messages* (SRMs) carried by newer devices and by video content. Once the authority of the receiving device has been established, the video is encrypted by an XOR operation with a stream cipher generated from keys exchanged during the authentication process. If a display device with no decryption ability attempts to display encrypted content, it appears as random noise.

Summary of Content Protection Schemes   All the various copy protection implementations are optional for the producer of a disc. If someone wishes to protect content on a disc, he or she can choose to apply APS, CSS, CPPM, or watermarking; a combination of more than one is usually applied. It is possible to use APS (Macrovision) without CSS, but it is easy to circumvent the APS trigger bits if the content is not protected with CSS. Therefore, Macrovision is almost always combined with CSS.

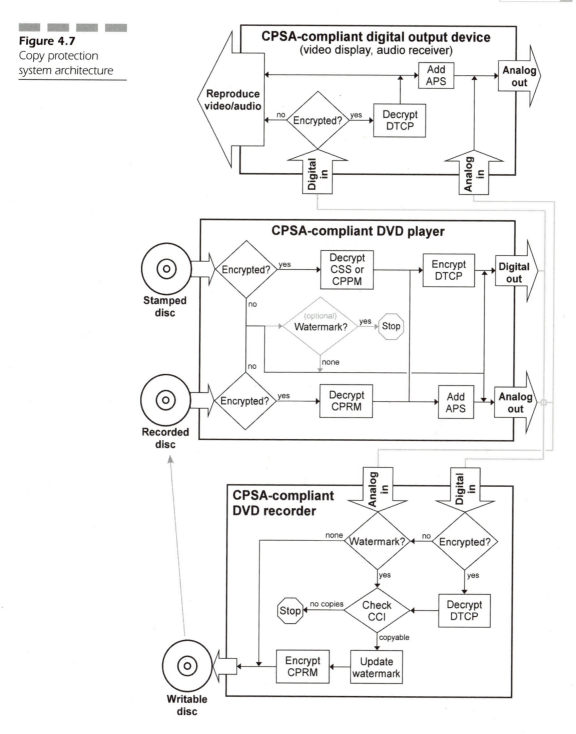

Figure 4.7
Copy protection
system architecture

Figure 4.8
Scope of copy
protection system
architecture

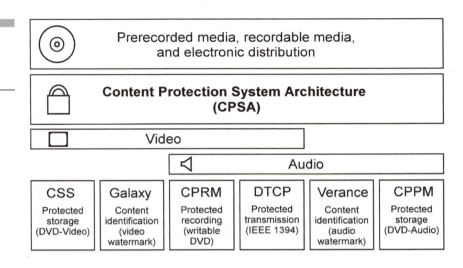

CPRM encryption is performed automatically by DVD recorders. DTCP and HDCP are handled by the DVD player or computer, not by the disc developer. As long as the developer has included CMI with the content, it will be honored or passed along by the digital output protection system.

CSS, CPPM, and CPRM decryption are optional for hardware and software player manufacturers; a player or computer without decryption capability can play only unencrypted discs.

Watermarking is being retrofitted into the CSS and CPPM licenses so that future compliant devices will be required to check for watermarking in unencrypted content. See Figure 4.9 for an overview of the content protection chain.

Ramifications of Copy Protection   All these copy protection schemes are designed to guard against casual copying, which the studios claim causes billions of dollars in lost revenue. The goal is to "keep the honest people honest." The people who developed the copy protection measures are the first to admit that they will not stop well-equipped pirates or even determined consumers. Video pirates have equipment that can easily make complete bit-by-bit copies of discs or create identical master copies for mass replication. Bit-by-bit copiers are available to computer owners who know where to look.

Movie studios began promoting legislation in 1994 that would make it illegal to defeat technical copy protection measures. The result is the *World Intellectual Property Organization* (WIPO) Copyright Treaty, the WIPO Performances and Phonograms Treaty (December 1996), and the

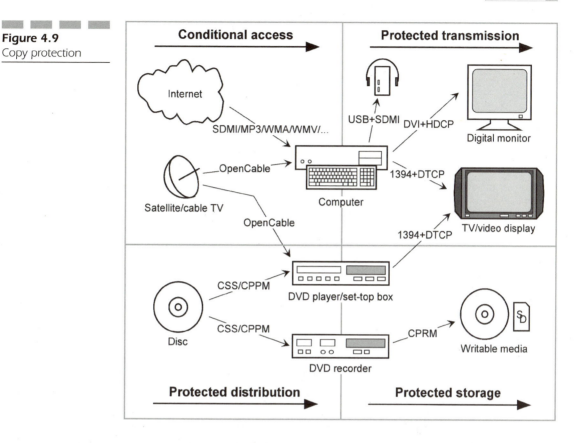

Figure 4.9
Copy protection

compliant U.S. *Digital Millennium Copyright Act* (DMCA), passed into law in October 1998. Processes or devices intended specifically to circumvent copy protection are now illegal in the United States and many other countries. As the legislation was being developed, a cochair of the CPTWG stated, ". . . in the video context, the contemplated legislation should also provide some specific assurances that certain reasonable and customary home recording practices will be permitted, in addition to providing penalties for circumvention." It is not at all clear how this may be "permitted" by a player or by studios that set the "no copies" flag on all their discs.

Although CPSA promotes compatibility and consistency between various content protection schemes, presumably improving the customer experience, many people regard it as an Orwellian nightmare of big brother intrusion. It can be argued that CPSA and associated content protection measures do away with the time-honored (and court-honored) doctrines of first sale rights and fair use. You can buy a DVD, but unless you have a playback device that is sanctioned by a licensing entity, you may not be able to play it. If the copyright owner chooses to not allow any copying of the disc, then reasonable copying is denied for educational use or for personal use, such as compiling a disc of favorite songs or making a copy to play in the car. Access to content is being tied to a monopolistic cabal of copyright owners and player manufacturers.

Effect of Copy Protection on Computers   DVD-ROM drives and computers, including DVD-ROM upgrade kits, are required to support Macrovision, CGMS, and CSS. Computers that play DVD-Audio discs or record video onto writable DVDs must support CPPM and CPRM, respectively. PC video cards with TV outputs that do not support Macrovision will not work with CSS-encrypted movies. Computers with IEEE 1394/FireWire connections must support the DTCP standard in order to work with other DTCP devices. New computers with DVI outputs must support HDCP. Every DVD-ROM drive includes CSS/CPPM circuitry to establish a secure conntection to the decoder hardware or software in the computer, although CSS can only be used on DVD-Video content and CPPM can only be used on DVD-Audio content. Writable DVD drives must include support for CPRM. The various protection systems are only used for audio and video data, not for other types of computer data. Of course, since DVD-ROM can hold any form of data, any desired encryption scheme can be implemented beyond those which are part of CPSA.

Operating systems such as Windows and Mac OS are beginning to integrate security and copy protection measures into the core of the system. More robust hardware-based security, such as encryption/decryption chips proposed by Intel, is also on the horizon. This kind of support from the computer industry makes content owners more willing to trust their content to computer and Internet environments, but it also makes playback much more complicated and often interferes with the use and creation of nonprotected content. See Chapter 11 for additional discussion of copy protection on computers.

Licensing

No single company "owns" DVD. The official format specification was developed initially by a consortium of ten companies: Hitachi, JVC, Matsushita, Mitsubishi, Philips, Pioneer, Sony, Thomson, Time Warner, and Toshiba. Various working groups within the DVD Forum are responsible for different parts of the DVD specification, and representatives from many other companies have contributed in various working groups since the original DVD-ROM and DVD-Video specifications were produced. Although a wide range of people have contributed to the DVD format, it is not an international standard. Some of the DVD physical format specifications were submitted to ECMA and ISO for international standardization, but they are only a small part of the complete, multivolume DVD specification.

The DVD format specification books are available from the DVD Format and Logo Licensing Corporation (FLLC)[12] only after signing a nondisclosure agreement and after payment of a $5000 fee for the first book, plus $500 for each additional book. Manufacture of DVD products and use of the DVD logo for nonpromotional purposes requires an additional format and logo license, with a $10,000 fee for each format. For example, a combination DVD-Video and DVD-Audio player requires a license for the base DVD-ROM format, plus a DVD-Video format license and a DVD-Audio format license, for a total of $30,000. The books do not include information about copy protection systems, which are licensed and documented separately.

The term *DVD* is too common to be trademarked or owned. Time Warner originally trademarked the DVD logo and has since assigned it to the DVD Format and Logo Licensing Corporation. Hardware manufacturers must license the DVD logo and certify compatibility of their products. Authoring tool developers must license the logo. Logo licensing is not required for certain promotional uses. Hardware distributors and retailers may use the DVD logo without a license if their product is manufactured by a licensee. System integrators may use the DVD logo without a license if the DVD components are manufactured by licensees and no additional logo is added (for example, an integrator cannot add a DVD-Video logo to a system that includes a DVD-ROM drive unless he or she signs a DVD-Video license). Content providers, title distributors, and retailers are allowed to use the DVD logo without a license if the disc is replicated by a licensee.

The format and logo license does not convey any patent royalties, which are claimed by dozens of different companies. Some of them have banded together and pooled their patents to make licensing easier, but there is still a bewildering array of companies holding out their hands for their slice of royalties (Table 4.10). Essential DVD technology patents must be licensed from a Philips/Pioneer/Sony pool, a Hitachi/Matsushita/Mitsubishi/Time Warner/Toshiba/Victor pool, and from Thomson. Patent

---

[12]Before April 14, 2000, logo and format licensing was administered by Toshiba in an interim capacity.

TABLE 4.10

DVD Patent
Licensing

Licensing Entity	License	Cost	Who Pays
Philips/Pioneer/ Sony pool	DVD and optical disc technology patents	3.5% per player, minimum $5; additional $2.50 for Video CD compatibility	Player manufacturers, software player developers
		$0.05 per disc	Disc replicators
Hitachi/ Matsushita/ Mitsubishi/ Time Warner/ Toshiba/ Victor pool	DVD technology patents	4% per player or drive, minimum $4; 4% per DVD decoder, minimum $1	Player manufacturers, software player developers
		$0.075 per disc	Disc replicators
Thomson	DVD technology patents		Player manufacturers
Discovision	Optical disc technology patents		Disc replicators
DVD CCA	CSS, CPRM, CPPM	$10,000 initial (no per-product fees)	Player manufacturers, software player developers, disc replicators, large content developers
Macrovision	Macrovision APS	$30,000 initial charge; $15,000 yearly renewal	Hardware manufacturers (players, graphics cards)
		$0.04-$0.10 per disc	Content developers
Dolby	Dolby Digital decoding patents	$0.26 per channel (maximum of $0.60 per player)	Player manufacturers, software player developers
	MLP technology	$0.003 per disc	Disc replicators

continues

TABLE 4.10 cont.

DVD Patent Licensing

Licensing Entity	License	Cost	Who Pays
Philips	Dolby Digital decoding patents		DVD-Audio player manufacturers and software player developers, A/V receiver manufacturers
		$0.20 per channel (maximum of $0.60 per player)	Player manufacturers, software player developers
MPEG LA	MPEG-2 patents	$0.003 per disc	Disc replicators
		$4 per player	Player manufacturers, software player developers
Verance	Watermarking patents	$0.04 per disc or per program	Disc replicators or content developers
		$25,000 per year (for source code) or $10,000 per year (for object code)	Player manufacturers, software player developers
		$50 per watermarked track	Audio production houses
		25% of revenue on watermarking equipment	Manufacturers of authoring or mastering systems
Nissim	Parental management patents	$0.25 per player	Player manufacturers, software player developers
Various companies and licensing entities	Packaging patents		Disc replicators and fulfillment houses

royalties also may be owed to Discovision Associates, which owns about 1300 optical disc patents. The licensor of CSS, CPPM, and CPRM encryption technology is DVD CCA (Copy Control Association).[13] Macrovision licenses its analog antirecording technology to hardware makers. There are no royalty charges for player manufacturers. Macrovision charges a per-disc royalty to content publishers. Dolby licenses Dolby Digital decoders on a per-channel basis. Philips, on behalf of CCETT and IRT, also charges per player and per disc for patents underlying Dolby Digital. MPEG-LA (MPEG Licensing Administrator) represents most MPEG-2 patent holders, with licenses per player and per disc, although there seems to be disagreement on whether content producers owe royalties for discs. Nissim claims 25 cents per player for parental management patents, but there is disagreement on whether the patents apply to DVD and if they are valid. Hewlett Packard and Philips state that there are no licensing costs for manufacturers who include DVD+RW read capability in their units. Implementation of the DVD-RAM specification incurs no royalties so long as no patented technologies are used.

The various licensing fees add up to over $30 in royalties for a $300 DVD player and about 20 cents per disc. Disc royalties are paid by the replicator.

Packaging

When DVD was introduced, manufacturers worried that customers might assume that a DVD would play in their CD player because it looks the same as a CD. Accordingly, the *Video Software Dealers Association* (VSDA) recommended a new-sized package 5 5/8 inches wide, 7 3/8 inches high, and between 3/8 and 5/8 inches deep, which is as wide as a CD jewel box and as tall as a VHS cassette box (Figure 4.10). There are many varieties on the theme, such as the popular Amaray plastic clamshell keep case, the polycarbonate super jewel box, and Time Warner's plastic-and-paperboard Snapper package.

As DVD-Audio comes on the scene, a similar concern is leading dealers to recommend a jewel case that is 1 inch taller than the standard jewel case (also called a *super jewel box,* but not the same as the existing movie-sized super jewel box). Since it will not fit in many dealer racks or customer storage cases, it may not fare well.

There is no official requirement for DVD package size, and many companies simply use standard CD jewel cases or sleeves, especially for DVD-ROMs.

---

[13]Before December 15, 1999, CSS licensing was administered on an interim basis by Matsushita.

DVD Overview

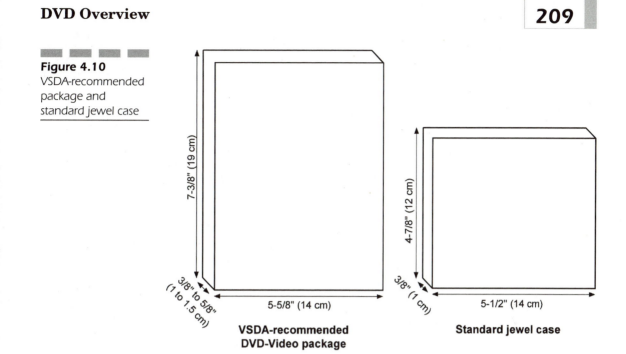

Figure 4.10
VSDA-recommended
package and
standard jewel case

7-3/8" (19 cm)

3/8" to 5/8"
(1 to 1.5 cm)

5-5/8" (14 cm)

**VSDA-recommended
DVD-Video package**

4-7/8" (12 cm)

3/8" (1 cm)

5-1/2" (14 cm)

Standard jewel case

CHAPTER **5**

Disc and Data Details

Introduction

This chapter covers the details of the physical DVD formats and data recording formats. A basic understanding of terms and concepts from Chapters 3 and 4 is assumed.

DVD-ROM is the foundation of DVD: the physical layer and the file system layer. DVD-Video and DVD-Audio are applications of DVD-ROM and are also applications of MPEG-2, Dolby Digital, MLP, and other standardized formats. That is, the DVD-Video and DVD-Audio formats define subsets of other standards to be applied in practice for audio and video playback. DVD-ROM can contain any desired digital information, but DVD-Video and DVD-Audio are limited to certain data types designed for video and audio reproduction. The DVD-Video data types are matched to NTSC and PAL television formats, since NTSC and PAL televisions are the primary target systems.

The DVD family includes recordable versions of DVD-ROM. Application specifications for recording are built on top of the recordable format specifications. These are DVD Video Recording (DVD-VR), DVD Audio Recording (DVD-AR), and DVD Stream Recording (DVD-SR). The DVD application specifications, particularly DVD-Video and DVD-Audio, are covered in detail in Chapter 6.

Physical Composition

Table 5.1 lists the physical characteristics of DVD. Most of these characteristics are shared by all the physical formats (read-only, writable, 12 centimeter and 8 centimeter).

Substrates and Layers

A DVD disc is composed of two bonded plastic substrates (Figure 5.1). Each substrate can contain one or two layers of data. The first layer, called *layer 0,* is closest to the side of the disc from which the data is read. The second layer, called *layer 1,* is further from the readout surface. A disc is read from the bottom, so layer 0 is below layer 1. The layers are spaced very close together; the distance between layers is 55 microns, which is less than one-tenth of the thickness of the substrate (less than one-twentieth of the thickness of the disc).

TABLE 5.1

Physical Characteristics of DVD

Thickness	1.2 mm (+0.30/−0.06) (two bonded substrates)
Substrate thickness	0.6 mm (+0.043/−0.030)
Spacing layer thickness	55 μm (±15)
Mass	13 to 20 g (12-cm disc) or 6 to 9 g (8-cm disc)
Diameter	120 or 80 mm (±0.30)
Spindle hole diameter	15 mm (+0.15/−0.00)
Clamping area diameter	22 to 33 mm
Inner guardband diameter	33 to 44 mm
Burst cutting area diameter	44.6 mm (+0.0/−0.8) to 47 (±0.10) mm
Lead-in diameter	45.2 to 48 mm (+0.0/−0.4)
Data diameter	48 mm (+0.0/−0.4) to 116 mm (12-cm disc) or 76 mm (8-cm disc)
Lead-out diameter	Data + 2 mm (70 mm min. to 117 mm max. or 77 mm max.)
Outer guardband diameter	117 to 120 mm or 77 to 80 mm
Radial runout (disc)	<0.3 mm, peak to peak
Radial runout (tracks)	<100 μm, peak to peak
Index of refraction	1.55 (±0.10)
Birefringence	0.10 μm max.
Reflectivity	45 to 85% (SL), 18 to 30% (DL)[a]
Readout wavelength	650 or 635 nm (640 ±15) (red laser)
Polarization	Circular
Numerical aperture	0.60 (±0.01) (objective lens)
Beam diameter	1.0 mm (±0.2)
Optical spot diameter	0.58 to 0.60 μm
Refractive index	1.55 (±0.10)
Tilt margin (radial)	±0.8°
Track spiral (outer layer)	Clockwise
Track spiral (inner layer)	Clockwise or counterclockwise
Track pitch	0.74 μm (±0.01 avg., ±0.03 max.)

continues

TABLE 5.1 cont.

Physical
Characteristics
of DVD

Pit length	0.400 to 1.866 μm (SL), 0.440 to 2.054 μm (DL) (3T to 14T)
Data bit length (avg.)	0.2667 μm (SL), 0.2934 μm (DL)
Channel bit length (avg.)	0.1333 (±0.0014) μm (SL), 0.1467 (±0.0015) μm (DL)
Jitter	<8% of channel bit clock period
Correctable burst error	6.0 mm (SL), 6.5 mm (DL)
Maximum local defects	100 μm (air bubble), 300 μm (black spot), no more than six defects between 30 and 300 μm in an 80-mm scanning distance
Rotation	Counterclockwise to readout surface
Rotational velocity[b]	570 to 1630 rpm (574 to 1528 rpm in data area)
Scanning velocity[b]	3.49 m/s (SL), 3.84 m/s (DL) (±0.03)
Storage Temperature	−20 to 50°C (−4 to 112°F), ≤ 15°C/h variation (59°F/h)
Storage humidity	−5 to 90% relative, 1 to 30 g/m^3 absolute, ≤ 10%/h variation
Operating temperature	−25 to 70°C (−13 to 158°F), ≤ 50°C sudden change (122°F)
Operating humidity	−3% to 95% relative, 0.5 to 60 g/m^3 absolute, ≤ 30% sudden change

[a]SL = single layer; DL = dual layer.

[b]Reference value for a single-speed drive.

Figure 5.1
Disc structure

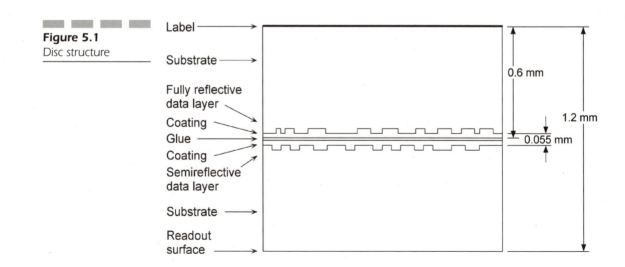

Mastering and Stamping

DVDs are usually made from polycarbonate, but they also can be made from acrylic, polyolefin, or similar transparent material that can be stamped in molten form in a mold. Material with lower birefringence such as acrylic (PMMA) or lexan is better but will not be used commonly until overall costs are lower.

The physical format of a DVD disc is close to that of a CD. In fact, most of the production process is the same as for a CD. More care needs to be taken to properly replicate tinier pits in thinner substrates, and two substrates need to be bonded together. For discs with only one data layer per substrate, the process often can be accomplished with minor modifications and additions to existing CD replication machinery. This is true as long as the equipment can be adjusted to meet the tighter tolerances of DVD, such as finer cutting laser focus, lower jitter, smoother stamper surface, proper thickness, reduced eccentricity, less warpage, more balanced label printing, and so on.

Mastering refers to the process of creating the physical stamping molds used in the replication process (Figure 5.2). The formatted data signal is used to modulate the cutting laser of a laser beam recorder machine (LBR), which creates pits in a glass disc that has been covered with a thin photoresist coating. The laser does not actually cut the glass but rather exposes sections of the photosensitive polymer. The coating is then developed and etched to remove the exposed areas. A vacuum evaporation or sputtering process applies an electrically conductive layer that is then thickened by an electrochemical plating process to create the stamping master, or *father*. The stamping master, usually made of nickel, contains a mirror image of the pits that are to be created on the disc. The raised bumps on the stamper are pressed into the liquid plastic during the injection molding process to create the pits. In some cases, the father is used to create *mother* discs, which can then be used to create multiple stampers called *sons*. Masters also can be made using a dye polymer and metalization process.

NOTE: *The term* mastering *is often misused to refer to the authoring or premastering process.*

In the case of a single layer, the stamped plastic surface is covered with a thin reflective layer of aluminum, silver, or other reflective metal by a

Figure 5.2
Mastering and
stamping

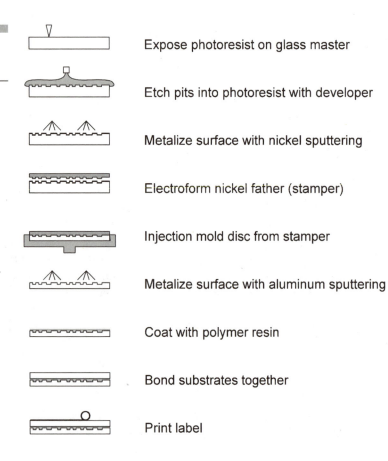

Expose photoresist on glass master

Etch pits into photoresist with developer

Metalize surface with nickel sputtering

Electroform nickel father (stamper)

Injection mold disc from stamper

Metalize surface with aluminum sputtering

Coat with polymer resin

Bond substrates together

Print label

sputtering process. This creates a metallic coating between 60 and 100 angstroms thick. Silicon also can be used for the reflective layer. For two layers, the laser must be able to reflect from the first layer when reading it but also focus through it when reading the second layer. Therefore, the first layer has a semitransparent coating. In many cases, gold is used for the semireflective layer, giving dual-layer discs their characteristic gold tint. The first layer is about 20 percent reflective, whereas the second layer is about 70 percent reflective. The top reflective layer is optionally given a protective overcoat before being bonded to the other substrate.

The confusing part is that there are two ways to make a single-sided, dual-layer disc. The first method, developed by Matsushita, puts one layer

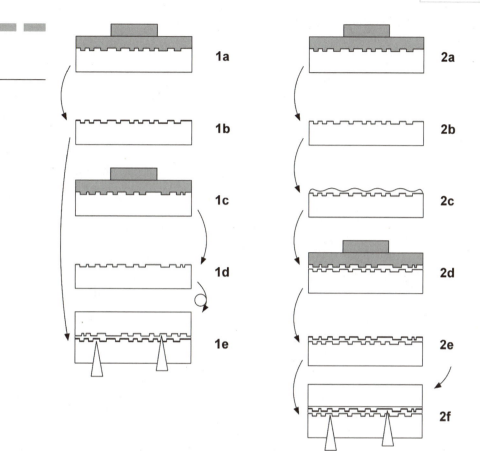

Figure 5.3
Dual-layer
construction

on each substrate, separated by a very thin, transparent adhesive film. The second method, developed by 3M (now Imation), puts one layer on a substrate and then adds a second layer to the same substrate using a photopolymer (2P) material.

A dual-layer disc can be made in two different ways (Figure 5.3):

1. One layer on each side

    a. Stamp the outer data layer into the first substrate.

    b. Add a semireflective coating.

    c. Stamp the inner layer "upside down" on the second substrate.

    d. Add fully reflective coating.

    e. Flip the second substrate over and glue with transparent adhesive to the first.

2. Two layers on one side

   a.   Stamp the outer data layer into first substrate.
   b.   Add a semireflective coating.
   c.   Cover with molten transparent material (ultraviolet resin).
   d.   Stamp the inner layer into the transparent material[1]
   e.   Add a fully reflective coating.
   f.   Glue to the second substrate.

Clearly, the first method is simpler. The second method is required for double-sided, dual-layer discs, which have four layers. These discs are slightly thicker than other discs.

In either case, the result is essentially the same, since the data layers end up at the proper distance from the outer surface. In the first case, the laser focuses through the first side and through the transparent glue onto the layer at the inner surface of the second side. In the second case, the laser focuses through the added transparent material onto the layer at the inner surface of the first side.

Bonding

There are various methods of bonding DVD substrates. Some roll the adhesive on, some "print" it with a screen, some spin the disc to spread the adhesive across the surface, and some rely on capillary flow. The most popular adhesives are hot-melt glue and ultraviolet (UV)-cured photopolymers. Tolerances are extremely tight (on the order of 30 microns), so this is one of the most demanding parts of the DVD replication process and the one that sets it apart from CD replication.

The hot-melt method rolls a thin coat of melted thermoplastic resin onto each substrate and then presses the substrates together with a hydraulic ram. Once the glue has set, a strong bond is formed that remains stable at temperatures up to 70°C (150°F). The hot-melt bonding technique is cheaper and easier than the UV technique.

Variations of UV bonding use a photopolymer (2P) lacquer that is spread across the bonding surface by spinning the disc (radical adhesive), applying a silk-screened layer (cationic adhesive), or attaching a UV film in a vac-

---

[1]The second layer may be transferred to the adhesive from an intermediate surface stamped into a temporary substrate made of a more "slippery" material such as PMMA or polyolefin.

uum. The lacquer is hardened by UV light. This forms an extremely strong bond that is unaffected by temperature. A transparent adhesive such as 2P is required for dual-layer discs, since the laser must be able to read through it. The silk-screening method is more appropriate for writable discs because the cationic adhesive curing process generates less heat, which is less likely to affect the dye or phase-change layers.

Similar bonding techniques are used with both *magneto-optical* (MO) discs and laserdiscs. Laserdiscs sometimes suffer from deterioration of the aluminum layer, which is colloquially referred to as "laser rot" even though it has nothing to do with the laser beam or with rotting. Laserdiscs are molded from acrylic (PMMA), which absorbs about 10 times more moisture than the polycarbonate used in DVDs and can lead to oxidation of the aluminum. The flexibility of laserdiscs allows movement along the bond, which also can damage the aluminum layer. Laser rot problems were more prevalent in the past when material purity was lower and the interaction between the materials that make up a disc was not as well understood as it is today. Since DVDs are molded from improved materials and are more rigid than laserdiscs, they are not as susceptible to laser rot.

Burst Cutting Area

A section near the hub of the disc, called the *burst cutting area* (BCA), optionally can be used for individualizing discs during the replication process. A strong laser is used to cut a series of stripes, somewhat like a bar code, to store up to 188 bytes of information such as ID codes or serial numbers. The BCA sits between 44.6 and 47 millimeters from the center of the disc (Figure 5.4). The BCA can be read by the same laser pickup head that reads the disc. BCA information can be used for inventory purposes or by storage systems such as disc jukeboxes to quickly identify discs.

The BCA is also used by copy protection systems to uniquely encode the data on recordable discs so that they can be decoded only with a key stored in the BCA. Manufacturers of recordable media write a unique media ID in the BCA of each disc. During the development of this system, the DVD Forum realized that data in the BCA needed to be organized because the existing free-for-all would cause problems. They defined a BCA pack format, where the first 2 bytes of each pack hold an application ID to identify the chunk of data stored in the pack. The third byte is a version number, the fourth byte indicates the length of the data in the pack, and remaining bytes of the pack are the data. This allows multiple packs to be placed in the

Figure 5.4
Burst cutting area

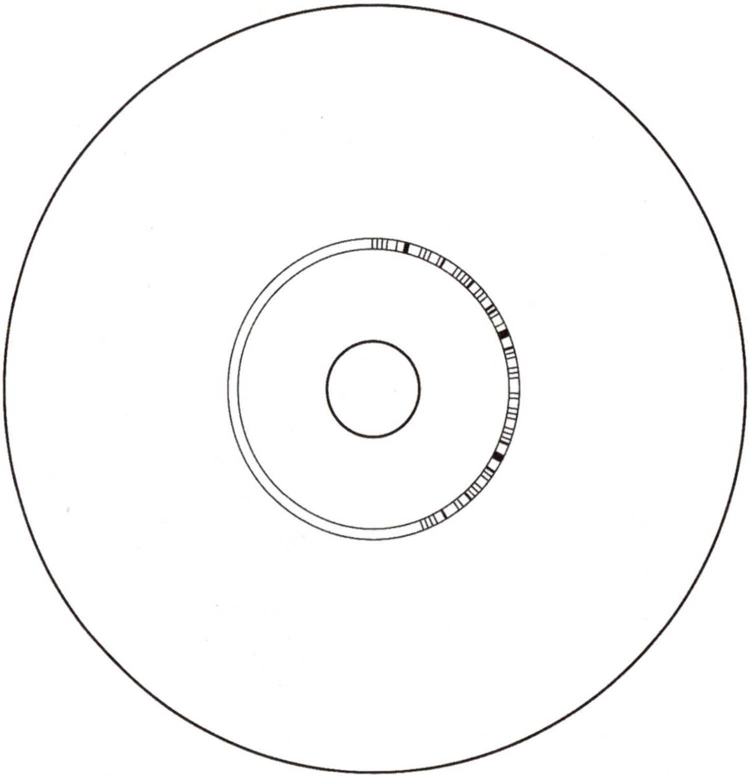

BCA. Various applications can read each pack until they find one that has the desired application ID.

A variation called *narrow burst cutting area* (NBCA) is used by DVD-RW and possibly other writable DVD formats. NBCA barcodes are cut in the 22.7- to 23.5-millimeter-radius section.

Optical Pickups

A critical component of DVD units is the optical pickup, which houses the laser (or lasers), optical elements, and associated electronic circuitry. The optical pickup is mounted on an arm that physically positions it under the disc, moving in and out to read different tracks. DVD readers use semiconductor red laser diodes at a wavelength of 635 or 650 nanometers.

Because of the requirement to read DVDs (which have data layers recorded at two different distances from the surface of the disc) as well as CDs (which have the data recorded at yet a third level), many different techniques have been developed for controlling focus depth. This is also complicated by the fact that CD-R discs require a different wavelength laser (780 nanometers) than is needed to read DVD discs.

Holographic pickups use aspherical lenses with annular rings that can focus reflected laser light from two different depths, one for DVDs and one for CDs. These pickups are unable to read CD-R discs unless used with dual lasers. Twin-lens pickups use two different lenses designed to focus at DVD or CD depth. The appropriate lens is physically moved into the path of the laser beam by a magnetic actuator. These pickups are unable to read CD-R discs. Twin-laser pickups use two separate laser and lens assemblies, with one laser wavelength tuned for CD and CD-R and the other for DVD. In some cases these pickups include holographic lenses. Other approaches are possible, such as twin-wavelength laser diodes that generate two wavelengths of light.

The scenario will become even more complicated when high-density DVD is developed, since it probably will require a third laser. The blue laser wavelength of around 400 nanometers will not be properly reflected by CD-R media and probably will not be reflected by recordable DVD media either.

In all cases, the focusing control is used to change the focus depth between DVD layers, which are spaced about 55 microns apart. The DVD data layers are approximately 0.55 millimeter from the surface of the disc, and the CD data layer is approximately 1.15 millimeters deep.

Media Storage and Longevity

DVDs can be stored and used in a surprisingly wide range of environments (Table 5.2). You can keep them in your refrigerator or in your hot water heater, although neither of these storage methods is recommended in place of a shelf in the living room or office. A cool, dry storage environment is best for long-term data protection. If a disc has been in an environment different from the operating environment, it should be conditioned in the operating environment for at least 2 hours before use.

DVDs are quite stable, and if they are treated well, they usually will last longer than the person who buys them. Estimating the lifetime of a storage medium is a very complex process that relies on simulated aging and statistical extrapolations. Based on accelerated aging tests and past experience with optical media, the general consensus is that replicated discs will last anywhere from 50 to 300 years.

DVD-R discs are expected to last from 40 to 200 years, about as long as CD-R discs.[2] Shelf life before recording is about 10 years. The primary factor in the lifespan of DVD-R discs is aging of the organic dye material, which can change its absorbance properties. Long-term storage of DVD-Rs should be in a relatively dark environment, since the dyes are photosensitive, especially to blue or UV light (both of which are present in sunlight). Cyanine media are more susceptible than phthalocyanine media. There are anecdotal reports of CD-R and DVD-R discs "wearing out" after being played for long periods of time, ostensibly because of alteration in the dye caused by reading it with a laser, but there are counterreports of discs that play 24 hours a day in kiosk installations for months or years without a hitch.

The erasable formats (DVD-RAM, DVD-RW, and DVD+RW) are expected to last from 25 to 100 years. The primary factor in the lifespan of these discs is the chemical tendency of the phase-change alloys to separate and aggregate, thus reducing their ability to hold state. This also limits the number of times a disc can be rewritten.

For comparison, magnetic media (tapes and disks) last 10 to 30 years; high-quality, acid-neutral paper can last a hundred years or longer; and archival-quality microfilm is projected to last 300 years or more.

Note that computer storage media usually become technically obsolete within 20 to 30 years, long before they physically deteriorate. In other words, before the discs become nonviable, it will become difficult or impossible to find equipment that can read them.

As discussed earlier under "Bonding," DVDs may be subject to "laser rot" —oxidation of the reflective layer. Normally, laser rot is not visible, although there have been a few anecdotal reports of cloudiness in discs that correlates with playback errors. Media types such as dual-layer discs and DVD-Rs that use gold for the reflective layer are not susceptible to oxidation because gold is a stable element.

---

[2]Kodak officially states that its CD-R media will last 100 years if stored properly. Testing by Kodak engineers indicates that 95 percent of the discs should last 217 years.

TABLE 5.2

Environmental
Conditions

	Ambient Temperature	Maximum Variation	Relative Humidity	Maximum Variation
Storage				
DVD-ROM, DVD-R	−20 to 50°C (−4 to 122°F)	15°C (59°F) per hour	5 to 90%	10% per hour
DVD-RAM	−10 to 50°C (14 to 122 °F)	10°C (50°F) per hour	3 to 85%	10% per hour
DVD+RW	−10 to 55°C (14 to 131°F)	15°C (59°F) per hour	3 to 90%	10% per hour
Operation				
DVD-ROM, DVD-R	−25 to 70°C (−13 to 158°F)	15°C (59°F) per hour	3 to 95%	10% per hour
DVD-RAM	5 to 60°C (41 to 140°F)	10°C (50°F) per hour	3 to 85%	10% per hour
DVD+RW	−5 to 55°C (23 to 131°F)	10°C (50°F) per hour	3 to 85%	10% per hour

Writable DVD

There are five recordable versions of DVD-ROM: DVD-R for General, DVD-R for Authoring, DVD-RAM, DVD-RW, and DVD+RW.[3] The drive units of each format can read DVD-ROM discs, but each uses a different type of disc for recording. DVD-R records data only once (sequentially), whereas DVD-RAM, DVD-RW, and DVD+RW can be rewritten thousands of times. All writable DVD formats can be used for computer data storage as well as for recording video or audio in consumer devices.

It is also possible to create a hybrid, sometimes called *DVD-PROM*,[4] where a portion of the disc is read-only (stamped in a replication line) and

---

[3]*R* stands for "recordable." *RAM* stands for "random-access memory." Officially, DVD-RAM is "DVD rewritable," whereas DVD-RW, which ought to be "rewritable," is officially "DVD rerecordable."

[4]*PROM* stands for "programmable read-only memory." The acronyms *ROM, RAM,* and *PROM* are all borrowed from roughly analogous solid-state memory devices.

another portion is writable. This could be a disc with two layers, one prere-corded and one writable, or it could be a disc where part of one layer is stamped and the rest is writable.

The three erasable formats (DVD-RAM, DVD-RW, and DVD+RW) are essentially in competition with each other. DVD-RAM has a head start of about 2 years. A major problem is that none of the writable formats are fully compatible with each other or even with existing drives and players. As time goes by, they will become more compatible and more intermixed. For example, combination DVD-ROM/CD-RW drives were released near the end of 1999. Video recorders from Pioneer released near the end of 2000 combine DVD-RW and DVD-R. Future DVD-RAM drives will write to DVD-R discs as well as CD-R/RW. See Table 5.3 for compatibility details.

Each writable DVD format is detailed in the next sections, followed by a summary and comparison.

DVD-R

DVD-R is the record-once version of DVD-ROM, similar in function to CD-R. Data can only be written permanently to the disc, which makes DVD-R well suited for archiving, especially for data that requires an audit trail. DVD-R is compatible with most DVD drives and players. First-generation

TABLE 5.3

Compatibility of Writable DVD Formats

	DVD Unit	DVD-R(G) Unit	DVD-R(A) Unit	DVD-RW Unit	DVD-RAM Unit	DVD+RW Unit
DVD-ROM disc	Reads	Reads	Reads	Reads	Reads	Reads
DVD-R(G) disc	Usually reads	Reads, writes	Reads, does not write	Reads, writes	Reads	Reads
DVD-R(A) disc	Usually reads	Reads, does not write	Reads, writes	Reads, does not write	Reads	Reads
DVD-RW disc	Usually reads	Reads	Reads	Reads, writes	Usually reads	Usually reads
DVD-RAM disc	Rarely reads	Does not read	Does not read	Does not read	Reads, writes	Does not read
DVD+RW disc	Usually reads	Usually reads	Usually reads	Usually reads	Usually reads	Reads, writes

capacity was 3.95 billion bytes (3.68 gigabytes), later extended to 4.7 billion bytes (4.37 gigabytes) in version 2.0. Matching the 4.7-gigabyte capacity of DVD-ROM was crucial for title development and testing.

DVD-R originally was a single format, but in early 2000 it was split into an "authoring" version and a "general" version. The general version uses a 650-nanometer laser (instead of 635 nanometers) for the ability to also write DVD-RAM. DVD-R(G) is intended for home use, whereas DVD-R(A) is intended for professional development. DVD-R(A) media are not writable in DVD-R(G) recorders, and vice versa, but both kinds of media are readable in most DVD players and drives. DVD-R(A) retains the characteristics of the original DVD-R format. The changes made to DVD-R(G), in addition to recording wavelength, are decrementing prepit addresses, a prestamped or prerecorded control area, and double-sided discs (Table 5.4). Wobble frequency is the same for both. A major addition to DVD-R(A) 2.0 is the cutting master format (CMF) feature, which specifies how a disc description protocol (DDP) file is to be written in the lead-in area. The inclusion of a DDP file enables a DVD-R to be used in place of a DLT for submission to a replicator (see the replication section of Chapter 12).

DVD-R uses organic dye technology. A dye material such as cyanine, phthalocyanine, or azo coats a wobbled groove molded into the polycarbonate substrate. The dye is backed with a reflective metallic layer—usually gold for high reflectivity. The wobbled pregroove provides a self-regulating clock signal to guide the laser beam of the DVD-R recorder as it "burns" pits into the photosensitive dye layer. The recorder writes data by pulsing the laser at high power (6 to 12 milliwatts) to heat the dye layer. The series of pulses avoids an overaccumulation of heat, which would create oversized marks. The heated area of the dye becomes dark and less transparent, causing less reflection from the metallic layer underneath. The disc can be read by standard DVD readers, which ignore the wobble. Wobble frequency is eight times the size of a sync frame. The wobble is not modulated—that is, it does not contain addressing information. Addressing and synchronization information is prestamped into the *land prepit* area at the beginning of each sector of the disc. As the recorder's laser beam follows the pregroove, the land prepits are contacted peripherally and create a second pattern of light reflected back to the photodetector. Since the land prepits generate a different signal frequency than the pregroove wobble, the encoded information can be extracted to tell the recorder what sector it is writing to. The land prepits are offset in alternating directions to avoid conflict when they coincide on each side of the groove.

It is theoretically possible to make dual-layer DVD-R discs, but it will take years for the technology to make it out of the laboratory.

Since sectors are already physically allocated on the disc, no physical formatting is required before recording. DVD-Rs usually are recorded in a single session, called *disc at once* (DAO). This is widely supported by DVD-R formatting tools and provides the most compatibility with readers. Incremental writing, where data is appended to the end of a DVD-R in multiple sessions, can be done in two ways. The first uses the multisession features of UDF similar to the packet writing method developed for CD-R. This method is supported by only a few formatting tools. The second method is called *border zone recording,* where a small buffer zone called a *border-out* area is written at the end of each session. Border zone recording was adapted from multisession CD-R designs. The border-out acts as a temporary lead-out area that a compatible reader can ignore once more data is written and a new border zone is added at the end. Reading a disc written with border zone recording requires drives or players that fully support the DVD-R specification. However, many do not support it, although the DVD Multi program requires it. As with CD-R, the price of DVD-R will drop to the point where it is often easier to burn a new disc rather than go through the complications of incremental writing.

The first area on the disc is the *power calibration area* (PCA), which is used by the recorder to perform tests to optimize the laser power for the inserted disc. This process is called *optimal power calibration* (OPC) and is done automatically by the drive before recording. Power calibration can be performed about 7088 times on a single disc. The PCA extends from physical sectors 20800h to 223AFh in version 1.0 and from sectors 1E800h to 203AFh in version 2.0.

The PCA is followed by the *recording management area* (RMA), where the recorder stores *recording management data* (RMD) such as power calibration information, recorder ID, recording history, incremental recording details, and so on. Beyond the RMA is the *information area,* which contains the lead-in, the data recording area, and the lead-out. Only the data recording area is user-accessible. In addition to recording user data, the drive can record *copyright management information* (CPM and CGMS) in sector headers of the data area. DVD-R(A) recorders can write to the control data in the lead-in area using special commands, although only 0s can be written in the part used for storing copy protection keys. DVD-R(G) media are required to have prestamped or prerecorded control areas so that copy protection keys from read-only media cannot be written. Copy protection is provided by CPRM (see Chapter 4).

DVD-R is the most compatible writable format. Version 2.0 media are slightly less compatible with existing readers than the older version 1.0

	DVD-R General	DVD-R Authoring
Recording wavelength	635 nanometers	650 nanometers
Prepit addressing	Decrementing (from sector FFCFFFh)	Incrementing (from sector 003000h)
Sides	One or two	One
Control area content protection)	Prestamped with 0s	0s written by drive
Content protection	None required	CPRM

TABLE 5.4

Differences Between DVD-R Subformats

because the added capacity requires tighter track spacing. The formulation of the media can have an effect on compatibility. Discs from some manufacturers work better than others. It is expected that compatibility will improve slightly as media manufacturers improve the quality of their blank discs. Compatibility also will improve as new players are released because only a few firmware "tweaks" are needed to ensure that DVD-Rs can be read. The DVD-R 1.0 format is standardized in ECMA-279.

DVD-RW

DVD-RW, officially known as DVD rerecordable, is a rewritable version of DVD-R. The name is pronounced as DVD "dash" RW, although many in the opposing camps pejoratively refer to it as DVD "minus" RW. It was briefly called DVD-ER and then DVD-R/W before being dubbed DVD-RW. The format is intended for use in authoring and consumer video and audio recording. Because of the physical similarities between DVD-R and DVD-RW, most DVD-RW recorders also will write to DVD-R(G) media.

DVD-RW, along with the other two rewritable formats, uses phase-change media (see the following section for more details). Compatibility of DVD-RW with most DVD drives and players is slightly below that of DVD-R. The media has a low reflectivity, which confuses some readers into thinking it is dual-layer media. Simple firmware upgrades can solve the problem, since all DVD drives already have *automatic gain control* (AGC) circuitry to compensate for reflectivity differences between single-layer and dual-layer discs. DVD-RW uses wobbled-groove recording with address info on land prepit areas for synchronization at write time. Readers ignore the wobble and the land area. DVD-RW capacity is 4.7 billion bytes (4.37 gigabytes). DVD-RW media can be rewritten about 1000 times. DVD-RW media typically comes bare, but it can use the same cartridges as DVD-RAM. See the following section for details.

Extra information, called a *linking sector,* is recorded between each ECC block on the disc. This extra padding allows for "write splices" to occur at the start and end of each write to the disc. Linking sectors have the same length and format as a normal sector, but they contain information that identifies them as not containing user data. A reader must allow for these gaps in the data by resynchronizing the data clock and otherwise ignoring them.

Unlike DVD-RAM and DVD+RW, DVD-RW is designed for sequential access, using *constant-linear-velocity* (CLV) rotation control. Sequential access improves storage capacity and compatibility. DVD-RW uses a restricted overwrite mode, which reduces unrecorded gaps on the media. This is designed to limit extra seeks during recording (other than defect management sparing) in order to improve real-time recording performance. The alternative to restricted overwrite mode is sequential recording mode, which is similar to that of DVD-R. As with DVD-R, border zone recording can be done to provide compatibility with standard DVD-ROM drives and players. Defect management is performed in software, using features of UDF 2.0.

For restricted overwrite mode, ECC blocks must be formatted in advance. Since formatting the entire disc can take a long time (usually more than an hour), special quick format operations are specified for formatting small areas at a time and for adding or enlarging border zones.

The *power management area* (PMA) and *recording management area* (RMA) are the same as for DVD-R. The control area is preembossed to prevent copy protection keys being written. Copy protection is provided by CPRM (see Chapter 4).

DVD-RAM

DVD-RAM, officially known as DVD rewritable, is an erasable, rerecordable version of DVD-ROM. Version 1.0 of DVD-RAM had a storage capacity of 2.58 billion bytes (2.4 gigabytes). It was in increased in version 2.0 to 4.7 billion bytes (4.37 gigabytes). DVD-RAM media can be rewritten more than 100,000 times. DVD-RAM version 2.0 doubled the recording data rate to 22.16 Mbps (equivalent to 18X CD recording). Version 2.1 added an 80-millimeter disc size designed for use in portable devices such as digital camcorders.

DVD-RAM uses phase-change (PD) technology with some CD-RW and MO features mixed in. A wobbled groove provides clocking data with a

1.33-micron period. Marks are written in both the groove and the land between grooves. The alternating groove and land recording inverts the tracking signal once per revolution. The grooves and preembossed sector headers are molded into the disc during manufacturing.

Sectors on DVD-RAM are arranged in a *zoned CLV* layout, a hybrid of CLV and *constant angular velocity* (CAV). Rather than using the same angular velocity for the entire disc, which makes for fast access but is an inefficient use of recording surface, the disc is divided into multiple concentric rings, called *zones* (Figure 5.5). The first sector of each revolution in the zones is always aligned. This allows for faster access than a traditional CLV layout. The speed of rotation remains constant within a zone. That is, CAV recording is used so that data density in each zone is the same. A 120-millimeter version 2.0 disc is partitioned into 35 zones. Each zone has a fixed radius in width so that each successive zone, moving out from the hub, contains more sectors. One sector per revolution is added to each zone out from the center of the disc, starting with 17 sectors per track, or a total of 39,200 sectors (2450 ECC blocks) in zone 0 and 105,728 sectors (6608 ECC blocks) in zone 34. Version 1.0 discs have 24 zones. A version 2.1 80-millimeter disc has 14 zones.

Figure 5.5
Zoned CLV

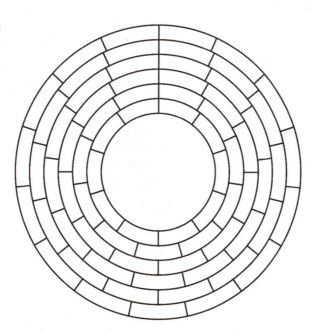

DVD-RAM is the best suited of the writable DVD formats for use in computers because of its hardware-based defect management, zoned CLV format for rapid access, and high number of rewrites. DVD-RAM is also intended for consumer audio and video recording. A major difference between DVD-RAM and all other physical DVD formats is that embossed headers are used to identify the physical sectors. The address used by the drive to read or write sectors is the physical address, not the logical sector number. Because of this difference, along with low reflectivity, tracking changes for land and groove recording, modified access control for zoned CLV, defect management, and a different starting location of the data area, DVD-RAM discs cannot be read by DVD drives and players built before 1998, as well as many built since then.

Defect management is handled automatically by the drive. Two replacement methods are used. *Slipping replacement* is used first, where a defective physical sector is replaced by the first nondefective following physical sector. The *primary defect list* (PDL) records information about defective sectors found during manufacturing, as well as those found during later formatting. The PDL can only be changed by re-formatting. The number of sectors in a group to be listed in the PDL shall not exceed the number of sectors in the *spare area* in that group. *Linear replacement* is the second defect-management method, where a defective physical sector causes the containing ECC block (16 sectors) to be replaced by a block of sectors from a spare area. Defective sector blocks are listed in the *secondary defect list* (SDL) recorded on the disc.

DVD-RAM defines two types of spare areas from which to allocate replacement sectors. The *primary spare area* (PSA), consisting of 12,800 sectors, is preassigned during formatting. The *supplementary spare area* (SSA) is also assigned during formatting but can be expanded later. The supplementary spare area can range from 0 to 97,792 sectors. Once the primary spare area is used up, the supplementary spare area is used and enlarged as needed.

Slipping replacement is quick because defective sectors are replaced by nearby sectors. Linear replacement is slower because it requires a head seek operation to get to the spare area to write or read the replacement block. However, linear replacement is the only way to handle defects during recording without formatting the media.

Each time a data sector is overwritten, the recording start position is randomly shifted by 0 to 15 channel bits, and the lengths of the guard fields are changed by 0 to 7 bytes to accommodate. This random placement of the data keeps the recording surface from being stressed when the same data is written multiple times.

DVD-RAM media comes preformatted. A disc can be re-formatted to erase the contents and to reorganize defect-management information for better performance. A full format—which takes about an hour—verifies and initializes each sector, clears all but the manufacturer defect information from the primary defect list, and creates new spare areas. A *quick improvement* format—which takes only seconds—changes linear replacement sectors to slipped sectors. No data is erased. A *quick clearing* format—which takes only seconds—initializes the media for use, by clearing the PDL, as with full formatting, but without testing any sectors. All data on the disc is erased. Version 1.0 defined *zoned formatting,* which performs the equivalent of a quick clearing format to a single zone, erasing only the data in the zone.

In order to protect the media and make recording more reliable, DVD-RAM media usually comes in cartridges. Once the disc is written, it can be removed from the cartridge to be used in standard drives and players. There are three types of cartridges: type 1 is sealed, type 2 allows the disc to be removed and reinserted, and type 3 is an empty cartridge that a disc can be inserted into for more writing. Single-sided DVD-RAM discs come with or without cartridges. Double-sided DVD-RAM discs are only available in sealed type 1 cartridges. Special cartridges are available for 80-millimeter DVD-RAM; the two-part cartridge includes a disc holder that makes it easy to put the disc in and take it out without touching the recording surface. Adapter cartridges let 80-millimeter cartridges be used in recorders that only accept 120-millimeter cartridges. Standard cartridge dimensions are 124.6 × 135.5 × 8.0 millimeters (Figure 5.6).

Information placed in the embossed lead-in area on the disc specifies if it can be written without a cartridge. If the field is set to 00h, the recorder will not write to the disc unless it is mounted in a cartridge. Cartridges have write-inhibit holes. If the hole is open, the recorder will not write on the disc. A write-inhibit flag (in version 2.0 only) can be set or cleared by the recorder. If the flag is set, the recorder will not write on the disc. Each cartridge also has a sensor hole that indicates if the disc has been taken out at least once. If the sensor hole indicates that the disc has been removed, the recorder may reject certain operations, such as writing with verification turned off.

The DVD-RAM 1.0 format is standardized in ECMA-272 and ECMA-273. DVD-RAM 1.0 drives were able to read older 650-MB PD discs.[5] Compatibility with PD media was dropped in version 2.0 drives. DVD-RAM 2.0 drives can read and write 1.0 media.

---

[5]Largely because many of the features of DVD-RAM are based on technology developed for PD.

Figure 5.6
DVD cartridges

Images courtesy Panasonic and Marken Communications

DVD+RW

DVD+RW is an erasable format based on CD-RW technology. It is not supported by the DVD Forum, even though the companies backing it (primarily Hewlett Packard, Philips, and Sony) are members of the DVD Forum.

The DVD+RW format uses phase-change media with a high-frequency wobbled groove that provides highly accurate sector alignment during recording, thus eliminating linking sectors. Marks are written in the groove only. Unlike the other writable formats, there is no embossed addressing information. The groove wobble is frequency modulated to carry addressing information called *address in pregroove* (ADIP). DVD+RW discs can be recorded in either CLV format for sequential video access (read at CAV speeds by the drive) or CAV format for random access. CAV-formatted discs are not compatible with most standard DVD readers. CLV-formatted discs hold 4.7 billion bytes (4.37 gigabytes). CLV formatting is used for audio and video recording, whereas CAV formatting is designed for computer data storage. DVD+RW media can be rewritten about 1000 times. Most DVD+RW drives also will write to CD-R and CD-RW media.

The lack of linking sectors, the option of no defect management, and CLV formatting make DVD+RW discs compatible with many existing DVD readers. An early 1.0 version of DVD+RW, which held 3 billion bytes (2.8 gigabytes) and was not compatible with anything else, was abandoned in late 1999 in favor of the improved 2.0 version.

Optional defect management is similar to that of DVD-RAM, with primary and secondary defect lists and slipping and linear replacement.

Phase-Change Recording

Phase-change recording technology, used for CD-RW, DVD-RW, DVD-RAM, DVD+RW, and other optical technologies, depends on changes in reflectivity between the amorphous and crystalline states of special alloys. When the alloy is heated by a laser at low power (*bias*) to reach a temperature around 200°C (400°F), it melts and crystallizes into a state of high reflectivity. When the alloy is heated by the laser at high power (*peak*) to a temperature between 500 and 700°C (900 and 1300°F), it melts and then cools rapidly to an amorphous state in which the randomized atom placement causes low reflectivity. As the disc rotates under the laser, it writes marks with high-power pulses and "erases" between the marks with low-power pulses. The marks can then be read with a much lower power setting to sense the difference in reflectivity (Figure 5.7).

Phase-change discs are made of a polymer substrate to which a recording layer and a protective lacquer are applied. The recording layer is a sandwich of four thin films: the lower dielectric film, the recording film, the upper dielectric film, and the reflective film. The upper and lower dielectric films rapidly draw heat away from the recording layer to create the 10-second supercooling effect needed to keep it from returning to a crystalline state. The recording film is made up of an alloy such as germanium, antimony, and tellurium (Ge-Sb-Te) or indium, silver, antimony, and tellurium (In-Ag-Sb-Te). The dielectric layers are made of a material such as zinc sulfide and silicon dioxide ($ZnS-SiO_2$). The reflective film is usually aluminum or gold.

Real-Time Recording Mode

The version 2.0 file system specifications for the various rewritable formats define a special real-time recording mode. Real-time stream recording and playback are critical applications, yet the drive units, especially those used

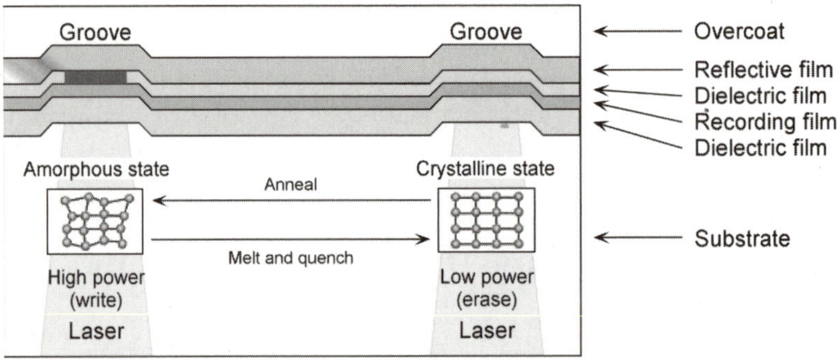

Figure 5.7
Phase-change
recording

in consumer players, have low performance compared with hard-disk drives. In addition to slow seek times and lower data rates, dispersion of data across the medium can degrade performance. A practical solution to the problem is to make sure that streaming data is arranged in continuous order on the disc so that no head seeks are required during recording or playback. This helps guarantee a minimum data rate. The real-time stream model specifies new methods to handle defective sectors on writable discs.

When a player or operating system specifies that a file containing real-time streaming content is being recorded, data is placed in the largest unallocated extents that provide a physically contiguous space. Since linear replacement causes discontinuities in the data, it is disabled for streaming write operations (slipping, which is performed at formatting time, remains in effect). Even if the recorder encounters a defective block, it will not replace it, since the highest priority is given to continuity of data. Using real-time stream operations thus may result in erroneous data. Minor errors in an MPEG-2 video stream, for example, are often undetectable or not objectionable during playback and are preferable to loss of recording or playback continuity.

Summary of Writable DVD

Many differences and many similarities exist among writable DVD formats (Tables 5.5 and 5.6). Although some of the writable formats are slightly more suited for one application than for another, when it comes down to it, they all do a similar job of recording data. Some have speed advantages, although faster drives and the use of CAV speeds when reading or writing CLV discs will make most of the write/record speeds moot. There will be

TABLE 5.5

Writable DVD Overview

	DVD-RAM 1.0	DVD-RAM 2.0	DVD-R 1.0	DVD-R 2.0	DVD-RW 1.0	DVD+RW 2.0
First available in U.S.	Q3 1998	Q2 2000	Q4 1997	Q2 1999	Q4 2000	Q1 2001
Capacity	2.6 GB/side	4.7 GB/side	3.95 GB/side	4.7 GB/side	4.7 GB/side	4.7 GB/side
Rewrites	~100,000			0	~1000	~1000
Write method	Wobbled land and groove		Wobbled groove		Wobbled groove	High-frequency wobbled groove
Recording technology (similar to)	Phase change (PD & MO & CD-RW)		Organic dye (CD-R)		Phase change (CD-RW)	Phase change (CD-RW)
Formatting	Zoned CLV		CLV		CLV	CAV or CLV
Cartridge	Optional		None		None	None
Applications	Computer data storage backup, audio/video recording		Development testing and premastering short-run duplication, copying, archiving		Audio/video recording	Computer data storage, backup, audio/video recording
Backers	Hitachi, Matsushita, Samsung, Toshiba		Pioneer		Pioneer Ricoh, Sharp, Sony, Yamaha	Hewlett Packard, Phillips, Sony, Thomson, Mitsubishi, Ricoh, Yamaha

endless arguments about which technology is better, just as there will be endless arguments about the benefits and drawbacks of cartridges.

The bottom line, however, is that people prefer bare discs. The question of whether to use caddyless media was a point of some contention when deciding on the DVD read-only specs, but the consensus was to use bare discs like CDs. Early CD-ROM drives used caddies, but they were soon

TABLE 5.6

Comparison of Physical Formats

	Capacity (12 cm, billion bytes)	Capacity (8 cm, billion bytes)	Recording Wavelength (nm)	Reflectivity (%)	Data Bit Length (μm)	Channel Pit Length (μm)	Min. Pit Length (μm)	Max. Pit Length (μm)	Track Pitch (μm)	Scanning Velocity[a] (m/s)
DVD-ROM (single layer)	4.70	1.46	n/a	45–85	0.267	0.133	0.400	1.866	0.74	3.49
DVD-ROM (dual layer)	8.54	2.66	n/a	18–30	0.293	0.147	0.440	2.054	0.74	3.84
DVD-R 1.0	3.95	1.23	635	45–85	0.293	0.147	0.440	2.054	0.80	3.84
DVD-R 2.0	4.70	1.46	635/650[b]	45–85	0.267	0.133	0.400	1.866	0.74	3.49
DVD-RW 1.0	4.70	1.46	650	18–30	0.267	0.133	0.400	1.866	0.74	3.49
DVD-RAM 1.0	2.6	n/a	650	15–35	0.409–0.435[c]	0.205–0.218[c]	0.614–0.653[c]	2.863–3.045[c]	0.74	5.96–6.35[c]
DVD-RAM 2.0	4.70	1.46	650	15–35	0.280–0.291[c] (0.280–0.295[d])	0.140–0.146[c] (0.140–0.148[d])	0.420–0.437[c] (0.420–0.443[d])	1.960–2.037[c] (1.960–2.065[d])	0.615	8.16–8.49[c] (8.16–8.61[d])
DVD+RW	4.70	1.46	650	10–20		0.176			0.80	4.9–6.25[e] 3.02–7.35[f]
AS-MO	6.00	n/a	635		0.235	0.235	0.235		0.6	4.5–10.9[e]
CD-R	0.65	0.19								
CD-RW	0.65	0.19								

[a]Reference rate. Many drives run faster.
[b]650 nanometers for DVD-R General, 630 nanometers for DVD-R Authoring.
[c]Range for zoned CLV format.
[d]Values for 8-centimeter media.
[e]Range for CLV format.
[f]Range for CAV format.

dropped in favor of drives that could accept bare discs. The customers had spoken. The recommendations of the DVD Forum working groups charged with designing rewritable DVD formats also were for bare discs. DVD-RAM and DVD-RW use optional cartridges (not caddies). With DVD-RAM, the engineers put more emphasis on protecting the media.

Data Format

On top of the physical formats of DVD-ROM and the various writable versions sit the data and file formats. They are generally similar, with minor differences for writable discs to deal with error correction and rewriting of data (Table 5.7).

Sector Makeup and Error Correction

Each user sector of 2048 bytes is scrambled with a bit-shifting process to help spread the data around for error correction. Sixteen extra bytes are added to the beginning: 4 bytes for the sector ID (1 byte of sector information, 3 bytes for the sector number), 2 bytes for ID error detection, and 6 bytes of copyright management information (CPR_MAI). Four bytes of payload error detection code are also added to the end. This makes a data sector of 2064 bytes, called a *data unit 1* (Figure 5.8). Each data sector is arranged into 12 rows of 172 bytes, and the rows of 16 data sectors are interleaved together (spreading them apart to help with burst errors) into error correction code blocks (192 rows of 172 bytes) (Figure 5.9). For each of the 172 columns of the ECC block, a 16-byte outer-parity Reed-Solomon code is calculated, forming 16 new rows at the bottom. For each of the 208 (192 + 16) rows of the ECC block, a 10-byte inner-parity Reed-Solomon code is calculated and appended. The ECC block is then broken up into recording sectors by taking a group of 12 rows and adding 1 row of parity codes. This spreads the parity codes apart for further error resilience. Each recording sector is 13 rows (12 + 1) of 182 bytes (172 + 10).

Each recording sector is split down the middle, and 1-byte sync codes are inserted in front of each half-row (Figure 5.10). This results in a recording data unit of 2418 bytes, called a *data unit 3,* which is identical for all DVD physical formats (DVD-ROM, DVD-R, DVD-RW, DVD-RAM, and DVD+RW). DVD-RAM media adds data between each data unit 3 before recording (Figure 5.11). The data unit is processed with 8/16 modulation

TABLE 5.7

Data Storage
Characteristics
of DVD

Modulation	8/16 (EFMPlus)
Sector size (user data)	2048 bytes
Logical sector size (data unit 1)	2064 bytes (2048 + 12 header + 4 EDC)
Recording sector size (data unit 2)	2366 bytes (2064 + 302 ECC)
Unmodulated physical sector (data unit 3)	2418 bytes (2366 + 52 sync)
Physical sector size	4836 (2418 × 2 modulation)
Error correction	Reed-Solomon product code (208,192,17) × (182,172,11)
Error correction overhead	15% (13% of recording sector: 308/2366)
ECC block size	16 sectors (32678 bytes user data, 37856 bytes total)
Format overhead	16% (37856/32678)
Maximum random error	≤ 280 in 8 ECC blocks
Channel data rate[a]	26.16 Mbps
User data rate[a]	11.08 Mbps
Capacity (per side, 12 cm)	4.38 to 7.95GB (4.70 to 8.54 billion bytes)
Capacity (per side, 8 cm)	1.36 to 2.48GB (1.46 to 2.66 billion bytes)

[a]Reference value for a single-speed drive.

(doubling each 8-bit byte to 16 bits). This creates a physical sector of 4836 bytes, which is written out row by row to the disc as channel data. Data is written using NRZI format (nonreturn to zero, inverted), where each transition from pit to land represents a one, and the lack of a transition represents a string of 2 to 10 zeroes.

The 8/16 modulation process replaces each byte with a 16-bit code selected from a set of tables (or a four-state machine). These codes have been chosen carefully to minimize low-frequency (DC) energy, and they also incorporate sync and merge characteristics. The codes guarantee at least 2 and at most 10 zeroes between each pair of ones. 8/16 modulation is sometimes called *EFMPlus*. This is a very odd name because EFM is the method used for channel data modulation on CDs and stands for "eight-to-fourteen modulation." Apparently "plus" means to add two to the second number.

The additional error detection and correction information takes up approximately 13 percent of the total data: 308 bytes out of 2366 bytes for

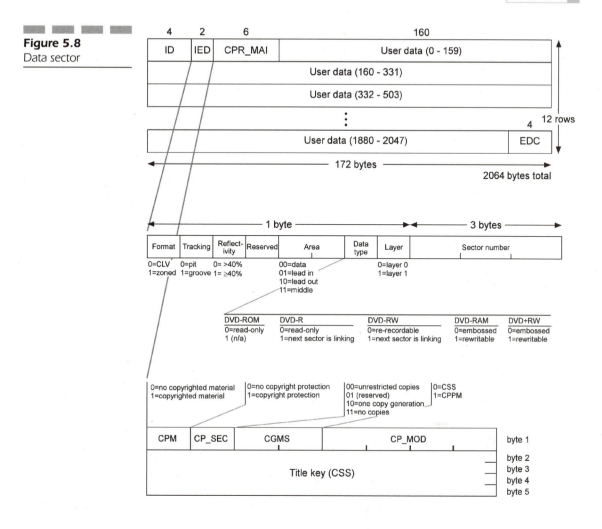

Figure 5.8
Data sector

each sector. Six error detection bytes are included with the sector data (2 for the ID and 4 for the complete sector). There are 302 bytes for RS parity codes (120 for the 12 rows, 172 for the 172 columns, and 10 for the thirteenth row of column codes). The product code part of the error correction system refers to calculating parity codes on top of parity codes, where the rows of outer parity cross the columns of inner parity codes. This creates an extra level of detection and correction. RS-PC can reduce a random error rate from 2×10^{-2} (1 error in 200) to less than 1×10^{-15} (1 error in 1 quadrillion, or 1 million-billion). This is an error correction efficiency approximately 10 times that of CD.

Unlike one-dimensional error correction systems such as CIRC, which treat the data as a continuous stream, RS-PC works on relatively small

Figure 5.9
ECC block and
recording sectors

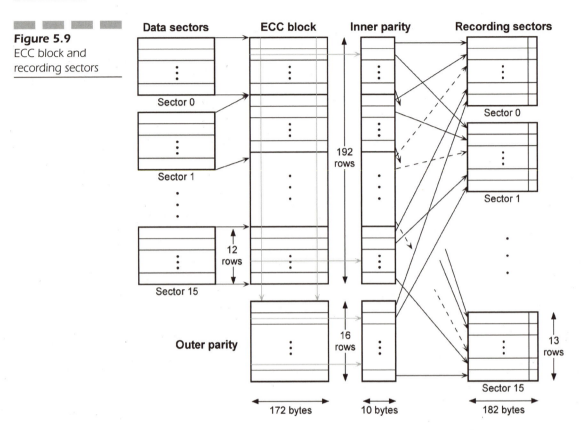

Figure 5.10
Physical sector

sync 0	row 1, bytes 0-90	sync 5	row 1, bytes 91-181
sync 1	row 2, bytes 182-272	sync 5	row 2, bytes 273-363
sync 2		sync 5	
sync 3		sync 5	
sync 4		sync 5	
sync 1		sync 6	
sync 2		sync 6	
sync 3		sync 6	
sync 4		sync 6	
sync 1		sync 7	
sync 2		sync 7	
sync 3	row 12, bytes 2002-2092	sync 7	row 12, bytes 2093-2183
sync 4	row 13, bytes 2184-2274	sync 7	row 13, bytes 2275-2365

←2→◄——— 91 bytes ———►◄2►◄——— 91 bytes ———►

Inner parity

**Outer
parity**

(186 x 13 = 2418 bytes)

Channel bits:

sync 0	0-90	sync 5	91-181	sync 1	182-272	. . .	sync 7	2275-2365
32	1456	32	1456	32	1456		32	1456 = 38688

(4836 bytes)

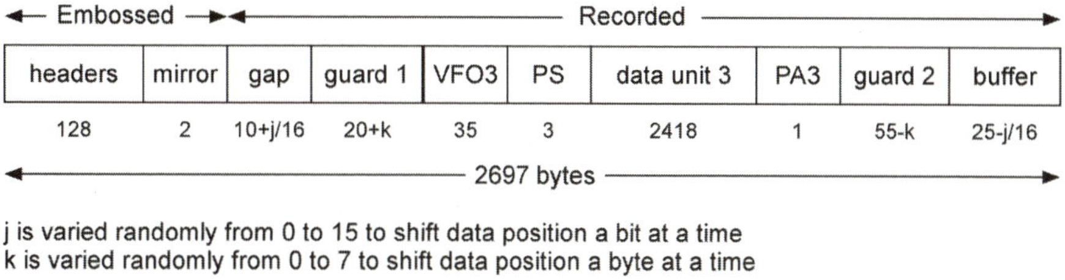

Figure 5.11
DVD-RAM physical sector.

	headers	mirror	gap	guard 1	VFO3	PS	data unit 3	PA3	guard 2	buffer
	128	2	10+j/16	20+k	35	3	2418	1	55-k	25-j/16

← Embossed → ← ———————————————— Recorded ———————————————— →

← ———————————————— 2697 bytes ———————————————— →

j is varied randomly from 0 to 15 to shift data position a bit at a time
k is varied randomly from 0 to 7 to shift data position a byte at a time

blocks that are more appropriate for computer data. A disadvantage of RS-PC is that the matrix structure requires twice the buffer size, but given the ever-falling cost of memory, this is no longer a critical issue.

The maximum correctable burst error length for DVD-ROM is approximately 2800 bytes, corresponding to physical damage of 6 millimeters in length (6.5 millimeters for dual layers). In comparison, the CIRC structure of CDs can only correct burst errors of approximately 500 bytes, corresponding to 2.4 millimeters of physical damage. The writable formats each define slightly different correctable burst error lengths.

Writable media sector size is 2048 bytes. The writable formats cannot record encryption keys in the extra 6 bytes of copyright management information that is present on DVD-ROM discs. The MPAA has asked for a version of DVD-R that would be able to include CSS and CPPM information. The DVD Forum has been considering how to implement this request (which would create a third DVD-R subformat), but it may not ever see the light of day.

Data Flow and Buffering

Raw channel data is read off the disc at a constant 26.16 Mbps. The 16/8 demodulation process reduces the data by half to 13.08 Mbps. After error correction, the user data stream runs at a constant 11.08 Mbps (see Figure 6.1).

The track is broken into sectors of 2048 bytes (2 kilobytes). A 37,856-byte ECC block is made up of 16 sectors; 5088 bytes of this is overhead: 4832 bytes of RS-PC error correction, 96 bytes of sector error detection and correction, and 160 bytes of ID and copy protection information. Because error correction is applied to an entire block, data must be read and written in complete blocks. Each ECC block contains 32,767 (32 kilobytes) of user data.

DVD drives, sometimes referred to as DVD *logical units,* transfer data in 2048-byte chunks—one sector's worth. An internal cache parcels out these units from an ECC block. This presents a special problem for rewritable drives that receive data from the host in 2-kilobyte chunks but must write it to the drive in 32-kilobyte chunks. Drives must implement an internal process for maintaining an ECC block that can be read, modified to add a new sector's worth of data, and then rewritten. This read-modify-write process may use an internal cache rather than writing to the disc.

If the data is not cached, more than one rotation is needed to write the data. First, the ECC block must be read from the drive, the new data must be written and new ECC information calculated, and then the new ECC block must be written back to the same place on the disc. A technique to provide better performance with rewritable media is to write data in sizes that are a multiple of 32,768 bytes starting at logical block addresses that are a multiple of 16, which results in a one-pass direct overwrite operation.

Disc Organization

A DVD contains data written in a continuous spiral track. The track in layer 0 travels from the inner to the outer part of the disc. If layer 1 exists, its track can travel in either direction. *Opposite track path* (OTP) is designed for continuous data from layer to layer. The laser pickup assembly begins reading layer 0 at the center of the disc, starting with the lead-in area; when it reaches the outside of the disc, it enters the middle area and refocuses to layer 1 and then reads back toward the center of the disc until it reaches the lead-out area. *Parallel track path* (PTP) is designed for special applications where it is advantageous to switch between layers. In this case, each layer is treated separately, and each has its own lead-in and lead-out area. There is no middle area. (Figure 5.12).

The lead-in area contains information about the disc (Table 5.8 and Figure 5.13). Copy protection information and disc keys or media key blocks are also contained in the lead-in area.

The middle area and lead-out area each provide a guide to the readout system that it has reached the middle of a two-layer extent or the end of a layer. They are simply sectors full of zeros.

Unlike CD, DVD has no specially defined low-level data formats for audio or video. In addition, no TOC or subchannel data exist. There is only the concept of general data.

On most units, the rotational velocity of the disc varies; the farther from the center the laser pickup is, the slower the disc spins. This creates a CLV, keeping the current track surface moving at a constant speed under the pickup head so that the data density remains the same across the entire

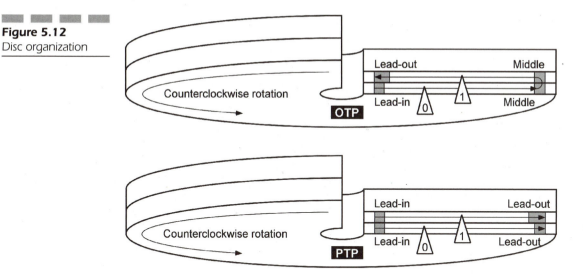

Figure 5.12
Disc organization

surface. Accordingly, there are more data sectors per revolution further from the center of the disc. Some units spin the disc at a CAV, keeping the revolutions per minute the same regardless of the position of the pickup head. They use more advanced circuitry to deal with the difference in data readout rates. These units only achieve their maximum rated speed when reading the outermost section of the disc.

DVD-ROM, DVD-R, and DVD-RW media are formatted for CLV reading. DVD+RW media can be formatted in either CLV or CAV mode. In CAV mode, there are always the same number of sectors per revolution. Since data toward the outer edge of the disc is spread across more surface area in CAV mode, the capacity of the disc is reduced. The advantage of CAV is that seeking to a specific sector is faster because the motor does not have to speed up or slow down as the head travels in or out.

DVD-RAM uses a combination of CLV and CAV formatting, called *zoned CLV* (ZCLV). Details are in the previous DVD-RAM section.

File Format

Data sectors on a DVD-ROM can contain essentially any type of data in any format. Officially, the OSTA UDF file format standard is mandatory. UDF defines specific ways in which the ISO 13346 volume and file structure standard is applied for specific operating systems.

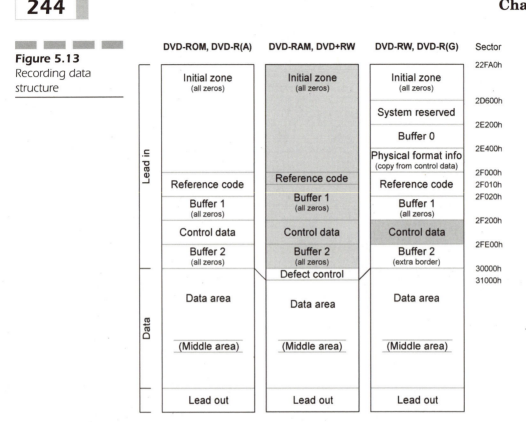

Figure 5.13
Recording data
structure

shaded areas are pre-embossed or pre-recorded and can't be changed

UDF limits ISO 13346 by making multivolume and multipartition support optional, defining filename translation algorithms, and defining extended attributes such as Mac OS file/creator types and resource forks. UDF provides information specific to DOS, OS/2, Macintosh, Windows 95, Windows NT, and UNIX. On top of the UDF limitations, the DVD-ROM standard requires that the logical sector size and logical block size be 2048 bytes. UDF includes an appendix defining additional restrictions for DVD-Video. See "File Format" under "DVD-Video" in Chapter 6 for details.

UDF also defines an extended attribute for storing DVD copyright management information. It is an implementation use extended attribute with the implementation identifier set to "*UDF DVD CGMS Info."

TABLE 5.8		PHYSICAL DATA	
Contents of Lead-in Area	Format version	4 bits (B0:b_0–b_3)	0 = 0.9 1 = 1.0 4 = 1.9 5 = 2.0
	Disc type	4 bits (B0:b_4–b_7)	0 = DVD-ROM 1 = DVD-RAM 2 = DVD-R 3 = DVD-RW 9 = DVD+RW
	Maximum data rate	4 bits (B1:b_0–b_3)	0 = 2.52 Mbps[a] 1 = 5.04 Mbps[a] 2 = 10.08 Mbps 255 = none specified
	Disc size	4 bits (B1:b_4–b_7)	0 = 120 mm, 1 = 80 mm
	Layer type	4 bits (B2:b_0–b_3)	b_0: 1 = read-only b_1: 1 = write-once b_2: 1 = write-many
	Track path	1 bit (B2:b_4)	0 = PTP, 1 = OTP
	Number of layers	2 bits (B2:b_5–b_6)	1, 2
	Track density	4 bits (B3:b_0–b_3)	0 = 0.74 μm/track 1 = 0.80 μm/track 2 = 0.615 μm/track
	Linear density	4 bits (B3:b_4–b_7)	0 = 0.267 μm/data bit 1 = 0.293 μm/data bit 2 = 0.409–0.435 μm/data bit 4 = 0.280–0.291 μm/data bit 8 = 0.353 μm/data bit
	Starting sector number	3 bytes (B5–B7)	030000h (031000h for DVD-RAM and DVD+RW)
	Ending sector number (main)	3 bytes (B9–B11)	
	Ending sector number (layer 0)	3 bytes (B13–B15)	
	BCA flag	1 bit (B16:b_7)	Does not exist, exists

continues

TABLE 5.8 cont.

Contents of Lead-in
Area

	COPYRIGHT DATA	
Copyright protection system	1 byte	0 = none 1 = CSS or CPPM 2 = CPRM
Region management flags	1 byte	1 bit per region (8)
Encryption Data	2048 Bytes	
Manufacturing Data	2048 Bytes	1 per layer
Content Provider Information	28672 Bytes	

[a]Slow rotation for battery-operated drives. Not appropriate for video or audio playback.

DVD-ROM Premastering

The premastering process for a DVD-ROM is very similar to the premastering process for a CD. The computer directory structure and data files are written to a UDF, UDF Bridge (UDF/ISO 9660), or other format disc image. Some premastering systems allow the disc image to be used to simulate an actual disc. The disc image is copied to DLT or DVD-R for creating a one-off test disc or for mastering and replication. The DVD-Video premastering process is described in Chapter 6.

Improvement over CD

The storage capacity of a single-layer DVD is seven times higher than that of a CD. This is accomplished by a combination of smaller physical bit size, slightly larger data area, and improved logical channel coding (Table 5.9).

TABLE 5.9

Incremental
Improvements of
DVD over CD-ROM

Factor	DVD	CD	Gain
Smaller pit length	0.400 μm	0.833 μm	2.08x
Tighter tracks	0.74 μm	1.6 μm	2.16x
Slightly larger data area	8605 mm^2	8759 mm^2	1.02x
More efficient modulation	16/8	17/8	1.06x
More efficient error correction	13% (308/2366)	34% (1064/3124)	1.32x
Less sector overhead	2.6% (62/2418)	8.2% (278/3390)	1.06x
Total increase			~7x

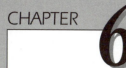

Application Details: DVD-Video and DVD-Audio

Introduction

The DVD-Video format specification is an application of DVD-ROM intended to provide high-quality audio and video in a format supported worldwide by consumer electronics manufacturers, movie studios, video publishers, and computer makers. Likewise, the DVD-Audio specification is an application of DVD-ROM focused on high-fidelity multichannel audio. In addition, the DVD Forum has defined recording formats for video, audio, and generic streaming data. This chapter covers the technical details of these application formats.

It is worth noting that the DVD application specifications are not formal standards—that is, they have not gone through an open review process under the auspices of a standards organization. The physical format specifications for DVD-ROM and writable DVDs were submitted to the ECMA and have become ISO/IEC standards, but the application formats are controlled by the DVD Forum—a consortium of companies with vested interests.

Variations of DVD for professional use, video games, Web-connected content, and other areas also may beget application specifications based on DVD-ROM or writable DVD. The DVD Forum may define some of these, whereas other organizations may define additional applications, such as Philips and Sony's SACD audio format for DVD and Sony's own PlayStation format.

Data Flow and Buffering

Raw channel data is read off a disc at a constant 26.16 Mbps. The 16/8 demodulation process cuts it in half to 13.08 Mbps. After error correction, the user data stream goes into the track buffer at a constant 11.08 Mbps. Data search information (DSI) is copied from the data stream before it reaches the track buffer. The track buffer feeds the program stream data out at a variable rate of up to 10.08 Mbps. If the buffer fills, the input stream is halted (while the disc continues to spin) until there is room in the buffer for more incoming data. The program stream contains various packetized elementary streams (PESs): video, audio, subpicture, still picture, real-time text presentation control information (PCI), and data search information (DSI). PCI and DSI together constitute DVD system overhead.

The maximum combined rate of the remaining streams is 10.08 Mbps, with a per-stream limit of 9.8 Mbps (see Figure 6.1).

The MPEG-1 or MPEG-2 video stream is a standard MPEG elementary stream. MPEG-1 or MPEG-2 audio, if any, is also presented as a standard MPEG audio stream. An MPEG private stream is used to hold PCI and DSI, and a second private stream holds subpicture and all other audio data.

The video stream is limited to 9.8 Mbps but must be lower to allow for audio. MPEG-1 video is limited to 1.856 Mbps.

A pulse-code modulated (PCM) audio stream on a DVD-Video disc is limited to 6.144 Mbps (sufficient for 8 channels at 16/48 sampling). A PCM audio stream on a DVD-Audio disc is limited to 9.6 Mbps. Dolby Digital is limited to 448 kbps, MPEG-1 audio is limited to 384 kbps, the MPEG-2 audio extension stream is limited to 528 kbps (giving a combined limit of 912 kbps), DTS is limited to 1536 kbps, and SDDS is limited to 1280 kbps. See Tables 6.1 and 6.2, and Figures 6.2 and 6.3.

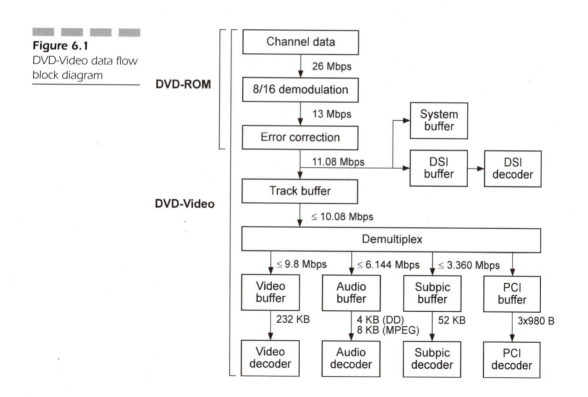

Figure 6.1
DVD-Video data flow block diagram

TABLE 6.1

Stream Data Rates

	Minimum (Mbps)	Typical (kbps)	Maximum (kbps)
MPEG-2 video	1500[a]	3500	9800
MPEG-1 video	900[a]	1150	1856
PCM (DVD-Video)	768	1536	6144
MLP/PCM (DVD-Audio)	n/a	6900	9600
Dolby Digital	64	384	448
MPEG-1 audio	64	192	384
MPEG-2 audio	64	384	912
Subpicture	n/a	10	3360

[a]Not an absolute limit but a practical limit below which video quality is too poor.

Figure 6.2

Relative sizes of DVD-Video data streams

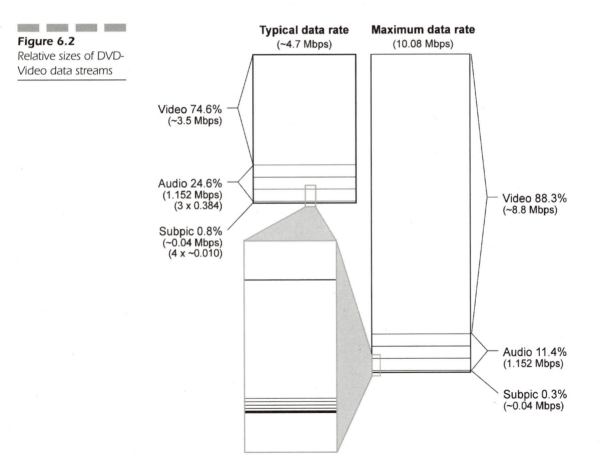

TABLE 6.2

Playing times at various data rates

Video (Average)	Audio (Tracks & Format)	Total[a]	DVD-5	DVD-9	DVD-10	DVD-14	DVD-18
	Data Rate (Mbps)			**Playing Time per Disc, minutes (hours)**			
3.5	1.344 (3 DD5.1)	4.88	128 (2.1)	233 (3.8)	256 (4.2)	361 (6)	466 (7.7)
3.5	0.896 (2 DD5.1)	4.44	141 (2.3)	256 (4.2)	282 (4.7)	397 (6.6)	513 (8.5)
3.5	0.448 (1 DD5.1)	3.99	157 (2.6)	285 (4.7)	314 (5.2)	442 (7.3)	571 (9.5)
3.5	3.072 (8 DD5.1)	6.61	94 (1.5)	172 (2.8)	189 (3.1)	266 (4.4)	344 (5.7)
3.5	1.536 (1 16/48 PCM)	5.08	123 (2)	224 (3.7)	246 (4.1)	347 (5.7)	448 (7.4)
8.7	1.344 (3 DD5.1)	10.08	62 (1)	112 (1.8)	124 (2)	175 (2.9)	225 (3.7)
9.6	0.448 (1 DD5.1)	10.08	62 (1)	112 (1.8)	124 (2)	175 (2.9)	225 (3.7)
7.0	0.896 (2 DD5.1)	7.94	78 (1.3)	143 (2.3)	157 (2.6)	222 (3.7)	286 (4.7)
6.0	0.896 (2 DD5.1)	6.94	90 (1.5)	164 (2.7)	180 (3)	254 (4.2)	328 (5.4)
6.0	0.384 (2 DD2.0)	6.42	97 (1.6)	177 (2.9)	195 (3.2)	274 (4.5)	354 (5.9)
5.0	0.896 (2 DD5.1)	5.94	105 (1.7)	191 (3.1)	211 (3.5)	297 (4.9)	383 (6.3)
4.0	0.896 (2 DD5.1)	4.94	126 (2.1)	230 (3.8)	253 (4.2)	357 (5.9)	461 (7.6)
3.0	0.896 (2 DD5.1)	3.94	159 (2.6)	289 (4.8)	318 (5.3)	448 (7.4)	578 (9.6)
2.0	0.192 (1 DD2.0)	2.23	280 (4.6)	510 (8.5)	561 (9.3)	790 (13.1)	1020 (17)
1.86[b]	0.192 (1 DD2.0)	2.09	299 (4.9)	544 (9)	599 (9.9)	843 (14)	1088 (18.1)
1.5[b]	0.192 (1 DD2.0)	1.73	361 (6)	657 (10.9)	723 (12)	1019 (16.9)	1314 (21.9)
1.15[c]	0.224 (1 Layer II)	1.41	443 (7.3)	805 (13.4)	886 (14.7)	1248 (20.8)	1610 (26.8)
1.15[b]	0.064 (1 DD1.0)	1.25	499 (8.3)	908 (15.1)	999 (16.6)	1407 (23.4)	1816 (30.2)
1.0[b]	0.064 (1 DD1.0)	1.10	567 (9.4)	1031 (17.1)	1135 (18.9)	1599 (26.6)	2062 (34.3)
0.7[d]	0.064 (1 MP3)	0.80	779 (12.9)	1416 (23.6)	1558 (25.9)	2195 (36.5)	2832 (47.2)

Note: DD = Dolby Digital.
[a]Total data rate includes four subpicture streams (0.04 Mbps).
[b]MPEG-1 video.
[c]Video and audio rates equivalent to video CD.
[d]MPEG-4 video. Will not play on a standard DVD-Video player.

Figure 6.3
Data rate versus
playing time

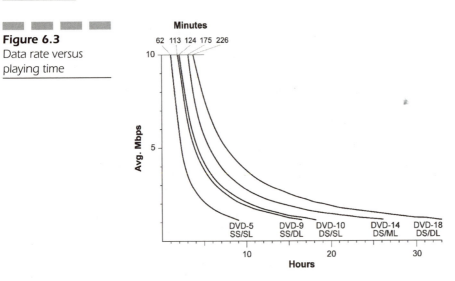

File Format

In order to limit the computing requirements for a home consumer DVD player, additional constraints were applied to the OSTA UDF file format. This constrained format is outlined in Appendix 6.9 of the UDF standard and is commonly referred to as *MicroUDF*. Constraints include the following:

- A DVD player is recommended to support UDF. ISO 9660 eventually will be phased out.
- There should be one logical volume per side, one partition, and one file set.
- Individual files must be less than 1 gigabyte in length ($2^{30} - 1$).
- Each file must be a contiguous extent.
- File and directory names must use only 8 bits per character.
- Only short allocation descriptors are allowed.
- Fields such as OS and implementation use must be zero.
- No symbolic links (aliases) can be used.
- ICB strategy 4 is used.
- No multisession format is allowed.
- No boot descriptor is allowed.

■ A specific directory (VIDEO_TS or AUDIO_TS) must be present and must contain a specific file (VIDEO_TS.IFO or AUDIO_TS.IFO).

These constraints apply only to the directory and files that the DVD player needs to access. Other files and directories can exist on the disc that are not intended for the DVD player and do not meet the preceding listed constraints. Such directories and files must be physically located on the disc following the audio zone and video zone and are ignored by DVD players (unless the DVD player employs extensions to the DVD standard). This allows many kinds of data and applications to coexist with DVD-Video, DVD-Audio, or data.

The DVD-Video specification prescribes a directory and set of files within the UDF file format. All files are stored in one directory (VIDEO_TS). DVD-Audio information is stored in a homologous directory (AUDIO_TS). An optional, root-level directory (JACKET_P) can contain identifying images for the disc in three sizes, including thumbnails for graphic directories of disc collections.

For a DVD-Video volume, the required VIDEO_TS.IFO file contains the video manager title set (main menu) information (VMGI). Additional .IFO files hold other title set information (VTSI), with backup copies in .BUP files. For each video title set on the disc, there can be up to 10 .VOB files that hold one chunk from video object set blocks. Likewise for DVD-Audio volumes, the required AUDIO_TS.IFO file contains the audio manager title set (main menu) information (AMGI). Each audio object set is broken into .AOB files (see Figures 6.4 and 6.5, as well as "Physical Data Structure," later in this chapter).

DVD-Video

DVD-Video provides one stream of MPEG-2 variable-bit-rate video with up to 9 interleaved camera angles; up to 8 streams of Dolby Digital multi-channel audio, MPEG-2 multichannel audio, or linear PCM audio; and up to 32 streams of full-screen graphic overlay subpictures supplemented with navigation menus, still pictures, seamless branching, parental management, and rudimentary interactivity. The DVD-Video format is designed for playback on standard 525-line (NTSC) and 625-line (PAL) television displays via analog video connections. It also can be played in standard resolution in interlaced or progressive format on digital displays and high-definition televisions via analog or digital connections.

Figure 6.4
DVD-Video and DVD-
Audio file structure

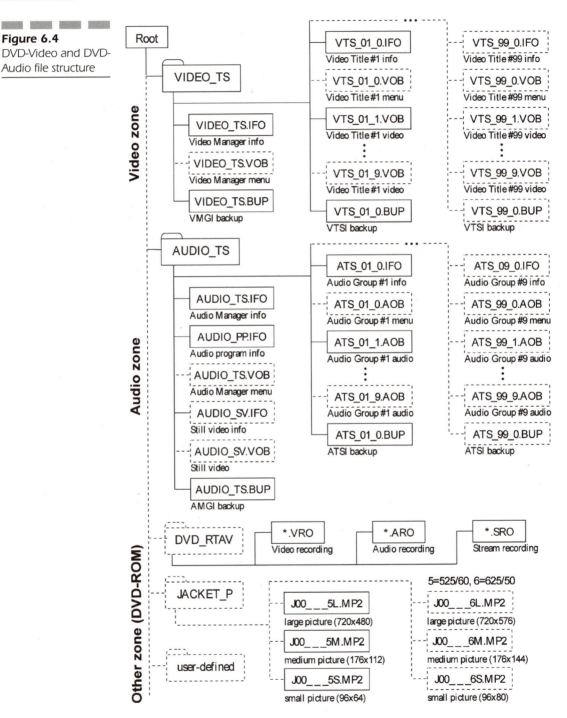

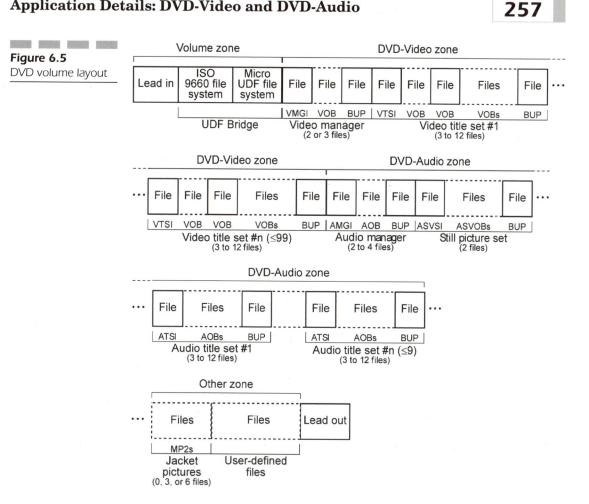

Figure 6.5
DVD volume layout

DVD-Video incorporates multichannel digital audio for compatible digital audio systems, and also provides digital or analog audio to support standard stereo audio systems (see Figure 6.6). DVD-Video also can be played on compatible computer systems (see Chapter 11 for details). DVD-Video incorporates video and audio formats from various standards (see Appendix B).

Navigation and Presentation Overview

DVD players (including software players) are based on a presentation engine and a navigation manager. The *presentation engine* uses the

Figure 6.6
DVD-Video player
block diagram

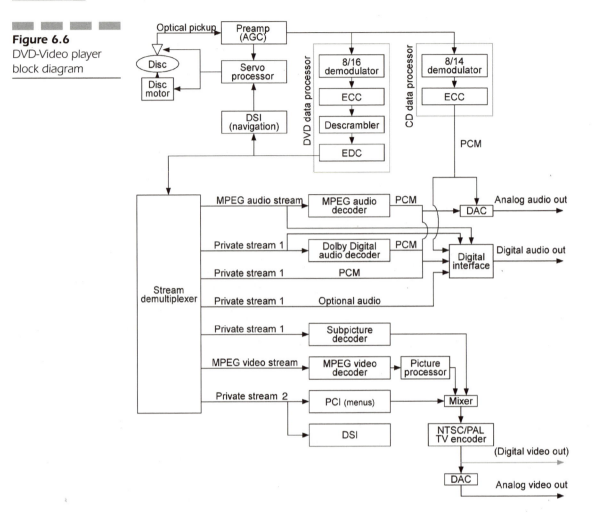

information in the presentation data stream from the disc to control what is shown. The *navigation manager* uses information in the navigation data stream from the disc to provide a user interface, create menus, control branching, and so on. In a general sense, the user input determines the path of the navigation manager, which controls the presentation engine to create the display (see Figure 6.7).

Navigation data includes information and a command set that provides rudimentary interactivity. Menus are present on almost all discs to allow

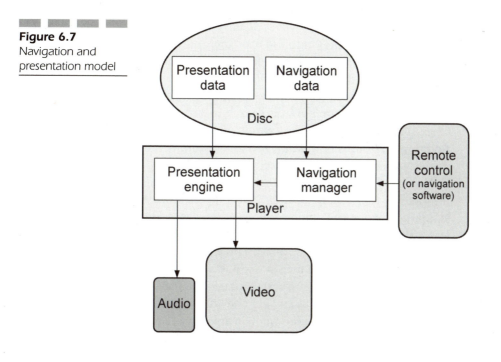

Figure 6.7
Navigation and
presentation model

content selection and feature control. DVD-Video content is broken into titles (movies or other programs) and parts of titles (chapters). For example, a disc containing four television episodes could present each episode as a title. A disc with a movie, supplemental information, and a theatrical preview might be organized into three titles, with the long movie title arrayed in chapters. A disc can have up to 99 titles, but in many cases there will be only one. This would be straightforward if it were not for the industry practice of calling discs titles. For example, *The Matrix* was the first title to ship a million copies. This particular title (disc) has 35 titles (video segments) on it, ranging from a title that holds the entire movie to titles containing short behind-the-scenes clips, logos, and the FBI warning.

Most discs have a main menu for the entire disc, from which titles are selected. Each title or group of titles can have its own menu. Depending on the complexity of the title, additional menus can be provided at any point. Each menu has a still or moving video background and onscreen buttons. It is possible to use menus anywhere on the disc, such as to provide pop-up features over the top of any video. Remote-control units have four directional arrow keys for selecting onscreen buttons, plus numeric keys, a "Select" or "Enter" key, a "Title" or "Top Menu" key, a "Menu" key, and a "Return" key. Standard playback functions are play/pause, fast forward, fast reverse, next, and previous. Additional remote functions may include step, slow, fast, multispeed scan, audio select, subtitle select, camera-angle select, play mode select, search to title, search to part of title (chapter), and search to time. Any of these features can be disabled by the producer of the disc. Additional features of the navigation and command set are covered in the navigation and presentation sections that follow.

Material for camera angles, seamless branching, and parental control is interleaved together in chunks. The player jumps from chunk to chunk, skipping over unused angles, branches, or scenes to stitch together a seamless video presentation. These chunks are not multiplexed as individual streams of video and audio[1], but are interleaved separately (see Figure 6.8). Each video angle has its own set of audio and subpicture tracks, although they are generally the same across all angles. The interleaved chunks have no direct effect on the bit rate, but they do reduce the maximum allowable rate because the track buffer must be able to continue supplying data while intervening chunks are skipped over. Adding one camera angle for the duration of a title roughly doubles the amount of space it requires (and cuts the playing time in half).

DVD application data are organized into a complex structure representing the physical location of the data on the disc. Since data may be shared among different titles and programs, logical data structures are overlaid on the physical structure. The logical structures contain navigation information and determine the presentation order of information, which is independent of the physical order.

Physical Data Structure

The physical data structure determines the way data is organized and placed on the disc.[2] The standard specifies that data must be stored sequentially—physically contiguous—according to the DVD-Video physical struc-

---

[1]Everything turns out to be interleaved if you dissect it far enough. At the MPEG packet level, the different program streams are interleaved, but the player sees them as individual streams coming out of the demultiplexer.

[2]*Physical* in this sense refers to the positional ordering of the data, not to the physical characteristics of the underlying storage medium.

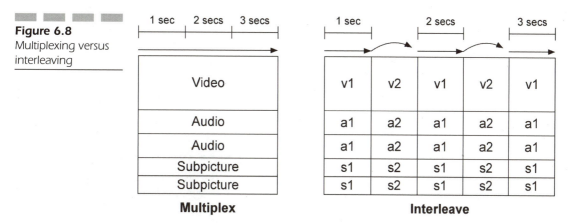

Figure 6.8
Multiplexing versus interleaving

ture. The top line of Figure 6.9 represents the order in which data is stored on the disc. The structure is hierarchical; each block can be broken up into component blocks, which can be further subdivided, as illustrated in Figure 6.9 (also see Table 6.7).

The primary block is a video title set (VTS), which carries internal information about the titles it contains (menu pointers, time maps, cell addresses, etc.), followed by video object sets (VOBSes) shown in Figures 6.10 and 6.11. Since the VTS information applies to all the titles in the set, the titles must contain the same number, format, and order of audio tracks and subpicture tracks. MPEG-1 and MPEG-2 video cannot be mixed within a VOBS. The first video object set may be an optional menu (VTSM), called a *root menu,* followed by video object sets that contain the actual title content.

The video manager (VMG) is a special case of a video title set that optionally can contain a main menu for the disc, called the *title menu* or *top menu.* This is the table of contents for the disc. If this menu is present, it is usually the first thing the viewer sees after inserting the disc. Alternately, there can be a special autoplay piece that automatically begins playback when the disc is inserted.

Data at the VOBS level includes attributes for video, audio, and subpicture (see Tables 6.3-6.5). The language of audio and subpictures can be identified with ISO 639 codes (see Table A.24) and also can be identified with an extension code as commentary, simplified audio, and so forth. Identification codes are stored on the disc as binary representations of two-letter (lowercase) codes. For example, the code for English is *en,* and the code for Zulu is *zu.* Different authoring systems and production tools show these codes in different ways. The ASCII hexadecimal representation for *e* is 65

Figure 6.9
DVD-Video physical
data structure

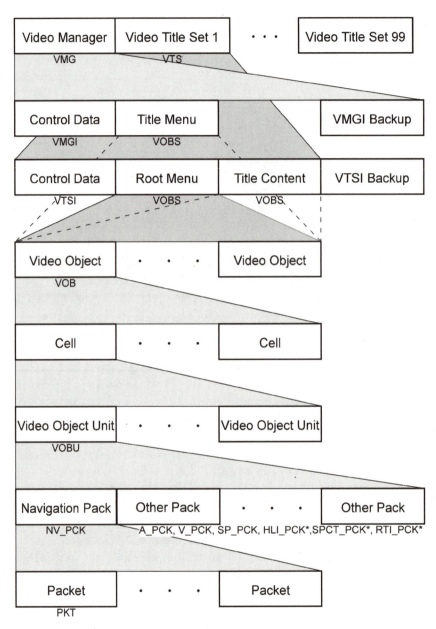

*DVD-Audio only

and for *n* is 6E, so the code might be displayed as 656E or as the decimal equivalent 25966. A few odd systems use an alternate code in which each letter is assigned a decimal value beginning with 1, so *en* would be 0114.

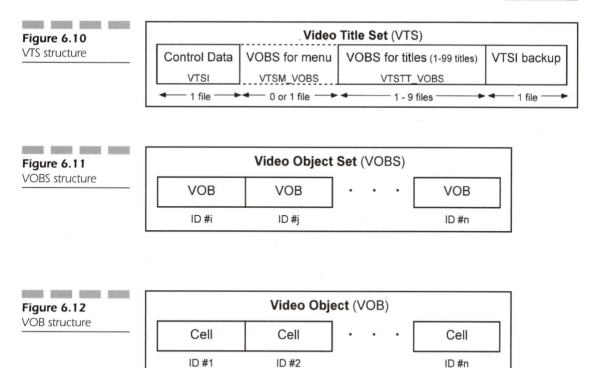

Figure 6.10
VTS structure

Video Title Set (VTS)			
Control Data	VOBS for menu	VOBS for titles (1-99 titles)	VTSI backup
VTSI	VTSM_VOBS	VTSTT_VOBS	
◄— 1 file —►	◄— 0 or 1 file —►	◄— 1 - 9 files —►	◄— 1 file —►

Figure 6.11
VOBS structure

Video Object Set (VOBS)			
VOB	VOB	• • •	VOB
ID #i	ID #j		ID #n

Figure 6.12
VOB structure

Video Object (VOB)			
Cell	Cell	• • •	Cell
ID #1	ID #2		ID #n

Each video object set (VOBS) is composed of one or more video objects (VOBs) (see Figure 6.12). A video object is part or all of an MPEG-2 program stream. The picture resolution and display rate are the same for all video objects in a video object set. Granularity at the VOB level is designed to group or interleave blocks for seamless branching and camera angles. Each VOB set contains one or more VOB blocks. A contiguous block contains one VOB in contiguous sectors on the disc. An interleaved block contains multiple VOBs broken into interleaved units (ILVUs) that are physically interleaved on the disc to enable seamless presentation. The size of the ILVUs is determined by the data rate of the streams, with the size kept small enough that the pickup head has time to jump over the intervening units of other video objects to get the next unit in sequence before the track buffer is depleted.

A VOB is made up of one or more cells (see Figure 6.13). A *cell* is a group of pictures or audio blocks and is the smallest addressable chunk. A cell can be as short as a second or as long as a movie. Some authoring systems call cells *scenes*. Cell IDs are relative to the video object in which they reside, so a cell can be uniquely identified with its cell ID and enclosing video object ID. Note that the player may need to get additional presentation information about the cell from the program chain (PGC, defined later). A cell cannot be spread across both layers of a disc. This rule is also supposed to apply to a pair of cells intended to play seamlessly, but so-called seamless layer changes can be created by violating this rule.

Each cell is further divided into video object units (VOBUs) as shown in Figure 6.14. A VOBU is the smallest unit of playback. Despite its name, a VOBU does not always contain video. A VOBU is an integer number of video fields and is from 0.4 to 1 second long, unless it is the last VOBU of a cell, in which case it can be up to 1.2 seconds long. A VOBU contains zero or more GOPs. A VOBU usually contains one GOP. If this is the case, it must begin with an MPEG sequence header followed by a GOP header followed by an I frame. If the last GOP in the VOBU does not align with the presentation end time of the VOBU, or if the VOBU is followed by a VOBU with no video data, then there must be an MPEG sequence_end code. Only one

Figure 6.13
Cell structure

Cell				
VOBU	VOBU	· · ·	VOBU	

Figure 6.14
VOBU structure

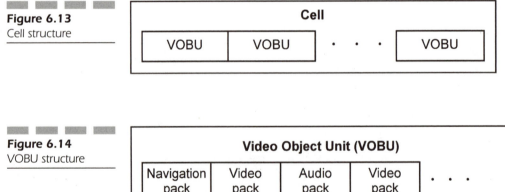

Video Object Unit (VOBU)

Navigation pack	Video pack	Audio pack	Video pack	· · ·
NV_PCK	V_PCK†	A_PCK†	V_PCK	
Required (1 only)	Optional (repeated in any order)			

· · ·	Subpicture pack	Highlight* pack	Still* pack	RT Text* pack
	SP_PCK†	HLI_PCK	SPCT_PCK†	RTI_PCK
	Optional (repeated in any order)			

*DVD-Audio only (Audio Object Unit)
†Encryptable with CSS or CPPM

sequence_end code is allowed per VOBU. Analog protection system (APS) data is stored at the video object unit level and specifies whether analog protection (Macrovision) is off or is one of three types (see Table 6.6). See the copy protection section of Chapter 4 for details.

Finally, at the bottom of the heap, VOBUs are broken into packs of packets (see Figures 6.15 and 6.16). The format of packs and packets is compliant with the MPEG program stream standard. Packs include system clock reference (SCR) information for timing and synchronization. Each packet identifies which stream it belongs to and carries a chunk of data for that stream. Packs are stored in recording order, interleaved according to the different streams that were multiplexed together. Different packs contain data for navigation, video, audio, and subpicture. Additional packs are used by the DVD-Audio and DVD recording formats, as detailed later in this chapter. DVD-Video physical units are outlined in Table 6.7.

TABLE 6.3

VOBS Video Attributes

Compression mode	MPEG-1 or MPEG-2
TV system	525/60 (NTSC) or 625/50 (PAL/SECAM)
Aspect ratio	4:3 or 16:9
Display mode permission[a]	Letterbox, pan and scan, both
4:3 source is letterboxed	Yes or no

[a]When 16:9 aspect ratio is presented on a 4:3 display.

TABLE 6.4

VOBS Audio Attributes

Audio coding mode[a]	AC-3, MPEG-1/MPEG-2, MPEG + extension, LPCM, DTS, or SDDS
DRC (dynamic range)	On, off
Quantization (PCM)	16, 20, or 24 bits
Sampling rate (PCM)	48 or 96 kHz
Number of channels[a]	1 to 8
Film/camera mode (625/50)	Film or camera
Application mode	Surround, karaoke, or unspecified
Code	ISO 639 language code or unspecified
Code extension	Unspecified, caption, for visually impaired, director comments 1, director comments 2

[a]A VOB containing a menu can only use 2-channel PCM audio.

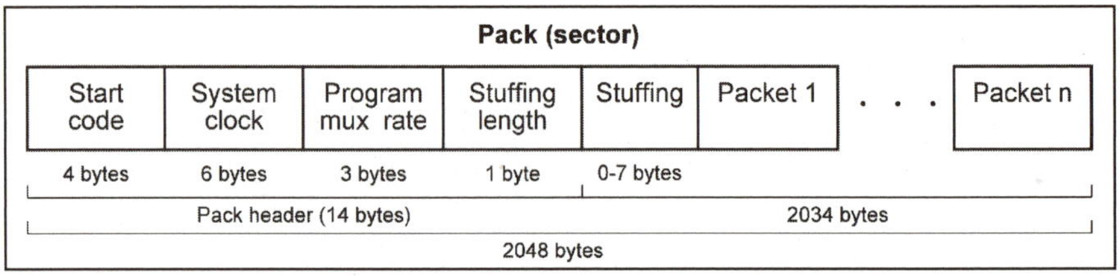

Figure 6.15
Pack structure

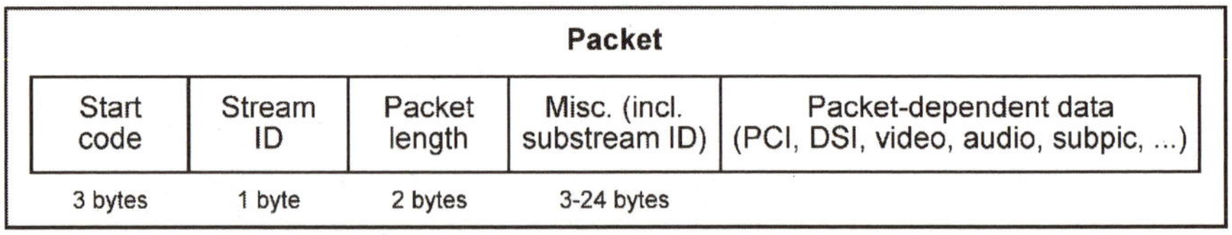

Pack (sector)							
Start code	System clock	Program mux rate	Stuffing length	Stuffing	Packet 1	. . .	Packet n
4 bytes	6 bytes	3 bytes	1 byte	0-7 bytes			

Pack header (14 bytes)   2034 bytes

2048 bytes

Figure 6.16
Packet structure

Packet				
Start code	Stream ID	Packet length	Misc. (incl. substream ID)	Packet-dependent data (PCI, DSI, video, audio, subpic, ...)
3 bytes	1 byte	2 bytes	3-24 bytes	

TABLE 6.5

VOBS Subpicture Attributes

Coding mode	2-bit RL (no other options)
Code	ISO 639 language code or unspecified
Code extension	Unspecified, normal caption, large caption, children's caption, normal closed caption, large closed caption, children's closed caption, forced caption, director caption, large director caption, director caption for children
Channel mixing info	Contents, mixing phase, mixed flag, mix mode

TABLE 6.6

VOBU APS (Macrovision) Attributes

0	Off
1	Pseudosync and AGC
2	Pseudosync and AGC, plus Colorstripe type 1 (inverted split burst on 2 lines of every 17)
3	Pseudosync and AGC, plus Colorstripe type 2 (inverted split burst on 4 lines of every 21)

TABLE 6.7

DVD-Video Physical Units

Unit	Maximum
Video title set (VTS)	99 per disc
Video object set (VOBS)	99 per VTS
Video object (VOB)	32767 per VOBS
Cell	255 per VOB
Video object unit (VOBU)	
Pack (PCK)	2048 bytes
Packet (PKT)	

Domains and Spaces

The concepts of *domains* and *spaces* relate to DVD navigation. Domains and spaces are not physical structures, nor are they logical structures defined by information on the disc. Rather, they are abstract groupings of data structures. Technically, a domain is a set of similar *program chains* (PGCs), although it is better to think of a DVD domain as a state that the player monitors in order to keep track of the type of content being displayed (see Table 6.8). The stop state is not actually a domain, since it is the absence of PGCs, but it is useful to classify it together with the other four domains. Navigation commands are constrained by domains to be appropriate for the current state. For example, a "select button" command is only meaningful if a menu is being displayed, and a fast-forward command makes no sense if the player is stopped or in a menu with a still image. Some authoring systems define subdomains for each menu language in the VMGM and VTSM domains.

Spaces are a way of grouping domains into functionally similar units, as shown in Table 6.9 (also see Table 6.14, which shows how the spaces overlap).

TABLE 6.8

Domains

Domain	Abbreviation	State	Valid commands	Number
First play domain	FP_DOM	Preparing to play	Non-playback commands	None or one
Video manager menu domain	VMGM_DOM	Displaying the title menu	Menu-related commands	None or one for each title menu language
Video title set menu domain	VTSM_DOM	Displaying a root menu or one of its submenus (subpicture, language, audio, or angle)	Menu-related commands	None or one for each menu and each language
Title domain	TT_DOM	Playing video content in a title	Playback commands	One for each title
Stop state		Stopped. The head is retracted from the disc	Play	n/a

TABLE 6.9

Spaces

Space	Scope	Number
System space	Everything other than individual titles FP_DOM + VMGM_DOMs + VTSM_DOMs	None or one
Menu space	All menus VMGM_DOM + VTSM_DOMs	None or one
VMG space	Intro and title menu FP_DOM and VMGM_DOMs	None or one
VTS space	Menus and titles of one VTS VTSM_DOMs + TT_DOMs	One for each VTS

Presentation Data Structure

The presentation data structure is a logical hierarchy overlaid on the physical data structure (see Figure 6.22 and Table 6.10). The presentation data structure determines the grouping of video sequences and the playback order of each block of video in a sequence (see Figures 6.17 and 6.18). The top level comprises titles. Each title contains up to 999 program chains (PGCs). The first PGC in a title is called the *entry PGC*. Linking between PGCs is not seamless. The title menu and each root menu are entry PGCs. A program chain contains 0 to 99 *programs* (PGs), which are groupings of cells. A PGC with no programs (no VOBs), called a *dummy PGC,* contains only navigation commands. The physical data and the logical presentation data structure converge at the cell level. Each PGC contains one *program control block* (PCB), which is an ordered list of pointers to cells that indicates in what order the programs and cells are to be played. Programs within a PCB can be flagged for sequential play, random play (programs are selected randomly and may repeat), or shuffle play (programs are played in random order without repeats). Individual cells may be used by more than one PGC, which is how parental management and seamless branching are accomplished—different program chains define different sequences through mostly the same material.

The presentation data structure includes additional groupings at various levels to provide additional organization (see Figure 6.18). Groupings include parental blocks for parental management, angle blocks for multiple camera angles, and language units for menus in multiple languages.

There is also the part-of-titles (PTT) construct, commonly called a *chapter.* A *part of title* is a marker or branch point, not a container. A PTT marker can only go at the beginning of a program.[3] For a multi-PGC title, the user may take different paths from PGC to PGC, so chapter 2 via one path may be made up of different PGCs than Chapter 2 via another path (see Figure 6.19).

Three types of titles exist: a monolithic title meant to be played straight through (one_sequential_PGC title), a title with multiple PGCs for varying program flow (multi_PGC title), and a title with multiple PGCs that are automatically selected according to the parental restriction setting of the player (parental_block title) (see Figures 6.20 and 6.21). One_sequential_ PGC titles are the only kind that have time maps for timecode display and searching.

Figure 6.22 shows the relationship between presentation data and physical data.

Navigation Data

Navigation data is a collection of information that determines how the physical data is accessed. In a sense, navigation data is built on top of presentation data. Navigation data controls access and interactive playback. It is grouped into four categories: control, search, user interface, and navigation commands (see Figure 6.23). There are five levels at which navigation data exists: *video manager information* (VMGI), stored in the VMG, which controls the video title sets and the title menu; *video title-set information* (VTSI), stored in each VTS, which controls the titles and menus in a VTS; *program chain information* (PGCI), stored in the PGC, which controls access to components of a PGC such as audio and subpicture streams; *presentation control information* (PCI), stored in packets spread throughout the data stream (one per VOBU), which controls menu display and program presentation in real time; and *data search information* (DSI), also stored in packets spread throughout the data stream (one per VOBU), which controls forward/reverse scanning and seamless branching (see Table 6.11).

Previous, next, and return (go up) PGC information is stored in the PGCI. The return link corresponds to the "Return" button on the remote control. The "Previous" (I◄◄) and "Next" (►►I) buttons on the remote control jump between programs until there is no previous or next program, in

---

[3]In general, because chapters must start on a program boundary, a chapter is equivalent to a program. Specifically, every chapter is a program, but not every program is a chapter. The "Next" and "Previous" keys on remote controls jump between programs, not chapters.

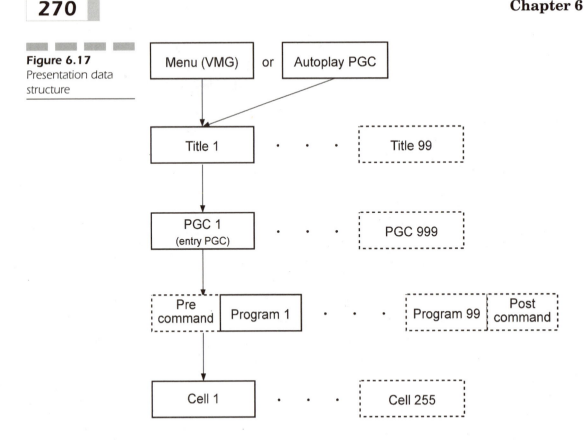

Figure 6.17
Presentation data
structure

which case they follow the previous and next links (if present) in the PGCI (see Figure 12.4 for an example remote control).

Control Information Control information describes the data. This includes characteristics such as format (525/60 or 625/50), aspect ratio (4:3 or 16:9), language, audio and subpicture selection, and moral codes for parental management. These are described elsewhere in more detail.

Search Data Search data defines a navigational structure that can be steered through with the remote control or by program control. There are seven search types, described in Table 6.12, which are associated with a key or combination of keys on the remote control. These searches are associated with a PGC; the search map is stored in the PGCI. Additional search types are provided for parts of titles (chapters), time, angles, and VOBUs (for trick play modes such as slow and fast).

Figure 6.18
Presentation data
structure groupings

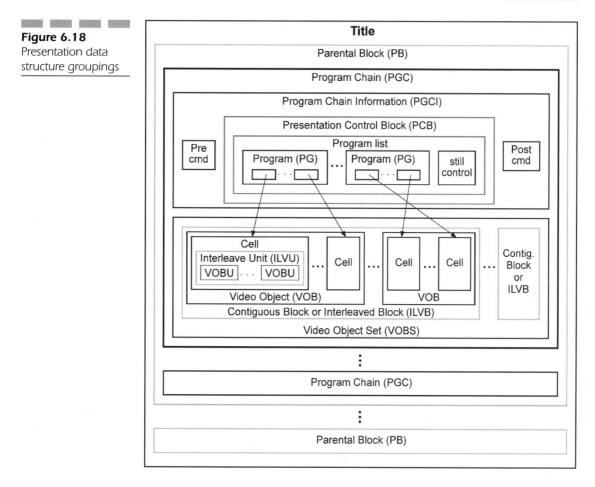

Figure 6.19
Example of part-of-
title markers

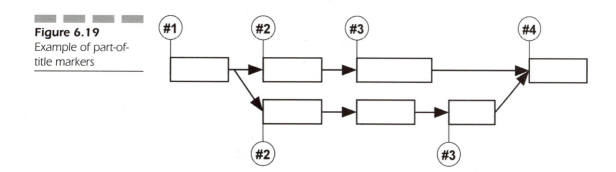

Figure 6.20
Title structures

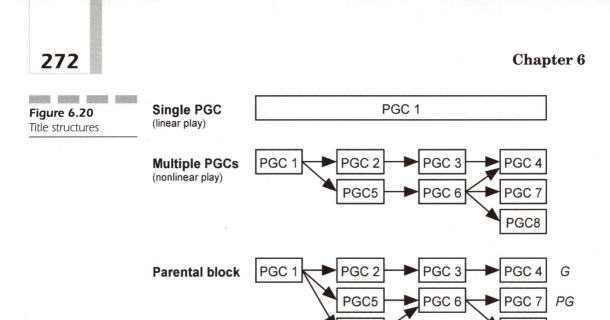

Single PGC
(linear play)

Multiple PGCs
(nonlinear play)

Parental block

Figure 6.21
Example presentation
structures

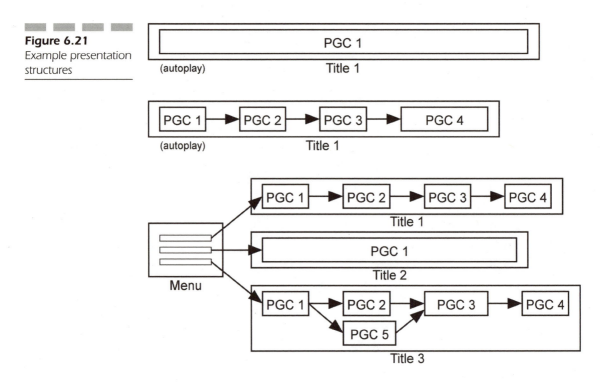

	Physical data	Logical data (presentation)

Figure 6.22
Relationship of
presentation data to
physical data

Physical data

Video Title Set (VTS)	**Title** (TT)
Video Object Set (VOBS)	Program Chain (PGC)
Video Object (VOB) ...	Program (PG)
Cell	Cell (cell pointer)
Video Object Unit (VOBU)	

TABLE 6.10

DVD-Video Logical Units

Unit	Maximum
Title	99 per disc
Parental block (PB)	
Program chain (PGC)	999 per title, 16 per parental block
Part of title (PTT)	999 per title, 99 per one-sequential-PGC title
Program (PG)	99 per PGC
Angle block (AB)	
Interleave block (ILVB)	
Interleave unit (ILVU)	
Cell pointer	255 per PGC

Figure 6.23
Navigation data
structure

Control
- Format (NTSC/PAL)
- Language
- Audio selection
- Subpicture selection
- Parental management
- Karaoke
- Display mode and aspect

Navigation
- General parameters
- System parameters
- Navigation timer
- Buttons

Search
- PGCI search (jump to menu)
  - Title (disc menu; VMG)
  - Root (local menu; VTS)
  - Audio (audio submenu)
  - Subpicture (subpicture submenu)
  - Angle (angle submenu)
  - Part of Title (chapter submenu)
- Presentation data search
  - Title
  - Part of Title (chapter)
  - Program (next/previous)
  - Time
  - Angle
  - VOBU (trick play)

If no search data is present for a particular search type, the player is unable to provide that function. It is possible to create a disc that plays from beginning to end and cannot be interrupted by the user (other than by ejecting the disc).

Navigation Commands and Parameters Commands provide the interactive features of DVD. Each PGC optionally can begin with a set of precommands, followed by cells that can each have one optional command, followed by an optional set of postcommands. In total, a PGC cannot have more than 128 commands, but since commands are stored in a table at the beginning of the PGC and referenced by number, they can be reused many times within the PGC. Cell commands are executed after the cell is presented. PGC commands are stored in the PGCI. Each menu button also can contain a single command, stored in the PCI (see "Buttons" later in this chapter).

The commands are similar to computer CPU instruction words and are in fact instructions to the processing unit of the player. Each command is up to 8 bytes long and can consist of one, two, or three instructions. Instructions include

- Math operations: add, subtract, multiply, divide, modulo, random

- Logical (bitwise) operations: AND, OR, XOR

- Comparisons: equal, not equal, greater than, greater than or equal to, less than, less than or equal to

- Register operations: load, move, swap

TABLE 6.11

Navigation Data

Name	Abbreviation	Important Content
Video manager information	VMGI	Number and attributes of title sets (VTS_ATRT); pointers to titles (VTS_PTT_SRPT); parental management table (PTL_MAIT); text data (TXTDT_MG); attributes of title menu (VMGM) video stream, audio streams, and subpictures; title menu (VMGM) cell pointers and VOBU maps
Video title set information	VTSI	Pointers to chapters (TT_SRPT); pointers to program chains; time maps; attributes of root menu (VTSM) video stream, audio streams, and subpictures; root menu (VTSM) cell pointers and VOBU maps; video title set (VTS) cell pointers and VOBU maps
Program chain information	PGCI	Number and length of programs; permitted user operations (UOPs); links between program chains (previous, next, return/go up); playback mode (sequential, random, shuffle); PGC still time; commands (pre, cell, and post); program maps and cell pointers; cell still times
Presentation control information	PCI	For each VOBU: APS, user operations (UOPs), nonseamless angle jump pointers, button information (rectangle, color, highlight, directional links, associated aspect ratio, command, and so forth), presentation times, ISRC for video/audio/subpicture streams
Data search information	DSI	For each VOBU: Reference picture pointers (I or P frames), link to next interleaved unit (ILVU), seamless angle jump pointers, presentation times, audio gap lengths, VOBU pointers for forward/reverse scanning, video synchronization pointers to audio and subpicture packs

- Program flow control: goto, break
- Video presentation control: link, jump, call, resume, exit

See Table 6.13 for a complete list of commands.

Video presentation control commands (the link group and the jump group) are limited by the current domain and the destination domain. Table 6.14 shows the commands necessary to move from one domain to another. Of course, there are myriad restrictions and gotchas. For example, a PGC cannot link directly to a PGC in a different domain, and one VTS cannot

TABLE 6.12

Search Types

Search Type	Example Keypresses (Remote Control)
PGCI search	
Title menu (top menu; VMG)	Top (or Title)
Root (menu; VTS)	Menu
Audio (title audio menu)	Audio+Menu
Subpicture (title subpicture menu)	Subtitle+Menu
Angle (title angle menu)	Angle+Menu
Part of title (title chapter menu)	Chapter+Menu
Data search	
Title	Search+Title+2+Enter
Part of title (chapter search)	Search+Chapter+8+Enter
Time	Search+Time+1+2+0+0+Enter
Angle	Angle

link directly to another VTS. Many discs are authored with dummy PGCs in the VMGM that serve as switching points between PGCs and VTSs. Presentation control with cell commands is not seamless. Branching within a title can be done with link commands. The JumpVTS_PTT command is different from LinkPTTN in that it always executes the precommands. LinkPG... commands are limited to the current PGC, except that LinkPrevPG can go to the beginning of the previous PGC.

There are 24 system-use 16-bit registers (called *system parameters,* or SPRMs; affectionately known as "sperms") that hold information such as language code, audio and subpicture settings, and parental level (see Table 6.15). Some SPRMs can only be used to determine the state of the player (read-only), whereas others can be set by commands (read/write), allowing programs to control presentation of audio, video, subpicture, camera angles, and so on. SPRMs cannot be set directly. They can only be set by special commands such as SetNVTMR that sets both SPRM 9 (countdown timer) and SPRM 10 (timer jump destination) or by commands that move to a different place on the disc, potentially causing SPRM 4 (title number in volume), SPRM 5 (title number in VTS), SPRM 6 (PCG), and SPRM 7 (chapter) to change. SPRMs 11 through 20 are called *player parameters* because

TABLE 6.13

Navigation
Commands

GoTo group	*Commands for controlling program flow within a pre or post sequence*	
GoTo	Jump to a specified command number	
Break	Stop executing commands in pre or post section	
NOP	No operation (do nothing)	
SetTmpPML	Ask user to confirm temporary parental level change. If ok, change level and go to specified command	
Link group	*Commands for moving within the current domain*	
LinkPGCN	Start at program chain number	
LinkPTTN	Start at part-of-title (chapter) number	
LinkPGN	Start at program number	
LinkCN	Start at cell number	
LinkTopPGC	Restart current PGC	
LinkTailPGC	Go to end of current PGC (execute postcommands)	
LinkPrevPGC	Start at previous PGC	
LinkNextPGC	Start at subsequent PGC	
LinkGoUpPGC	Start at higher PGC	
LinkPGCN	Start at specific PGC number	
LinkPTTN	Start at specific chapter number	
LinkTopPG	Restart current program	
LinkPrevPG	Start at previous program	
LinkNextPG	Start at next program	
LinkPGN	Start at specific program number	
RSM	Resume at location where playback was suspended by CallSS or MenuCall user operation. (Destination not limited to current domain.)	
Jump group	*Commands for moving out of the current domain*	
JumpTT	Start at title number (from VMG)	
JumpVTS_TT	Start at title number (in same VTS)	
CallSS	Start at a menu PGC number in system space,[a] saving resume state	
JumpSS	Start at PGC number in system space[1] (from system space)	
JumpVTS_PTT	Start at part-of-title (chapter) number in title number (in same VTS)	
Exit	Stop (enter stop state)	
Compare group	*Commands for comparing values and parameters*	
BC	Compare bitwise (logical and)	
EQ	Test if equal	
NE	Test if not equal	
GE	Test if greater than or equal	
GT	Test if greater than	
LE	Test if less than or equal	
LT	Test if less than	

continues

TABLE 6.13 cont.

Navigation Commands

SetSystem group	Commands for setting system parameters and general parameters
SetSTN	Set audio, subpicture, or angle number (SPRM 1, 2, and 3)
SetNVTMR	Set navigation countdown timer (SPRM 9 and 10)
SetHL_BTNN	Set selected button (SPRM 8)
SetAMXMD	Set karaoke audio mixing mode (SPRM 11)
SetGPRMMD	Set general parameter value and mode (register or counter) (GPRM 0 to 15)
Set group	Commands for setting and manipulating values in general parameters
Mov	Set GPRM value (from a constant, GPRM, or SPRM)
Swp	Exchange the values in two GPRMs
Add	Add a value (constant or GPRM) to a GPRM
Sub	Subtract a value (constant or GPRM) from a GPRM
Mul	Multiply a GPRM by a value (constant or GPRM)
Div	Divide a GPRM by a value (constant or GPRM)
Mod	Take the remainder of dividing a GPRM by a value (constant or GPRM)
Rnd	Set GPRM to a random number between 1 and a value (constant or GPRM)
And	Take the bitwise product of a GPRM and a value (constant, GPRM, or SPRM)
Or	Take the bitwise sum of a GPRM and a value (constant, GPRM, or SPRM)
Xor	Exclusive or a GPRM with a value (constant, GPRM, or SPRM)

[a]The command can jump to the first play PGC, the entry PGC (or entry PGC block) of a menu (in VMGM or VTSM domain) or a title (in VTS domain), or any PGC in the VMGM.

they reflect the settings of the player, although SPRM 0 ought to be in the same group.

There are 16 general-use 16-bit registers (called *general parameters,* or GPRMs; also known as "germs") that can be used by on-disc programs for such things as keeping score, storing viewer responses, or tracking what sections of the disc have been seen. Each parameter holds an unsigned integer, with a value from 0 to 65,535, but with clever programming, any number of smaller values can be combined in a single register; up to 16 one-bit flags. GPRMs also can be used in counter mode, where the value increases by one each second. A limitation is that all GPRMs are cleared when a title search or title play command is used, when the stop (or eject) command is used, and when the player is turned off.

Summary of Data Structures

To recap, building up in order from the lowest level (see Figure 6.18): video, audio, subpicture, and presentation/control information is divided up and interleaved into packets. Each chunk of content, usually 0.5 seconds long, is organized as an MPEG *group of pictures* (GOP).[4] GOPs are arranged in a *video object unit* (VOBU), with usually one GOP per VOBU. VOBUs are collected into cells. A sequence of cells and cell commands forms a program, whose audio/video content is stored in a *video object* (VOB). A program usually corresponds to one scene. Simple programs are usually held in a single cell. Chapter (part-of-title, PTT) markers are added to create access points, usually at program and cell boundaries. Programs are linked in order into a *presentation control block* (PCB). The PCB and additional command and *control information* (PGCI) make up a *program chain* (PGC). The audio/video content of a PGC is stored in a *video object set* (VOBS). PGCs contain video for the first play sequence, menus, and titles. PGCs can be grouped into logical parental blocks. PGCs are grouped conceptually into

TABLE 6.14

Domain Transitions

From	To			
	System Space			
		Menu Space		
	VMG Space		**VTS Space**	
	FP_DOM	**VMGM_DOM**	**VTSM_DOM**	**TT_DOM**
FP_DOM	n/a	JumpSS	JumpSS	JumpTT
VMGM_DOM	JumpSS	n/a	JumpSS	JumpTT, Link (GoUpPGC or RSM)
VTSM_DOM	JumpSS	JumpSS	JumpSS	JumpVTS_TT, JumpVTS_PTT, Link (GoUpPGC or RSM)
TT_DOM	CallSS	CallSS	CallSS	JumpVTS_TT, JumpVTS_PTT

[4]GOP headers are optional in MPEG-2 MP@ML. Entry points occur at MPEG-2 sequence headers.

TABLE 6.15

Player System
Parameters (SPRMs)

No.	Description	Access	Values	Default Value
0	Preferred menu language	Read-only	Two lowercase ASCII letters (ISO 639)	Player-specific
1	Audio stream number	Read/write	0–7 or 15 (none)	15 (Fh)
2	Subpicture stream number and on/off state	Read/write	b0–b5: 0–31 or 62 (none) or 63 (forced subpicture) b6: display flag (0 = do not display)	62 (3Eh)
3	Angle number	Read/write	1–9	1
4	Title number in volume	Read/write	1–99	1
5	Title number in VTS	Read/write	1–99	1
6	PGC number	Read/write	1–32,767	Undefined
7	Part of title number	Read/write	1–99	1
8	Highlighted button number	Read/write	1–36	1
9	Navigation timer	Read-only[a]	0–65,536 (seconds)	0
10	PGC jump for navigation timer	Read-only[a]	1–32,767 (PGC in current title)	Undefined
11	Karaoke audio mixing mode	Read/write	b2: mix ch2 to ch1 (0 = do not mix) b3: mix ch3 to ch1 b4: mix ch4 to ch1 b10: mix ch2 to ch0 b11: mix ch3 to ch0 b12: mix ch4 to ch0	0
12	Parental management country code	Read-only	Two uppercase ASCII letters (ISO 3166) or 65,535 (none)	Player-specific
13	Parental level	Read/write	1–8 or 15 (none)	Player-specific
14	Video preference and current mode	Read-only	b10–b11: preferred display aspect ratio 0 (00b): 4 × 3 2 (01b): not specified 3 (10b): reserved 4 (11b): 16 × 9 b8–b9: current video output mode 0 (00b): normal (4:3) or wide (16:9) 1 (01b): pan-scan (4:3)	Player-specific

continues

TABLE 6.15 cont.

Player System Parameters (SPRMs)

No.	Description	Access	Values	Default Value
			2 (10b): letterbox (4:3) 3 (11b): reserved	
15	Player audio capabilities	Read-only	b2: SDDS karaoke (0 = cannot play) b3: DTS karaoke b4: MPEG karaoke b6: Dolby Digital karaoke b7: PCM karaoke b10: SDDS b11: DTS b12: MPEG b14: Dolby Digital	Player-specific
16	Preferred audio language	Read-only	Two lowercase ASCII letters (ISO 639) or 65,535 (none)	65,535 (FFFFh)
17	Preferred audio language extension	Read-only	0 = not specified 1 = normal audio 2 = audio for visually impaired 3 = director comments 4 = alternate director comments	0
18	Preferred subpicture language	Read-only	Two lowercase ASCII letters (ISO 639) or 65,535 (none)	65,535 (FFFFh)
19	Preferred subpicture language extension		0 = not specified 1 = normal subtitles 2 = large subtitles 3 = subtitles for children 5 = normal captions 6 = large captions 7 = captions for children 9 = forced subtitles 13 = director comments 14 = large director comments 15 = director comments for children	0
20	Player region code (mask)	Read-only	One bit set for corresponding region (00000001 = region 1, 00000010 = region 2, etc.)	Player-specific
21	Reserved			
22	Reserved			
23	Reserved for extended playback mode			

[a]Bits within the word are referred to as b0 (low order bit) through b15 (high order bit).

domains with related navigational features. Domains are grouped conceptually into spaces. Similar titles are grouped logically along with their menus into a *video title set* (VTS).

Menus

Each disc can have one main menu, called the *title* (or *top*) *menu*. The title menu is optional. There also can be additional menus, called *root menus*. A disc may have no root menus, or it may have hundreds. Menus can be in a simple flat structure, organized into a hierarchy, or linked willy-nilly into any tortuous structure the author desires. There is nothing in the DVD specification to support hierarchies of menus and submenus—they are created only by linking buttons to menus and assigning destinations for MenuCall and GoUp commands.

There are four other kinds of menus that are essentially submenus of a root menu: part-of-title (chapter) menu, audio menu, angle menu, and subpicture menu. These menus reside in a VTS with the root menu. Most players do not provide a way to access these submenus directly, so many discs include buttons on the root menu that go to the submenus, when they exist. For example, if a title has multiple audio language tracks and includes an audio menu to select one, the root menu usually includes a button to get to the audio menu. A disc can be authored to automatically show a menu for available angles when it enters an angle block.

The buttons and selection highlights used to create menus are actually subpictures, which can be put anywhere in any video program. In-play menus offer countless possibilities for interactive video because they can appear, disappear, and change during playback. The reason for a designated title menu (VMGM) and root menus (VTSMs) is to provide menu access with the press of a single remote control key.

DVD menu nomenclature is confusing, to say the least. The title menu really should be called the *disc menu* or *top menu* because it is the top-level menu for the disc, not the menu for a title. Root menus (which are not at the root) are the menus for various titles and title sets. Pressing the "Top" ("Title") key on a remote control accesses the main menu, and pressing the "Menu" key accesses the root menu of the current title.[5] When the "Top" ("Title") or "Menu" key is pressed a second time (after going to the menu), playback resumes at the point where it was interrupted. There is also a "Return" or "Go Up" button that—sometimes—

returns from a submenu to its parent menu or to the current title if the menu is a top-level menu. Some remotes or software players use other labels such as "Guide" or "Setup" for the "Top" ("Title") key and "Root" or "Digest" for the "Menu" key. To make things worse, many Hollywood movies do not have a title menu, so the "Top" ("Title") key has no effect, and the "Menu" key acts like the "Top" ("Title") key. Confused? So are the owners of most DVD players!

The DVD Forum issued a recommendation for standard remote control key names in hopes of clarifying things.[6] See Table 6.16 for a summary. More consistent use of menus by disc producers also will help a great deal. Chapter 12 discusses consistency of menu navigation design.

TABLE 6.16

Remote Control Labeling

Internal Name	Technical Name	Recommended Key Label	Other Labels in Use
Title menu	VMGM (video manager menu)	Top Menu or Top	Title, Title Menu, Guide, Info, Setup
Root menu	VTSM (video title set menu)	Menu	Menu, Root, Root Menu, Digest
GoUp	MenuCall (GoUp)	Return	Back, Previous

[5]Some Pioneer remote controls require two keypresses to get to the title menu—the "Title" key and then the "Menu" key—unless the disc is stopped, in which case only the "Title" key is needed. The "Menu" key works with a single press, except that when the disc is stopped it goes to a player-generated menu that is neither the title menu nor the root menu.

[6]Unfortunately, in typical style of the DVD specification, the "clarification" is almost impenetrable. Here is an example, complete with typos, of a definition of button behavior; one of 18 paragraphs defining four rules for the "Top Menu" and "Menu" buttons: "When [MEMU] is pressed, Menu_Call() (the transition to Root menu) is executed and then the function of [TOP MENU] is changed to Menu_Call() and the function of [MENU] is changed to RSM. In case that no entry of Root menu exists, the transition to Root menu is not actually executed although Menu_Call() is executed."

Buttons

Information to create onscreen buttons is included in the navigation data. Up to 36 highlightable, rectangular buttons can be positioned on the screen. In the case of widescreen content (with anamorphic, automatic letterbox, or automatic pan and scan modes), only 18 buttons are allowed per screen when two modes are used, and only 12 buttons are allowed per screen when all three modes are used. In this case, a separate set of buttons for each display mode is required so that the highlights are drawn in the correct area of the modified picture. Button locations are relative to the final displayed picture, after possible formatting with pan and scan or letterbox, not to the anamorphic picture stored in the video stream. For example, many movie discs use a single, widescreen picture for the menu, but also enable pan and scan mode to crop to the center of the menu for 1.33 televisions. Since the picture is cropped, there must be a second set of buttons with different screen positions.

The display and highlighting of menu buttons are done with subpictures. The foreground pixel types are used to draw the buttons on the video background (which can be still or moving). The background pixel type is invisible so that the video will show through. Invisible buttons can be created by setting the pixel contrast to 0. In this case, there are usually high-quality buttons rendered into the menu video instead of low-quality subpicture graphics, and subpictures are used only to create the highlight art that indicates which button is selected.

The arrow keys on the remote control (see Figure 12.4) or software player interface are used to highlight buttons by jumping from one to another. Each button includes four directional links that determine which button on the screen is selected when the corresponding arrow keys are pressed. This creates a complex web of links between buttons that may or may not correspond to their physical arrangement. When a button is selected (highlighted), its color and contrast values (four each) are changed to those defined for the *select* state. The selected button is activated by pressing the "Enter" or "Select" key on the remote control. Alternatively, a button can be activated by pressing the corresponding number keys on the remote control. Some remotes activate buttons 1 through 9 with a single keypress; others require multiple keypresses. When the button is activated, its pixels are momentarily displayed in a new set of color and contrast values defined for the *action* state. See the section on subpictures for details about subpicture pixels, colors, and contrasts.

Each button has one command associated with it. This is generally a flow-control command that links to a title or a PGC. In many cases, a button is linked to a dummy PGC that contains no programs (physically does not contain any VOBs) but is just a set of precommands and postcommands, usually culminating in a jump to a title or PGC with video in it. PGCs can be linked together, allowing a button to trigger an arbitrarily large sequence of commands.

A button can be set for *auto action,* which means that selecting (highlighting) it will activate it immediately. This is useful to create menus where things change when buttons are selected or for creating menu navigation implementations where merely pressing a directional arrow will move to a new place, without requiring that the "Enter" key be pressed.

Each menu can have one button designated for *forced selection* and one button designated for *forced activation.* These can be set to occur after a specific amount of time. This feature is commonly called *idle out,* where something happens if the user does nothing for a certain period of time. The time is specified in increments of 1/90,000 of a second, up to about 795 minutes (2^{32}/90,000).

Stills

Still frames (encoded as MPEG-2 I frames) are supported and can be displayed indefinitely. Still frames can be accompanied by audio for a slideshow type of presentation. The DVD-Video presentation format allows automatic freeze frames at the end of any video segment (a PGC, a cell, or a VOBU).

Most still images on DVDs are authored as menus, and in fact, this is a good way to put still images on a disc. Each still video image or graphic is encoded as a single MPEG video frame. Hundreds or thousands of still images can be linked together with small or invisible menu buttons.

There are three types of stills: a PGC still, a cell still, and a VOBU still. Each has about the same functionality, with the primary difference being the way it is intended to be used and the location where the still information is stored. In all cases, the still occurs at the last PTM of a VOBU. A VOBU still causes a still at the end of a specific VOBU, a cell still causes a still at the last VOBU of the cell, and a PGC still causes a still at the last VOBU of the PGC. VOBU stills and cell stills occur before the cell command is executed. PGC stills, which can be used in random or shuffle PGCs but not sequential PGCs, occur after cell commands and PGC looping but before the postcommands. During a still, the navigation countdown timer and any GPRMs in counter mode continue to count. PGC and cell stills can be held indefinitely or from 1 to 254 seconds. VOBU stills are always indefinite.

Stills are a handy way to show nonmotion video (such as a logo) for a period of time without wasting space encoding hundreds of frames of the same image.

The still feature is not the same as the pause or freeze-frame feature. Stills are implemented automatically by the navigation system, whereas pauses are the result of the viewer pressing the "Pause" key. Depending on player design, the user may be able to continue past a still by pressing "Play," "Pause," or "Next."

User Operations

User operations are the low-level functions defined for a player. A player can implement any sort of remote control design or user interface, but it must then translate all user input into one of the defined user operations. Some operations are mandatory for all players, whereas others are optional (see Table 6.17). Since user operations are defined in Annex J of the DVD specification, they are sometimes referred to as *annex J operations*.

User operation controls (UOPs) are flags that the DVD author sets to restrict a viewer's navigation options at any particular place on the disc. For example, many discs do not allow the viewer to fast forward or jump to a menu at the beginning of the disc (for example, the FBI warning). Each disc can enable or disable a user operation at any point, even if the operation would otherwise be valid within the domain. For example, a disc may be authored to disallow fast forwarding in certain places or to prevent jumping to a menu once a title begins playing.

User operation controls are implemented as 1-bit flags, each identified with a bit number from 0 to 25 (see Table 6.18). These are identified as UOP0, UOP1, and so on. If the bit is set, the operation is prohibited. Some UOP flags control more than one user operation. User operation controls can be placed at the title level, PGC level, and VOBU (PCI) level. This creates nested scopes where lower-level prohibitions take precedence. In other words, if a user operation is not prohibited, it can be prohibited at lower levels, but once a user operation is prohibited, it cannot be "unprohibited" by UOPs at a lower level.

Video

See Table 6.19 for a summary of DVD-Video format details.

TABLE 6.17

User Operations

User Operation	Equivalent Navigation Command	Control	Mandatory in in Player
Title play	JumpTT	UOP2	Mandatory
Time play	None	UOP0	
Time search	None	UOP0, UOP5	
PTT play	JumpVTS_PTT	UOP1	
PTT search	LinkPTTN	UOP1, UOP5	
Stop	Exit	UOP3	Mandatory
GoUp	LinkGoUpPGC	UOP4	Mandatory
PrevPG search	LinkPrevPG	UOP6	Mandatory
Top PG search	LinkTopPG	UOP6	
NextPG search	LinkNextPG	UOP7	Mandatory
Forward scan	None	UOP8	
Backward scan	None	UOP9	
Menu call (Title)	JumpSS or CallSS	UOP10	Mandatory
Menu call (Root)	JumpSS or CallSS	UOP11	Mandatory
Menu call (Subpicture)	JumpSS or CallSS	UOP12	
Menu call (Audio)	JumpSS or CallSS	UOP13	
Menu call (Angle)	JumpSS or CallSS	UOP14	
Menu call (PTT)	JumpSS or CallSS	UOP15	
Resume	RSM	UOP16	Mandatory
Upper button select	SetHL_BTNN	UOP17	Mandatory
Lower button select	SetHL_BTNN	UOP17	Mandatory
Left button select	SetHL_BTNN	UOP17	Mandatory
Right button select	SetHL_BTNN	UOP17	Mandatory
Button activate	None	UOP17	Mandatory
Button select and activate	None	UOP17	
Still off	None	UOP18	Mandatory
Pause on	None	UOP19	
Pause off	None	None	
Menu language select	None	None	Mandatory
Audio stream change	SetSTN	UOP20	Mandatory
SP stream change	SetSTN	UOP21	Mandatory
Angle change	SetSTN	UOP22	Mandatory
Parental level select	None	None	
Parental country select	None	None	
Karaoke audio presentation mode change	SetAMXMD	UOP23	
Video presentation mode change	None	UOP24	Mandatory

TABLE 6.18

User Operation
Control

User Operation	Bit	Controlled in		
		Title	**PGC**	**VOBU**
Time play or seach	UOP0	√	√	
PTT play or search	UOP1	√	√	
Title play	UOP2		√	
Stop	UOP3		√	√
GoUp	UOP4			√
Time or PTT search	UOP5		√	√
TopPG or PrevPG search	UOP6		√	√
NextPG search	UOP7		√	√
Forward scan	UOP8		√	√
Backward scan	UOP9		√	√
Menu call (Title)	UOP10		√	√
Menu call (Root)	UOP11		√	√
Menu call (Subpicture)	UOP12		√	√
Menu call (Audio)	UOP13		√	√
Menu call (Angle)	UOP14		√	√
Menu call (PTT)	UOP15		√	√
Resume	UOP16		√	√
Button select or activate	UOP17		√	
Still off	UOP18		√	√
Pause on	UOP19		√	√
Audio stream change	UOP20		√	√
SP stream change	UOP21		√	√
Angle change	UOP22		√	√
Karaoke audio presentation mode change	UOP23		√	√
Video presentation mode change	UOP24		√	√

TABLE 6.19

DVD-Video Format

Multiplexed data rate	Up to 10.08 Mbps
Video data	One stream
Video data rate	Up to 9.8 Mbps (typical avg. 3.56)
TV system	525/60 (NTSC) or 625/50 (PAL)
Video coding	MPEG-2 MP@ML/SP@ML VBR/CBR or MPEG-1 VBR/CBR
Coded frame rate	24 fps[a] (film), 29.97 fps[b] (525/60), 25 fps[b] (625/50)
Display frame rate	29.97 fps[b] (525/60), 25 fps[b] (625/50)
MPEG-2 resolution	720 × 480, 704 × 480, 352 × 480 (525/60); 720 × 576, 704 × 576, 352 × 576 (625/50)
MPEG-1 resolution	352 × 240 (525/60); 352 × 288 (625/50)
MPEG-2 GOP max.	36 fields (525/60), 30 fields (625/50)
MPEG-1 GOP max.	18 frames (525/60), 15 frames (625/50)
Aspect ratio	4:3 or 16:9 anamorphic[c]
Pixel aspect ratio	Refer to Table 6.17

[a]Progressive (decoder performs 3-2 or 2-2 pulldown).

[b]Interlaced (59.94 fields per second or 50 fields per second).

[c]Anamorphic only allowed for 720 and 704 resolutions.

Video Stream DVD-Video is based on a subset of MPEG-2 (ISO/IEC 13818) Main Profile at Main Level (MP@ML) or Simple Profile at Main Level (SP@ML). Constant and variable bit rates (CBR and VBR) are supported.

DVD adds additional restrictions, which are detailed in Table 6.20. DVD-Video also supports MPEG-1 video at constant and variable bit rates (see Table 6.20).

Before MPEG-2 compression occurs, the video is subsampled from ITU-R BT.601 format at 4:2:0 sampling with 8 bits of precision, which allocates an average of 12 bits per pixel. The actual color depth of the samples is 24 bits (1 byte for Y, 1 byte for C_b, and 1 byte for C_r), but the C samples are shared by 4 pixels.

The uncompressed source data rate is 124.416 Mbps (720 × 480 × 12 × 30 or 720 × 576 × 12 × 25). For 24-fps film, the source is typically at video frame rates of 30 fps (a telecine pulldown process has added duplicate fields). The MPEG encoder performs an inverse telecine process to remove

TABLE 6.20

Differences between DVD and MP@ML

MPEG Parameter	DVD	MP@ML
Display frame rate (525/60)	29.97	23.976, 29.97, 30
Display frame rate (625/50)	25	24, 25
Coded frame rate (525/60)	23.976, 29.97	23.976, 24, 29.97, 30
Coded frame rate (625/50)	24, 25	24, 25
Data rate	9.8 Mbps	15 Mbps
Frame size (horizontal size × vertical size) (525/60)	720 × 480, 704 × 480, 352 × 480, 352 × 240	From 16 × 16 to 720 × 480
Frame size (horizontal size × vertical size) (625/50)	720 × 576, 704 × 576, 352 × 576, 352 × 288	From 16 × 16 to 720 × 576
Aspect ratio	4:3, 16:9	4:3, 16:9, 2.21:1
Display horizontal size	540 (4:3), 720 (16:9)	Variable
GOP maximum (525/60)	36 fields/18 frames (30/15 recommended)	No restriction
GOP maximum (625/50)	30 fields/15 frames (24/12 recommended)	No restriction
GOP header	Required (first GOP in VOBU)	Optional
Audio sample rate	48 kHz	32, 44.1, 48 kHz
Packet size	2048 bytes (one logical block)	Variable
Color primaries and transfer characteristics (525/60)	4 (ITU-R BT.624 M) 6 (SMPTE 170 M)	1, 2, 4, 5, 6, 7, (8)
Color primaries and transfer characteristics (625/50)	5 (ITU-R BT.624 B or G)	1, 2, 4, 5, 6, 7, (8)
Matrix coefficients (RGB to YC_bC_r)	5 (ITU-R BT.624 B or G), 6 (SMPTE 170 M)	1, 2, 4, 5, 6, 7
Low delay	Not permitted	Permitted

the duplicate fields. Therefore, it is appropriate to consider the uncompressed film source data rate to be 99.533 Mbps (720 × 480 × 12 × 24) or 119.439 Mbps (720 × 576 × 12 × 24).

The maximum video bit rate is 9.8 Mbps (but must be less to allow for audio). The "typical" bit rate varies from 3.5 to 6 Mbps, but the rate depends entirely on the length of the original video, the quality and complexity of the video, the amount of audio, and so forth. The canonical 3.5 Mbps average video data rate is a 36:1 reduction from uncompressed video source (124 Mbps) or a 28:1 reduction from film source (100 Mbps). Video running near 9 Mbps is compressed at less than a 14:1 ratio (refer to Table 3.1).

Variable-bit-rate (VBR) encoding allows more data to be allocated for complex scenes and less data to be used during simple scenes. By lowering the average data rate, longer amounts of video can be accommodated, but the extra headroom allows the encoder to maintain video quality when a higher data rate is needed. See Figure 6.24 for an illustration. It is like riding a bicycle up and down hills. If you pedal at the same rate going up and down hills, you will run out of energy sooner than if you coast downhill and pedal easier when it's flat. VBR is in contrast to the statistical multiplexing scheme used by most digital video transmission systems (digital satellite, digital cable), which allocates bits across multiple channels into a fixed transmission rate. When one channel demands more quality, bits are stolen from other channels.

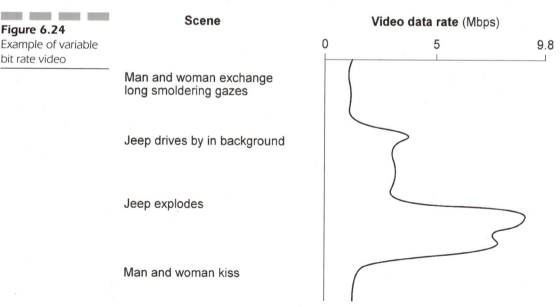

Figure 6.24
Example of variable bit rate video

Scene

Man and woman exchange long smoldering gazes

Jeep drives by in background

Jeep explodes

Man and woman kiss

Video data rate (Mbps)

0   5   9.8

Scanning and Frame Rates DVD-Video supports two display television systems of 525/60 (NTSC, 29.97 interlaced fps) and 625/50 (PAL/SECAM, 25 interlaced fps). Internal coded frame rates are typically at 29.97 fps interlaced scan from NTSC video, 25 fps interlaced scan from PAL video, and 24 fps progressive scan from film. In the case of 24 fps, the MPEG-2 encoder adds repeat_first_field flags to the data to make the decoder perform 2–3 pulldown for 60 (59.94) Hz displays or 2–2 pulldown for 50 Hz displays. For 50 Hz (PAL) display, the change from 24 to 25 fps causes a 4 percent speedup. Audio must be adjusted to match, resulting in a pitch shift (one semitone sharp) if it is not digitally readjusted.

A total of 480 lines of active video from a 525/60 video source are encoded and regenerated. Video encoding starts at line 23. For a 625/50 source, 576 lines of active video are encoded.

Very few DVD players convert from PAL video or film rates to NTSC display format or from NTSC video or film rates to PAL. Almost all PAL DVD players are able to produce video in NTSC scanning format transcoded to PAL color format for display on televisions supporting 4.43 NTSC signals (also called 60 Hz PAL). Some PAL players convert NTSC discs to standard PAL output, and a few NTSC players convert PAL discs to NTSC output. Computers are not tied to TV display rates, so most DVD computer software and hardware can play both NTSC and PAL. Some DVD computers can only display the converted video on the computer monitor, but others can output it as a video signal for a TV.

The actual coded picture rate in the MPEG-2 stream does not have to be exactly 24 or 30. Other coded picture rates or even varying rates will work, as long as the MPEG-2 repeat_first_field and top_field_first flags are set properly to produce either 25 or 29.97 fps display rates.

Resolution DVD-Video supports numerous resolutions designed for the NTSC and PAL television display systems (see Table 6.21). The lower resolutions are intended primarily for compatibility with MPEG-1 video formats. Most material uses a raster of 720 $\times$ 480 for 525/60 display and a raster of 720 $\times$ 576 for 625/50 display. The MPEG-2 display_horizontal_size value is set to 720 for 16:9 display mode and 540 for 4:3 display mode.

DVD pixels are not square; 525/60 pixels are tall, whereas 625/50 pixels are short. There are eight pixel aspect ratios (see Table 6.22) depending on the raster size and the picture aspect ratio. These ratios vary from the tallest of 0.909 to the widest of 2.909 (almost 3 times as wide as it is tall). Obviously, the pixels are wider for 16:9 anamorphic form than for normal 4:3 form. The 720- and 704-pixel rasters produce identical pixel aspect

TABLE 6.21		**(16:9 aspect ratio allowed)**		**(16:9 aspect ratio not allowed)**	
DVD-Video Resolutions (Rasters)	525/60 (NTSC)	720×480	704×480	352×480	352×240
	625/50 (PAL/SECAM)	720×576	704×576	352×576	352×288

ratios because the 720-pixel version includes more of the horizontal over-scan area (with a scanning line period of 53.33 microseconds), whereas the 704-pixel version is a tight scan (with a line period of 52.15 microseconds).

There is an alternate way of calculating pixel aspect ratios that divides the horizontal count by the vertical count and then divides the result by the picture aspect ratio. Confusingly, this gives a height-width aspect ratio rather than a consistent width-height aspect ratio. Table 6.22 includes values from the alternate method in parentheses as reciprocals and also adjusts them to match television scanning rates. The table also shows the integral conversion ratios used by most video digitizing hardware that converts between square and nonsquare pixels. Of course, there are no horizontal pixels in analog signals, although scan lines correspond to vertical pixels.

Using the traditional (and rather subjective) television measurement of *lines of horizontal resolution per picture height* (TV lines, or TVL), DVD has a theoretical maximum of 540 lines on a standard TV [720/(4/3)] and 405 on a widescreen TV [720/(16/9)]. Lines of horizontal resolution also can be approximated at 80 per MHz. DVD's MPEG-2 luma component is sampled at 6.75 MHz, which results in 540 lines. The actual observable lines of horizontal resolution may be closer to 500 on a standard TV due to low-pass filtering in the player. Typical luma frequency response maintains full amplitude to between 5.0 and 5.5 MHz. This is below the 6.75 MHz native frequency of the MPEG-2 digital signal. In other words, most players fall short of reproducing the full quality of DVD. Chroma frequency response is half that of luma.

For comparison, video from a laserdisc player has about 425 lines of horizontal resolution (5.3 MHz), SuperVHS and Hi8 have about 400 (5 MHz), broadcast television has about 335 (4.2 MHz),[7] and video from a VHS VCR has about 240 (3 MHz).

---

[7]Measurements for broadcast are usually tighter than those for recorded media, so broadcast quality is closer to SuperVHS and Hi8 than the numbers would indicate.

TABLE 6.22

DVD-Video Pixel Aspect Ratios

Resolution	4:3 Display	Standard Integral Ratio	16:9 Display
$720 \times 480,$[a] $704 \times 480,$[b] 352×240[b]	0.909 (1/1.095)	10/11	1.212 (1/0.821)
$720 \times 576,$[a] $704 \times 576,$[b] 352×288[b]	1.091 (1/0.9157)	59/54	1.455 (1/0.687)
352×480[b]	1.818 (1/2.19)	20/11	2.424 (1/0.411)
352×576[b]	2.182 (1/1.831)	118/54	2.909 (1/0.343)

[a]Overscan

[b]Exact scan

TABLE 6.23

Resolution Comparison of Different Video Formats

Format	VCD (1.78)	VCD (1.33)	VHS (1.78)	VHS (1.33)	LD (1.78)	LD (1.33)	DVD (1.78/1.33)	DTV3 (1.78)	DTV4 (1.78)
Horizontal pixels	352	352	333	333	567	567	720	1280	1,920
Vertical pixels	180	240	360	480	360	480	480	720	1,080
Total pixels	63,360	84,480	119,880	159,840	204,120	272,160	345,600	921,600	2,073,600
x VCD (16:9)		*4:3*	1.89	*2.52*	3.22	*4.30*	5.45	14.55	32.73
x VCD (4:3)			*1.42*	1.89	*2.42*	3.22	4.09	*10.91*	*24.55*
x VHS (16:9)				*4:3*	1.70	*2.27*	2.88	7.69	17.30
x VHS (4:3)					*1.28*	1.70	2.16	*5.77*	*12.97*
x LD (16:9)						*4:3*	1.69	4.51	10.16
x LD (4:3)							1.27	*3.39*	*7.62*
x DVD (16:9/4:3)								2.67	6.00
x DTV3 (16:9)									2.25

Note: 16:9 aspect ratios for VHS, LD, and VCD are letterboxed in a 4:3 picture. Comparisons between different aspect ratios are not as meaningful. These are shown in italics. Comparisons at 1.85 or 2.35 letterbox aspect ratios are essentially the same as at 1.78 (16:9).

DVD resolution in pixels can be roughly compared with analog video formats by considering pixels to be formed by the intersections of active scan lines and lines of horizontal resolution adjusted for aspect ratio. Table 6.23 shows how the resolution of DVD compares with other formats, including the two high-definition formats of the U.S. ATSC proposal (labeled DTV3 and DTV4), which also correspond to the H0 and H1 levels of the Microsoft/Intel/Compaq "Digital TV Team" proposal. Also see Table A-17 for additional video resolution figures.

Widescreen Format   Video can be stored on a DVD in aspect ratios of 4:3 or 16:9. The 16:9 format is anamorphic, meaning the picture is squeezed horizontally to fit a 4:3 rectangle and then unsqueezed during playback. DVD players can produce video in essentially four different ways:

■ 4:3 normal (for 4:3 or 16:9 displays)

■ 16:9 letterbox (for 4:3 displays)

■ 16:9 pan and scan (for 4:3 displays)

■ 16:9 widescreen (anamorphic, for 16:9 displays)

DVD-Video segments can be marked for the following display modes:

■ 4:3 full frame

■ 4:3 letterboxed (for automatically setting display mode on widescreen TVs)

■ 16:9 letterbox only (player not allowed to pan and scan)

■ 16:9 pan and scan allowed (viewer can select pan and scan or letterbox on 4:3 TV)

Some players send a signal to the television indicating that the picture is in anamorphic widescreen form so that widescreen televisions can adjust automatically. In some cases, the player also can inform the television that the 4:3 picture was transferred to video in letterbox format so that the television can expand the picture to remove the mattes. In Europe, the widescreen signaling system (WSS) may be used to convey this type of information (Tables 6.24 and 6.25). This standard is recommended for use by NTSC systems as well.

There is also a convention for signaling widescreen format by adding a 5-V direct current (DC) component to the chroma (C) line of the Y/C (s-video) output. This tells the widescreen equipment to expect video in anamorphic form. Most DVD players and many new video displays support this technique. Unfortunately, much of the existing equipment that the video signal might

TABLE 6.24

Widescreen
Signaling
Information

Aspect Ratio	Range	Format	Position	Active Lines
4:3 (1.33)	≤ 1.46	Full	Center	576 (480)
14:9 (1.57)	>1.46, ≤ 1.66	Letterbox	Top	504 (420)
14:9 (1.57)	>1.46, ≤ 1.66	Letterbox	Center	504 (420)
16:9 (1.78)	>1.66, ≤ 1.90	Letterbox	Top	430 (360)
16:9 (1.78)	>1.66, ≤ 1.90	Letterbox	Center	430 (360)
>16:9 (>1.78)	>1.90	Letterbox	Center	Undefined
14:9 (1.57)	>1.46, ≤ 1.66	Full[a]	Center[a]	576 (480)
16:9 (1.78)	>1.66, ≤ 1.90	Full (anamorphic)	n/a	576 (480)

[a]Shoot and protect 14:9. Soft matte format intended to be displayed with top and bottom cropped on a 16:9 display.

TABLE 6.25

Widescreen
Signaling on SCART
Connectors

SCART Pin 8 Voltage	Meaning
0 V	Normal
+6 V	16:9
+12 V	4:3 letterbox

pass through on its way to the display is designed to filter out such DC "noise." New A/V receivers and video switchers are being designed to recognize the widescreen signal and either pass it through or recreate it at the output.

In anamorphic mode, the pixels are fatter (see Table 6.22). Because of the high horizontal resolution of DVD, the wider anamorphic pixels are not objectionably noticeable.

Video in anamorphic form causes no problems with line doublers because they simply double the lines on their way to the widescreen display that then stretches out the lines.

Letterbox Conversion   For automatic letterbox mode, the player uses a letterbox filter that creates mattes at the top and the bottom of the picture (60 lines each for NTSC, 72 for PAL). This leaves three-quarters of the height remaining, creating a shorter but wider rectangle for the image. In order to fit the shape of this shorter rectangle, the player squeezes the picture vertically by combining every four lines into three. The vertical downsampling compensates for the anamorphic horizontal distortion and results in the movie being shown in its full width but with a 25 percent loss of vertical resolution. Some players simply throw away every fourth line. Better players use weighted averaging to give smoother results, albeit with a softer picture (see Figures 6.25 and 6.26). Some DVD players do a better job of letterbox filtering than others, perhaps by compensating for interlace jitter effects.

Pan and Scan Conversion   For automatic pan and scan mode, a portion of the image is shown at full height on a 4:3 screen by following a center-of-interest offset that is encoded in the video stream according to the preferences of the people who transferred the film to video. The pan and scan window is 75 percent of the full width, which reduces the horizontal resolution from 720 to 540 (see Figures 6.27 and 6.28), causing a 25 percent loss of horizontal resolution. Expanding the window by 33 percent compensates for the anamorphic distortion and achieves the proper 4:3 aspect ratio. The

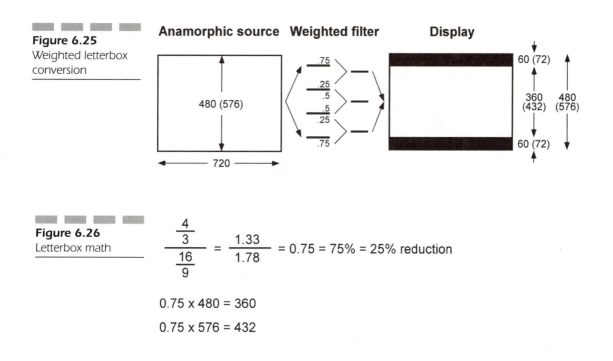

Figure 6.25
Weighted letterbox conversion

Figure 6.26
Letterbox math

$$\frac{\dfrac{4}{3}}{\dfrac{16}{9}} = \frac{1.33}{1.78} = 0.75 = 75\% = 25\% \text{ reduction}$$

0.75 x 480 = 360

0.75 x 576 = 432

Figure 6.27
Weighted pan and
scan conversion

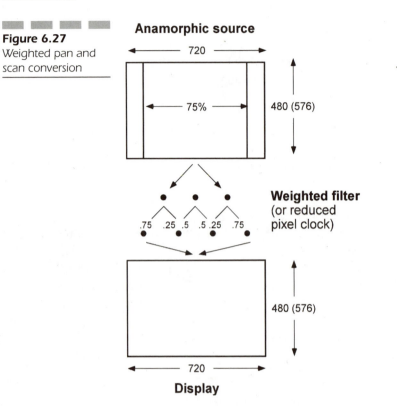

Anamorphic source

720

75%

480 (576)

Weighted filter
(or reduced
pixel clock)

.75 .25 .5 .5 .25 .75

480 (576)

720

Display

Figure 6.28
Pan and scan math

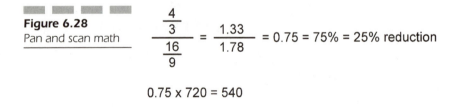

$$\frac{\frac{4}{3}}{\frac{16}{9}} = \frac{1.33}{1.78} = 0.75 = 75\% = 25\% \text{ reduction}$$

$$0.75 \times 720 = 540$$

$$\frac{\frac{16}{9}}{\frac{4}{3}} = \frac{1.78}{1.33} = 1.33 = 133\% = 33\% \text{ increase}$$

$$1.33 \times 540 = 720$$

540 extracted pixels on each line are interpolated by creating four pixels from every three to scale the line back to the full width of 720. Weighted averaging gives smoother results.

Unlike letterbox conversion, which is new to DVD, pan and scan conversion is part of the MPEG-2 standard. Most MPEG-2 decoder chips include a pan and scan conversion feature. The offset is specified in increments of one-sixteenth of a pixel. DVD allows only horizontal adjustments; that is, MPEG-2 frame_center_vertical_offset must be 0, while frame_center_horizontal_offset can vary from −1440 to +1440 for 720-pixel frames and from −1312 to +1312 for 704-pixel frames. It is also possible to stretch out the pixels by increasing the pixel clock in the TV encoder chip.

Video Interface   The MPEG video stream is decoded into 4:2:0 digital component format. This format can be sent directly from the player (with accompanying copy protection information) to a digital connection such as IEEE 1394/FireWire or SDI. However, in most cases the player is connected to an analog video display or recording system and requires that the signal be converted to analog form through a digital-to-analog converter. Although the original BT.601 video values are sampled with 8 bits of precision, better players use 10 bits or more in the digital-to-analog converter to provide headroom for more accurate calculations, which can help produce a smoother picture.

For analog component output, the digital signals are scaled and offset according to the desired output format(s) of YP_bP_r or RGB, blanking and sync information is added, and the digital values are converted to analog voltage levels. Some players label the analog component output as YC_bB_r, which is incorrect.

For analog s-video and composite baseband video, the digital signal is sent through a TV encoder to produce an NTSC signal from 525/60 format data or a PAL or SECAM signal from 625/50 format data. Almost all TV encoders support both NTSC and PAL, but most NTSC players do not enable PAL video signal output (see Figure 6.29). See "How to Hook Up a DVD Player" in Chapter 9 for more information about video connections, and see "Digital Connections" in Chapter 13 for information about future interface options.

Audio

DVD-Video supports three primary audio standards—Dolby Digital, MPEG-2, and linear PCM (LPCM). Two optional audio formats are included

Figure 6.29
Video block diagram

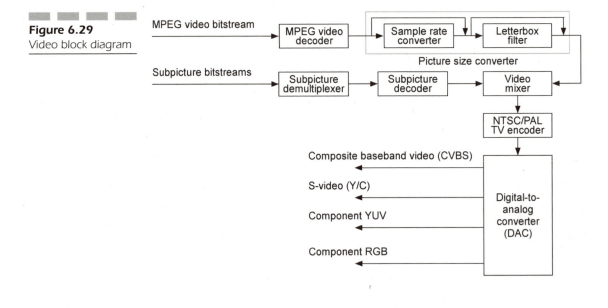

TABLE 6.26

DVD-Video
Format, Audio
Details

Audio	0 to 8 streams
Audio coding	Dolby Digital, MPEG-1, MPEG-2, LPCM
Audio bit rate	32 kbps to 6.144 Mbps, 384 typical

—DTS (Digital Theater Sound) and SDDS (Sony Dynamic Digital Sound)—but players are not required to support either one (see Table 6.26).

Dolby Digital and MPEG-2 can provide discrete multichannel audio. This gives clean sound separation with full dynamic range from each speaker. The result is a more realistic soundfield in which sounds can travel left to right and front to back.

Discs containing 525/60 (NTSC) video are required to include at least one audio track in Dolby Digital or PCM format. After that, any combination of formats is allowed. Discs containing 625/50 (PAL/SECAM) video are required to have at least one track of Dolby Digital, MPEG, or PCM audio. Since Dolby Digital decoders greatly outnumber MPEG-2 audio decoders, most disc producers also use Dolby Digital audio tracks on PAL discs instead of MPEG audio tracks.

Audio streams are encoded at various data rates, depending on the number of channels. See Table 6.27 for playing times.

TABLE 6.27

Approximate Audio Playing Times at Various Data Rates

| | | Playing Time per Disc (hours) | | | | | | | | | |
| | | No Video | | | | | +4 Mbps Video (Avg.) | | | | |
Format	kbps	DVD-5	DVD-9	DVD-10	DVD-14	DVD-18	DVD-5	DVD-9	DVD-10	DVD-14	DVD-18
DD 1.0	64	163.1	296.5	326	459.7	593	2.5	4.6	5.1	7.2	9.3
DD 2.0	192	54.3	98.8	108.6	153.2	197.6	2.4	4.5	4.9	7	9
DD 5.1	384	27.1	49.4	54.3	76.6	98.8	2.3	4.3	4.7	6.7	8.6
DD 5.1 max	448	23.3	42.3	46.5	65.6	84.7	2.3	4.2	4.6	6.6	8.5
2 DD 5.1	768	13.5	24.7	27.1	38.3	49.4	2.1	3.9	4.3	6.1	7.9
2 DD 5.1 max	896	11.6	21.1	23.2	32.8	42.3	2.1	3.8	4.2	6	7.7
MPEG 7.1 max	912	11.4	20.8	22.8	32.2	41.6	2.1	3.8	4.2	5.9	7.7
3 DD 5.1	1152	9	16.4	18.1	25.5	32.9	2	3.6	4	5.7	7.3
3 DD 5.1 max	1344	7.7	14.1	15.5	21.8	28.2	1.9	3.5	3.9	5.5	7.1
PCM 48/16 stereo	1536	6.7	12.3	13.5	19.1	24.7	1.8	3.4	3.7	5.3	6.8
PCM 48/20 stereo	1920	5.4	9.8	10.8	15.3	19.7	1.7	3.2	3.5	4.9	6.4
8 DD 5.1	3072	3.3	6.1	6.7	9.5	12.3	1.4	2.6	2.9	4.1	5.3
PCM 96/20 stereo	3840	2.7	4.9	5.4	7.6	9.8	1.3	2.4	2.6	3.7	4.8

Note: DD = Dolby Digital.

Dolby Digital Audio Details   Dolby Digital (AC-3) is the format used for audio tracks on almost all DVDs. It is a multichannel digital audio format, lossily compressed using perceptual coding technology from original PCM with a sample rate of 48 kHz at up to 20 bits of precision. The Dolby Digital standard provides for other sampling rates of 32 and 44.1 kHz, but these are not allowed with DVD. Frequency response is 3 Hz to 20 kHz for the main five channels and 3 to 120 Hz for the low-frequency effects (LFE) channel (Table 6.28).

The Dolby Digital bit rate is 64 to 448 kbps, with 384 or 448 kbps being the typical rate for 5.1 channels. The 448 kbps rate results in a compression ratio of 10:1 (90 percent) from 5.1 channels of 48/16 PCM. The typical bit rate for stereo (with or without Dolby Surround encoding) is 192 kbps. Monophonic audio is usually at 96 kbps for music or 64 kbps for voice.

There can be 1, 2, 3, 4, or 5 channels. The subwoofer (.1) channel can be added optionally to any combination. Two-channel mode can either be stereo or dual mono (where each channel is a separate track). An extra rear center channel is possible with the Dolby Digital Surround EX format, which is compatible with all DVD discs and players and with existing Dolby Digital decoders. The added channel is not an additional full-bandwidth discrete channel; rather, it is phase matrix encoded into the two rear channels in the same way Dolby Surround is matrixed into standard stereo channels. A new (or additional) decoder is needed to extract the rear center channel.

All Dolby Digital decoders are required to perform a downmixing process to adapt 5.1 channels to 2 channels for stereo PCM and analog output. The downmixing process matrixes the center and surround channels onto the main stereo channels in Dolby Surround format for use by Dolby Pro Logic decoders. The LFE channel is omitted from the downmix because most audio systems without six speakers cannot reproduce the low bass. This can help keep sound from becoming "muddy" on average home audio systems.

TABLE 6.28

Dolby Digital Audio Details

Sample frequency	48 kHz
Sample size	Up to 24 bits
Bit rate	64 to 448 kbps; 384 or 448 kbps typical
Channels (front/rear)[a]	1/0, 2/0, 3/0, 2/1, 2/2, 3/1, 3/2, 1+1/0 (dual mono)
Karaoke modes	L/R, M, V1, V2

[a]LFE channel can be added to all variations.

When the audio is encoded, the downmixed output is auditioned by using a reference decoder. If the quality is not adequate, the encoding process can be tweaked, the 5.1-channel mix can be tweaked, or a separate Dolby Surround track can be added (in either Dolby Digital or PCM format). Most modern action movies require minor adjustments to the 5.1-channel mix to make sure the dialogue is audible. Some disc producers prefer to create a separate two-channel surround mix rather than letting the decoder do the downmixing. Two-channel Dolby Surround streams are also used when only the Dolby Surround mix is available or where the disc producer does not want to remix from the multitrack masters.

Dolby Digital is not synonymous with 5.1 channels. A Dolby Digital soundtrack can be mono, dual mono, stereo, Dolby Surround stereo, and so forth. For example, older movies have only monophonic soundtracks encoded as Dolby Digital 1.0.

Dolby Digital also provides dynamic range compensation. DVDs have soundtracks with much wider dynamic range than most recorded media. This improves performance for a quality home theater setup but may make dialogue and other soft passages too low to hear clearly on less than optimal audio systems. Information is added to the encoded data to indicate what parts of the sound should be boosted when the player's dynamic range compression setting is turned on.

Additional information can be carried in the Dolby Digital stream, such as a copyright flag and a flag that identifies when Surround EX encoding has been used. See Chapter 3 for more details on Dolby Digital encoding.

MPEG Audio Details MPEG audio is multichannel digital audio, lossily compressed using perceptual coding from original PCM format with a sample rate of 48 kHz at 16 bits. MPEG-1 Layer II and MPEG-2 backward-compatible (BC) are supported. The variable bit rate is 64 to 912 kbps, with 384 kbps being the normal average rate. An MPEG-1 stream is limited to 384 kbps (see Table 6.29).

There can be 1, 2, 3, 4, 5, or 7 channels. The subwoofer (.1) channel is optional with any combination. The 7.1-channel format adds left-center and right-center channels but is not intended for home use. Stereo channels are provided in an MPEG-1 Layer II stream. Surround channels are matrixed onto the MPEG-1 stream and duplicated in an extension stream to provide discrete channel separation. The LFE channel is also added to the extension stream. An additional extension layer can be added for 7.1 channels. The extension streams are carried by MPEG packets. This layering and packetizing process makes MPEG-2 audio backward-compatible with MPEG-1 hardware; an MPEG-1 decoder will see only packets containing the two main channels.

TABLE 6.29

MPEG Audio
Details

Sample frequency	48 kHz only
Sample size	Up to 20 bits
MPEG-1	Layer II only
MPEG-1 bit rate	64 to 192 kbps (mono), 64 to 384 kbps (stereo)
MPEG-2	BC (matrix) mode only
MPEG-2 bit rate[a]	64 to 912 kbps
Extension streams[b]	5.1-channel, 7.1-channel
Channels (front/rear)[c]	1/0, 2/0, 2/1, 2/2, 3/0, 3/1, 3/2, 5/2 (no dual channel or multilingual)
Karaoke channels	L, R, A1, A2, G
Emphasis	None
Prediction	Not allowed

[a]MPEG-1 Layer II stream + extension stream(s).

[b]AAC (unmatrix, NBC) not allowed.

[c]LFE channel can be added to all variations.

Stereo output includes surround channel matrixing for Dolby Pro Logic processors. The MPEG signal already has the center and surround channels matrixed onto the main stereo channels, so no special downmixing is required.

Since MPEG-2 decoders were unavailable at the introduction of DVD, first-generation PAL players contain only MPEG-1 decoders.

The MPEG-2 standard includes an AAC (advanced audio coding) mode. In order for DVD-Video discs to be playable in players with only MPEG-1 decoders, the AAC mode is not allowed for the primary MPEG audio track, which always must be compatible with MPEG-1. The AAC mode originally was known as *NBC* (non-backward-compatible) and also is referred to as *unmatrix mode*. MPEG Layer II audio mode, also known as *MP3,* is also not allowed. Of course, AAC and MP3 tracks can be recorded on a disc outside the DVD-Video or DVD-Audio zones. See Chapter 3 for more details of MPEG audio encoding.

PCM Audio Details Linear PCM (pulse-code modulation) is lossless, uncompressed digital audio. The same format is used on CDs. DVD-V supports sampling rates of 48 or 96 kHz with 16, 20, or 24 bits per sample.

(Audio CD is limited to 44.1 kHz at 16 bits.) There can be from 1 to 8 channels in each track (see Table 6.30).

The maximum PCM bit rate is 6.144 Mbps, which limits sample rates and bit sizes with 5 or more channels (see Table 6.31). It is generally felt that the 96 dB dynamic range of 16 bits or even the 120 Db range of 20 bits combined with a frequency response of up to 22,000 Hz from 48 kHz sampling is adequate for high-fidelity sound reproduction. However, additional bits and higher sampling rates are useful in studio work, noise shaping, advanced digital processing, and three-dimensional sound field reproduction.

DVD players are required to support all the variations of PCM, but most models subsample the 96 kHz rate down to 48 kHz, and some may truncate extra bits above 16 or 20. High-end players pass 96 kHz audio to the digital audio outputs, but only from tracks that are not CSS encrypted, because the CSS license limits protected output to 48 kHz.

DTS Audio Details Digital Theater Systems Digital Surround is an optional multichannel (5.1 or 6.1) digital audio format, lossily compressed

TABLE 6.30

PCM Audio Details

Sample frequency	48 or 96 kHz
Sample size	16, 20, or 24 bits
Channels	1, 2, 3, 4, 5, 6, 7, or 8
Karaoke channels	L, R, V1, V2, G

TABLE 6.31

Allowable PCM Data Rates and Channels

		Number of Channels							
		1	2	3	4	5	6	7	8
kHz	Bits	Data Rate (Kbps)							
48	16	768	1536	2304	3072	3840	4608	5376	6144
48	20	960	1920	2880	3840	4800	5760	n/a	n/a
48	24	1152	2304	3456	4608	5760	n/a	n/a	n/a
96	16	1536	3072	4608	6144	n/a	n/a	n/a	n/a
96	20	1920	3840	5760	n/a	n/a	n/a	n/a	n/a
96	24	2304	4608	n/a	n/a	n/a	n/a	n/a	n/a

from PCM at 48 kHz and up to 24 bits. The data rate is from 64 to 1536 kbps, with 768 or 1536 kbps being the typical rates for 5.1-channel audio. The 768 kbps rates results in a compression ratio of 6:1 (83 percent) from 5.1 channels of 48/16 PCM. The 1536 kbps rate results in a compression ratio of 3:1 (67 percent) from 5.1 channels of 48/16 PCM. DTS supports up to 4096 kbps variable data rate for lossless compression, but DVD does not allow this. DVD also does not allow sampling rates other than 48 kHz (see Table 6.32).

Channel combinations are (front/surround): 1/0, 2/0, 3/0, 2/1, 2/2, 3/2, and 3/3. The LFE channel is optional with all seven combinations. A rear center channel is supported in two ways: (1) a Dolby Surround EX matrixed channel that is compatible with existing decoders (*DTS-ES Matrix 6.1*) or (2) a discrete seventh channel (*DTS-ES Discrete 6.1*). DTS also has a 7.1-channel mode (*DTS-ES Discrete 7.1*, with 8 discrete channels), but it is not used commonly. The 7-channel and 8-channel modes work with existing DTS decoders but require a new decoder to take advantage of all channels.

The DVD standard includes an audio stream format reserved for DTS, but many older players ignore it and are thus unable to recognize and output the DTS stream. These players play only the Dolby Digital or PCM stream that is mandatory on all discs, including DTS discs.

The encoding system used on DVDs (DTS Coherent Acoustics, a psychoacoustic transform coder) is different from the one used in theaters (Audio Processing Technology's apt-X, an ADPCM coder). All DVD players

TABLE 6.32

DTS Audio Details

Sample frequency	48 kHz
Sample size	Up to 24 bits
Bit rate	64 to 1536 kbps; 768 or 1536 kbps typical
Channels (front/rear)[a]	1/0, 2/0, 3/0, 2/1, 2/2, 3/2, 3/3 (no multilingual)
Karaoke modes	L/R, M, V1, V2

[a]LFE channel can be added to all variations.

can play DTS audio CDs because the standard PCM stream holds the DTS code. See Chapter 3 for more details on DTS encoding.

SDDS Audio Details   SDDS is an optional multichannel (5.1 or 7.1) digital audio format, lossily compressed from PCM at 48 kHz. The data rate can go up to 1280 kbps. SDDS is a theatrical film soundtrack format based on the ATRAC transform coder compression format that is also used by Sony's MiniDisc.

Karaoke Modes   All five DVD-Video audio formats support karaoke mode, which has two channels for stereo (L and R) plus an optional melody channel (M) or a guide channel (G), and two optional vocal channels (V1 and V2). The karaoke modes are generally implemented only by DVD player models with karaoke features for mixing audio and microphone input. Karaoke mixing is implemented as a multichannel track that is selectively mixed to two output channels. Source channels 1 and 2 are usually the background instrumentals, channels 3 and 4 are typically vocal channels (usually a harmony channel and a melody channel), and channel 5 is the guide or melody helper channel (usually a single instrument playing the melody).

Codes are included to identify audio segments such as master of ceremonies intro, solo, duet, male vocal, female vocal, climax, interlude, ending, and so on.

Audio Interface   The digital audio signal from the currently selected audio track is sent to the audio subsection. Multichannel (Dolby Digital or MPEG-2) audio signals are directed to the digital audio output jacks for decoding by external equipment. The multichannel signals are also sent to the appropriate decoder where they are downmixed to two channels and converted to PCM signals. PCM signals from the decoder or directly from the disc are also routed to the digital audio output jacks for processing by external equipment with digital-to-analog converters. (See Figures 6.30 and 6.31.) Some players also include built-in digital-to-analog audio converters and external jacks for discrete output from the decoder. Most players do not include built-in DTS decoders, but instead pass it directly out the digital audio output jacks for processing by external equipment. Built-in DTS decoders operate the same as Dolby Digital decoders.

Figure 6.30
Dolby Digital audio
block diagram

Figure 6.31
MPEG-2 audio block
diagram

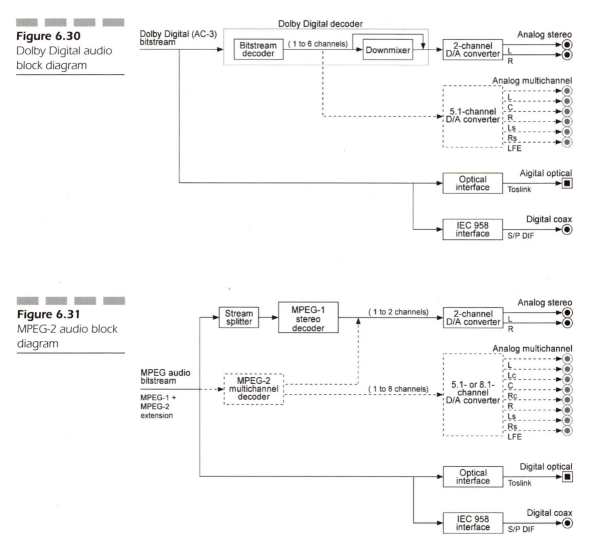

Some players provide PCM audio and multichannel audio signals on separate connectors; others provide dual-purpose connectors that must be switched between PCM and multichannel output. A standard extension to IEC 958, designed by Dolby and described in ATSC A/52, is used for Dolby Digital audio signals. This extension, with additions from Philips and others, is being formalized as IEC 1937.

PCM audio signals are also routed to a digital-to-analog converter for standard analog stereo output. See "How to Hook Up a DVD Player" in Chapter 9 for more information about audio connections. See "Digital Connections" in Chapter 13 for information about future interface options.

Subpictures

DVD-Video supports up to 32 subpicture streams that overlay the video for subtitles, captions, karaoke lyrics, menus, simple animation, and so on. These are part- to full-screen, run-length-encoded bitmaps limited to four pixel values (see Table 6.33).

Each pixel is represented by 2 bits, allowing four types. The four pixel types are officially designated as *foreground* (also called *pattern*), *background, emphasis-1,* and *emphasis-2,* but are not strictly tied to these functions. The highlight functions are part of the menu button feature (described earlier). Each pixel type is associated with one color from a palette of 16 and one contrast or transparency level. The 24-bit YC_bC_r color palette entries provide selections from more than 11 million colors. The

TABLE 6.33

Subpicture Details

Data	0 to 32 streams
Data rate	Up to 3.36 Mbps
Unit size	53,220 bytes (up to 32,000 bytes of control data)
Coding	RLE (max. 1440 bits/line)
Resolution[a]	Up to 720 × 478 (525/60) or 720 × 573 (625/50)
Bits per pixel	2 (defining one of four types)
Pixel types	Background, foreground, emphasis-1, emphasis-2
Colors[a]	4 of 16 (from 4-bit palette,[b] one per type)
Contrasts[a]	4 of 16 (from 4-bit palette,[b] one per type)

[a]Area, content, color, and contrast can be changed for each field.

[b]Color palette and contrast can be changed every PGC.

transparency is set directly, from invisible (0), through 14 levels of transparency (1–14), to opaque (15).

The display area (starting coordinates, width, and height) can be specified up to almost a full-screen rectangle (0,0,720,478 or 0,0,720,573). The display area and content (bitmap) can be changed for each frame or field. The color and contrast of the four pixel types can be changed for each frame or field; the palettes can be changed for each PGC.

The maximum data rate for a single subpicture stream is 3.36 Mbps, with a maximum size per frame of 53,220 bytes. Each run-length-encoded line can be up to 1440 bits (720 × 2). Associated display control sequences (DCSQs) can be up to 32,000 bytes (see Figure 6.32).

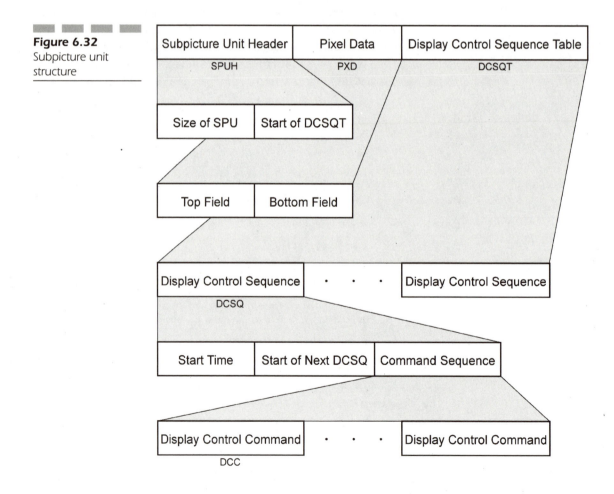

Figure 6.32
Subpicture unit
structure

A typical set of subtitles are presented at about 14 per minute, for around 1500 per movie.

The subpicture display commands (DCCs) are used to change the location, scroll position, transparency, and so on. A sequence of commands can be used to create effects such as color changes, moving highlights, fades, crawls, and so on.

Closed Captions

DVD includes support for NTSC Closed Captions. Closed Captions (CC) are a standardized method of encoding text into an NTSC television signal. The text can be displayed by a TV with a built-in decoder or by a separate decoder. All TVs larger than 13 inches sold in the United States since 1993 have Closed Caption decoders.

Even though the terms *caption* and *subtitle* have similar definitions, *captions* commonly refer to onscreen text specifically designed for hearing-impaired viewers, whereas *subtitles* are straight transcriptions or translations of the dialogue. Captions are usually positioned below the person who is speaking, and they include descriptions of sounds and music. Closed captions are not visible until the viewer activates them. Open captions are always visible, such as subtitles on foreign videotapes.

Closed Captions on DVDs are carried in the MPEG-2 video stream and are sent automatically to the TV. They cannot be turned on or off from the DVD player. They are stored as individual character codes (one character per field). Subtitles, on the other hand, are DVD subpictures. Subpictures also can be used to create captions. To differentiate NTSC Closed Captions from subtitles, captions created as subpictures are usually called "captions for the hearing impaired." Some DVD players do not output NTSC Closed Captions.

DVD does not support PAL Teletext, the much-improved European equivalent of Closed Captions. The Advanced Television Enhancement Forum (ATVEF) recommends the use of EIA-768 TV links, which are URLs stored in the line 21 T2 service, to add Web links to video. See Chapter 11 for more information.

Camera Angles

A DVD video program can have up to nine camera angles, which are essentially interleaved video tracks. The viewer can switch among the angles by pressing the "Angle" key on the remote control. The angle also can be changed with commands on the disc. The alternate chunks of video must all be the same length. The viewer cannot switch to a different camera angle and have the action be out of sync (unless the video itself was not produced in sync). Each angle is technically required to have identical audio tracks, but it is possible to get around this requirement and assign a different audio track to each angle.

Video objects, one for each angle, are interleaved into an interleaved block (ILVB). ILVBs are required for camera angles and optionally can be used for parental branching and seamless branching (see Figure 6.33).

A major advantage of camera angles is that they are supported directly by the buffer management feature of DVD, which automatically leapfrogs through the interleaved video objects, reading only the ones assigned to the selected angle. This means that adding an angle does not reduce the data rate limit, although it does reduce playing time. For example, a single-angle program that runs for 5 minutes at 7 Mbps would be about 262 megabytes long. If it were changed to a two-angle program, each angle could still run at 7 Mbps, but the program would take up 524 megabytes. Limitations are imposed on the maximum data rate depending on the number of angles (see Table 6.34). The data-rate limit applies to the angle block and for 2.5 seconds preceding the angle block.

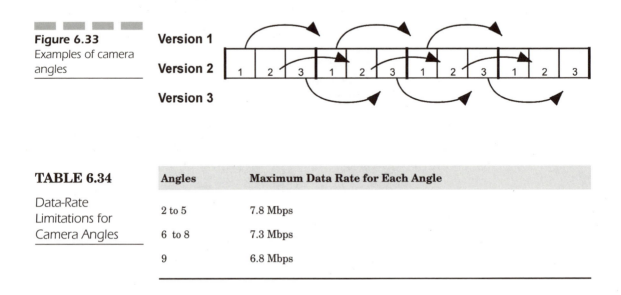

Figure 6.33
Examples of camera angles

TABLE 6.34	Angles	Maximum Data Rate for Each Angle
Data-Rate Limitations for Camera Angles	2 to 5	7.8 Mbps
	6 to 8	7.3 Mbps
	9	6.8 Mbps

Parental Management

Playback restrictions can be placed on a player by choosing a parental level from the settings menu. There are eight levels, each less restrictive of content than the level below. The Motion Picture Association of America (MPAA) standard movie ratings are mapped to specific parental levels (see Table 6.35). Some players may be set to any of the eight levels, others limit the selection to the five MPAA levels, and still others offer three levels: all titles, no adult titles, only children's titles. The DVD Forum has not established an official correspondence to classification systems of other countries, such as the United Kingdom's BBFC, Germany's FSK, or France's FSS. In any case, most discs released outside the United States show a classification code on the package but do not set a corresponding parental level on the disc.

Parental level support is optional for players, but most support it. Parents can set a password or code in the player to prevent the setting from being changed. However, resetting the player generally clears the password. A disc with no parental management information on it is treated by some players as being at the highest level of restriction.

Parental levels restrict the playback either of an entire disc or of certain scenes. Parental codes are placed on the disc in each parental block (a group of PGCs) so that the player can automatically select the proper path from

TABLE 6.35	**MPAA Rating**	**Parental Level**	**General Description**
MPAA Ratings and Corresponding Parental Levels		8	Unrated (most restricted audience)
	NC-17	7	Adult audience
	R	6	Mature audience
		5	Mature teenage audience
	PG-13	4	Teenage audience
	PG	3	Mature young audience
		2	Most audiences
	G	1	General (unrestricted audience)

scene to scene. This allows multiple ratings versions of a movie to be put on a single disc. For this to work, the video must be broken carefully into scenes. Objectionable scenes must be coded so that they can be skipped over, or alternate versions of the scenes must be provided and coded appropriately. Video, audio, and subpictures need to be orchestrated so that there is no discontinuity across scene splices. Obviously, this is a significant undertaking. Few discs are produced with multiple versions for different parental levels.

Because chapters can exist within a parental block, there may be two versions of the same chapter in a title, each assigned a different parental management level and in a different parental block. For example, a child who logs in and plays the disc would see one version of chapter 3, and an adult who logs in would see a different version, assuming that the application supports parental management levels.

A command can be inserted in the middle of a video sequence to temporarily change the parental management level. If the current level is too low for the new level, the player will pause and ask the user to change the level. If the user is unable to raise the level, playback will stop. Temporary parental levels generally are authored as angle blocks, so a scene in a film may have two versions, one rated for younger viewers and one for adults.

Seamless Playback

Seamless presentation, also called *multistory,* allows multiple paths through mostly the same material by jumping from place to place without a break in the video (see Figure 6.34). This is useful for movies with alternate endings, directors' cuts, and so on. Seamless branching is accomplished in a manner similar to parental management. In fact, since commands exist to set the parental level, it is possible to use the parental features to control seamless branching. Camera angles can be included in the seamless branching process. For example, a dual-language video that shows a scene of a book can be authored with multiangle sections wher-

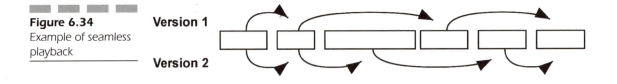

Figure 6.34
Example of seamless playback

Version 1

Version 2

ever the book is shown. The angle can be set to match the language, and as playback moves from a single-angle section to a multiangle section, it will seamlessly select the correct angle. Alternatively, the video can be authored in *partial interleave* mode, where one PGC is used for one language and a second PGC is used for the other language. The different book scenes are interleaved together so that the PGC can jump over one set of cells while showing the other.

Seamless branching can be accomplished within a PGC. Cells may be contiguous or noncontiguous, but the distance between them is restricted depending on the data rate (see Table 6.36). At higher data rates, the distance must be smaller so that the pickup head can jump to the new position and begin reading data before the track buffer runs out. Seamless branching (and camera angles) generally impose a combined stream data-rate limit of 8 Mbps or lower. Also, video objects for seamless playback must be within a single PGC and on the same layer.

Seamless branching generally cannot be done on the fly in response to user interaction. Each path is encapsulated by a PGC, which contains cell pointers to link together the desired chunks of video. Each PGC must be created during the authoring process and stored on the disc. Cell commands cannot be used for seamless branching. Jumping between PGCs is also not guaranteed to be seamless.

Text

A disc can contain a wide variety of text data (TXTDT) that is identified for its purpose and the part of the content to which it applies. This includes information to be displayed for the user, such as title names, song names, and production credits; identifying information from the production process, such as version numbers and product codes; and extra information

TABLE 6.36

Allowable Distance Between Cells for Seamless Playback

	Bit Rate (Mbps)			
	8.5	8	7.5	7
Maximum jump sectors	5000	10,000	15,000	20,000
Minimum buffer sectors	201	221	220	216

for enhancing use of the disc, such as Web links. DVD text is organized in a way that mirrors the logical hierarchy of the DVD volume (see Figure 6.35).

Text is stored in language blocks so that the contents can be described in different languages. There is no requirement that the information be the same in both languages. For example, detailed information could be presented about every title and chapter in English but only title information in other languages.

DVD has two types of text items: *structure identifiers* and *application content items*. Each text item includes a numerical code that identifies it. Text items with a code of 1 through 32 are structure identifiers. They have no associated text. The numerical code identifies the logical structure to which any following content items belong (see Table 6.37). A hierarchical structure is used that corresponds closely to the logical structure of a DVD disc contents: volume, title, chapter, and so on. Codes above 32 identify content items that hold text information. The exact way in which content

Figure 6.35
DVD text data
example

Text	Identifiers	Structure
	\<Volume\>	Volume
DVD Demystified 2	Name	
Omnibus	General category	
	\<Title\>	Title 1
About DVD	Name	
	\<Title\>	Title 2
Recipe 4 DVD	Name	
	\<Chapter\>	Chapter 1
Introduction	Name	
	\<Chapter\>	Chapter 2
Plan	Name	
	\<Chapter\>	Chapter 3
Encode	Name	
	\<Chapter\>	Chapter 4
Author	Name	
	\<Chapter\>	Chapter 5
Multiplex	Name	
	\<Chapter\>	Chapter 6
Deliver	Name	
	\<Title\>	Title 3
Motion Menu Tutorial	Name	
	\<Title\>	Title 4
In A New York Minute	Song	
Eagles: Hell Freezes Over	Album	
Music	General category	
Pop Music	Music category	
	\<Audio stream\>	Audio 1
DTS 5.1	Name	
	\<Audio stream\>	Audio 2
Dolby Digital 2.0	Name	

\<\> = structure identifier

TABLE 6.37

DVD Text Data
Codes

Item	Code (hex)	Code (decimal)	Description
General Structure Identifiers			
Volume	01h	1	Indicates that following items pertain to the entire disc side
Title	02h	2	Indicates that following items pertain to a title
Parental ID	03h	3	Indicates that following items pertain to a particular part of a parental block
Chapter	04h	4	Indicates that following items pertain to a chapter
Cell	05h	5	Indicates that following items pertain to a cell (usually one scene from a movie)
Stream Structure Identifiers			
Audio	10h	16	Indicates that following items pertain to an audio stream
Subpicture	11h	17	Indicates that following items pertain to a subpicture stream
Angle	12h	18	Indicates that following items pertain to an angle block
Audio Channel Structure Identifiers			
Channel	20h	32	Indicates that following items pertain to one channel in an audio stream
General Content			
Name	30h	48	The most common name for the volume, title names, chapter names, song names, and so on
Comments	31h	49	General comments about the title, chapter, song, and so on
Title Content			
Series	38h	56	Title of the series that the volume, title, or chapter is part of
Movie	39h	57	Title of the content if it is a movie
Video	3Ah	58	Title of the content if it is a video
Album	3Bh	59	Title of the content if it is an album
Song	3Ch	60	Title of the content if it is a song
Other	3Fh	63	Other title of the volume, title, or chapter if it belongs to some other genre or category

continues

TABLE 6.37 cont.

DVD Text Data Codes

Item	Code (hex)	Code (decimal)	Description
Secondary Title Content			
Series	40h	64	Secondary or alternate title of the series that the volume, title, or chapter is part of
Movie	41h	65	Secondary or alternate title of the content if it is a movie
Video	42h	66	Secondary or alternate title of the content if it is a video
Album	43h	67	Secondary or alternate title of the content if it is an album
Song	44h	68	Secondary or alternate title of the content if it is a song
Other	47h	71	Secondary or alternate title of the volume, title, or chapter if it belongs to some other genre or category
Original Title Content			
Series	48h	72	Original name of the series that the volume, title, or chapter is part of
Movie	49h	73	Original name of the content if it is a movie
Video	4Ah	74	Original name of the content if it is a video
Album	4Bh	75	Original name of the content if it is an album
Song	4Ch	76	Original name of the content if it is a song
Other	4Fh	79	Original name of the volume, title, or chapter if it belongs to some other genre or category
Other Info Content			
Other scene	50h	80	Additional information about a scene in a title or chapter
Other cut	51h	81	Additional information about a cut in a title or chapter
Other take	52h	82	Additional information about an alternate take in a title or chapter
Other label	53h	83	Other label for a volume, title, or chapter

continues

Item	Code (hex)	Code (decimal)	Description
TABLE 6.37 cont. DVD Text Data Codes			

Item	Code (hex)	Code (decimal)	Description
		Miscellaneous Other Content	
Language	58h–59h	88–89	No one seems to know what this is for, or why it is also labeled "stand in"
Work	5Ch–67h	92–103	Color, shooting location, production company, and so on
Character	6Ch–8Dh	108–141	Leading actor, leading actress, supporting actor, supporting actress, producer, conductor, orchestra, animator, and so on
Data	90h–92h	144–146	Production, award, historical background
Karaoke	94h–99h	148–153	Male melody, female harmony, and so on
Category	9Ch–9Fh	156–159	Category
Lyrics	A0h–A2h	160–162	Lyrics
Document	A4h	164	Liner notes
Other	A8h	168	Other
Administration	B0h–C9h	176–201	Product number, SKU, ISRC, copyright, release date, encoding information, master tape information, and so forth
Vendor unique	E0h–Efh	224–239	Available for private use by disc author
Sorting	F0h	240	Special extension code for adding a second entry that provides an alphabetically or numerically sortable variation of the previous entry

strings are used is not closely defined, so DVD authors can use them in various ways.

Correspondence of text items to logical objects on the disc is established by the number of times the structure identifier is repeated. For example, the first occurrence of a title structure identifier indicates that all following content items apply to title 1. The second occurrence of a title identifier indicates that all following content items apply to title 2. The first occurrence of a chapter identifier after a title identifier indicates that all following content items apply to chapter 1 of that title, and so on.

DVD text is optional. It does not have to be included on the disc, and players are not required to do anything with it. Historically, DVD text has been used almost exclusively on karaoke discs, and these discs mostly use structure identifiers 1 and 2 and content type 48. However, as players and jukeboxes become more sophisticated, and as computer-based DVD players proliferate, DVD text is becoming more important.

DVD-Audio

DVD-Audio is a separate application format from DVD-Video. The DVD-Audio specification includes different data types and features (see Table 6.38), with content stored in a separate audio zone on the disc (the AUDIO_TS directory) (see Figure 6.5). The DVD-Audio format includes a subset of DVD-Video features to provide onscreen menus and motion video. DVD-Audio discs that include video are sometimes called *DVD-AudioV* discs.

Understanding the DVD-Audio specification first requires understanding the basics of the DVD-Video specification, since DVD-Audio inherits much of the architecture and navigation features of DVD-Video. Any use of video with the audio also depends on the DVD-Video format. This section of the book generally describes the parts of DVD-Audio that are new or different. Other features, such as karaoke and text data, are essentially the same for both formats, so they are only described in the preceding DVD-Video section.

TABLE 6.38

Comparison of Audio Features

Feature	DVD-Video	DVD-Audio
Simultaneous streams	1 to 8	1 to 2
Channels/stream	1 to 8	1 o 6
PCM sample size	16, 20, 24 bits	16, 20, 24 bits
PCM sample frequency	48, 96 Hz	44.1, 48, 88.2, 96, 176.4, 192 Hz
Lossy compression (Dolby Digital, DTS, MPEG-2)	Standard	Optional
Lossless compression (MLP)	No	Yes

DVD-Audio players come in two flavors: *audio-only players* (AOPs), such as portable players, and *video-capable audio players* (VCAPs) that also can show onscreen information and play a subset of the DVD-Video format, as defined in the DVD-Audio specification. In addition, there are *universal players*, or *DVD-Audio / Video* players, that play both DVD-Video and DVD-Audio discs. Audio-only players may have a simple display, such as a one-line liquid-crystal display (LCD), that can display text information (song name, artist name, and so forth) and playback information (group number, track number, and so forth). Audio-only players are not required to display still pictures or motion video from DVD-Audio discs. Since visual menus are not always available, a simplified interface allows direct access to tracks and groups of tracks.

It is also possible to make a "universal disc" that plays in all DVD-Audio and DVD-Video players and computers by including a Dolby Digital version of the audio in the video zone. Anyone producing a DVD-Audio disc should strongly consider investing the small amount of extra work to make a disc that plays in all DVD players.

DVD-Audio players work with existing receivers. They output PCM and Dolby Digital, and some support the optional DTS and DSD formats. However, most current receivers are not designed with high-definition multi-channel PCM audio inputs, and even if they were, there is no standard for this type of digital audio connection. DVD-Audio players with high-end digital-to-analog converters (DACs) can be hooked up to receivers with two- or six-channel analog inputs, but some quality is lost if the receiver converts back to digital for processing. Improved receivers with advanced digital connections such as IEEE 1394 (FireWire) are needed to use the full digital resolution of DVD-Audio.

Data Structures

The DVD-Audio specification defines physical and logical data structures similar to those in DVD-Video (see Figure 6.36). The audio manager (AMG) is the top-level table of contents for a disc. It contains the navigation data for accessing smaller audio units. The audio manager optionally can contain a DVD-Video-style video object that provides a visual menu for the disc. One or more audio title sets (ATSs) hold audio objects (AOBs) and also can contain visual menus. Each audio object contains one or more tracks of audio, a set of still images, and text. In general, one AOB is used for each track on the disc. An AOB is limited to two audio streams, usually a multi-channel version and a stereo version. Still images related to an audio title

Figure 6.36
DVD-Audio logical
structure

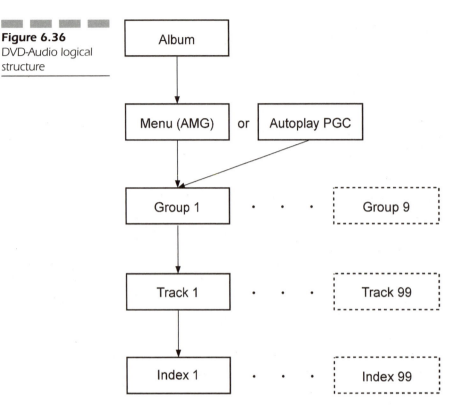

set are stored in an audio still video set (ASVS). Still pictures are stored in the multiplexed stream in SPCT packs. Real-time text, such as synchronized lyrics, are stored in real-time information (RTI) packs in the multiplexed stream (refer to Figure 6.15).

The structure of AOB streams is simplified from video object streams, with less interleaving of data in order to support continuous high-resolution audio playback. Links can be defined in the DVD-Audio structure that tie it to the DVD-Video content in the video zone.

DVD-Audio discs are broken into groups of tracks (songs). Tracks on a DVD-Audio disc are similar to tracks on a CD; each one is usually a single song or musical performance. If there are many tracks or tracks of similar nature (such as a set of stereo songs and a set of the same songs in multi-channel versions), they can be allocated into audio title groups (ATTGs). A disc (album) can have up to nine groups. Each group can have up to 99 tracks. As with CDs, optional indexes allow direct access to points within a track. Each track can have up to 99 indexes. Many discs have only one group. In DVD-Video terms, a group is a title, a track is a program (or a PGC), and an index is a cell (see Table 6.39).

A track may contain an audio selection block, which allows different channel configurations of the same music. The user can select stereo or multichannel, and the appropriate version of the track will be played. This is an alternative to including coefficients for downmixing a multichannel track to a stereo track.

DVD-Audio defines a *silent cell,* at index 0, that is the first cell in a program. Silent cells contain no audio data and are specially recognized by the player to provide a break between tracks. When tracks are played sequentially, the player "plays" the silent cell. When tracks are accessed randomly or skipped over, the silent cell is not played. If a still picture is associated with the silent cell, it will not be shown.

TABLE 6.39

DVD-Audio Terminology Recommendations

Friendly Term	Technical Term from DVD-Audio Specification	Example	Equivalent in DVD-Video
Album	Volume		Volume
Group	Title group	Stereo play list, multichannel play list	Title
Bonus group (enter four-digit key number to access)	Hidden group	Bonus tracks, hidden tracks	None
Track	Track		Program
Index	Index	Music marker	Cell
Spotlight	Spotlight	Highlighted lyrics	None
Album text	ATXTDT	Song, artist, liner notes	TXTDT
Track text	RTXTDT	Lyrics, song credits	Note
Slideshow	Slideshow	On-screen lyrics	Slideshow
Browsable pictures	Browsable pictures	Gallery	None

Navigation

Because DVD-Audio-only players have no video screen, and because users may wish to play DVD-Audio discs without turning on the television, a simplified navigation method is specified, called *simplified audio program play* (SAPP). The basic functions are play/pause, group access (jump to the first track of a group), track access (skip to the next or previous track), index access (skip to the next or previous index), and audio selection (choose between audio streams). Some of these functions, such as group access and index access, are optional. A set of still picture access commands also is defined: next, previous, and home. Next and previous are equivalent to "next page" and "previous page," whereas home is equivalent to "return." The disc author specifies which is the home image.

Bonus Tracks

The last group on a disc can be designated for conditional access. The group is hidden, not normally accessible to the user, unless the user knows a four-digit access code. Once the user enters the access code, the tracks in the last group become playable.

The idea behind this feature is to provide bonus tracks that can be used for prizes or incentives. There could be a series of puzzles in the disc packaging (or even in the disc menus) that have to be solved to reveal the access code. Or customers may need to visit a band or record label Web site to get the code. Of course, any secret such as this will not remain a secret for long, especially in the age of the Internet. It is inevitable that Internet sites will post lists of all known access codes for DVD-Audio discs.

Audio Formats

Linear PCM (LPCM) is mandatory on DVD-Audio discs, with up to 6 channels at sample rates of 44.1, 48, 88.2, 96, 176.4, and 192kHz with sample sizes of 16, 20, or 24 bits. This allows a theoretical frequency response of up to 96 kHz and a dynamic range of up to 144 decibels. Multichannel PCM tracks are downmixable by the player, although at 192 and 176.4 kHz, only two channels are available. The maximum data rate for a PCM stream is 9.6 Mbps.

To provide for longer playing times without compromising audio quality, *Meridian Lossless Packing* (MLP) can be used to compress the data. MLP removes redundancy from the signal to achieve a compression ratio of about 2:1 while allowing the PCM signal to be completely recreated by the MLP decoder. DVD-Audio players are required to include an MLP decoder.

The other audio formats of DVD-Video (Dolby Digital, MPEG audio, and DTS) are optional on DVD-Audio discs, although Dolby Digital is required for audio content that has associated video. A subset of DVD-Video features is allowed; those that are not are listed in Table 6.40. See Table 6.41 for compression ratios and Table 6.42 for playing times.

Sampling rates and sizes can vary for different channels by using a pre-defined set of two groups; as shown in Table 6.43. For mono and stereo configurations, there is only one group. For other configurations, there are two groups. Group 1 always has sampling rates and sizes that are higher than or the same as group 2. Note that channel configurations 8 to 12 are the same as 13 to 17, other than the channel groupings. Groups apply to PCM streams and MLP streams.

TABLE 6.40

DVD-Video
Features Not
Allowed in
DVD-Audio

Branching (multistory)
Camera angles
Parental management
Region control
User operation control

TABLE 6.41

MLP Compression

PCM Source	Minimum	Typical
44.1 kHz, 16 bits	25%	50%
48 kHz, 16 bits	10%	50%
96 kHz, 20 bits	40%	55%
96 kHz, 24 bits	38%	52%
192 kHz, 24 bits	43%	50%

	Audio Format	Playing Time (PCM)	Typical Playing Time (MLP)
TABLE 6.42 DVD-Audio Playing Times	6 channels, 96-kHz, 24-bit	45 minutes	85 minutes
	5.1 channels, 96-kHz, 24-bit	45 minutes	105 minutes
	5.1 channels + 2 channels, 96-kHz, 24-bit	33 minutes	75 minutes
	2 channels, 192-kHz, 24-bit	1 hour	2 hours
	2 channels, 96-kHz, 24-bit	2 hours	4 hours
	2 channels, 44.1-kHz, 16-bit	7 hours	12 hours
	1 channel, 44.1-kHz, 16-bit	14 hours	25 hours

High-Frequency Audio Concerns

Most existing equipment was designed with a 20 kHz frequency in mind, based on performance limitations of vinyl records, cassette tapes, CDs, and so on. With sampling frequencies of 176.4 and 192 kHz, DVD-Audio theoretically can reproduce frequencies of 88.2 or 96 kHz, with attendant supersonic signals. Existing equipment generally has not been tested for power-handling capacity when processing sustained supersonic signals. System components such as tweeters and supertweeters, passive speaker crossover networks, and power amplifier output stages are particularly vulnerable to overheating and damage from high-frequency signals. The DVD Forum, on advice from audio experts, has recommended that warnings be placed on test discs containing test signals with very high frequencies or very large amplitudes. The DVD Forum also recommends that DVD audio players include a protection device, such as a low-pass filter. The protection device can be turned off for use with equipment that has been adequately tested. Audio engineers and developers of audio mastering equipment also should take care not to introduce dangerous high-energy signal levels that might come from sources such as excessive noise shaping in signal processing or conversion, aliasing from upsampling without adequate filtering, equipment or hookup faults producing electronic oscillations in the supersonic region, and supersonic output from electronic instruments.

TABLE 6.43

DVD-Audio
Channel
Assignments

Channel Assignment	Ch 0	Ch 1	Ch 2	Ch 3	Ch 4	Ch 5
0	C					
1	L	R				
2	L	R	S			
3	L	R	Ls	Rs		
4	L	R	LFE			
5	L	R	LFE	S		
6	L	R	LFE	Ls	Rs	
7	L	R	C			
8	L	R	C	S		
9	L	R	C	Ls	Rs	
10	L	R	C	LFE		
11	L	R	C	LFE	S	
12	L	R	C	LFE	Ls	Rs
13	L	R	C	S		
14	L	R	C	Ls	Rs	
15	L	R	C	LFE		
16	L	R	C	LFE	S	
17	L	R	C	LFE	Ls	Rs
18	L	R	Ls	Rs	LFE	
19	L	R	Ls	Rs	C	
20	L	R	Ls	Rs	C	LFE
	Group 1				Group 2	

Note: L = left; R = right; C = center; LFE = low-frequency effects; S = surround; Ls = left surround; Rs = right surround.

Downmixing

DVD-Audio includes specialized downmixing features for PCM channels. PCM downmixing was included in DVD-Video but was never well defined or supported. DVD-Audio includes coefficient tables to control mixdown and avoid volume buildup from channel aggregation. Up to 16 tables can be defined by each audio title set (album), and each track can be identified with a table. Coefficients range from 0 to 60 decibels. This feature goes by the horribly contrived name of *SMART* (system-managed audio resource technique).

MLP includes a separate downmixing feature. MLP downmix coefficients can range from +6 to −24 decibels or 0 decibels (infinite attenuation). In place of downmixing, the coefficients can be use to control positive or negative phase.

Stills and Slideshows

DVD-Audio defines the oddly named term *audio still video* (ASV). An ASV is a still image (an MPEG I frame) that is displayed on a video screen while the audio plays. ASVs may include buttons, in which case they include subpicture and highlight information similar to that of DVD-Video. ASVs can be presented synchronously in slideshow mode or asynchronously in browse mode, depending on how the disc was authored. Slideshow mode uses predefined display times for each image, synchronized to the music. Browse mode allows the viewer to step forward and back through the images, independent of the music playback. In either mode, the images can be displayed either in sequential order or in random or shuffle order.

There can be up to 99 still images per audio track. The still pictures are loaded into a 2MB buffer in the player before the track begins playing, so the number of stills depends on how heavily compressed they are. At typical compression levels, there is room for about 20 stills per track. A predefined set of transitions can be used between stills: cut in, cut out, fade in, fade out, dissolve, and six wipe patterns.

Text

DVD-Audio provides two types of text: general text (*album text*) and real-time synchronized text (*track text*). General text data is stored in a single place on the disc and uses the same format as DVD-Video, covered earlier in this section. It is intended to hold general information about the content of the disc, such as album and track names, artists, producers, and ISRC or UPC_EAN codes. General text data also can hold URLs of related Web sites.

DVD-Audio real-time text is stored in packets along with the audio and video. It is intended to be displayed onscreen (or on the small display of an audio-only player) while the music plays, showing lyrics, credits, song notes, and other kinds of information directly related to the music. In a way, real-time text is like a character-based version of the DVD-Video subtitle feature.

Characters are encoded in ISO 8859-1 (Latin) for alphanumeric European languages or in Music Shift JIS for the Japanese language. Audio-only players can display European (single-byte) text using 4 lines of 30 characters each. Japanese (double-byte) text is displayed on 2 lines of 15 characters each.

Super Audio CD (SACD)

Sony and Philips (both members of the DVD Forum) developed an independent high-fidelity audio format for DVD media called *Super Audio CD* (SACD). SACD uses direct stream digital (DSD) encoding with sampling rates of up to 100 kHz. DSD is based on the *pulse-density modulation* (PDM) technique that uses single bits to represent the incremental rise or fall of the audio waveform. This supposedly improves quality by removing the "brick wall" filters required for PCM encoding. It also makes downsampling more accurate and efficient. Many modern analog-to-digital conversion systems use delta sigma oversampling, which is the technology used by DSD. In conventional digital audio systems, the bit stream is decimated to a PCM representation. Since DSD retains the 1-bit signal, it avoids decimation and interpolation steps, which can result in a more accurate and natural representation of the original waveform. DSD provides frequency response from DC to more than 100 kHz with a dynamic range exceeding 120 decibels. DSD includes a lossless encoding technique that produces approximately 2:1 data reduction by predicting each sample and then run-length encoding the error signal. Maximum data rate is 2.825 Mbps.

SACD is intended to provide "legacy" discs that have two layers: a CD-style layer that can be read by existing CD players and DVD-style layer for SACD players. A 0.6 mm substrate containing CD data at the outside surface is bonded to a 0.6 mm DVD substrate. This places the CD content at the standard 1.2mm distance from the read-out surface. The DVD layer is designed to be reflective at 650 nanometers (DVD laser wavelength) but transparent at 780 nanometers (CD laser wavelength). SACD players must be designed to recognize the DVD layer rather than first checking for CD data and assuming that the disc is a CD. Technical difficulties have limited the release of dual-format discs, especially considering that the initial price of these dual-layer discs was higher than for a standard CD plus a standard DVD.

Both stereo and multichannel versions of each track can be provided. SACD includes text and still graphics, but no video. SACD includes a physical watermarking feature called *pit signal processing* (PSP). PSP modulates the width of pits on the disc to store a digital watermark. Since normal user data are stored using the pit length, modifying the pit width has no effect on the data. The optical pickup must contain additional circuitry to read the PSP watermark, which is then compared with information on the disc to make sure that it is legitimate. Because of the requirement for PSP detection circuitry, protected SACD discs cannot be played in standard DVD-ROM drives.

SACD technology is available to existing Sony/Philips CD licensees at no additional cost. Pioneer, which released the first DVD-Audio players in Japan at the end of 1999, included SACD support in its DVD-Audio players. If other manufacturers follow suit, the entire SACD versus DVD-Audio format debate will be moot because DVD-Audio players will play both types of discs.

DVD Recording

The DVD recording application formats are designed to work on all the various writable DVD versions. Because the DVD Forum does not endorse DVD+RW, no effort has been made to ensure compatibility, but there is no reason the recording formats cannot be used with DVD+RW media.

Of course, compatibility issues do exist, both at the physical level and at the application level. More than 15 million DVD-Video players (more than 50 million, counting computers) were in the marketplace before the first wave of DVD recorders hit in late 2000 and early 2001. None of those 50 million existing players could read the DVD recording formats. As of this writing, it is too early to tell, but the recording formats may be supplanted by unofficial use of the DVD-Video format for recording. Clever formatting tricks allow recorders to use the DVD-Video format for real-time recording with only a quick fixup pass at the end to update content pointers. Some of the fancy features of the DVD recording formats such as custom playlists are not available when using the DVD-Video format, but this pales against compatibility with existing players and computers. The various compatibility issues are covered in more detail in the compatibility section of Chapter 5.

DVD-VR

The DVD Video Recording application format is designed to record video from analog and digital sources in real time. The original DVD-Video format requires that some information, such as time maps, pointers to video objects, and so on, be determined after the size and running time of the video are known. When recording real-time streaming data, this information is not available, so DVD-VR includes alternative ways to include navigation and search information. Instead of VTSI time maps and multiplexed navigation packs that DVD-Video and DVD-Audio use, DVD-VR defines a new VOBU map that stores timestamps along with the size of the video recorded since the last timestamp. The player can then calculate access points by adding up sizes. VOBU time maps are stored in video object information (VOBI) structures in a new object management layer that sits between the PGCI and the VOB levels. The object management layer also includes video object group information (VOGI) structures to keep track of still pictures.

New picture coding resolutions of 544×480 and 480×480 are defined for NTSC signals, and new still picture VOBs allow easy recording of MPEG I-frames for use with digital cameras. Audio recording in PCM is limited to one or two channels.

DVD-VR includes a data structure to record vertical blanking interval (VBI) signals that contain NTSC Closed Captions, PAL Teletext, aspect ratio signals, APS, CGMS, Web links, and so on. VBI data are recorded in real-time data information (RDI) packs in the multiplexed stream.

DVD-VR also supports user editing of recorded video. Users can delete recorded programs and create custom play lists. A *play list* is a collection of time pointers—a user-defined program chain—that allows video scenes to be presented in any order without modifying the original recorded data. After programs are recorded, DVD-VR recorders can create menus to provide instant access to each program. Using standard DVD-Video menu features, they can even create menus with thumbnail stills representing each video program.

DVD-AR

The DVD Audio Recording application format was still under development at the end of 2000. It is a modified version of the DVD-Video and DVD-Audio specs designed for real-time audio recording from analog and digital sources. The navigation structure is taken from DVD-Video, and the presentation structure is taken from DVD-Audio. Of course, it would be too much to hope that the discs would be compatible with DVD-Audio or DVD-Video players.

Incoming audio signals can be sampled and recorded at the same resolutions as DVD-Audio (44.1, 44, 88.2, 96, 176.4, or 192 kHz and 16, 20, or 24 bits). Only one audio stream can be recorded at a time. The audio can be stored as uncompressed PCM or in compressed formats such as Dolby Digital and MPEG audio.

Still pictures can be recorded along with the audio as MPEG-1 or MPEG-2 I-frames. A memory buffer as in DVD-Audio players allows a group of still pictures to be displayed without causing interruption of the audio. Text information (album name, track name, lyrics, and so forth) also can be recorded, along with the audio as album text or as real-time track text, similar to DVD-Audio.

As with DVD-VR, user-defined play lists can be created to customize playback of the recorded audio. Segments in a play list can be added, deleted, reordered, combined, and divided, all without changing the original recorded audio.

DVD-SR

The DVD Stream Recording application format was still under development at the end of 2000. The general idea of DVD-SR is to provide a standardized way to record streaming data from any source, such as a digital satellite receiver, digital cable tuner, digital video camera, or even streaming content from the Internet. The recorder does not "understand" the contents of what it is recording—it does not know if it is audio, video, stock tickers, or something else. Therefore, the recorder is incapable of directly playing back what it has recorded. It must pass the stream on to some other device that knows how to interpret, decode, and present the recorded content. DVD-SR players will use IEEE 1394/FireWire to connect to other devices that supply or play back streams of digital content. They also will implement copy protection using CPRM and DTCP.

DVD-SR uses the same general structure as DVD-Video but much simplified. Data is stored in stream recording objects (SROs). Instead of recording video packs, audio packs, subpicture packs, and so on, DVD-SR records only application data packs. Incoming data, such as MPEG transport streams, MPEG program streams, Microsoft ASF streams, and so forth, is chopped into 1998-byte chunks and wrapped in packs. If the recorder is designed to recognize high-level information in the incoming stream, such as timestamps, it can generate additional navigation data to support searching and trick play (fast forward, reverse, slow, and so on).

DVD-SR content can be mixed with DVD-VR content. They can share play lists and can have mixed program chains that include both VR cells and SR cells. Draft versions of the DVD-SR format defined streaming objects, or SOBs, causing much hilarity among those more familiar with the English language.

What's Wrong with DVD

Introduction

DVD-ROM is a well-designed update of CD-ROM. Storage space and speed are much improved, and many of the shortcomings of CD-ROM have been rectified. DVD-Video, on the other hand, is a collection of compromises in which commercially and politically motivated restrictions have outweighed technical limitations. It is unfortunate that in an overzealous attempt to protect their intellectual property, Hollywood studios have crippled the format that can give them the best venue of expression outside a theater.

Perhaps the biggest shortcoming of DVD-Video is ease of piracy, or, more accurately, the perception of potential losses from piracy. Ironically, the superior digital quality of DVD has proven to be its Achilles' heel. The ease with which unscrupulous or unthinking people may be able to make a perfect digital copy and spread it across the Internet has caused too many sleepless nights on the part of studio executives. The thought of discs, legal or not, available for sale anywhere in the world is also an apparent concern. The reaction has been to add restrictive regional codes, an analog copy protection system that can degrade the picture, and limits on digital audio output quality. These restrictions, along with encryption and watermarking, also complicate the otherwise simple use of DVDs in computers.

Regional Management

Movies are not released all over the world at the same time. A major motion picture may come out on video in the United States—months after its theatrical release—at the same time it is being shown on theater screens for the first time in Europe or Japan. Movie distributors are worried that if video copies of a movie were available before official release, it would reduce the success of the movie in the theaters. In addition, different distribution rights and licenses are established in different countries. For example, Disney may own distribution rights in the United States, but Studio Ghibli may own them in Japan.

Because of this, studios and distributors want a way to control distribution. Every DVD player has a code that identifies the geographic region in which it was sold. Discs intended for use only in certain countries have codes stored on them that identify which regions are allowed and which are not allowed. A DVD player will not play a disc that is not permitted to be played in the region in which the player was sold.

This regional management system wreaks havoc with import markets. An otherwise legally imported disc often will not work because of its regional codes. A worse problem is that discs and players purchased in one country may not work when taken to another. If someone in Japan moves to the United States and takes a DVD player along, that person will not be able to play most movies purchased in the United States. Likewise, if that person buys a new DVD player in the United States, he or she would not be able to play many of the discs brought from Japan.

Regional management was implemented to preserve a parochial system of distribution. As the Internet breaks down national boundaries of commerce, and as digital cinema allows movies to debut in theaters worldwide at the same time, region codes will become mostly irrelevant. Nevertheless, DVD movies undoubtedly will continue to be circumscribed geographically.

DVD regional locks are optional; any disc can be designed to play in all players. However, most movie releases—even those which have been available for years—include regional locks. As DVD was being developed and first released, some studios stated that older movies that had already made their theatrical run would not be restricted by region. However, none of the studios have yet lived up to this promise.

Because of the inconvenience of regional codes outside the United States, over 50 percent of players sold in most countries are modified (or modifiable) to disable region coding.

DVD-ROM drives also are regionally coded. Computer DVD-Video player applications check for regional codes before playing movies. Computer software on DVD-ROM does not use regional codes.

Details of the regional management system are covered in more detail in Chapter 4.

Copy Protection

The movie industry claims that over $700 million worldwide is lost each year to casual illegal copying of videotapes. Surveys of video rental outlets show that more than half believe that their business is hurt by consumers who make copies of rental tapes. It is debatable how many illegal copies would have been purchased or rented had the tapes been uncopyable. The industry's own data show that 30 percent of consumers who are thwarted by videotape copy protection subsequently rent or buy a legitimate copy, and video retailers believe that copy-protected tapes increase their revenue by 18 percent. Regardless of the effectiveness of copy protection schemes, DVD includes a number of them.

Every DVD-Video player must include Macrovision or similar analog copy protection technology to deter copying of DVD onto videotape. Macrovision adds pulses to the video signal that confuse the recording circuitry of VCRs and render a tape copy unwatchable, supposedly without affecting the picture when the DVD is played on a television. However, some equipment, especially line doublers and high-end televisions, is unable to cleanly display video that has been altered by the Macrovision process.

Analog and digital copying of DVD is controlled by information specifying whether the video can be copied and whether it can be copied only once or unlimited times. Digital recording equipment must respect this information. The "copy once" setting is intended to allow consumers to make copies for personal use but not allow copies of the copies (since the copy is marked for "no copies").

Digital copying is also prevented by encrypting critical sectors on the disc. Video data that has been copied from a disc using a DVD-ROM drive will not play without the decryption keys that were hidden on the original disc. During normal playback, the video data stays scrambled until just before being decoded and displayed, so it cannot be intercepted and copied easily within the computer.

Even the people who designed the copy protection techniques (the industry's Copy Protection Technical Working Group) freely admit that they do not expect copy protection to slow down professional thieves. The MPAA claims that over $2.5 billion a year is lost to professional piracy worldwide. Anyone equipped with the proper equipment to defeat Macrovision or make a bit-by-bit digital copy of a disc can get around the copy protection barriers. In fact, the targets of these measures are people who make a copy or two for friends. The stated goal is "to keep the honest people honest."

Movie studios and consumer electronics companies have promoted legislation to make it illegal to defeat DVD copy protection. The result is the *World Intellectual Property Organization* (WIPO) Copyright Treaty, the WIPO Performances and Phonograms Treaty (December 1996), and the compliant U.S. *Digital Millennium Copyright Act* (DMCA), passed into law in October 1998. Software or devices intended specifically and primarily to circumvent copy protection are now illegal in the United States and many other countries. A cochair of the legal group of the copy protection committee stated, "In the video context, the contemplated legislation should also provide some specific assurances that certain reasonable and customary home recording practices will be permitted, in addition to providing penalties for circumvention." Since most DVD movies are designed to prevent any sort of copying, even single copies for personal use, it is not at all clear how this might be "permitted" by a player.

All copy protection systems are optional for DVD publishers. Decryption in players is also optional, so it is possible that a small number of DVD-ROM computers and possibly even DVD-Video players will not be able to play protected movies.

Hollywood Baggage on Computers

Given the ability of many computers to play movies from DVD, the regional management and copy protection requirements apply to computers as well as home players. Manufacturers of hardware or software involved in the playback of scrambled movies are required to obtain a license from the DVD Copy Control Authority. The hardware and software must ensure that decrypted files cannot be copied, that digital outputs contain proper copy protection information, and that analog outputs are protected by Macrovision or a similar process.

The upside is that these safeguards assure Hollywood that its property will not be plundered and spread illegally from computer to computer. The downside is that most law-abiding computer owners are inconvenienced and may pay slightly more because of these protection measures. This is especially irritating to those who have no interest in watching commercial movies on their computer screens. In addition, copying and editing of legitimate video, such as from digital camcorders, is made difficult or impossible because of copyright safeguards. See Chapter 11 for more details of copy protection and regional management issues on computers.

NTSC versus PAL

Because DVD-Video is based on standard MPEG-2 digital video, it could have been a worldwide standard capable of working with both 525/60 (NTSC) and 625/50 (PAL/SECAM) television systems, allowing discs produced anywhere in the world to be played anywhere else in the world. Sadly, this is not the case. Partly in an attempt to limit the widespread distribution of discs, but mostly because it was easier, two different video formats are used for DVD discs and players. Two different surround audio standards are also used: Dolby Digital and MPEG-2 audio. Movies intended for distribution in the United States, Japan, and other countries using the NTSC system are encoded for display at 30 frames per second along with Dolby Digital audio, whereas movies intended for release in Europe and in countries using the PAL or similar SECAM standard are encoded for display at 25 frames per second. PAL discs optionally can use the MPEG audio format, which many NTSC players do not recognize. On top of the format differences, playback speed is slightly faster for movies in PAL format. Because of these differences, countries with NTSC televisions require NTSC DVD playback hardware, and countries with PAL televisions require PAL DVD playback hardware.

Since NTSC is the dominant standard, almost all DVD players released in PAL countries can play both types of discs as long as the right kind of television is connected. When playing NTSC discs, PAL players can output a 4.43 NTSC signal (called *60-Hz PAL*) that combines the NTSC scanning rate with the PAL color-encoding system. Most PAL televisions sold since mid 1990 can handle this type of signal. Some players also provide the option to produce a pure 525/60 signal from an NTSC disc, but this requires a fully NTSC-compatible TV.

There are a few standards-converting PAL players that completely convert 525/60 NTSC video to standard 625/50 PAL output. The conversion process is quite complex. Achieving high quality requires expensive hardware to handle scaling, temporal conversion, and object motion analysis. The audio speed also must be adjusted to conform to the different display rate, which can prevent the use of a digital audio connection. Because most DVD players use shortcuts and cheap circuitry to convert video, using 60-Hz PAL output with a PAL-compatible TV provides a better picture than converted PAL output.

Things are much less compatible the other way around. Most NTSC players cannot play PAL discs. A very small number of NTSC players can convert 625/50 PAL to 525/60 NTSC, but again, video quality suffers from low-cost conversion solutions.

The restrictions of television systems do not apply to computer monitors. Properly equipped computers can play movies from both NTSC and PAL discs and can decode both Dolby Digital and MPEG audio.

Tardy DVD-Audio

No standard for the DVD-Audio format was ready when DVD was introduced. In fact, it was not until 4 years later that DVD-Audio products finally became available. However, since the DVD-Video standard already included very high quality audio, and since DVD-Video players can play audio CDs, there has not been a big hurry or big demand for a new audio standard.

Incompatible Recordable Formats

Despite many obstacles, the companies backing DVD managed to come out with a single standard for DVD-ROM and DVD-Video. The incredible success of DVD clearly proves that customers will readily adopt a common format that has widespread support. Unfortunately, this lesson seems to have been lost on the very companies that should have learned it. There are no less than five different recordable variations of DVD, with other varieties looming. Each recordable format has different incompatibilities with the other formats as well as with existing players and DVD-ROM drives. On top of the physical incompatibilities, there are three separate recording application formats: one for video, one for audio, and one for "streaming data" such as from a camcorder or a digital video receiver. None of these recording file formats are readable by standard DVD video or audio players, and all require software upgrades on DVD computers. The details of incompatible formats are covered in Chapter 4.

Late-Blooming Video Recording

Recordable DVD arrived about a year after DVD-ROM and DVD-Video, but only for recording data on computers. It took another 3 years for DVD home video recorders to arrive, accompanied by $2000+ price tags and, as mentioned in the preceding section, plagued with compatibility problems.

Some people believe that DVD home recorders will never be much of a success because digital tape is more cost-effective. On the other hand, digital tape lacks many of the advantages of DVD, such as seamless branching, instant rewind/fast forward, instant search, and durability, not to mention the appeal of shiny discs. Therefore, once the encoding technology and recording hardware are fast and cheap enough and the blank discs are cheap enough, DVD recorders could be the VCRs of the new millennium. However, given the stupidity of not defining a recording format before millions of players and drives were sold, and given the competition between manufacturers releasing a jumble of proprietary recorders, it will take recordable DVD much longer to succeed than it should have.

Recordable DVD also may face competition from digital videotape (DV). However, DV is not currently intended for mass-market prerecorded video, and D-VHS can only record already-digitized video signals. See the "Digital Videotape" section of Chapter 8 for more information.

Playback Incompatibilities

The DVD-Video specification was put together like a patchwork quilt, with different parts coming from dozens of different engineers speaking Japanese, Dutch, and English. The books were assembled and produced in Japanese and then translated into English. Engineers from a large number of companies all over the world implemented the specs in player designs. It is no surprise, therefore, that the DVD-Video feature set is incompletely or improperly implemented in many players. Even then, many incompatibility problems have resulted simply from laziness or carelessness. Many early DVD title developers had to scale back their plans after discovering that their ingeniously designed discs worked differently or not at all on different players.

For example, the DVD specification allows 999 chapters per title, but some players cannot handle more than 511. Likewise, players are supposed to support 999 program chains, but some early players balk at 244. Some cannot handle full video data rates of 9.8 Mbps, whereas others cannot deal with extra files in the root directory. The release of *The Matrix* in 1999 brought wide publicity to the general malady. Because *The Matrix* was the first million-seller DVD, and because it was an unusually complex title that also contained PC enhancements, more people than ever before had problems playing the disc. Some problems were caused by authoring and formatting errors on the disc, but most problems were caused by flaws in a surprisingly high number of player models.

Taking the long view, *The Matrix* and other "problem discs" did the DVD world a favor by exposing flaws in players that failed to properly play discs authored according to the DVD specification. Although the backlash to the studios, production houses, and InterActual (the company providing the PCFriendly computer enhancement software) was painful, pushing the envelope early ensured that DVD manufacturers are now more responsible about making players that work right. It is better to deal with these "growing pains" early on when there are only a few million players than later when there are tens of millions.

The number of players that continue to have design flaws, even after many iterations, is inexcusable because they fail on discs that have been available for years. Some problems are caused by bugs in DVD authoring software, but as authoring software programs mature, most of them incorporate workarounds to avoid known errors and deficiencies in players. The situation is slowly improving with each new release of players, but producers who want their discs to work on their customers' players must still accommodate older models.

Synchronization Problems

When certain discs are played in certain players, there is a "lip sync" breakdown where audio lags slightly behind the video or in some cases slightly precedes the video. Perception of the sync problem is highly subjective—some people are bothered by it, whereas others cannot discern it at all. The cause is a complex interaction of as many as four factors:

1. Improper matching of audio and video tracks in the DVD encoding/authoring process
2. Poor sync techniques during film production or editing (especially postdubbing or looping)
3. Loose sync tolerances in the DVD player
4. Delay in the external decoder/receiver

Factor 1 or 2 usually must be present in order for factor 3 or 4 to become apparent. Some discs with severe sync problems have been reissued after being reencoded to fix the problem. In some cases the sync problem in the player (factor 3) can be fixed by pausing or stopping playback and then restarting or by turning the player off, waiting a few seconds, and then turning it back on. Unfortunately, there is no simple answer and no single, simple fix. Player manufacturers and disc producers need to cooperate to nail down and fix the various causes.

Feeble Support of Parental Choice Features

Hollywood was quick to request a parental management feature for DVD but very slow to actually use it. DVDs can be designed to play a different version of a movie depending on the parental level that has been set in the player. By taking advantage of the seamless branching feature of DVD, objectionable scenes can be skipped over automatically or replaced during playback. This requires that the disc be carefully authored with alternate scenes and branch points that do not cause interruptions or discontinuities in the soundtrack. Unfortunately, very few multirating discs have been produced. Hollywood studios are not convinced that there is a big enough demand to justify the extra work involved (shooting extra footage, recording extra audio, editing new sequences, creating branch points, synchronizing the soundtrack across jumps, submitting new versions for MPAA rating, dealing with players that do not properly implement parental branching, having video store chains refuse to carry discs with unrated content, and more). The result is that a distinctive feature of DVD is largely wasted.

Since studios are not supporting the built-in parental management feature, an alternative is to use a software player on a computer that can read a "play list" that tells it where to skip scenes or mute the audio. Play lists can be created for the thousands of DVD movies that have been produced without parental control information. Devices such as TV Guardian that connect between the player and the TV and read Closed Captions in order to filter out profanity and vulgar language also work with DVD, as long as the disc contains Closed Captions.

Not Better Enough

Part of the reason for the success of CD was that the technology was not overly advanced. Technical difficulties in production and playback were minimized, tolerances could be met easily, and costs could be kept low. DVD has taken the same approach, which will go far toward helping it become cheap and widespread, but it also means that the bar was not raised terribly high. High-definition television (HDTV) is just around the corner, and DVD could have been the first format to fully support it. There will be a second generation of DVD designed for HDTV, but in a way it is a shame that the designers did not try harder to achieve more in the first generation.

During the early years of HDTV (which was officially introduced in the United States in late 1998), DVD provides the best source for prerecorded video. Since HDTV sets include analog video connectors that work with all DVD players and other existing video equipment such as VCRs, existing DVD players and discs work perfectly with HDTV sets and provide a much better picture than any other prerecorded consumer video format. Progressive-scan DVD players provide an even better picture when connected to a progressive-scan display. Unfortunately, some DVDs contain video from interlaced sources, which is difficult to make look good on a progressive display. Even for progressive-source video, certain steps in the DVD production process are oriented toward optimizing video quality on interlaced displays, which means that detail may be reduced from what it could be in pure progressive format. The details of progressive scan are covered in Chapter 3.

No Reverse Gear

Because of the way MPEG-2 compressed video builds frames by using the differences from previous frames, it is impossible to play in reverse without a large memory buffer in which to store a set of previous frames or without a very complex high-speed process of jumping back and forth on the disc to build up a frame in order from a key frame. RAM for video buffers is expensive, so only DVD computers and very high-end DVD players can play backwards at normal speed. As memory gets cheaper, this feature will trickle down to cheaper players.

Some players can move backwards through a disc by skipping between key frames. Attempting to display these frames at the proper time intervals results in jerky playback with delays of about 1/2 second between each frame. Smooth scan can only be achieved by showing the frames at 12 to 15 times normal speed.

Only Two Aspect Ratios

DVD is limited to 1.33 (4:3) and 1.78 (16:9) aspect ratios, even though MPEG-2 allows a third aspect ratio of 2.21. Better yet would be the ability to support any aspect ratio. Because of the need for a standard physical shape, televisions essentially come in two shapes: 4:3 and 16:9. However, if the player or TV were able to unsqueeze an anamorphic source of any ratio, it would provide better resolution because pixels would not be wasted on letterbox mattes. Letterbox mattes would still need to be generated by the player or the display, but high-resolution displays would be able to make the most of every pixel of the anamorphic signal. The obvious disadvantage to this feature is that variable-geometry picture scaling circuitry is more expensive than the fixed-geometry scaling of DVD players and existing wide-screen TVs.

A related problem is that the DVD specification includes a provision for tagging preletterboxed video so that wide-screen displays can adjust automatically, but some discs are authored without the flag. That is, the television should automatically enlarge the picture to fill the screen and get rid of the letterbox mattes that were encoded with the video, but because the flag was not set properly during disc production, the player does not know to send the signal to the television. See Chapter 3 for details on aspect ratios.

Deficient Pan and Scan

DVD uses a feature of MPEG-2 to store picture offset information that allows anamorphic wide-screen video to be converted automatically to full-frame pan and scan. However, the horizontal-only limitation of this feature has caused it to be widely ignored. The full-frame conversion process in a studio not only pans from side to side but also moves up and down and zooms in and out and usually includes extra picture from above and below the wide-screen area. Disc producers are not happy with DVD's limited automatic pan and scan feature, so they either create a wide screen-only disc or they include a second full-frame pan and scan version of the movie on the disc. The designers of DVD could have included the option for full-frame video storage at high resolution to allow extraction of 4:3 pan and scan, 4:3 letterbox, and 16:9 anamorphic with no loss of resolution. This would require more pixels and thus more storage space (39 percent more for 720×666 anamorphic and 71 percent more for 888×666 full frame) but would be more efficient than encoding two (or three) separate versions on the disc. The added video processing complexity would have slightly increased player cost in early generations.

Inefficient Multitrack Audio

MPEG-2 audio is inefficient in MPEG-1–compatibility mode. Parts of the audio signal must be duplicated to achieve full channel separation, thus creating an overhead of about 10 percent (see Chapter 3 for details). MPEG-2 provides a non-backward–compatible format (AAC), but since this system was not finalized when DVD was introduced, and since MPEG-2 decoders were not available, DVD-Video discs intended for PAL players are saddled with a less-than-optimal algorithm on MPEG-2 multichannel tracks in order to support MPEG-1 decoders.

Other ideas for improving the general efficiency of multiple audio tracks were considered. Alternate language sound tracks contain the same music and sound effects, differing only in the dialogue. It could be possible to store all dialogue tracks in mono or stereo form and have the player use a two-track decoder to mix them together. The problem is that dialogue is not isolated to one or two tracks. Using a mono or nondiscrete dialogue track could detract from the audio presentation. Another possibility is differential encoding, which allows additional speech tracks to be encoded relative to the main sound track containing music and special effects, rather than recoding an entire track for each language. This results in a lower data rate for additional language tracks. These ideas have the benefit of reducing data rate (and thus increasing quality or playing time) but the drawback of adding complexity to the audio decoding component of players.

On a related note, there is no provision in DVD for automatically dealing with the difference in playback speed when a 24 frame/sec (fps) film is displayed at 25 fps (PAL). The 4 percent speedup must be taken care of before the audio is encoded. If Dolby Digital and DTS were able to automatically conform the audio at playback time for different speeds, production would be greatly simplified, as would the manufacture of players that convert from PAL to NTSC or vice versa.

Inadequate Interactivity

The ability of DVD-Video players to process and react to user control is limited. There is little beyond the bare basics of menus and branching. Features such as navigation, user input, score display, indexes, searching, and the like are possible with DVD-Video, but each permutation must be anticipated and put in place when the disc is produced. All video displayed by the player has to be created ahead of time, unlike other systems such as CD-i or video game consoles that can generate graphics and text on the fly.

Consider, for example, a DVD-Video quiz game designed for a standard DVD player. Commands on the disc can program the player to keep track of the player's score, but there is no way to directly display the score. Instead, a set of screens must be created ahead of time to cover all the possible results. To give a score to within 5 percent, 21 screens are needed (0, 5, 10, . . . , 90, and 100 percent). Given the amount of work required to create the screens and to design the code for displaying them, it is more likely that a compromise of 3 or 4 screens would be used, giving a coarse result such as "bad," "fair," "good," and "great." This type of compromise is likely to pervade all attempts at interactivity in DVD-Video.

Searches or lookups can only be done with DVD-Video in a hierarchical or sequential manner. For example, to look up a word in a small 1000-word glossary, the first screen would list the 26 letters of the alphabet (DVD allows up to 36 buttons on a screen). After the first letter was selected, additional screens of words would be displayed. Ten or more screens might be required to list all the words beginning with a common letter. The "Next" and "Previous" buttons on the remote control would be used to page through the screens one at a time. Finally, after a word was selected, a screen with the definition would be shown. This would require the laborious preparation of 1000 screens for the definitions and more than 100 menu screens to access them. A more powerful system than DVD simply could read the text of a definition from a database file and display it on a generic definition screen.

The use of DVD-Video for education, productivity, games, and the like is disappointingly limited. The feature set was clearly designed from the limited goals of on-screen menus, simple branching, and karaoke. Clever use of the rudimentary interactive features of DVD can accomplish a truly surprising amount, but most producers will be hard pressed to put in the extra time and effort required, thus resulting in reduced usability. A DVD-Video cookbook, for instance, may provide a simple index search of ingredients or dishes that would include chickpeas and stew but would frustrate someone looking for garbanzo beans or goulash and may not even include the carrots that are in the stew.

Limited Graphics

The subpicture feature of DVD could be extended far beyond simple captions, menus, and crude animation if it were not limited to four colors and four transparency levels at a time. Much richer visual interaction could be achieved with more colors. Advanced features such as sprites and rudimentary three-dimensional (3D) capabilities also would expand DVD's repertoire.

The designers obviously chose to limit the subpicture format to save cost and bandwidth, but even a small improvement of 16 simultaneous colors would have made for significantly better-looking subtitles, more sophisticated highlighting, and a superior graphic overlay environment.

Small Discs

Part of the appeal of DVD is its small, convenient size. Ironically, however, this is also a drawback. People are psychologically averse to paying the same amount for something smaller. Even though the smaller item may be better, the larger item somehow seems to be worth more.

The small size of DVDs also limits the cover art. Gone are the days of innovative art on LP jackets or good-sized movie art and descriptive text on laserdisc covers. Liner notes are also limited, but this shortcoming can be compensated for by including still pictures, short clips, and other material in video form on the disc itself.

False Alarms

Most DVD movie packages include security tags inserted at the factory. The added cost of these security measures is easily offset by a reduction in shoplifting costs, but a minor side effect is that when discs are purchased in a store that has not implemented security, the tags are not deactivated. Unsuspecting shoppers may get a rude surprise when they walk into another store and set off an alarm.

No Bar-Code Standard

One of the most powerful features of laserdisc players used in training and education is bar codes. Printed bar codes are scanned using a wand that sends commands to the player via the infrared remote interface, telling it to search to a specific picture or play a certain segment. A simple player becomes a powerful interactive presentation tool when combined with a bar-code reader. Bar codes can be added to textbooks, charts, posters, lesson outlines, storybooks, workbooks, and much more, enhancing them with quick access to pictures and movies.

Some industrial/educational DVD players from companies such as Pioneer and Philips support bar-code readers, but the lack of standardized support for bar-code readers, even as add-ons, ultimately denies the advantages to average player owners.

No External Control Standard

Most consumer laserdisc players include an external control connector, and all industrial laserdisc players include a serial port for connection to a computer. An entire genre of multimedia evolved during the 1980s using laserdisc players to add sound and video to computer software. Admittedly, this is less important today as the multimedia features of computers improve, but many applications of DVD such as video editing, kiosks, and custom installations are limited by lack of an external control standard. As with bar codes, some industrial/educational DVD player models include RS-232 or similar external control ports, but each player manufacturer uses a different proprietary command protocol.

Poor Computer Compatibility

The multimedia CD-ROM industry has long been plagued by incompatibility problems. In 1995, return rates of CD-ROMs were as high as 40 percent, mostly because customers were unable to get them to work on their computers. Compatibility problems were caused by incorrect hardware or software setups, defects in video and audio hardware, bugs in video and audio driver software, and the basic problem that hardware such as CD-ROM drives and microprocessors often were not powerful enough for the tasks demanded of them. The potential for problems with DVD-ROM is even worse. In addition to all the compatibility problems of CD-ROMs, DVD-ROMs will have to deal with defects in video and audio decoder hardware or software, incompatibilities of proprietary playback implementations, decoder software that cannot keep up with full-rate movies, DVD-Video navigation software that does not correctly emulate a DVD-Video player, and so on. Not all computers with a DVD-ROM drive will play movies from a DVD-Video disc, especially computers that are upgraded from a CD-ROM to a DVD-ROM drive. Someone buying or upgrading a computer may not understand this.

No WebDVD Standard

One of the most interesting uses of DVD is in combination with HTML and the Internet. Hollywood studios, corporations, educators, and thousands of other DVD makers are excited about the potential of combining the best of DVD with the best of the Internet. Unfortunately, the DVD-Video specification has no provisions to make this easier or standardized. The DVD-Audio specification does have a URL link feature, but it hardly scratches the surface. Other groups are working on standards for Web-connected DVDs, but most enhanced discs only work on certain platforms. See Chapter 11 for more on WebDVD.

Escalated Obsolescence

DVD is attached at the hip to the computer marketplace, which has a nasty habit of evolving faster than the average customer would prefer. For contrast, consider laserdisc, which is more than 20 years old. Laserdiscs made in 1978 still play on most players and look great. Compare this with computers; the first microcomputers came out only a few years before laserdisc, but they have already churned through many generations. How usable is most DOS software from the 1980s? How useful today are word-processing files from a TRS-80 or games from an Apple II?

The greatest advantage of digital information—its flexibility—is in a way its greatest shortcoming. It is too easy to tinker with the formula. TV, radio, and other systems have lasted for years with only minor improvements because it was too hard to make major improvements without starting over. The digital clay of the DVD format, however, is so malleable that in too short a time the temptation to revamp it becomes irresistible. This does not mean that existing discs will not play in future players, but it does mean that new features will be added that will require new or updated equipment, to the delight of those who sell new equipment, but to the dismay of their customers.

The next generation of DVD was under development even before the first generation was released. It will use blue lasers to achieve the higher data density required for HDTV. Discs in the new HD-DVD format will not play on older DVD players unless they are made as hybrids with an "old" side and a "new" side. Average American consumers replace their TVs every 7 to 8 years. Let us hope that the phases of DVD last at least that long.

Summary

The creators of DVD are aware of most of its limitations. Some were deliberate compromises to keep down cost and complexity. Some were simply beyond what they wanted to deal with.

There are solutions to most of DVD's shortcomings. Improvements perhaps will appear in proprietary enhanced versions of DVD or, better yet, be officially supported in the next generation of DVD. See Chapter 13 for more perspective on the future evolution of the format.

Considering that DVD was a compromise solution beaten into a form that was acceptable to hundreds of people from dozens of companies in the consumer electronics, movie, and computer industries, all with different priorities and different motivations—other than profit—it is amazing that DVD turned out as well as it did.

DVD is not the be-all and end-all medium for entertainment or computer data storage, nor was it ever intended to be. Technology had reached the point where it was time to introduce a new format, free of the drawbacks and dead ends of compact disc, videotape, and laserdisc. Detractors complain that DVD should have waited longer and taken advantage of new technology such as shorter-wavelength lasers in order to store even more data, or that DVD-Video is not sufficiently improved over laserdisc, or that it should have waited and fully supported HDTV. This kind of argument leaves one forever poised on the edge of a leap that is never taken, however, because something better is always around the corner. Once blue lasers are available commercially, ultraviolet lasers will be in the laboratories, promising even greater storage density. And how much better is "enough?" Realistically, DVD-Video is already capable of more than most televisions can deliver. Improving the quality would have made DVD more expensive and less reliable for little or no visible gain. And HDTV, as promising as it may be, is experiencing the most sluggish introduction ever seen; it will take many long years to establish an HDTV market able to sustain a major consumer electronics product such as DVD.

All in all, DVD is a major step forward in many areas. There is certainly room for improvement, but even the Brave Little Tailor killed only seven in one blow.

DVD
Comparison

Introduction

This chapter compares DVD systems with related consumer electronics and computer data storage products. Each section presents technical specifications as well as advantages and disadvantages. The charts are, of necessity, rather terse and technical, but most points are explained in the accompanying paragraphs or in Chapters 3 and 4. Most terms and acronyms are also defined in the Glossary.

Some specifications, such as *signal-to-noise ratio* (SNR) and *dynamic range,* are technical maximums that usually are lower in practice. For example, both DVD video and audio at 24 bits per sample have a theoretical SNR of 144 decibels, but MPEG compression creates variable video noise and most recording equipment cannot actually achieve an SNR of 144 decibels, and current digital-to-analog converters are incapable of reproducing a perfectly clean signal.

While some technologies may be considered competitors to DVD, they also may complement DVD, and vice versa. For example, VHS and DVD can coexist much like audiocassette tape and audio CD. *Digital videotape* (DV) is a popular recording source in producing video for DVD.

Laserdisc and CD-Video (CDV)

Laserdisc is the most obvious competitor to DVD-Video because it is a high-quality video format on optical disc. DVD player manufacturers found their initial primary customers to be videophiles and home theater aficionados, many of whom own laserdisc players.

Even before it came to market, DVD dealt laserdisc a mortal blow. Anticipation of DVD in 1996 drove laserdisc player sales down 37 percent, even though sales of VCRs and hi-fi/surround-sound systems were up. Disc sales also were down over 30 percent. Approximately 70 percent of early DVD buyers already owned laserdisc players. In July 1999, Pioneer Entertainment, the largest laserdisc distributor, announced that it had shifted focus to VHS and DVD to replace all its laserdisc business.[1] Image Entertainment, formerly the largest independent distributor of laserdiscs, released its last laserdisc titles in February 2000.

---

[1] Other arms of Pioneer continued to distribute laserdiscs for education, corporate training, and special applications such as museum kiosks.

DVD quickly displaced laserdisc as the premiere home entertainment format, but it will never achieve 100 percent replacement. About 10,000 laserdisc titles were released in the United States for a peak installed base of about 2 million players. Over 35,000 laserdisc titles were released worldwide into a market that reached approximately 7 million laserdisc players. DVD attained the same player base in less than three years, but it will take four or five years to build a similar library of titles. Laserdisc has the superiority of tenure and will continue to be a source of quality video, especially for rare titles that may not appear on DVD for a long while, if ever. Most laserdisc player owners bought DVD players shortly after they became available, but few have rushed to replace their laserdisc collection.

An important distinction between laserdisc and DVD is that laserdiscs do not contain digital video, and they do not always use digital audio. The laserdisc video format is analog pulse FM-encoded composite video.

CDV, sometimes called *Video Single* or *CD-Video* (not to be confused with Video CD), is actually a hybrid of CD and laserdisc. Part of a CDV contains 20 minutes of digital audio playable on any CD player, DVD player, or CD-compatible laserdisc player. The other part of a CDV contains 5 or 6 minutes of analog video and digital audio in laserdisc format, playable only on CDV-compatible laserdisc systems. Table 8.1 lists laserdisc and DVD-Video specifications.

Advantages of DVD-Video over Laserdisc

Features. DVD-Video has the same basic features as CLV laserdisc (such as, scan, pause, search) plus most of the added benefits of CAV laserdisc (such as, freeze, slow, fast). DVD goes beyond laserdisc with multistory branching, parental control, multiple camera angles, video menus, interactivity, and more. Level II laserdisc players had a command language similar to that of DVD, but level II discs and players never grew beyond the small niches of education and industrial training.

Capacity. Programs on DVD can be over four times longer than those on laserdisc at equivalent quality. A single-layer DVD-Video holds over two hours of material per side, and a dual-layer disc holds over four hours. A CLV laserdisc holds one hour per side, and a CAV laserdisc holds only one half hour.

DVD-Video supports still frames with audio, allowing for hundreds or thousands of pictures accompanied by hours of surround sound. Laserdisc still frames have no audio (unless specially produced discs are connected to expensive still-frame audio equipment).

TABLE 8.1

Laserdisc and
DVD-Video
Specifications

	Laserdisc	DVD-Video
Diameter	30 or 20 cm	12 or 8 cm
Thickness	2.4 mm	1.2 mm
Average pit length	1.3 m	1.2 m
Track pitch	1.6 m	0.74 m
Rotational velocity	600 to 1800 rpm	570 to 1600 rpm
Video	Composite analog NTSC	Component digital MPEG-2
Playing time	1 h/side (CLV), 0.5 h/side (CAV)	2+ h/side (1 layer), 4+ h/side (2 layer)
Widescreen support	Letterbox[a]	Anamorphic
Analog copy protection	None	Macrovision
Video SNR	~50 dB	~70 dB
NTSC resolution[b]	~272,160 pels (567 × 480); ~204,120 (567 × 360) letterboxed to 16:9	345,600 pels (720 × 480)
PAL resolution[b]	~326,592 pels (567 × 576); ~244,944 (567 × 432) letterboxed to 16:9	414,720 pels (720 × 576)
Audio	Two analog channels (FM), two digital channels (LPCM with optional Dolby Digital or DTS)	Eight digital tracks of up to eight channels each (LPCM/Dolby Digital/MPEG-2 with optional DTS or SDDS)
Uncompressed audio	16-bit 44.1-kHz PCM	16/20/24-bit 48/96-kHz PCM
Compressed audio	384 kbps Dolby Digital	64 to 448 kbps Dolby Digital
Optional compressed audio	1411 kbps DTS	32 to 1536 kbps DTS
Audio SNR	115 dB (PCM)	96 to 144 dB (PCM)
Dynamic range	96 dB (16-bit PCM)	96/120/144 dB (16/20/24-bit PCM), 120 dB (20-bit Dolby Digital)
Frequency (±0.3 dB)	4 to 20,000 Hz	4 to 22,000 Hz (48 kHz), 4 to 44,000 Hz (96 kHz)

[a]Rare anamorphic laserdiscs are available, but standard laserdisc players cannot format them for 4:3 televisions.

[b]Analog laserdisc video does not actually have pixels, but the count can be approximated using TV lines of horizontal resolution (4:3 aspect ratio) and scan lines.

Convenience. Laserdiscs are large and can be cumbersome to handle. The disc size also makes the players larger and noisier than DVD players. DVD discs can be handled easily and can be sent through the mail cheaply. DVD players can be portable—the same size as CD players. DVD discs fit into standard-width drives designed for computers. One drawback of the smaller disc size is less space on the package for art and information.

Because laserdiscs cannot hold more than one hour per side, the disc must be changed one or more times during a movie. Some laserdisc players can flip the disc automatically, which still causes a break of about ten seconds and does not help for movies that are more than two hours long (except in the case of exotic two-disc players designed for viewers who are obsessed with cinematic continuity). Laserdisc sides often end where it is technically convenient rather than where it is unobtrusive. In comparison, a DVD can hold a three- or four-hour movie on one side if both layers are used.

Audio. DVD-Video has up to eight audio tracks. Laserdisc has two stereo audio tracks: one analog and one digital.

DVD-Video uncompressed digital audio (*pulse-code modulation*, or PCM) enables sampling rates of 48 or 96 kHz with 16, 20, or 24 bits of precision. Laserdisc uncompressed digital audio uses 44-kHz sampling at 16 bits.

DVD-Video compressed audio uses Dolby Digital 5.1-channel surround sound at a typical data rate of 384 or 448 kbps. DVD-Video optionally can include compressed DTS or SDDS audio. DVD-Video also can use 5.1- or 7.1-channel MPEG-2 audio; although many players do not support it, and few home theaters have eight speakers. Laserdisc carries compressed Dolby Digital surround sound by preempting one of the analog audio channels. Laserdisc optionally can carry DTS surround sound by replacing both digital audio channels.

To be fair, it should be recognized that most movies on DVD use compressed Dolby Digital audio for the sound tracks, whereas most laserdiscs have an uncompressed PCM digital audio track. It is difficult to compare the two, given the subjective importance of 5.1-channel surround sound and the differences in audio levels, mixing, and EQ, but most tests show that the average listener cannot tell the difference between uncompressed PCM stereo audio and compressed Dolby Digital stereo audio.

Video. DVD almost always has better video than laserdisc. Technically, the resolution of DVD-Video is approximately one-third better than laserdisc and two-thirds better in widescreen mode. Laserdisc suffers from degradation inherent in its analog format and in the composite NTSC or PAL video signal. DVD uses component digital video;

even though it is heavily compressed, it is near studio master quality when encoded properly and carefully. Technically, analog laserdisc video is also compressed, since the color component (chroma) of the video is reduced to less than one-sixth the resolution of the brightness component (luma).

This does not mean that the video quality of DVD is always better than that of laserdisc—only that it can and should be better. Poorly made DVDs look worse than well-made laserdiscs, but any DVD that has had sufficient care taken in the process of film transfer and compression should look better than a laserdisc made with equal care.

It is worth noting that the average television under 25 inches does not have the precision to show much difference between laserdisc and DVD. Home theater systems using large or widescreen TV sets with s-video or component video inputs are needed to take best advantage of the improved picture quality of DVD.

Of course, just as with vinyl records and CDs, the arguments about analog laserdisc quality versus digital DVD quality will rage eternally. The only final answer is to compare them objectively, side by side, and form your own opinion.

Noise. Most laserdisc players make a whirring noise that can be heard during quiet segments of a movie. This does not bother some people but is quite annoying to others. This mechanical noise is due to the large size and higher spin rate of laserdiscs (600 to 1800 rpm compared with 600 to 1600 rpm for DVDs and 200 to 500 rpm for CDs), especially since the outer edge of a 30-centimeter laserdisc moves much faster than the outer edge of an 8-centimeter DVD spinning at the same rate. Most DVD players are as quiet as CD players.

Subtitles. When subtitles are included on a laserdisc, they must be added permanently to the video picture. On letterboxed movies, they can be placed in the matte area so as not to cover up the picture, but they are still obtrusive, especially to those who do not like subtitles. DVD allows up to 32 different subtitles or graphic overlays that can be turned on or off at any time.

Reliability. Laserdiscs were based on very advanced technology when they were introduced, and they still show signs of that heritage. Production is expensive, and stamping small, precise pits on such a large surface makes clean mold separation difficult. CDs and DVDs were based on better-established technology when they were introduced, leading to simpler and more reliable production.

Both laserdiscs and DVDs are made from two bonded substrates, but the thinner profile and smaller size of DVDs make them much more stable and subject to fewer warp problems. Laserdiscs are subject to what is commonly called "laser rot": the deterioration of the aluminum coating that can occur if the seal between disc sides is broken. Laserdiscs absorb moisture, which can penetrate the seal. The large size of laserdiscs makes them flexible, thus enabling more movement along the bond between sides, which can ruin the aluminum layer. This was mostly a problem during early days of laserdisc replication but still crops up from time to time. DVDs are usually molded from polycarbonate, which absorbs about ten times less moisture than the acrylic used for laserdiscs. DVDs are much more rigid, so less bond flexion occurs.

Laserdiscs incorporate no error correction, although many newer players have noise-reduction circuitry. Noise in analog video is expected and accepted. Laserdisc video often suffers from *dropouts,* small white specks caused by minor imperfections and scratches. Severe flaws or large scratches can cause playback to skip or may cause the disc to be unplayable. Because DVD is a digital medium, errors can have a more drastic result than on laserdisc. This is why the DVD format includes a very robust error-correction system that can compensate for scratches as wide as 6 mm. Most imperfections or scratches on a DVD disc will not affect playback. Large flaws or scratches may cause the picture to skip or break up and may even render the disc unplayable. On the whole, DVD is more tolerant of physical flaws and surface damage than laserdisc.

Availability and Support.   Many more manufacturers of DVD players exist than of laserdisc players, without counting computers that can play DVD-Video. Dealers were amazingly quick to clear out their inventories of laserdiscs and players and replace them with DVD discs and players. The rapid decline of the laserdisc market has meant that most new releases and rereleases are only available on DVD.

Price.   DVD players and discs are cheaper than laserdisc players and laserdiscs. The ability to use components made cheap by the huge DVD-ROM and CD markets and the rapidly expanding use of digital video and digital audio are driving down costs. DVDs are cheaper to replicate than laserdiscs, and production costs have dropped to the point where DVDs can be produced with a desktop computer and replicated for less than $2 a copy. Initial pricing for DVD movies was at the same level as VHS tapes, about 30 percent lower than laserdiscs. Some Hollywood studios have begun releasing older movies on DVD at even lower prices than videotape.

Laserdiscs have one advantage over new releases that are priced for rental on videotape and DVD, typically at $80 to $90. Laserdiscs are seldom released at rental prices and are thus available for about 60 percent less during the period before the videotape and DVD products are dropped to lower prices for retail sale.

Advantages of Laserdisc over DVD-Video

Established Market. Laserdisc enjoyed over 20 years of modest growth as the premiere videophile format. It will take the DVD market many years to catch up with laserdisc's 35,000 titles worldwide.

Reverse Play. Because of the way MPEG-2 compressed video builds pictures by using the differences from key frames, it is impossible to play in reverse at normal speed without a large amount of memory in which to hold the set of previous frames. RAM for video and decoding is expensive, so most DVD players can only play backward by jumping to key frames, which results in video playback about 15 times faster than normal or jerky playback at slower speeds. Laserdisc players cannot play CLV discs backward at all, but they can play CAV discs backward and forward at various speeds.

Still-Frame Capacity. A CAV laserdisc can hold 54,000 still images per side. Each frame can be accessed directly by entering a frame number into the remote control unit or by using a bar-code reader. DVDs can hold thousands of still images, but they are difficult to author onto the disc and cannot be accessed directly in most players.

No Regional Codes. DVD-Video discs can be coded so that they will not play in certain geographic regions. Laserdiscs have no such codes. Any NTSC laserdisc will play in any NSTC laserdisc player, and any PAL laserdisc will play in any PAL player.

No Copy Protection. Laserdiscs do not use Macrovision or any similar tampering with the video signal to prevent copying, unlike DVD, where the superior video signal can be distorted by copy protection schemes. Ironically, the lack of copy protection has hindered the growth of laserdisc because many studios are reluctant to release movies on laserdisc, especially new blockbuster hits, before they have achieved worldwide release.

Compatibility of Laserdisc and DVD-Video

No normal DVD player will play a laserdisc. No standard laserdisc player will play a DVD. Laserdisc uses analog video, whereas DVD uses digital video; they are very different formats. However, some manufacturers—most notably Pioneer—produced combination players that play DVDs and laserdiscs (as well as CDVs and audio CDs).

The ability to modify or upgrade a laserdisc player to play DVD will never be an option. DVD circuitry is completely different: the pickup laser is a different wavelength, the tracking control is more precise, the motor speeds are different, and so on. In any case, a hardware upgrade would be more expensive than buying a new DVD player to put next to the old laserdisc player. As for CDV discs, the audio portion will play on any DVD player, but the video portion is only viewable on players that include laserdisc compatibility.

Videotape

When DVD was introduced at the end of 1996, over 175 million VCRs had been purchased by U.S. households—coverage of over 86 percent—and over 400 million VCRs worldwide. In terms of market targets for DVD, this is the broad side of a barn. However, until DVD is recordable, it could miss the barn completely. Despite the fact that 70 percent of VCR owners have supposedly never recorded anything with their VCRs, consumers demand the ability to record. It is like buying a convertible automobile; people in sunny climates take down the top all the time, and people in rainy locales still want the option in case they get a few days of sun. Until DVD video recorders drop below the magic $500 price point, DVD players will largely be replacements for CD players, but not VCRs.

The Betamax videotape format provides a slightly better picture than VHS (around 250 lines of horizontal resolution compared with 240) but is not covered here because it is not used widely. *Super-VHS* (S-VHS) and *SuperBeta* (ED Beta) are even more improved but are likewise not covered here for the same reason. (S-VHS produces about 400 lines of horizontal resolution, whereas ED Beta produces close to 500).

In quantitative terms, DVD has over twice the resolution of VHS. In terms of perceived quality on a sufficiently good monitor, a stunning difference exists between the two, especially with progressive-scan DVD display, which effectively has over five times the resolution. Table 8.2 compares VHS with DVD.

Advantages of DVD-Video over Videotape

Capacity.   For typical use, such as prerecorded movies, both VHS and DVD have sufficient capacity. For anything requiring longer playing times, however, such as training videos or video libraries, a double-sided, dual-layer DVD can hold over 8 hours of high-quality video or 33 hours at VHS quality.

TABLE 8.2

VHS and DVD-Video Specifications

	VHS Videotape	DVD-Video
Video	Composite analog NTSC	Component digital MPEG-2
Playing time	2 h/tape (SP) or 6 h/tape (EP) (8 h EP with T-160 tapes)	2+ h/side (1 layer), 4+ h/side (2 layer)
Wide-screen support	Letterbox[a]	Anamorphic
Analog copy protection	Macrovision	Macrovision
Video SNR	~45 dB	~70 dB
NTSC resolution[b]	~153,600 pels (320×480)	345,600 pels (720×480)
PAL resolution[b]	~185,600 pels (320×580)	414,720 pels (720×576)
Audio	1 analog mono track or 1 analog hi-fi stereo track	8 digital tracks, each with up to 8 channels of surround sound
Audio SNR	~40 dB (mono), ~90 dB (hi-fi)	96 to 144 dB
Dynamic range	90 dB (hi-fi)	96/120/144 dB (16/20/24-bit PCM), 120 dB (Dolby Digital)
Frequency (±0.3 dB)	70 to 10,000 Hz (mono), 20 to 20,000 Hz (hi-fi)	4 to 22,000 Hz (48 kHz), 4 to 44,000 Hz (96 kHz)

[a]Anamorphic video can be recorded on any video system, but standard VCRs cannot format it for 4:3 televisions.

[b]Analog VHS video does not actually have pixels, but the count can be approximated using TV lines of horizontal resolution (4:3 aspect ratio) and scan lines.

Features.   Beyond the basic VCR features of play, pause, step, slow, fast, fast forward, and rewind, DVD adds instant rewind, high-speed scan, instant search, multistory branching, parental control, multiple camera angles, video menus, interactivity, and more. Not all discs include these features, but they are part of the basic DVD format.

Convenience.   Discs can be more compactly stored and can be sent through the mail more easily and cheaply. DVD players can be portable and battery-powered, at a size only slightly larger than the disc itself. DVD jukeboxes can put hundreds of discs at push-button access in a very small box.

Durability.   Videotapes are subject to degradation from wear and stretching, erasure from magnetic fields, and damage from heat. DVDs never wear out, are impervious to magnetic fields, and are less susceptible to heat damage. Discs can be scratched, but as with CDs, only large scratches will cause noticeable playback problems.

"Tape eating" VCRs eventually may become something to reminisce about, like the hazards of being covered with soot after a train ride behind a coal-fired engine. It is possible that loose material or a defect in a DVD player could scratch a DVD, but this has been very infrequent with millions of CD players.

VCR owners who rent videos subject their machines to tapes of dubious history covered with unknown substances, thus requiring more frequent head cleanings. Frequently rented tapes are recognized easily by their excessive glitches and tracking problems. Since the laser head in a DVD player never touches the surface of a disc, the condition of the disc does not affect the player so long as the disc is not broken or severely warped. DVDs are doing well in the rental market because their overall durability, ease of use, and superior video outweigh their susceptibility to scratches.

VCRs generally break down because of mechanical failure exacerbated by rewinding. DVD players have much simpler mechanisms and never need to rewind.

Audio.   Videotape audio is analog. The amount of tape dedicated to the audio track in monophonic VHS and linear stereo VHS is only one-half of one percent. Hi-fi VHS uses more tape and a helical scan to get close to CD-quality audio. Dolby Surround encoding can be used to matrix two surround channels into the videotape stereo signal. In comparison, DVD includes up to eight audio tracks of CD-quality audio with Dolby Digital or DTS discrete 5-channel surround sound and a subwoofer channel. PCM tracks can provide better-than-CD audio. The extra audio tracks can be used for foreign language, commentary, additional music, and more.

Video.   The video quality of VHS tape is much lower than broadcast or cable signals. Tape dropouts (poor or missing magnetic particles) cause dots and small flashes in the picture. Wear and stretching from kids playing their favorite tape twice a day cause the picture to degrade quickly, not to mention requiring more frequent head cleanings. Head alignment differences cause tracking problems and additional loss of quality. DVD digital video, even though it is compressed, can look almost as good as studio masters. Discs never wear out from repeated playing; servo-controlled laser tracking keeps everything in perfect alignment; and error-correction codes compensate for defects and damage.

Prerecorded videotapes are copied in high-speed duplicating machines; they do not nearly match the quality of the duplication master. DVDs are stamped into plastic using metal plates; they contain virtually the same data as the master.

Each generation of a videotape copy loses quality. A digital copy of a DVD is a perfect replica, no matter how many generations it is removed from the original. Of course, this assumes that the original copy is not protected. Copies of home videos sent to Grandma will no longer be so blurry that she cannot tell the grandkids apart.

Price.   Discs are cheaper than tapes and can be mass-produced faster and more easily. Whether this savings ever gets passed on to the consumer remains questionable, since it never seemed to happen with CDs. Some movie studio executives have stated that they plan to make DVDs as cheap or cheaper than videotape, especially older movies that have already made back their original cost.

Subtitles.   When subtitles are included on a videotape, they must be added permanently to the video picture. DVD allows up to 32 different subtitles or graphic overlays that can be turned on or off at will.

Advantages of Videotape over DVD-Video

Recordable.   Recordable DVD-Video will not be available in the home before the year 2001. Recordable DVD-Video technology will have to improve significantly and become much cheaper before it will become widespread. Copy protection issues, which have seriously hampered other digital recording formats such as *digital audiotape* (DAT) and *digital videotape* (DV), also may delay this vital feature.

Established Market.   VHS has been around for just over 20 years. Supposedly more than 30,000 different VHS titles are available in the United States and over 50,000 worldwide. It will take DVD a very long time, if ever, to reach this point. Many titles available on tape may never appear on DVD.

VHS also has a well-established rental market. Approximately 27,000 video stores in the United States are visited each week by over 65 million people, who, in 1996, rented more than 3 billion videotapes and purchased more than 580 million. DVD must penetrate this market while it is being eroded by video-on-demand and pay-per-view programming.

No Regional Codes.   A code can be added to a DVD disc so that it will not play in players or computers from certain geographic regions. Videotapes have no such codes. The only limitation is among television systems, but an NTSC videotape will play in any NSTC VCR, and a PAL videotape will play in any PAL VCR.

Multistandard Players.   Both DVD and videotape must support two incompatible television systems: 525/60 (NTSC) and 625/50 (PAL). The video signal is stored differently for each system, and they are not interchangeable. However, multistandard VCRs are available that can play either type of tape if connected to a multistandard monitor. Standards-converting VCRs are also available that use digital processing to convert between formats. The future widespread availability of multistandard and standards-converting DVD players is uncertain and is complicated by Japan (NTSC) and Europe (PAL) sharing the same DVD region code.

Availability and Support.   Until DVD-Video becomes well established in the home market, if ever, it may be difficult to buy or rent discs. Finding repair shops and qualified technicians also may be difficult at first.

Compatibility of VHS and DVD-Video

As ludicrous as it seems, people have done stranger things than to try to put a disc in a tape player. A videotape player will not play a DVD. No DVD player will play a videotape. However, some manufacturers have developed dual models that contain both a DVD deck and a VHS deck in a single unit.

Digital Videotape (DV, Digital8, and D-VHS)

Digital videotape systems generally are aimed at the professional production and "prosumer" markets. However, it is worth taking a look at them because they are the closest digital video competitors to DVD. Essentially two systems exist: DV (or DVC) and D-VHS. There is also a variation of DV that uses 8mm tapes. Purely professional systems such as Digital-S, Betacam SX, Digital Betacam, and D1 are not covered here.

Like DVD, DV has gained unified industry support. The original proposal from Matsushita, Philips, Sony, and Thomson was endorsed in 1993 by Hitachi, JVC, Mitsubishi, Sanyo, Sharp, and Toshiba and since then has been supported by many other companies and standardized as IEC-16884. Nonstandard enhancements have already been made to the format: Matsushita's DVCPro and Sony's DVCam use different physical formats or recording densities that make the tapes incompatible, but they use the same data formats in order to connect to other DV equipment.

DV camcorders appeared in 1995, but DV recording decks were delayed in the United States primarily by concerns that they might be able to make perfect copies of DVDs. DV recorders appeared in 1997 after a copyright protection system was added.

DV uses I-MPEG compression. I-MPEG is based on DCT and quantization similar to MPEG, but it compresses each frame separately (similar to MPEG's I frames or motion-JPEG) and uses interfield compression to take advantage of redundancy within a frame. Because no interframe compression is present, I-MPEG is less efficient than MPEG but is better for editing because frames can be moved and combined without requiring decoding and recoding of a group of dependent frames. As with DAT, the DV format specifies a computer data storage variation intended for backup and archiving.

Digital-VHS (D-VHS) was developed by JVC and is ostensibly supported by Hitachi, Matsushita, and Philips. D-VHS is oriented more toward the consumer market than DV, and it is backward-compatible with VHS, meaning that it can read and write VHS tapes as well as read and write digital data using special D-VHS tapes. D-VHS was announced in April of 1995 but has yet to appear as a commercial product outside Japan. D-VHS originally stood for Digital-VHS but now stands for Data-VHS. Unlike other digital tape formats, D-VHS does not convert or record analog video—it only records and reads bit streams. This means that D-VHS can record from a digital source such as DBS, digital cable, *high-definition TV* (HDTV), and DVD (assuming that each of these has a compressed digital bit-stream output), but the D-VHS player has to send the data back out to a compatible decoder in order to display it. This requires a television with a built-in decoder or a bit-stream input on a digital video device such as a DVD player. At the time DVD was introduced, such digital bit-stream connections were not available on any consumer products.

Because of its bit-stream capability, D-VHS is also being positioned for computer data backup in the home. Analog video sources are recorded and reproduced by D-VHS decks in analog form, not digital. D-VHS systems as planned are unable to do any conversion between analog and digital formats. Table 8.3 lists DV and DVD-Video specifications.

TABLE 8.3

DV and DVD-Video Specifications

	DV	D-VHS	DVD-Video
Video	Component digital	Composite analog or digital bit stream	Component digital
Playing time	0.5 to 1 h/tape (mini), 4.5 h/tape (standard)	2 to 6 h/tape (VHS), 2.5 (28.2 Mbps) to 49 (2 Mbps) h/tape (D-VHS)	2+ h/side (1 layer), 4+ h/side (2 layer)
Data capacity	5.5 to 50G bytes	31.7G bytes (300 m tape), 44.4G bytes (420 m tape)	4.7 to 17G bytes
Digital compression	I-MPEG, ~5:1 (from 8-bit 4:1:1 or 4:2:0)	None (external)	MPEG-1 or MPEG-2, ~30:1 (from 8-bit 4:2:0)
Data rate	25.146 Mbps CBR video, up to 1.536 Mbps audio, ~35.5 Mbps max.	28.2 Mbps (HD), 14.1 Mbps (STD), 2 to 7 Mbps (LP)	Up to 9.8 Mbps combined VBR/CBR video and audio
Error correction	RS-CIRC	RS (inner/outer)	RS-PC
Coded frame rate	29.97^a (525/60), 25^a (625/50)	n/a	29.97^a or 24^b (525/60), 25^a or 24^b (625/50)
Wide-screen support	Anamorphic^c	Letterbox (VHS)	Anamorphic
Analog copy protection	None	Macrovision (VHS)	Macrovision
Copy management	CGMS	CGMS	CGMS
Encryption	Yes	No (external)	Yes, CSS
525/60 resolution	345,600 pels (720 × 480)	~153,600 pels (320 × 480) (VHS)	345,600 pels (720 × 480)

continues

TABLE 8.3 cont.

DV and DVD-Video
Specifications

	DV	D-VHS	DVD-Video
625/50 resolution	414,720 pels (720 × 576)	~185,600 pels (320 × 580) (VHS)	414,720 pels (720 × 576)
Audio	Two tracks of two channels (32 kHz, 12-bit nonlinear PCM) or one track of two channels 32/44.1/48 kHz, 16-bit linear PCM)	Analog VHS stereo hi-fi or digital bit stream	Eight tracks of up to eight channels each (LPCM/Dolby Digital/MPEG-2)
Audio SNR	72 dB (12-bit), 96 dB (16-bit)	~40 dB (mono VHS), ~90 dB (hi-fi VHS)	96 to 144 dB
Dynamic range	96 dB	~90 dB (hi-fi VHS)	96/120/144 dB (16/20/24-bit PCM), 120 dB (20-bit Dolby Digital)
Frequency (±0.3 dB)	4 to 15,000 Hz (32 kHz), 4 to 20,000 Hz (44.1 kHz), 4 to 22,000 Hz (48 kHz)	70 to 10,000 Hz (mono VHS), 20 to 20,000 Hz (hi-fi VHS)	4 to 22,000 Hz (48 kHz), 4 to 44,000 Hz (96 kHz)

[a]Interlaced.
[b]Progressive.
[c]Special feature on some equipment only.

Advantages of DVD-Video over DV

Capacity. Mini-DV cassettes can hold 1 hour of video, and standard DV cassettes can hold 4 1/2 hours. An 8-centimeter DVD can hold approximately 3/4 to 2 1/2 hours; a 12-centimeter DVD can hold approximately 2 to 8 hours.

Features. Beyond DV and D-VHS features of play, pause, slow, fast, step, fast forward, and rewind, DVD adds instant access, search, menus, interactivity, and more. Not all discs include these features, but they are part of the basic DVD format.

Durability. Videotapes are subject to degradation from wear and stretching, erasure from magnetic fields, and damage from heat. Even though the very small tape width of DV makes it susceptible to stretching and dropouts, the digital data format and error correction largely compensate. In comparison, discs never wear out, are impervious to magnetic fields, and are less susceptible to heat damage. Discs can be scratched, but as with CDs, only large scratches will cause noticeable playback problems.

Even high-precision, semiprofessional videotape equipment occasionally "eats" a tape. It is possible that a DVD player could scratch a disc, but this is very rare.

Videotape equipment is quite complex, with DV heads spinning at 9000 rpm and hundreds of small components that can break down because of mechanical failure. DVD players have much simpler mechanisms.

Video. In VHS mode, D-VHS provides VHS video. In digital bit-stream mode, the video depends entirely on the digital source and may be better or worse than DVD.

Audio. DV provides one stereo track at high quality (48 kHz, 16-bit) or two stereo tracks at slightly lower quality (48 kHz, 12-bit). DVD provides one stereo track at superhigh quality (96 kHz, 24-bit), one 8-channel track at high quality (48 kHz, 16-bit), or up to 8 tracks of 5.1-channel Dolby Digital surround sound or 5.1/7.1-channel MPEG-2 surround sound.

In VHS mode, D-VHS provides standard VHS hi-fi audio. In digital bit-stream mode, the audio depends entirely on the digital source.

Price. Discs are cheaper than tapes and can be mass-produced faster and more easily than tapes. It is unlikely, however, that DV will ever be used for prerecorded commercial video.

At the time DVD was introduced, DV cameras and decks cost $3,000 to $5,000, and the professional DVCPro cameras and decks cost more than $15,000. The prices of DV cassette tapes were $10 to $25. Recordable DVD discs started at around $40 and eventually dropped to $5 or less. DV hardware and tape prices also will come down, but without the advantage of a mass-market computer counterpart such as DVD-ROM, they will not drop nearly as far as DVD hardware and disc prices.

Essentially no D-VHS equipment was available when this book was written. Decks are expected to be about the same price as DVD players, and tapes will be much cheaper when measured in cost per gigabyte (see the following section). Separate digital audio/video equipment with digital connections—such as IEEE-1394 FireWire—will be required to take advantage of D-VHS's digital bit-stream capability.

Advantages of DV over DVD-Video

Capacity. D-VHS has a much greater capacity than DVD (and DV), with a digital tape holding almost 50 gigabytes of data. Even though the real bit capacity of DV is significantly higher than D-VHS, DV tape is narrower.

Recordable. As with VCRs, this is the difference that makes all the difference. The high compression required by DVD-Video makes it an unsuitable candidate for shooting and editing. Recordable DVD more likely will be the final destination of digital video files from DV cameras and editing systems. Since it is possible to store DV-format (I-MPEG) video on a DVD-ROM, this may become popular for quick-access archiving. D-VHS can record from a digital source, but a digital decoder also must be available for playback.

Editing. A significant advantage of DV is that each frame is compressed individually, so they can be inserted, deleted, and combined in any order. This is especially useful for nonlinear digital editing systems. Because DVD's MPEG-2 P and B frames rely on nearby frames, they must be decompressed and recompressed to make an edit within a group of pictures. This can degrade picture quality.

Video. DV uses the same component digital format as DVD-Video, the only difference being the sampling system. DV uses 4:1:1 sampling for NTSC and 4:2:0 sampling for PAL. DVD-Video uses 4:2:0 sampling for both NTSC and PAL. Both sampling methods record the same amount of information but in slightly different ways, and the arguments are endless over which is better. DV's I-MPEG compression removes much less information than DVD's MPEG-2 compression but is correspondingly less efficient. DV video quality is superior to DVD's average video data rate of 3.5 Mbps but is not noticeably different when DVD rates are increased to more than 5 or 6 Mbps. D-VHS can record a digital video bit stream at up to 28 Mbps, which is more than adequate for even high-quality digital video formats such as those used for HDTV.

Price. D-VHS tapes may cost more than DVD discs, even more than recordable discs in the long term, but they are much more cost-effective when measured in price per gigabyte.

Compatibility of DV and DVD-Video

DV and DVD-Video are not directly compatible. DV's intraframe compression technique is similar to MPEG's but is technically not the same. Certain key, low-level differences make them incompatible and cause many headaches for implementors of hardware and software intended to support both formats. As long as these differences are accounted for, DV's I-MPEG compression format can be converted easily to MPEG-2. Both are component digital formats, and other than errors introduced in converting between sampling systems and possible minor artifacts introduced by the differences in compression techniques, almost no quality will be lost when transferring from one to the other.

D-VHS and DVD-Video are not directly compatible. However, D-VHS could potentially be used to record from DVD. In this case, both the DVD player and the D-VHS recorder must have a digital connection, such as FireWire. In addition, either the DVD player or some other device such as an HDTV display must be able to decode the DVD-specific MPEG-2/Dolby Digital bit stream when connected to the D-VHS deck. Features of DVD-Video such as seamless branching, camera angles, and so on would no longer be available from a sequential bit stream. It is expected that both DV and D-VHS recorders will respect the copy-generation management information sent by the DVD player.

Audio CD

Both DVD-Audio and DVD-Video format pack quite a wallop. DVD-Video discs can be produced with only a single still menu, leaving the remainder of the disc for audio. Table 8.4 presents a comparison of specifications.

Advantages of DVD over Audio CD

Quality. Many people feel that CD is near enough to perfect. Others claim it is not even close. In any case, DVD has it beat. DVD-Video specifies four different audio formats: linear PCM (pulse-code modulation, the same as audio CD), 5.1-channel Dolby Digital surround, 5.1/7.1 DTS Digital Surround, and 5.1/7.1-channel MPEG-2 surround. DVD-Audio adds even higher sampling rates and lossless compression for longer playing times.

TABLE 8.4

Audio CD and
DVD-Video
Specifications

	Audio CD	DVD-Video
Audio	16-bit 44.1-kHz PCM	16/20/24-bit 48/96-kHz PCM (DVD-Video), 44.1/48/88.2/96/176.4/192 kHz PCM with MLP lossless compression (DVD-Audio), 24-bit 48-kHz Dolby Digital/DTS/MPEG audio
Channels	2 (4 using Dolby Surround encoding, 5.1 using DTS encoding)	8 using PCM (DVD-Video), 6 using PCM (DVD-Audio), 5.1 using Dolby Digital, 5.1/71 using DTS or MPEG-2 multichannel
Audio SNR	105 dB	96 to 144 dB
Dynamic range	96 dB (16-bit PCM)	96/120/144 dB (16/20/24-bit PCM, Dolby Digital, DTS)
Frequency (±0.3 dB)	4 to 20,000 Hz (44.1 kHz PCM)	4 to 22,000 Hz (48 kHz), 4 to 44,000 Hz (96 kHz), 4 to 96,000 Hz (192 kHz)
Playing time	~80 min	~27 h of surround sound, ~7 h of 20-bit 48-kHz stereo, ~138 min of 24-bit 96-kHz stereo[a]

[a]All times are for a single 4.7-gigabyte layer. Times for a dual-layer disc are almost twice as long.

CD audio is sampled 44,100 times a second using 16 digital bits to hold each value. DVD-Video PCM audio is sampled at either 48,000 or 96,000 times a second and uses 16, 20, or 24 bits to hold the values. DVD-Audio PCM is sampled at CD multiples of 44,100, 88,200, and 176,400 Hz, as well as 48,000, 96,000, and 192,000 Hz, all with sample sizes of 16, 20, or 24 bits. The reproducible frequency is just below half the sampling rate, which gives CD a limit of about 20 kHz, DVD-Video a limit of about 44 kHz, and DVD-Audio a limit of about 94 kHz. Since the average human hearing range does not extend beyond about 20 kHz, it may seem that DVD-Video holds an advantage only for dogs and that DVD-Audio was designed for bats and dolphins, but higher sampling rates have been shown to result in improved sound reproduction. The sampling size determines the dynamic range (approximately 6 decibels per bit), but it also has benefits for noise shaping and other digital signal-processing techniques that can take advantage of the extra bits of precision.

CD audio uses two channels: left and right stereo. At lower sample rates and bit rates, DVD PCM can have up to eight channels. Dolby Digital uses 5.1 channels: left, right, center, left-rear, right-rear, and subwoofer. MPEG-2 optionally can add left-center and right-center channels.

Both of DVD-Video's surround-sound formats are compressed from 16-, 18-, or 20-bit 48-kHz sources. Psychoacoustic processing is used to remove imperceptible sounds and redundant information. Occasional compression artifacts may be heard, but the resulting surround-sound experience is similar to that of a theater—assuming the DVD player is hooked up to a sufficiently good home sound system.

A few CDs are available with DTS surround sound. These CDs are not playable on regular CD players. DVD offers DTS as an option, and with its larger capacity, DVD can contain standard PCM audio, a Dolby Digital version for those with Dolby Digital decoders, and a DTS version for those with DTS decoders.

Capacity. By squeezing the tracks tightly together, an audio CD can be made to hold as much as 84 minutes of audio rather than the specified 72 or 74 minutes. DVD, even at its highest level of uncompressed stereo sound quality on a single-layer disc, holds over two hours of audio. Using all 5.1 channels of Dolby Digital audio, a single-sided, single-layer DVD can play for over 27 hours. And using only two channels of Dolby Digital, more than 54 hours fit on a single layer, and a mind-boggling 197 hours can be packed onto a double-sided, dual-layer disc.

Video. It may seem strange to list video as an advantage of DVD over audio CD, but there have been many attempts to add video to music CDs, including CDV, CDG, and Enhanced CD (see the sections on laserdisc and other CD formats in this chapter for details).

Given the achievements of MTV, the success of music performance videos, and the amount of existing music video footage with no retail channel, it is expected that DVD-Video's combination of high-quality video with the convenience and audio quality of CD will tap a larger market of music listeners who are interested in the visual aspects of the performance. DVD music video also will appeal to the karaoke market, especially with its multilingual subtitle capabilities.

Advantages of Audio CD over DVD-Video

Established Base. By the end of 1999, over 800 million audio CD players and another 300 million or so CD-ROM drives were capable of playing audio CDs. Over 14 billion audio CDs have been produced since 1982. Since DVDs may not play in CD players but CDs will play in DVD players, no compelling reason justifies a publisher's decision to release music titles in DVD format—especially not before the DVD-Audio standard is established.

DVD-Video makes sense for music with video and for high-end music titles where the publisher wants to take advantage of higher PCM sampling rates and bit rates. For standard music titles, however, publishers will stick with a market of 700 million CD units that will continue to grow with CD-capable DVD units.

Compatibility of Audio CD and DVD-Video

In general, all DVD-Video players and DVD-ROM drives can play an audio CD. This is not actually required by the DVD standard, but so far all manufacturers have designed their DVD hardware to read CDs.

No CD player or CD-ROM drive can play audio from a DVD. It is possible that the DVD-Audio format may include a legacy variation containing music in CD format on one layer of the disc and in DVD format on another or CD data sandwiched between the DVD lead-in and an outer ring of DVD data. However, certain technical obstacles make either approach difficult or expensive. See "Other CD Formats" later in this chapter for details on compatibility between DVD and variations of the CD format that include music.

CD-ROM

DVD-ROM drives eventually will make CD-ROM drives extinct. Since DVD-ROM drives are backward-compatible and more expensive, CDs will continue to be used whenever the larger capacity of DVD is not needed. Table 8.5 presents a listing of CD-ROM and DVD-ROM specifications.

TABLE 8.5

CD-ROM and
DVD-ROM
Specifications

	CD-ROM	DVD-ROM
Capacity	650.4 MB	4.37 to 15.9 GB
Error correction	RS-CIRC	RS-PC
Error correction overhead	34 percent	13 percent
Modulation	8/14 (EFM)	8/16 (EFMPlus)
Transfer rate $(1\times)^a$	150 KB/s (1.23 Mbps)	1353 KB/s (11.08 Mbps)
File system	ISO 9660, HFS, other	UDF, UDF Bridge, ISO 9660, other

[a]Reference rates for single-speed drives. The transfer rate for double-speed drives is 300 KB/s for CD-ROM and 2705 KB/s for DVD-ROM, and so on.

Advantages of DVD-ROM over CD-ROM

Capacity.   A CD-ROM holds about 650 megabytes. This can be pushed to about 730 megabytes by squeezing the tracks tighter together. A single-layer DVD holds 4.4 gigabytes, which is seven times what a CD-ROM holds. A dual-layer DVD holds 8.0 gigabytes, 12.5 times more than a CD-ROM. A double-sided, single-layer DVD holds 8.8 gigabytes, 14 times more than a CD-ROM. A double-sided, dual-layer DVD holds 15.9 gigabytes, which is 25 times what a CD-ROM holds. (See page 133 for an explanation of the improvements from CD to DVD that account for the increased capacity.)

Speed.   The CD-ROM specification requires a minimum transfer rate of 150 KB/s (150×2^{10} bytes per second). Memory buffers and multispeed CD-ROM drives can raise the transfer rate much higher, above that of a single-speed (1X) DVD-ROM drive. Multispeed drives spin the disc at higher multiples of the standard velocity to increase the rate at which data are read off the CD. However, producers of CD-ROM–based software often must deal with the installed base of single-speed or double-speed CD-ROM drives.

The DVD-ROM specification requires a minimum transfer rate of 11.08 million bits per second (1353 kilobytes per second), which is nine times faster than a single-speed CD-ROM drive at 1.23 million bits per second or roughly equivalent to a 9X CD-ROM drive. As with CD-ROM drives, the manufacturers will continually increase the speed of DVD-ROM drives to improve performance. Double-speed and faster DVD-ROM drives came out so quickly after first-generation drives that it is quite reasonable to consider 22.16 Mbps (18X CD data rate) to be the minimum DVD data rate.

Some early DVD-ROM drives read CDs and DVDs at the same scanning velocity and therefore transfer CD-ROM data about as fast as a triple-speed (3X) CD-ROM drive. Most DVD-ROM drives increase the spindle speed to achieve higher data rates when reading CD-ROMs, usually 24X and higher.

Reliability. Even though DVD-ROM reduces the amount of space taken up for error correction to 13 percent versus 34 percent for CD-ROMs, DVD-ROM uses a more advanced technique that is about ten times more effective at correcting errors.

Because DVDs are made of two thin substrates glued together, they are more rigid than a solid CD-ROM. The improved rigidity makes them spin more smoothly, resulting in better reading accuracy. Disc bonding technology has become well developed for use with laserdiscs (which are also made of two sides glued together) and is quite reliable, although the hot-melt glue method may be inadequate for discs left on a car dashboard in July; the bond could begin to slip at temperatures above 150°F (70°C).

The data layer of a CD is at the top of the disc (the laser reads from underneath) and is protected by only a thin layer of lacquer. Deep scratches on the label side can physically damage the metallic layer or expose it to oxidation. The data layers of a DVD are at the center, on the inside surface of the substrates, and are consequently more protected.

Improved Standards. CD-ROM is an afterthought added three years later to the CD-Audio standard. Other variations such as CD-ROM XA, multisession, CD-R, Video CD, and CD-RW are afterthoughts built on afterthoughts. Many incompatibilities exist in the CD family. Enhanced CD (including CD Plus, which was a replacement for CD Extra, which enables computer data to be added to a disc with music tracks that can be played in an audio CD player) was not standardized until 1995 and is incompatible with early CD-ROM drives and driver software. The erasable CD-RW standard, established at the end of 1996, is not compatible with any CD-ROM drive built before 1997, nor with most built after 1997. The ISO 9660 file system went through several iterations under such names as High Sierra, Rockridge, and Frankfurt and did not adequately support Macintosh and UNIX file systems. Full support for these file systems had to be added as incompatible extensions.

DVD, on the other hand, was designed from the ground up to be computer-compatible. DVD-Video and DVD-Audio are layers on top of the DVD-'

ROM standard. The recordable and erasable versions of DVD were anticipated and partially accounted for. The UDF file system, developed by the *Optical Storage Technology Association* (OSTA) supports all major operating systems and provides for recordability and erasability. Unfortunately, DVD-RAM was not finalized until long after the first DVD-ROM drives were built and is incompatible with them.

Advantages of CD-ROM over DVD-ROM

Established Base. In 2000, a base of about 300 million CD-ROM drives was installed worldwide, compared with about 45 million DVD-ROM drives. Forecasts show the DVD-ROM installed base surpassing that of CD-ROM sometime between 2003 and 2005. In 2000, DVD-ROM drives were just reaching the point of critical mass where publishers begin to require them for their software. It will take much longer before publishers begin to abandon CD-ROM, especially since DVD-ROM drives are backward-compatible.

CD production systems are in place and well established, both at the desktop development level and at the manufacturing and replication level. Development of DVD-ROM content that does not rely on the higher data rate or specific video features of DVD is almost exactly the same as for CD-ROM, although actually getting the content onto DVD-ROM is harder than it should be, with surprisingly complex tools. DVD-R formatting and testing tools do not have near the polish and maturity of their CD-R counterparts, and development of DVD-Video content or DVD-Video–based multimedia software requires new equipment and techniques, especially for desktop production.

Recordable. Although DVD-R recorders were released in 1997, followed by DVD-RAM in 1998, the drives and blank discs remained at three to ten times the price of CD-R/RW drives and discs.

No Regional Codes or Encryption. Technically, regional codes, and encryption are not part of the DVD-ROM standard. However, the importance placed on computers being able to play or copy DVD-Video discs has dragged them into the equation. The result: long delays in the introduction of DVD, longer delays in the release of DVD-ROM drives, delays and complications in writable DVD formats, a small added cost to DVD computer hardware for regionalization and authentication (and potentially, watermark checking), added complexity for operating system and application support of regions and copy protection, and general muddling of the elementary distinction between computer data on DVD-ROM and movies on DVD-Video. Aside from feeble SDMI initiatives, CD-ROM has been largely free of this type of extraneous influence.

Compatibility of CD-ROM and DVD-ROM

In general, all DVD-ROM drives can read CD-ROMs. However, many variations of the CD format are detailed in the "Video CD and CD-i" and the "Other CD Formats" sections. Some of these variations are compatible with DVD-ROM drives and some are not. The most notable exception is that about half the first generation of DVD-ROM drives are unable to read CD-R discs. Most later-generation DVD-ROM drives have no problems reading CD-R discs.

No CD-ROM drive can read a DVD-ROM disc. Legacy discs, containing a CD data layer and a DVD layer, can be produced, but they are quite expensive and not completely compatible with CD-ROM drives.

Video CD and CD-i

Video CD, often called VCD but not to be confused with *CD-Video* (CDV), is a standard for storing audio and video using the CD format. In many ways it can be considered the precursor to DVD. Because of the capacity and speed limitations of the CD format, Video CD quality is limited; it is considered to be near that of VHS videotape. Super VCD, a major enhancement to Video CD, is covered in the next section.

The Philips CD-i format was created as a specialized interactive (that's what the *i* stands for) application of CD for custom players that connect to a TV. A Digital Video feature based on MPEG-1 was later added to CD-i. In 1993 it was revised slightly to form the Video CD 1.0 standard. New models of CD-i players can play Video CDs; older models require a motion video adapter. Video CD was improved in 1995 with menus, fast forward/rewind, and high-resolution still frames. VCD-Internet (1997) standardized the way to link the discs to Web pages. Table 8.6 presents a listing of the Video CD and DVD-Video specifications.

TABLE 8.6

Video CD, Super
Video CD (SVCD),
and DVD-Video
Specifications

	White Book (Video CD 2.0)	Super Video CD (IEC-62107)	DVD-Video
Video	Component digital MPEG-1	Component digital MPEG-2	Component digital MPEG-2
Playing time	74 min	~About 70 min	2+ h/side (single layer), 4+ h/side (dual layer)
Widescreen support	Letterboxed[a]	Letterboxed[a]	Anamorphic
Compression	~90:1 (from 8-bit 4:2:0)	~40:1 (from 8-bit 4:2:0)	~30:1 (from 8-bit 4:2:0)
Analog copy protection	None	CGMS-A	Macrovision, CGMS-A
Digital copy management	None	None	CGMS-D
525/60 (NTSC) resolution	84,480 pels (352 × 240)	230,400 pels (480 × 480)	345,600 pels (720 × 480)
625/50 (PAL) resolution	101,376 pels (352 × 288)	276,480 pels (480 × 576)	414,720 pels (720 × 576)
Display frame rate[b]	29.97 (525/60), 25 (625/50)	29.97 (525/60), 25 (625/50)	29.97 (525/60), 25 (625/50)
Graphic overlay	Not available	2-bits/pel; 4-color/ 4-transparency CLUTs	2-bits/pel; 16-color/ 16-transparency CLUTs
Still picture (I frame)	704 × 480 (525/60), 704 × 576 (625/50)	704 × 480 (525/60), 704 × 576 (625/50)	720 × 480 (525/60), 720 × 576 (625/50)
Audio	One track of mono, dual-mono, or stereo	Two tracks of mono, dual-mono, stereo, or 5.1-channels	Eight tracks of up to 8 channels each
Uncompressed audio	16-bit 44.1- kHz LPCM (if no video)	Not available	16/20/24-bit 48/96-kHz LPCM
Compressed audio	224 kbps MPEG-1 Layer II (44.1 kHz)	32 to 384 kbps VBR MPEG-1 Layer II or MPEG-2 multichannel (44.1 kHz)	64 to 448 kbps Dolby Digital (48 kHz) or 64 to 912 kbps MPEG-2 (48 (kHz) or 64 to 384 kbps MPEG-1 Layer II (44.1 kHz)

continues

	White Book (Video CD 2.0)	Super Video CD (IEC-62107)	DVD-Video
Karaoke features	Two channels	Two channels, lyrics overlay	Five channels, lyrics overlay
Audio SNR	96 dB	96 dB	96 to 144 dB
Dynamic range	96 dB	96 dB	96/120/144 dB (16/20/24-bit PCM), 120 dB (Dolby Digital)
Frequency (±0.3 dB)	4 to 20,000 Hz (44.1 kHz)	4 to 20,000 Hz (44.1 kHz)	4 to 22,000 Hz (48 kHz), 4 to 44,000 Hz (96 kHz)
Data rate	Max. 1.1519291 kbps	Max. 2.2 Mbps	Max. 9.8 Mbps

[a]Anamorphic video could be put on video CDs, but the players cannot re-format it for 4:3 televisions.

[b]MPEG-1 enables any coded picture rate to displayed at either 29.97 or 25 fps when decoded. MPEG-2 requires the pulldown rate to be determined at encoding time.

Advantages and Disadvantages of CD-i

Compared with DVD, CD-i has a few advantages and disadvantages not shared by Video CD. These are covered first, followed by the remaining comparisons with Video CD.

Features.   The interactivity of CD-i goes far beyond that of DVD-Video. CD-i players are actually special-purpose computers running a real-time operating system (a derivative of OS-9). They contain a Motorola 68000 processor, the same chip that powered early Macintosh computers. Additional specialized processors provide support for graphics and video. CD-i players have memory for data storage, optional keyboards, and even optional modems for connecting to the Internet. CD-i supports fully interactive multimedia programs and can generate text and graphics in response to user input. DVD-Video has limited interactivity using only premade text and graphics. CD-i was the first consumer device to use MPEG-1 video. CD-i players can decode multiple MPEG-1 video channels from a single multiplexed stream for multiple video windows of selectable sizes and aspect ratios. CD-i has up to 16 audio tracks and four audio formats, for up to 18 hours of audio per disc. Full- or partial-screen images can be overlaid with transparency on motion video with hardware-rendered transition effects such as dissolves, wipes, slides, and scrolls. Text overlay can be generated on the fly, including up to 32 stored subpicture streams, memory-resident pop-up menus, and real-time graphics. A computerlike graphic interface with an onscreen cursor is controlled by a wireless remote control.

CD-i features are impressive. On the other hand, a DVD designed for use in a PC or video game console can exceed the level of interactivity and video quality provided by CD-i.

Availability and Support. CD-i hardware and software was almost entirely produced or funded by Philips and Korea's LG Electronics with a handful of independent hardware and software licensees on the side. About half a million players were sold in the United States and about 2 million worldwide.

More than 70 manufacturers produce DVD-Video players, not counting hundreds of manufacturers of DVD-capable computers. Hundreds of companies and independent developers are creating interactive DVD-Video titles despite the format's limited capabilities.

Advantages of DVD-Video over Video CD

Features. The DVD-Video format includes the same basic features as Video CD (menus, pause, search, freeze, slow, fast, scan) and adds seamless branching, parental control, multiple camera angles, and more. Not all discs include these features, but they are part of the basic DVD format.

Capacity. Programs on DVD can be over four times longer than on Video CD. A single-layer DVD-Video holds over two hours of material per side, and a dual-layer disc holds over four hours. A Video CD holds 74 minutes and has only one side. Both DVD and Video CD support still frames with audio, but DVD has the potential for hundreds more pictures and hours more surround sound.

Convenience. Since Video CDs cannot hold much more than an hour, the disc must be changed one or more times during a movie, unless the player holds more than one disc at a time. A DVD-Video can easily hold a four-hour movie on one side.

Audio. DVD-Video has up to eight digital audio tracks. Each track can hold uncompressed audio with quality better than audio CD or compressed

5.1-channel surround sound in Dolby Digital or MPEG-2 format. Video CD normally has one track of compressed 2-channel audio in MPEG-1 format.

Video. The quality of DVD-Video vastly exceeds that of Video CD, with four times the resolution in normal mode and 5.5 times in widescreen mode (compared with letterboxed Video CD). DVD uses MPEG-2 compression, which is more efficient than the MPEG-1 compression used by Video CD. Video in MPEG-2 looks better than in MPEG-1, even at the same resolution and data rate and especially if a variable bit rate is used. MPEG-1 video is better only at data rates below 2 Mbps. DVD supports both MPEG-1 and MPEG-2.

Advantages of Video CD over DVD-Video

Worldwide Standard (Sort of). Video CD solved the problem of compatibility between 525/60 (NTSC) television systems and 625/50 (PAL) television systems in a rather procrustean manner. Video is stored internally in one of two formats, depending on which kind of players it is expected to be played on primarily. NTSC players convert PAL discs by repeating fields and chopping lines off the top and bottom. PAL players convert NTSC discs by adding extra black lines to the top and bottom.

DVD likewise stores video in a format corresponding to the intended display system, but only multistandard DVD players can play both types of disc, and only if connected to the right kind of television. This not a problem on computers because they already have to manipulate the video in order to display it on the monitor.

The DVD-Video standard includes region codes that producers can use to prevent playback in certain geographic regions. Video CDs have no such codes.

Established Base. At the end of 1999, over 40 million Video CD players were distributed, mostly in Asia. Video CDs also can be played on most CD-i players and on almost any computer with a CD-ROM drive and the proper MPEG-1 decoding hardware or software. Over 100 million computers in the United States alone are capable of playing Video CDs.

Over 8,000 Video CD titles exist worldwide, with about 700 available in the United States. Video CDs generally cost about the same or slightly less than videotapes.

Price. Currently, Video CD has the price advantage over DVD. Video CD movies sell for an average of $20 in the United States and for as little as

$3 in Asian countries. Video CD players typically cost $100 to $250, compared with DVD players at $150 to $500. DVD movies may become cheaper than Video CDs (outside Asia) because the production and manufacturing costs will be equivalent and the market will be larger. DVD-Video player prices will likewise be lowered by the larger market.

Compatibility of CD-i and DVD

No DVD drive or player can play a CD-i disc (Green Book format). A DVD-ROM drive can read data only from a CD-i bridge disc and requires special hardware and a specific operating system to make use of the CD-i data.

No CD-i player can play a DVD. Philips, the inventor of CD-i, once announced that it would make a DVD player with a CD-i adapter for playing existing CD-i discs. Some people also expect Philips to create a "DVD-i format" in attempt to breathe a little more life into CD-i (and recoup a bit more from its investment of more than $1 billion), but with the rapid decline of CD-i, Philips does not seem inclined to make a CD-i–adaptable player or a DVD-i variation.

Compatibility of Video CD and DVD

Many DVD-Video players and DVD-ROM drives can play audio and video from a Video CD. Compatibility with the White Book Video CD standard is not required by the DVD specification, but it is trivial from an engineering standpoint because any MPEG-2 decoder also can decode MPEG-1. About 50 percent of DVD players will play Video CDs, but anyone purchasing a DVD player with Video CDs in mind is advised to check the player specifications because some players do not play Video CDs. This is more of a marketing decision than a technical decision and may be influenced by the problem of pirated movies being cheaply available on Video CD.

Super Video CD

Super Video CD (SVCD) is an enhancement to Video CD that was developed by a Chinese government–backed committee of manufacturers and researchers, partly to sidestep DVD technology royalties and partly to create pressure for lower DVD player and disc prices in China. The final SVCD specification was announced by the China National Committee of

Recording Standards in September 1998. It was essentially a combination of the Chinese government's original SVCD proposal, the HQ-VCD proposal from the developers of the original Video CD (Philips, JVC, Sony, and Matsushita), and C-Cube's China Video CD (CVD) format. Although the final SVCD format borrowed from the CVD format, they were incompatible. Since C-Cube had a head start and widespread manufacturer support, which threatened the new standard, a new specification called Chao-Ji VCD (the Chinese equivalent of *super*) was quickly produced, requiring support for both formats. Almost all SVCD players are actually Chao-Ji players, although most discs are now produced in the more capable SVCD format. SVCD was later standardized as IEC-62107.

In terms of video and audio quality, SVCD is in between Video CD and DVD, using a 2X CD drive to support 2.2-Mbps variable-bit-rate MPEG-2 video (at 480 × 480 resolution for NTSC and 480 × 567 resolution for PAL), 2-channel MPEG-2 Layer II audio, text and graphic overlays, and other DVD-like features. Because the data rate is twice as fast, a disc that maximizes the quality will only play half as long: 37 minutes. Therefore, most SVCD players are three-disc changers that automatically switch from one disc to the next to achieve playing times long enough for a feature-length movie. In 2000 the unofficial DSVCD format appeared, which squeezes the tracks tighter together to achieve longer playing times. The *D* stands for "double," although typical capacity increases are only 1.5. Table 8.6 also presents a technical comparison of SVCD and DVD-Video specifications.

Advantages of DVD-Video over SVCD

Features. The DVD-Video format includes the same basic features as SVCD (menus, subpictures, pause, search, freeze, slow, fast, scan) and adds seamless branching, parental control, multiple camera angles, and more.

Capacity. Programs on DVD can be over eight times longer than on SVCD or even longer at equivalent quality levels. A single-layer DVD-Video holds over 2 hours of material per side, and a dual-layer disc holds over four hours. An SVCD holds 35 to 70 minutes and has only one side. Both DVD and SVCD support still frames with audio, but DVD has the potential for hundreds more pictures and hours more surround sound.

Convenience. Since SVCDs cannot hold much more than one half hour at high quality, the disc must be changed one or more times during a movie, unless the player holds more than one disc at a time. The DSVCD format bumps the playing time up to around one hour. A DVD-Video can easily hold a 4-hour movie on one side.

Audio.   DVD-Video has up to eight digital audio tracks. Each track can hold uncompressed audio with quality better than audio CD or compressed 5.1-channel surround sound in Dolby Digital or MPEG-2 format. SVCD has one or two tracks of compressed 2-channel audio in MPEG-1 Layer II format or one track of MPEG 5.1-channel audio.

Video.   The quality of DVD-Video exceeds that of SVCD, with one and a half times the resolution in normal mode and two times in widescreen mode (compared with letterboxed SVCD). DVD uses an average MPEG-2 data rate that is two or three times higher than SVCD.

Advantages of SVCD over DVD-Video

Price.   Currently, SVCD has a price advantage over DVD. SVCD movies sell for as little as $3 in Asian countries, although generally they are not available outside Asia. SVCD players typically cost $150 to $250, compared with DVD players at $150 to $500. DVD-Video player prices will drop as its market becomes bigger than the SVCD market, especially since DVD players do not require three-disc mechanisms.

Compatibility of SVCD and DVD

A few DVD-Video players, particularly those made in Asian countries, can play SVCDs because the DVD player already has MPEG capability. Fast Pentium II or Mac G3 computers can play SVCDs as long as they have the necessary software.

Other CD Formats

Many variations of the basic CD standard have been created to address limitations of the original design or to add new features. Most of these features have been integrated into DVD. DVD-ROM drives and DVD-Video players generally are backward-compatible with common mutations of CD. The details follow:

Compatibility of CD-R and DVD

A few DVD-ROM drives, usually early models, cannot read recordable CDs (Orange Book Part II). Many DVD-Video players cannot read CD-Rs. CD-RW compatibility is better. The primary reason to use a CD-R or CD-RW in a DVD-Video player is for homemade music CDs or homemade Video CDs.

The problem is that CD-Rs are invisible to the wavelength of laser required by DVD because the dye used in CD-Rs does not reflect the laser beam properly. This problem has been addressed in many ways. A number of manufacturers use twin-laser pickups in which one laser is used for reading DVDs and the other for reading CDs and CD-Rs. Sony developed a new dual-wavelength laser diode for the DVD drive in PlayStation 2. These kinds of solutions provide complete backward-compatibility with the millions of extant CD-R discs. An attempt to create a new "type II" CD-R medium was abandoned in part because of technical difficulties but also as a result of the realization that new discs would cost more than CD-Rs and would take far too long to supplant the existing CD-R supply.

Compatibility of CD-RW and DVD-ROM

DVD drives and players usually can read rewritable CDs (Orange Book Part III). CD-RW discs do not work in CD-ROM readers manufactured before 1997, and many CD audio players still do not read CD-RW discs.

CD-RW has a lower reflectivity difference than specified by the CD standard, thus requiring *automatic gain control* (AGC) circuitry. Since DVD-ROM drives and DVD players already contain AGC, they are usually able to read CD-RW discs because the CD-RW format does not have the invisibility problem of CD-R (see the preceding section).

The OSTA MultiRead logo guarantees compatibility with CD-R and CD-RW, but it is not widely used, even on hardware that is compatible.

Compatibility of Photo CD and DVD

Most DVD-ROM drives read Kodak Photo CDs because the drives already support the requisite CD-ROM XA and multisession standards. However, the operating system or a software application must specifically support the Photo CD file format in order to view the pictures. Since Photo CDs are usually produced using CD-R media, they suffer from the CD-R problem (see the preceding sections).

Three years after DVD's release, with over 70 brands of DVD players, none were able to play Photo CDs. The CD-R problem aside, DVD-Video players could support Photo CDs with an extra chip and a license from Kodak, but it seems that no manufacturer is interested.

Compatibility of Enhanced CD (CD Extra) and DVD

DVD-Video players can play music from Enhanced CDs (which contain music tracks followed by computer data tracks). Most DVD-ROM drives will play music and read data from Enhanced CDs, unless they are of the track-zero (pregap) type that is not properly supported by some DVD/CD-ROM drivers.

Compatibility of CDG and DVD

A few DVD-Video players support CD+G (music with additional graphics), specifically players with karaoke features. Most players sold in the United States do not. Most DVD-ROM drives can read CD+G discs but require proper software to reconstruct the graphics.

MovieCD

MovieCD is a computer movie playback system introduced in February 1997 by Sirius Publishing, based on its own proprietary video-encoding technology: Motion Pixels. MovieCDs are intended solely for playback on computers—stand-alone players do not exist.

MovieCD titles are stored on standard CD-ROMs and can be played back on a 486 DX2 66-MHz or faster computer using the Microsoft Windows operating system. At least 8 megabytes of RAM and a double-speed CD-ROM drive are required. Video quality, according to MovieCD's creators, is slightly better than VHS videotape. MovieCD's primary competitor is Video CD (see the preceding sections), which never succeeded in the United States. Reviewers consider MovieCD video quality to be just below that of Video CD. In early 1997, about 50 movies were released in MovieCD format, none of which were new releases and many of which were not big sellers in theaters or on videotape. Few titles have been added to the catalog since then because MovieCD was completely eclipsed by DVD.

Since most new computers with DVD-ROM drives are capable of playing audio and video at much higher quality, MovieCD can be thought of as a "poor person's DVD" and has been outclassed as more computers become DVD-capable. It is possible that a higher-quality DVD-based version of MovieCD will be developed, but it will still be confined to playback on computers.

MiniDisc (MD) and DCC

MiniDisc was introduced in 1991 by Sony as a replacement for cassette tapes (not CDs, as many have mistakenly assumed). MiniDiscs are 64-millimeter erasable *magneto-optical* (MO) discs in a plastic shell similar to that of small floppy disks. Prerecorded discs can be stamped like CDs or DVDs. The MiniDisc format uses ATRAC audio compression, a 52-band perceptual coding technique also used by Sony's theatrical SDDS system. A variation of MiniDisc for computer use, called MD-Data, is capable of holding 140 megabytes.

MiniDisc competed fiercely with *Digital Compact Cassette* (DCC), a format that was introduced in 1992 by Philips. DCC had the advantageous capability of playing and recording standard compact cassette tapes but did not do well in the marketplace and was officially abandoned by Philips in late 1996. DCC used PASC compression, which is essentially the same as MPEG-1 Layer II. Table 8.7 lists the specifications of MD and DVD-Video.

Advantages of DVD-Video over MiniDisc

Quality. MiniDisc was rushed to market because of rival DCC, and the flaws in the first version of its audio compression system were not well received. Since then, the ATRAC compression technology has gone through four generations and has improved significantly, slowly overcoming initial bad impressions. Still, many audio purists feel that MiniDisc's compression adversely affects audio quality. DVD-Video's PCM format uses no compression and is technically much superior. DVD-Video's compressed audio formats, Dolby Digital, DTS, and MPEG-2, use a higher data rate than

TABLE 8.7

MiniDisc and
DVD-Video
Specifications

	MiniDisc	DVD-Video
Diameter	64 mm	80 or 120 mm
Thickness	1.2 mm	1.2 mm (two 0.6-mm substrates)
Audio	20-bit 44.1-kHz ATRAC compressed, 2 channels	16/20/24-bit 48/96-kHz PCM, 16-bit 48-kHz Dolby Digital/MPEG-2 5.1/7.1-channel surround sound
Data rate	292 kbps CBR	64 to 448 kbps CBR (Dolby Digital), 64 to 912 kbps VBR (MPEG)
Error correction	RS-CIRC	RS-PC
Modulation	EFM (8/14)	EFMPlus (8/16)
Audio SNR	120 dBb	96 to 144 dB
Dynamic range	105 dB (20-bit PCM)	96/120/144 dB (16/20/24-bit PCM), 120 dB (20-bit Dolby Digital)
Frequency ($\pm$0.3 dB)	4 to 20,000 Hz (44.1 kHz)	4 to 22,000 Hz (48 kHz), 4 to 44,000 Hz (96 kHz PCM)
Playing time	74 min	~27 h of surround sound, ~7 h of 20-bit 48-kHz stereo, ~138 min of 24-bit 96-kHz stereo[a]
Data capacity	140 MB	4.4 to 15.9 GB

[a]All times are for a single 4.4-GB layer. Times for a dual-layer DVD are almost twice as long.

MiniDisc (usually 384 to 768 kbps average compared with 292 kbps), but they contain six channels of audio as compared with MiniDisc's two channels. Of course, the same audio purists who dislike MiniDisc's compression turn up their noses at Dolby Digital, DTS, and MPEG-2 audio as well.

Capacity. A MiniDisc holds 74 minutes of audio, or 140 million bytes of data. DVD, using uncompressed stereo sound on a single-layer disc, holds over two hours of audio. Using all 5.1 channels of Dolby Digital audio, a single-sided, single-layer DVD can play for over 27 hours. DVD holds over 8.5 billion bytes of data per side, over 60 times more than MD-Data.

Support. MiniDisc is supported primarily by Sony, whereas DVD had already garnered the support of over 100 companies when it was introduced.

Advantages of MiniDisc over DVD-Video

Durability. The optical disc used by the MiniDisc system is more vulnerable than a DVD-ROM disc, but it is encased in a plastic shell to protect it. DVDs are unprotected and subject to scratches.

The MiniDisc shell adds significantly to the production cost. The designers of DVD rejected a mandatory shell, but offer caddies or cartridges as an option. DVD-RAM discs, specifically, can be protected in cartridges.

Convenience. Because MiniDiscs are in a shell, they can be handled easily without fear of fingerprints and scratches.

MiniDiscs are about 3 inches across, making them very easy to use and to store. Standard DVD discs are over 5 inches across, but the DVD format also includes a 3-inch version, which may become the preferred size for some video and audio recording devices.

Compatibility of MiniDisc and DVD-Video

No DVD player can play a MiniDisc, and no MiniDisc player can play a DVD. It is not likely that a dual-format player will be developed because the systems are quite different.

The ATRAC encoding system used to store audio on a MiniDisc is proprietary to Sony. The SDDS format, which is an option on DVD-Video discs, is based on ATRAC, but no current DVD players support SDDS, no discs have been announced with SDDS audio tracks, nor are any external SDDS decoders available to consumers.

Digital Audiotape (DAT)

DAT was developed by Sony and Philips for digital storage of music and data. Sadly, DAT was an early casualty of copy protection battles and was delayed for years while serial copy management schemes and legislation were produced. The format achieved little success in the home audio market but has become a standard for professional audio recording and computer data archiving.

Most computer DAT drives include lossless hardware data compression in order to store more data on a tape. Table 8.8 lists DAT specifications.

TABLE 8.8

DAT and
DVD-Video
Specifications

	DAT	DVD
Audio	16-bit 44.1/48-kHz linear PCM, 12-bit 32-kHz nonlinear PCM; 2 channels	16/20/24-bit 48/96-kHz linear PCM, 20-bit 48-kHz Dolby Digital/MPEG-2; multichannel
Audio SNR	~92 to 96 dB	~96 to 144 dB
Dynamic range	92 to 96 dB (16-bit PCM)	96/120/144 dB (16/20/24-bit PCM), 120 dB (20-bit Dolby Digital)
Frequency (±0.3 dB)	4 to 14,500 kHz (32 kHz), 4 to 22,000 Hz (44.1/48 kHz)	4 to 22,000 Hz (48 kHz), 4 to 44,000 Hz (96 kHz PCM only)
Playing time	2 h (44.1/48 kHz), 4 h (32 kHz)	~27 h of surround sound, ~7 h of 20-bit 48-kHz stereo, ~138 min of 24-bit 96-kHz stereo[a]
Copy management	SCMS	CGMS, CSS
Data capacity	2 to 4 GB[b]	4.4 to 15.9 GB
Data rate	~2 to 5 Mbps	11.08 Mbps

[a]All times are for a single 4.4-GB layer. Times for a dual-layer disc are almost twice as long.
[b]Uncompressed.

Advantages of DVD over DAT

Cost. The cheapest DAT decks cost around $500, with professional units beginning in the $1000 price range. Computer DAT drives generally cost $700 to $1200. DVD players and DVD-ROM drives are cheaper. Recordable DVD drives (DVD-RAM and DVD+RW) are cheaper than DAT drives. DVD audio and video recorders were initially priced above $2000 but should fall to the same price levels as DAT recorders and drives within a year or two. DVD-ROM discs, of course, are much cheaper to produce than DAT cassettes. Writable DVD discs are also cheaper.

Speed. All versions of DVD have higher data transfer rates than DAT. Obviously, random access times are not even close.

Durability. Tape is subject to degradation from wear and stretching, dropouts, erasure from magnetic fields, and damage from heat. DAT error correction can compensate for most raw data errors. In comparison, discs never wear out, are impervious to magnetic fields, and are less susceptible to heat damage. DVD error correction is much more robust. Discs can be scratched, but only large scratches will cause noticeable playback problems or data loss.

DAT decks and drives "eat" tapes. It is possible that a DVD player could scratch a disc, but this is rare.

DAT equipment includes helical scanning heads, tape-loading mechanisms, and hundreds of small components that can easily break down. DVD players have much simpler mechanisms.

Advantages of DAT over DVD

Capacity. DAT capacities continue to grow from the original 4 gigabytes to 20 gigabytes. DVD-18 discs are roughly equivalent, but recordable DVD capacities are lower than what most DAT tapes and drives provide.

Support. In the audio recording industry, DAT equipment is used widely and well supported. In the computer industry, DAT is popular for data backup, but the various writable DVD formats eventually will eclipse it.

Compatibility of DAT and DVD

DAT tapes and decks are not physically compatible with DVD discs and players, but DAT and DVD-Video share the same 48-kHz PCM digital audio signal format. In fact, most PCM audio on DVD-Video discs was recorded or archived using DAT.

Both DAT and DVD-ROM carry digital data. A computer with a DAT drive and a DVD-ROM drive connected to it can interchange data between the two media.

Magneto-Optical (MO) Drives

An MO drive, also called a *rewritable optical drive,* is essentially a hybrid of optical disc technology and magnetic disc technology. MO systems rely on a property of certain materials that enables their magnetic state to be changed when heated to a certain temperature. A low-power polarized laser is used to read the data by detecting the magnetic orientation of spots on the disc. A high-power laser is used to heat spots that are changed by a local magnetic field.

Essentially, two standard formats of MO discs and drives exist. The 3.5-inch format first appeared with 128 megabytes of capacity, moved up to 230 megabytes, and recently became available in 640-megabyte sizes. The 5.25-inch format currently is available in 1.2- and 2.6-gigabyte capacities. The 2.6-gigabyte version holds 1.3 gigabytes on each side of the disc.

Proprietary versions of MO drives can store as much as 4.6 gigabytes, but these do not follow the ISO standards. Although they can read discs from other drives, their high-density formatted discs do not work in other drives. Table 8.9 lists MO specifications.

Advantages of DVD-ROM over MO

Cost. A 3.5-inch MO drive costs $500 to $700, with blank cartridges priced around $30. The 5.25-inch drives cost around $2000, with blank discs in the $40 to $60 range. DVD drives and discs, including blank writable discs, are cheaper, and prices continue to drop.

DVD discs can be stamped quickly and cheaply. MO discs generally are not used for prerecorded data because they are expensive and must be recorded individually.

Capacity. DVD-ROM discs hold much more than MO discs. Writable DVD discs hold 4.7 billion bytes, more than the largest standard MO disc. The next generation of MO technology is expected to have storage capacities of about 7 billion bytes per side, close to the 8.5 billion bytes of dual-layer DVD-ROM discs, although writable DVD is not available in dual-layer form. MO development has slowed down considerably since recordable DVD was released.

TABLE 8.9		MO	DVD
MO and DVD-ROM Specifications	Capacity (10×)	128M bytes to 2.6G bytes	4.7 to 17G bytes
	Sector size	512 or 1024 bytes	2048 bytes
	Transfer rate[a] (2×)	Up to 3.9 MB/s	1.3 MB/s

[a]Reference rates for single-speed drives.

Support.   MO technology is supported by several companies such as Sony, Fujitsu, and Pinnacle. DVD-ROM has much broader industry support, and even the various writable DVD formats have more companies behind them.

Advantages of MO over DVD-ROM

Durable.   MO discs are more vulnerable than DVD discs, but they are encased in a plastic shell for protection. DVD-RAM discs optionally use shells. DVDs are unprotected and subject to scratches. Shells add to the cost.

Compatibility of MO and DVD-ROM

Since MO discs are encased in a cartridge, they are incompatible with DVD. A new version of MO technology—ASMO—is under development as well as promises to provide drives that can read DVD discs and MO cartridges. Both media store digital information. A computer with both types of drives connected can transfer data between the two formats.

Other Removable Data Storage

Recordable DVD formats, particularly DVD-RAM, compete with traditional "removable hard drives" such as Jaz drives. Recordable DVD also competes with Zip, SuperDisc, HiFD, and similar new-generation floppy drives.

In general, DVD has the advantage. DVD-ROMs can be mass-produced, whereas recordable DVD has much lower cost per megabyte. DVD discs are not susceptible to magnetic fields.

Most removable-disk formats have faster seek times and write times than recordable DVD, although newer-generation DVD-RAM drives are just as fast as Jaz drives. SuperDisc and HiFD drives, at around $150, cost more than DVD-ROM drives but less than DVD-RAM drives or other recordable DVD drives. SuperDisc and HiFD drives have the advantage of reading old-style 3.5-in floppies. Of course, this must be weighed against the ability of DVD drives to read CDs and DVDs.

DVD at Home

Introduction

The primary focus of DVD marketing and sales is home entertainment. Even though there were seven times as many DVD PCs as home DVD players in 2000, there were about 100 times as many video titles as computer software titles.

This chapter talks about how to use a DVD player at home—how to choose a player and how to hook it up to a television and stereo system for the best sound and picture. If you are not sure if DVD is for you, a buying-decision quiz will help you make the decision.

Choosing a DVD Player

Just as most CD players meet the needs of the average buyer, most DVD players have the features and quality that most buyers are looking for. Some low-cost players may have problems playing more advanced discs, but overall, many good players are available. Video and audio performance in most modern DVD players is excellent. Unless you have a high-end home theater setup, a player that costs under $400 should be completely adequate. A higher price will buy a slight improvement in video quality (or a big improvement if you buy a progressive-scan player), high-end audio features, or more reliability and sturdiness.

Personal preferences, your budget, and your existing home theater setup play the most important roles in what player is best for you. Make a list of things that are important to you (see the checklist in Table 9.1). Make a list of players in your price range, and then eliminate or downgrade the ones that do not match your list. Try out a few of the players that remain, focusing on ease of use (remote control design, user interface, and front-panel controls). Because there is not much variation in picture quality and sound quality within a given price range, convenience features play a big part. Pay special attention to the remote control. Are the buttons easy to find, with clear labels? Is it comfortable to hold, and can you reach the important buttons with your thumb? Are the controls illuminated? Do you prefer a jog-shuttle knob? You will use the remote control all the time, and it will drive you crazy if it does not suit your style.

You may wish to consider buying a DVD computer instead of a standard DVD player, especially if you want progressive video or if you enjoy PC-enhanced DVDs (Table 9.2; see also Chapter 11).

Table 9.1

Checklist of Player
Features

Do I want selectable sound tracks and subtitles, multiangle viewing, aspect ratio control, parental/multirating features, fast and slow playback, digital video, multichannel digital audio, compatibility with Dolby Pro Logic and Dolby Digital receivers, on-screen menus, dual-layer playback, and ability to play audio CDs?	This is the wrong question to ask because all DVD players have all these features.
Do I want 96-kHz, 24-bit audio decoding?	For high-end audio systems (analog connection).
Do I want 96-kHz, 24-bit PCM digital out?	For high-end audio systems (digital connection).
Do I need an internal 6-channel Dolby Digital or DTS decoder?	Important if you have a multichannel ("Dolby Digital ready") amplifier.
Do I want DTS audio output?	Look for a player with the "DTS Digital Out" logo.
Does my receiver have only optical or only coaxial digital audio inputs?	Make sure the player has outputs to match.
Do I want virtual surround?	Useful if you only have two speakers.
Do I want progressive-scan video?	Gives the best-quality picture, but only if you have a progressive-scan TV.
Do I want to play video CDs?	Check the player specs for video CD compatibility.
Do I want to play HDCDs?	Check for the "HDCD" logo.
Do I need a headphone jack?	Handy for late-night viewing.
Do I want on-screen player setup menus and displays in languages other than English?	Look for multilanguage setup feature. (Note: All players support multilanguage menus when provided on the disc.)
Do I want to play homemade CD-R audio discs?	Look for the "dual laser" feature.
Do I want to replace my CD player?	You might want a changer model that can hold three, five, or even hundreds of discs.
Do I want to control all my entertainment devices with one remote control?	Look for a player with a programmable universal remote, or make sure your existing universal remote will run the new DVD player.

continues

Table 9.1 cont.

Checklist of Player
Features

Do I want a player that holds more than one disc?	Three- or five-disc players are useful for music. Players that hold 100 or more discs allow you to make your entire collection quickly available.
Am I bothered by bright front panel displays when watching movies?	Check for an option to dim or turn off the lighted display.
Do I want to zoom in to check details of the picture?	Look for players with picture zoom.
Do I care about black-level adjustment?	Check for a 0/0.7 IRE setup option.
Do I value special deals?	Some players come with free DVD coupons and free DVD rentals.

How to Hook Up a DVD Player

The DVD format is designed around some of the latest advances in digital audio and video, yet the players are also designed to work with TVs and video systems of all varieties. Therefore, the back of a DVD player can have a confusing diversity of connectors producing a potpourri of signals.

This section discusses the advantages and disadvantages of each kind of output signal, what the connectors are like, and how to hook them up to typical audio/video equipment. Refer to "Video Interface" and "Audio Interface" in Chapter 6 for technical details.

Signal Spaghetti

Most DVD players produce the following output signals:

- *Analog stereo audio.* This standard two-channel audio signal can include Dolby Surround encoding.
- *Digital audio.* This raw digital signal can connect to an external digital-to-analog converter or digital audio decoder. There are two different signal interface formats for digital audio: S/PDIF and Toslink. Both formats can carry *pulse-code modulated* (PCM) audio, multichannel Dolby Digital (AC-3) encoded audio, multichannel DTS encoded audio, and multichannel MPEG-2 encoded audio. The digital audio output on DVD-Audio-capable players also can carry MLP encoded audio.

Table 9.2

Pros and Cons of
DVD Players, DVD
PCs, and VCRs

Capability	DVD Player	DVD Computer	VCR
Plays VHS tapes	No	No	Yes
Records video	Usually not	Maybe (with right hardware or software)	Yes
Plays DVD-Video	Yes	Yes (with right hardware or software)	No
Plays Video CD	Yes	Yes	No
Plays audio CD	Yes	Yes	No
Plays DVD-ROM	No	Yes	No
Plays CD-ROM	No	Yes	No
Connects to TV	Yes	Usually	Yes
Connects to VGA monitor	No	Yes	No
Progressive video output	Usually not	Yes	No
Dolby Surround audio	Yes	Yes	Yes (if stereo model)
Dolby Digital audio	Yes	Maybe (depends on decoder and sound card)	No
DTS audio	Usually	Maybe (depends on decoder and sound card)	No
Connects to Internet	Usually not	Yes (with modem)	No
Plays games	Usually not	Yes	No
Can be updated	Not easily	Yes, especially if software decoder	Not necessary
Input devices	Remote	Mouse, keyboard, remote	Remote
Noise level	Very low	Low to very noisy	Low
Portable players	Yes	Yes (laptops)	No
Typical price	$150-$500	$400-$2000	$75-$400

- *Composite baseband video.* This is the standard video signal for connecting to a TV with direct video inputs or an *audio-visual* (A/V) receiver.
- *S-video (Y/C).* This is a higher-quality video signal in which the luma and chroma portions travel on separate wires.

Some players may produce additional signals:

- *Six-channel analog surround.* These six audio signals from the internal audio decoder connect to a multichannel amplifier or a Dolby-Digital-ready (AC-3-ready) receiver.

- *AC-3 radio-frequency (RF) audio.* The Dolby Digital FM audio signal from a laserdisc, connects to an audio processor or receiver with an AC-3 demodulator and decoder.

- *Radio-frequency (RF) audio/video.* These old-style combined audio and video signals are modulated onto a VHF RF carrier for connecting to the antenna leads of a TV tuner.

- *Component analog video (interlaced scan).* These three video signals (RGB or YP_bP_r) can connect to a high-end TV monitor or video projector.

- *Component analog video (progressive scan).* These three video signals (RGB or YP_bP_r) can connect to a progressive-scan monitor or projector.

- *Digital audio/video.* This is IEEE 1394/FireWire or other digital interconnect.

Connector Soup

The different audio and video signals may be presented on the following types of connectors:

- *RCA phono* (Figure 9.1). This is the most common connector, used for analog audio, digital audio, composite video, and component video. The term *cinch* is also used for this connector.

- *BNC* (Figure 9.2). This connector carries the same signals as RCA connectors but is more popular on high-end equipment.

- *Phono or miniphono* (Figure 9.3). This connector carries stereo analog audio signals and may be used by portable DVD players. It also may appear on the front of a DVD player for use with headphones.

- *S-video DIN-4* (Figure 9.4). This connector, also called *Y/C,* carries separated chroma and luma video signals on a special four-conductor cable.

- *Toslink fiberoptic* (Figure 9.5). This connector, developed by Toshiba, uses a fiberoptic cable to carry digital audio. One advantage of the fiberoptic interface is that it is not affected by external interference and magnetic fields. The cable should not be more than 30 to 50 feet (10 to 15 meters) long.

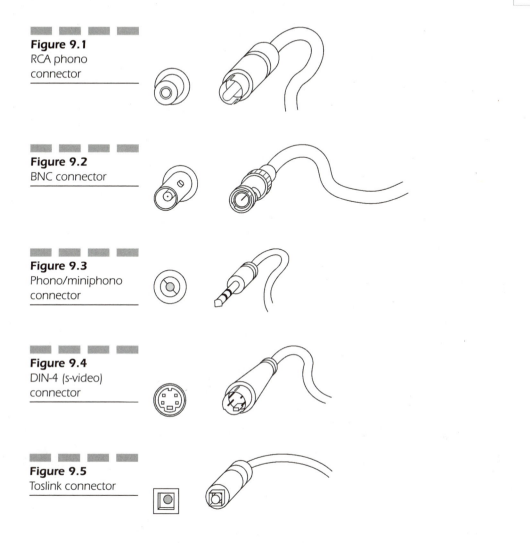

Figure 9.1
RCA phono
connector

Figure 9.2
BNC connector

Figure 9.3
Phono/miniphono
connector

Figure 9.4
DIN-4 (s-video)
connector

Figure 9.5
Toslink connector

■ *IEEE 1394* (Figure 9.6). Also known as *FireWire* or *i.Link*. This connector, not available on most DVD players, is actually an external bus, carrying all types of digital signals. Once copy protection issues are resolved, more consumer electronics devices will use IEEE 1394 connectors, making component hookup a breeze, with only one cable for all audio and video signals.

■ *DB-25* (Figure 9.7). This 25-pin connector, adapted from the computer industry, is used by some audio systems for multichannel audio input.

■ *SCART* (Figure 9.8). This 21-pin multipurpose connector, used primarily in Europe, carries many audio and video signals on a single cable: analog audio, composite RGB video, component video, and RF. Also called a *Euro* or *Peritel* connector.

■ *Type F* (Figure 9.9). This connector typically carries a combined audio and video RF signal over a 75-ohm cable. A 75- to 300-ohm converter may be required.

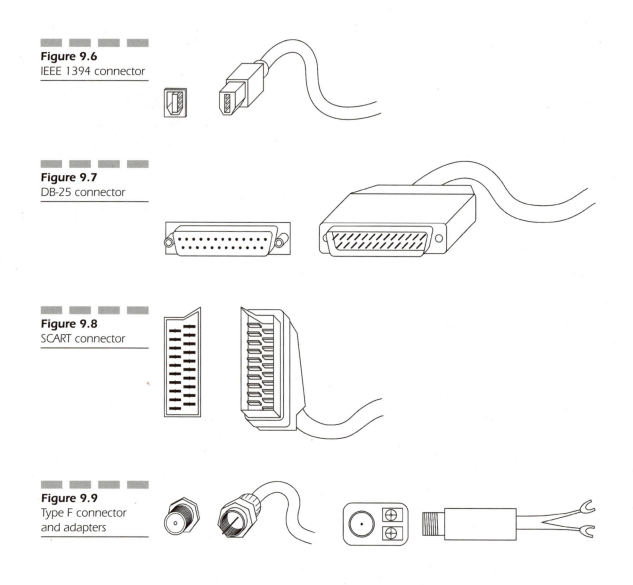

Figure 9.6
IEEE 1394 connector

Figure 9.7
DB-25 connector

Figure 9.8
SCART connector

Figure 9.9
Type F connector
and adapters

Audio Hookup

A DVD player must be connected to an audio system: a receiver, a control amp or preamp, a digital-to-analog converter, an audio processor, an audio decoder, an all-in-one stereo, a TV, a boombox, or other equipment designed to process or reproduce audio. For simplicity, the following sections occasionally will use the term *multichannel audio* to refer to Dolby Digital (AC-3) audio, DTS audio, and MPEG-2 audio. The term is not used to refer to Dolby Surround audio, which has only a two-channel signal. If your audio system provides both Dolby Digital and Dolby Pro Logic, connect the DVD player to the Dolby Digital inputs for the best result.

Discs for NTSC players are required to provide at least one audio track using either Dolby Digital or PCM audio. Discs for PAL players are required to provide at least one Dolby Digital, MPEG, or PCM audio track. Not all NTSC players are able to play MPEG audio tracks. All PAL players are able to play both MPEG and Dolby Digital audio tracks. Dolby Digital is becoming the dominant multichannel audio standard for DVD and other digital video formats, and it is not likely that MPEG-2 will ever see much use. DVD also supports optional multichannel formats, including DTS and SDDS. See Chapter 6 for details on different audio formats, including the difference between Dolby Surround/Pro Logic and Dolby Digital.

Some players also provide the Dolby Headphone feature, which processes the multichannel audio to recreate a more three-dimensional sound field for headphones. The Dolby Headphone feature works with all standard headphones.

If the amplifier has a subwoofer output and your subwoofer has a direct (coaxial) audio input, use that rather than connecting the speaker outputs from the subwoofer. Unless the subwoofer is much more expensive than the amplifier, the amplifier will do a better job of bass management (see the following "Bass Management"). Most DVD players also provide two or three audio hookup options. These options are detailed in the following sections.

Digital Audio The digital audio outputs provide the highest-quality audio signal. This is the preferred connection for audio systems that have them. Almost all DVD players have digital audio outputs for PCM audio and multichannel audio. These outputs carry either the raw digital audio signal directly from the digital audio track or the two-channel downmixed PCM signal from the internal multichannel decoder.

For multichannel audio output, the encoded digital signal bypasses the player's internal decoder. The appropriate decoder is required in the receiver or as a separate audio processor. For PCM audio output, the PCM signal from an audio track is sent directly to the digital audio output. Or alternatively, the multichannel decoder in the player produces a PCM signal. In either case, a receiver with a built-in digital-to-analog converter (DAC) or an outboard digital-to-analog converter is required. Some players provide separate outputs for multichannel audio and for PCM audio. Other players have either a switch on the back or a section in the onscreen setup menu where you can choose between PCM output and multichannel output (undecoded Dolby Digital, DTS, or MPEG audio). The multichannel output menu option is usually labeled "AC-3" or "Dolby Digital."

All NTSC DVD players include a two-channel Dolby Digital decoder, so they can produce PCM audio from Dolby Digital audio. Most PAL DVD players include both two-channel Dolby Digital and two-channel MPEG audio decoders, so they can produce PCM output from either format. Many players (NTSC and PAL) also provide DTS audio output, but only for connection to an external DTS decoder. Few players have built-in DTS decoders.

The digital audio output is also used for PCM audio from a CD. Players that can play Video CDs also may produce PCM audio output converted from the MPEG-1 audio signal. Combination laserdisc/DVD players also use this output for the laserdisc's PCM audio track (but not the AC-3 track; see the following).

The direct output from PCM tracks on a DVD is at a 48 or 96 kHz sampling rate with 16, 20, or 24 bits. The converted PCM output from multichannel audio tracks is at 48 kHz and up to 24 bits. The PCM output from a CD is at 44.1 kHz and 16 bits. The PCM output from a laserdisc player is also at 44.1 kHz and 16 bits. The connected audio component does not need to be able to handle all these variations, but the more the better. A system capable of 16 and 20 bits at sampling rates of 48 and 96 kHz is recommended. Some DVD players are incapable of properly formatting a PCM signal for output at high sampling rates or bit sizes. If you have an external system capable of 24 bits or 96 kHz, make sure the player can correctly produce the digital audio signal. Also be aware that players are required by the CSS/CPPM license to restrict digital output of 96 kHz audio when the disc is encrypted. In this case, most players downsample to 48 kHz.

The digital audio output must be connected to a system designed to accept either PCM digital audio, Dolby Digital (AC-3), or both. Most modern digital receivers can automatically sense the type of incoming signal.

Some players include a dynamic range control setting (also called *midnight mode*) that boosts soft audio and reduces loud sound effects. This setting should be turned off for best effect with a home theater system, but it should be turned on for environments where the dialogue cannot be clearly heard, such as when everyone else in the house has gone to bed and the volume is down low.

Dolby Digital also has a feature called *dialog normalization,* which is designed to match the volume level from various sources. Each encoded audio source includes information about the relative volume level. Dialog normalization automatically adjusts the playback volume so that the overall level of dialog remains constant. It makes no other changes to the audio, including dynamic range; it is equivalent to manually turning the volume control up or down when a new program is too soft or too loud. Usually, only one setting exists in a program; that is, the volume control does not change in the middle. Dialog normalization is especially useful with a digital television source, to handle variations in volume when changing channels. It is less useful for DVDs, unless they have many separate programs on them.

DTS is not used on most DVD discs. SDDS is not used on any discs. They are optional multichannel surround formats that are allowed in the DVD format but not directly supported by most players. Each requires the appropriate decoder in the receiver or a separate audio processor.

Connecting Digital Audio Two different standards govern the digital audio connection interface: coaxial and optical. The arguments are many and varied as to which one is better, but since they are both digital signal transports, good-quality cables and connectors will deliver the exact same data. Some players have only one type of connector, although many have both.

Coaxial digital audio connections use the IEC-958 II standard, also known as *S/PDIF* (Sony/Philips Digital Interface Format). Most players use RCA phono connectors, but some use BNC connectors. Use a 75-ohm cable to connect the player to the audio system. Multichannel connectors are usually labeled "Dolby Digital" or "AC-3." PCM audio connectors usually are labeled "PCM," "digital audio," "digital coax," "optical digital," or so on. Dual-purpose connectors may be labeled "PCM/AC-3," "PCM/Dolby Digital," or something similar. Make sure you use a quality cable; a cheap RCA patch cable may degrade the digital signal to the point that it does not work.

Optical digital audio connections use the EIAJ CP-340 standard, also known as *Toslink.* Connect an optical cable between the player and the audio receiver or audio processor. The connectors are labeled "Toslink," "PCM/AC-3," "optical," "digital," "digital audio," or the like.

If the connection (either coaxial or optical) is made to a multichannel audio system, select Dolby Digital/AC-3 (or DTS or MPEG multichannel) audio output from the player's setup menu or via the switch on the back of the player. If the connection is to a standard digital audio system (including one with a Dolby Pro Logic processor), select PCM audio output instead. In cases where a player has an optical (Toslink) connection but the audio system has a coaxial (S/PDIF) connection, or vice versa, a converter can be purchased for a few hundred dollars.

When you have used a digital connection, there is no need to make an analog audio connection. The exception is when certain players are set to output PCM audio at a 96 kHz sampling rate. When an encrypted disc prohibits 96 kHz output, some players will only produce analog audio output. In this case, a separate analog connection is needed.

Multichannel Analog Audio A component multichannel audio connection can be as good as a digital audio connection. However, such outputs use the digital-to-analog converters that are built into the player, and they may not always be of the best quality, especially on a low-cost player. The analog signal also must be converted back to digital when connected to a digital receiver. If you have an amplifier with multichannel inputs or a Dolby-Digital-ready receiver, six-channel audio connections are an appropriate choice, since a Dolby Digital decoder is no longer required.

All DVD players include a built-in two-channel Dolby Digital audio decoder. Only some players include a full six-channel decoder along with the multichannel digital-to-analog converters and external connectors necessary to make the decoded audio available. Some players support the Dolby EX or DTS ES formats, which add a rear center channel. In this case, a seventh audio connection is necessary to use the added channel.

DVD-Audio, with more emphasis on multichannel PCM audio, makes a multichannel amplifier more important. Unless the amplifier has MLP inputs (most do not), six analog connections or three pairs of two-channel digital audio connections are needed to carry all six PCM audio signals. Otherwise, the player must downmix to two-channel Dolby Surround.

Connecting Multichannel Analog Audio The player typically has six RCA or BNC jacks (or seven for EX/ES formats), one for each channel. A receiver/amplifier with six audio inputs—or more than one amplifier—is needed. Hook six audio cables to the connectors on the player and to the matching connectors on the audio system. The connectors typically are labeled for each speaker position: L, LT, or Left; R, RT, or Right; C or Center; LR, Left Rear, LS, or Left Surround; RR, Right Rear, RS, or Right Surround; Subwoofer or LFE; and sometimes, CR, Center Rear, CS, or Center Surround. Some receivers use a single DB-25 connector instead of separate connectors. An adapter cable is required to convert from DB-25 on one end to multiple RCA connectors on the other.

Stereo/Surround Analog Audio   A two-channel audio connection is the most widely used option, but it does not have the quality and discrete channel separation of a digital or multichannel audio connection. All DVD players include at least one pair of RCA (or sometimes BNC) connectors for stereo output. Any disc with multichannel audio will be downmixed automatically by the player to Dolby Surround output for connection to a regular stereo system or a Dolby Surround/Pro Logic system.

Connecting Stereo/Surround Analog Audio   Connect two audio cables with RCA or BNC connectors to the player. Connect the other ends to a receiver, an amplifier, a TV, or other audio amplification system. Connectors may be labeled "audio," "left," or "right." The connector for the left channel is usually white, and the connector for the right channel is usually red.

In some cases, the audio input on the stereo system (such as a boom box, if it can be rightly called a stereo system) will be a phono or miniphono jack instead of two RCA jacks. An adapter cable must be used. If the player is a portable player with a miniphono connector, a phono-to-RCA adapter cable is usually required to connect the player to the audio system. If the player includes a phono or miniphono connector for headphones, it is generally not recommended that the headphone output be used to connect the player to a stereo system because the line levels are not appropriate.

AC-3 RF Digital Audio   This digital audio output is provided only by combination laserdisc/DVD players. The AC-3 digital audio signal from the FM audio track of a laserdisc is presented at this output. Laserdisc AC-3 audio does not appear at the standard PCM/AC-3 output (although the stereo PCM audio track does). Audio from a DVD does not come out the AC-3 RF output. In other words, this is a special output designed solely for the AC-3 signal from a laserdisc, which is in a different format from DVD's AC-3 signal.

Hook a coaxial cable from the AC-3 RF output of the player to the AC-3 RF input of the receiver or AC-3 processor. Make sure the receiving end is set to rf mode or can automatically adapt to an RF signal.

In order to receive all audio signals from a combination laserdisc/DVD player, three separate audio hookups are required: a PCM/AC-3 connection (for DVD digital audio and laserdisc PCM digital audio), an AC-3 RF connection (for laserdisc AC-3 audio), and an analog stereo connection (for the laserdisc analog channels, which often contain supplemental audio).

A Bit About Bass

The heart-thumping, seat-shaking excitement of action movies relies heavily on deep, powerful low-frequency audio effects. DVD provides audio quality that is actually better than what comes on film for theaters—it is up to the home theater owner (subject to the spousal approval factor and the neighbor tolerance factor) how close he or she wants to get to a theater sound system.

All the ".1" sound encoding formats on DVD provide special channels for low-frequency effects (LFE). Despite becoming a standard feature, the LFE channel is misunderstood by people producing DVDs and by people listening to them. Part of the problem is that the LFE channel is overused by audio engineers. It possible and quite normal for all the bass in a movie to be mixed in the five main channels, since they are all full frequency. The LFE channel should be reserved for extraordinary bass effects, the type that only work well in a full discrete surround system with at least one subwoofer. Again, however, the same bass effects could be mixed in a 5.0 configuration with no loss or compromise. The reason is that modern receivers, particularly those with Dolby Digital and DTS decoders and a separate subwoofer output, have integrated bass management. Depending on the speaker configuration, the receiver automatically filters and routes bass below a certain frequency to the speakers that can reproduce it. For example, if an audio system has five small bookshelf speakers and a subwoofer, the receiver should send all the bass below 80 Hz or so to the subwoofer. In an audio system with a few large speakers, some smaller speakers, and subwoofer or two, the receiver will route low-frequency audio from all channels to the large speakers and the subwoofers. It does not matter what channel the bass comes from; all low frequencies from the main channels and the LFE channel will be sent to every speaker that can handle them.

This is the key to understanding why certain complaints about bass and LFE are groundless. A 5.0 mix does not compromise the audio or cheat owners of high-end audio systems, since all the necessary bass is still in the mix. Omitting the LFE channel when downmixing to two channels is not a terrible thing, since nothing important is in the LFE channel—only extra "oomph" effects that few two-channel systems can do justice to. This does assume that the engineer creating the audio mix understands the purpose of the LFE channel and does not blindly move all low frequencies into it. See Chapter 12 for more about bass allocation and LFE mixing.

Video Hookup

A DVD player must be connected to a video system: a television, a video projector, a flat-panel display, a video processor, an audio/visual (A/V) receiver or video switcher, a VCR, a video capture card, or other equipment capable of displaying or processing a video signal. If you have a widescreen TV, the details can be confusing. See Chapter 3 for information about aspect ratios and wide-screen display modes.

For all but RF video, audio cables are also required, since the video connection does not carry audio. Things get a bit tricky when there are multiple devices fighting over a single TV. For example, a DVD player, VCR, cable box, and video game console may all need to be connected to a TV with only one video input. In this case, the best option is to use an A/V receiver, which will switch the video along with the audio. When buying a new A/V receiver, get as many video inputs as you can afford—you will always end up with more video sources than you think you will. If an A/V receiver is out of your price range, get a new TV with more video inputs or get a manual video switching box. If you have only a DVD player and a VCR (or cable box), you can hook the VCR or cable box to the antenna input of the TV and hook the DVD player to the video input. Using the remote control, switch between channel 3 (or 4) and the auxiliary video input.

Do not connect the DVD player through the VCR. Most movies use Macrovision protection (see Chapter 4), which affects VCRs and causes problems such as a repeated darkening and lightening of the picture. You also may have problems with a TV/VCR combo, since many of them route the video input through the VCR circuitry. In this case, the only solution is to get a device that removes Macrovision from the signal.

Most DVD players provide two or three video hookup options, detailed in the following sections.

Component Video Until digital connections are available, this is the preferred method of connecting a DVD player to a video system. Component video output provides three separate video signals in RGB or YP_bP_r (Y, B-Y, R-Y) format. These are two different formats that are not directly interchangeable.

Unlike composite or s-video connections, component signals do not interfere with each other and are thus not subject to the slight picture degradation caused by crosstalk. Since the video is stored in three component parts on the disc, this provides the cleanest path from disc to display.

Two versions of component video exist: interlaced-scan and progressive-scan. Progressive-scan component video produces a picture with significantly more detail than interlaced-scan component video and requires a progressive-scan display.

Not all DVD players provide component video output, very few televisions have component video connections, and very few receivers or A/V controllers can switch component video inputs. There are even fewer progressive players and progressive TVs.

Connecting Component Video Some DVD players, notably U.S. and Japanese models, have YP_bP_r component video output in the form of three RCA or BNC connectors. The connectors may be labeled "Y," "U," and "V," or "Y," "P_b," and "P_r," or "Y," "B-Y," and "R-Y."[1] The connectors may be colored green, blue, and red, respectively.

Some DVD players, notably European models, have RGB component video output via a SCART connector or via three RCA or BNC connectors. The RGB connectors are generally labeled "R," "G," and "B" and may be colored to match. Hook a SCART cable from the player to the video system, or hook three video cables from the three video outputs of the player to the three video inputs of the video system.

S-Video All DVD players have s-video (Y/C) output, which generally gives a better picture than composite video output unless the s-video cable is very long. In most cases, the picture from an s-video connection is distinctly better than from a composite connection and is only slightly inferior to a component connection. S-video provides more detail, better color, and less color bleeding than composite video. The advantage of s-video is that the luma (Y) and chroma (C) signals are carried separately. Because they are stored independently on the disc, the best results are achieved when they are not combined by the player and then reseparated by a comb (or similar) filter in the TV. S-video is sometimes erroneously referred to as S-VHS because it was popularized by S-VHS VCRs. Note that most low-end A/V receivers are unable to switch s-video signals, or only provide it on a few inputs.

Connecting S-Video Hook an s-video cable from the player to the video system. The round, 4-pin connectors maybe labeled Y/C, s-video, or S-VHS.

---

[1]Many DVD players label the YP_bP_r connectors as YC_bC_r. This is incorrect because YC_bC_r refers only to digital component video signals, not analog component video signals.

Composite Video   This is the most common but lowest-quality connection. All DVD players have standard baseband video connectors. This is the same type of video output provided by most VCRs, camcorders, and video game consoles. This signal is also called *composite video baseband signal* (CVBS).

Connecting Composite Video   Hook a standard video cable from the player to the video system. The connectors are usually yellow and may be labeled "video," "CVBS," "composite," "baseband," and so on.

RF Audio/Video   This is the worst way to connect a DVD player to a television and is only provided by a few players for compatibility with older televisions that have only an antenna connection. The RF signal carries both audio and video modulated onto a VHF carrier frequency. This type of output is provided by many VCRs and cable boxes.

Connecting RF Audio/Video   Connect a coaxial cable with type F connectors from the player to the antenna input of the TV. The connectors may be labeled "RF," "TV," "VHF," "antenna," "Ch. 3/4," or something similar. If the TV antenna connection has two screws rather than a screw-on terminal, a 75- to 300-ohm adapter is needed. Set the switch near the connector on the back of the player to either channel 3 or channel 4, whichever is not used for broadcast in your area. Tune the TV to the same channel. If you have a TV with only RF antenna inputs, you will need to either get a DVD player with RF output or an RF modulator ($20 to $30).

Digital Hookup

When DVD players were introduced, none included digital video connections or digital bitstream connections. This is partly because the copyright protection and encryption systems for digital video had not yet been established and partly because almost no other consumer equipment provided digital connections at that time. There is also no standard format for streaming compressed audio/video over digital connections. Once the standards are developed, new DVD players must convert their playback information into the proper format.

It is expected that the copyright and standards issues will be resolved and that eventually most consumer electronics equipment will be interconnected via IEEE 1394/FireWire or a similar format. See "Digital Connections" in Chapter 13 for more information.

How to Get the Best Picture and Sound

If possible, use component video connections. If this is not an option, an s-video connection is better than a composite video connection. The improvement from composite to s-video is much greater than the improvement from s-video to component. For a significant improvement over component video, use a progressive-scan player or DVD computer connected to a progressive-scan display.

The most important step is to adjust the television properly. Turn the sharpness control on the TV all the way down. Video from DVD is much clearer than from traditional analog sources. The TV's sharpness feature adds an artificial high-frequency boost. If the sharpness control is not turned down, it exaggerates the high frequencies and causes distortion, just as the treble control set too high for a CD causes it to sound harsh. This can create a shimmering or ringing effect. The brightness control is usually set too high as well. Many DVD players output video with a black-level setup of 0 IRE (Japanese standard) rather than 7.5 IRE (U.S. standard). On TVs that are not adjusted properly, this can cause some blotchiness in dark scenes. DVD video has exceptional color fidelity, so muddy or washed-out colors are almost always a problem in the display, not in the DVD player or disc.

If you get audio hum or noisy video, it is probably caused by interference or a ground loop. Interference can be reduced with an adequately shielded cable. The shorter the cable, the better is the result. You may be able to isolate the source of the interference by turning off all equipment except the pieces you are testing. Try moving things farther apart. Try plugging them into a different circuit. Wrap your entire house in tinfoil. Make sure all equipment is plugged into the same outlet.

It may be hard to believe, but televisions are not necessarily adjusted properly in the factory. Get your TV professionally calibrated, or calibrate it yourself. A correctly calibrated TV is adjusted to proper color temperature, visual convergence, and so on, resulting in accurate colors and skin tones, straight lines, and a more accurate video reproduction than is generally provided by a television when it comes out of the box. Organizations such as the Imaging Science Foundation (ISF) train technicians to calibrate televisions using special equipment. They usually charge between $175 and $600. Another option is to use Joe Kane's *Video Essentials* DVD, based on the *Video Essentials* laserdisc. The *AVIA* and *The Ultimate DVD: Platinum* discs also provide instructions, test pictures, and other resources to calibrate audio and video systems.

Connect the DVD player to a good sound system. This may sound like a strange way to improve the picture, but numerous tests have shown that when viewers are presented with identical pictures and two different quality levels of audio, they perceive the picture that is accompanied by high-quality audio to be better than the picture associated with low-quality audio.

Viewing Distance

Much research has gone into human visual acuity and how it relates to viewing distance, information size, and display resolution. Viewing distance is usually measured in display screen heights. In general, the best viewing distance is about 3 to 5 times the screen height. Industry recommendations vary from 2 to 10 times the screen height. If you sit too close to the screen, you will see video scanline structures and pixels. If you sit too far away, you will lose visual detail and will not engage enough of your field of view to draw you into the experience. The optimal viewing experience is to fill at least 35 degrees of your field of vision while staying beyond the limit of picture resolution. The THX recommends a 36-degree viewing angle, whereas Fox recommends a 45-degree viewing angle.[2] There is disagreement about the dimensions of the average human field of view, especially as it relates to watching video, but the vertical range is about 60 to 90 degrees, whereas the horizontal angle is around 100 to 150 degrees. However, fields broader than about 40 degrees can make the viewer uncomfortable after a period of time because human vision tends to focus toward the center.

Often the resolution of the display determines the distance at which viewers naturally sit. Psychophysical studies have shown that human viewers tend to position themselves relative to a scene such that the smallest detail of interest subtends an angle of about 1 minute of arc, which is the limit of angular discrimination for normal vision. For the 480 visible lines of 525-line (NTSC) television, this produces a viewing distance of about 7 times picture height, with a resulting horizontal viewing angle of about 11 degrees for a 4:3 picture and 15 degrees for a 16:9 picture. To deliver the optimal one-pixel per arc-minute viewing experience at DVD resolution requires a viewing distance of 7 times screen height for NTSC, 6 times screen height for PAL.

---

[2]The vertical visual angle is calculated as arctan(height/2/distance) $\times$ 2. The horizontal viewing angle is arctan(width/2/distance) $\times$ 2. Dividing height or width by 2 and multiplying the final result by 2 centers the viewpoint in the middle of the screen.

THX Certification

Lucasfilm THX (the letters come from Tomlinson Holman experiment) was developed in 1983 with the goal of ensuring that theatergoers experience quality picture and sound presentation as intended by the filmmaker. THX technologies and standards were then extended to the living room. The Home THX program works with leading manufacturers to incorporate proprietary designs into certified home theater components such as receivers, speakers, and laserdisc players. The THX Digital Mastering program offers studios the certification of picture and audio quality during film-to-video transfer, mastering, and replication of laserdiscs and tapes. Lucasfilm has adapted these two programs for DVD.

THX is not an audio format like Dolby Digital or DTS. It is a certification and quality control program that applies to sound systems and acoustics in theaters, home equipment, and digital mastering processes.

Software Certification

A large part of the THX software certification process for DVD is the same as for laserdisc. This includes calibrating the video and sound equipment, monitoring the transfer from film to video master tape, and adding a special THX test signal in the vertical blanking interval. The same video master generally is used for laserdisc, videotape, and DVD.

The new and critical aspect of THX certification for DVD involves the MPEG-2 video encoding process. Part of the task is simply understanding the potential problems and paying attention to the many details such as video noise, picture detail, color balance, black level, white level, and correct audio level. The MPEG encoding process attempts to reduce redundant information; the lower the entropy of the signal, the better is the compression. Entropy can be increased by many characteristics of the source video such as lack of stability (weave), random noise (film grain), edge sharpness, and so on. These cause the encoder to require a higher bit rate and may affect the quality if the needed bit rate is higher than can be allowed. For example, the first DVD movie to be THX-certified—*Twister*—contains considerable hand-held camera work. The shakiness of the picture required a higher bit rate and careful attention to difficult sequences to make sure they were encoded cleanly.

The THX staff also works to request the best possible transfer, using D1 tapes with minimal image processing and noise reduction, and to request an audio remix if needed. THX does not interfere with choice of aspect ratio or anamorphic/nonanamorphic mode (although many DVD fans point out that only anamorphic mode can provide the highest picture quality for widescreen video). THX specifies multipass variable-bit-rate encoding with manual correction of trouble spots and limits on the number of audio tracks (to keep the video data rate above 5 Mbps). The final steps are a check of the physical stampers and thorough testing of check discs and final replicated discs.

The primary difficulty is that few tools exist for objectively measuring the output of an MPEG encoder. No machines or software algorithms can simulate human visual response. Therefore, THX relies on trained viewers to monitor the output from a reference decoder connected to the MPEG encoder. The viewers are trained to look for digital compression artifacts such as macroblocking (visible squares) and "mosquitoes" (fuzzy dots around sharp edges). The encoding process is adjusted and repeated as many times as needed to eliminate problem areas. In some cases where the nature of the transfer or print causes special encoding problems, a new video transfer or even a different print may be requested.

THX verifies that a preliminary check disc plays as expected on a wide range of players. One advantage of DVD is that there is little chance for errors to be introduced after the digital premaster is created. The mastering and replication operations deal with a bitstream, not an analog video signal that might be degraded by improperly adjusted equipment.

Because multichannel theater audio has only recently become widespread, many older movies were released in stereo even though they had four- or six-track masters. To make the most of DVD's capabilities with these older films, a new sound track is mixed from the original multitrack masters. THX was developed initially for audio, and this is still its forte. THX engineers supervise the mixing process, test the equipment, and even adjust the equalization where necessary to make up for the difference between the prevailing standards when the program material was made and modern systems.

Although DVD supports 5.1-channel Dolby Digital audio, the vast majority of viewers will hear the downmixed stereo on a standard stereo system or a system based on Dolby Pro Logic surround sound. Therefore, the audio encoding process is monitored with a decoder connected to a Dolby Pro Logic processor. Trained listeners audition the result in a calibrated THX listening room, checking primarily that the dialogue—which was designed originally for a discrete center speaker—remains audible, even when played on only two speakers or on multiple speakers through Pro Logic decoding. Older action movies require that the audio mix be adjusted slightly to boost the dialogue or reduce the sound effects. This is a delicate process, since the balance and accuracy of the 5.1-channel version must be preserved along with the artistic decisions of the original audio editor. In a few cases, no amount of tweaking can achieve a satisfactory result, so a separate Dolby Surround track is added to the disc.

Hardware Certification

THX continues to develop criteria for THX-certified DVD players. These players will be held to a high level of performance, which can be measured using test signals in the same way THX-certified laserdisc players are tested. The most critical component is the analog video encoder. The numerical output of MPEG decoders varies little from player to player. On the other hand, the circuits that produce the composite and s-video signals have the most significant effect on the visual signal quality and therefore must be given the closest attention.

The original THX specification for home theater equipment based on studio reference equipment is now called *THX Ultra. THX Select* is a simplified specification designed for audio systems in the average living room, allowing smaller speakers and more flexibility in speaker design.

THX-certified amplifiers enhance Dolby Pro Logic and Dolby Digital: bass managements sends low-frequency signals from all channels (or front channels for Dolby Surround) to the subwoofer; front channel reequalization compensates for the high-frequency boost in theater mixes designed for speakers behind the screen; rear channels are timbre matched and are decorrelated when both channels have the same audio; low frequencies are emphasized; and signal processing is used to account for the distance of each speaker from the viewing location.

THX's quality assurance and certification programs for audio and video help bring out the best of DVD. Many people feel that THX engineers have raised the quality of laserdisc close to the level of studio master tapes that were the standard only 10 years ago. THX is likewise working to help DVD live up to its potential to approach the quality of today's D-1 studio masters.

Understanding Your DVD Player

This section explains a few of the more confusing features that can be found on DVD players.

Remote Control and Navigation

The "Title" key, which is more often being called the "Top Menu" key, should take you to the main menu for the disc. The "Menu" key is intended to take you to the menu that is most appropriate for your current location on the disc. Discs with only one program on them often disable the "Title" key and use the "Menu" key to get to the main menu.

Menus can have submenus for audio, subtitle, and chapter selection. Some remote controls let you directly access these submenus by pressing a combination of keys. Without direct access from the remote control, you must use the onscreen buttons to get to the submenus.

The "Return" key should perform the valuable function of taking you up a level, somewhat like the "Back" button on a Web browser, but unfortunately, it is not understood by most disc authors and is not well supported.

Player Setup

Field/Frame Still   Some players let you choose between showing fields or frames when paused. Frame stills look better because they show all lines from both fields rather than half the lines from one field. However, with an interlaced video source, a frame still can produce odd twitter effects. Some players have an auto setting where the player attempts to determine which type of still is better for the current video.

Digital Audio Output   Most players give you a choice of PCM or Dolby Digital output on the digital audio connectors. The Dolby Digital setting also selects DTS, when available. Some players let you specify that DTS be the primary choice when available. Some players can automatically choose the DTS track; others require you to choose it manually.

Some players can automatically choose the first 5.1-channel soundtrack, since many discs come with both 2-channel Dolby Digital and 5.1-channel Dolby Digital. This solves the annoying problem of being partway into a movie with a nagging thought in the back of your mind that it does not sound quite right, only to realize that the 2-channel Dolby Digital track is selected instead of the 5.1-channel track.

Care and Feeding of Discs

Since DVDs are read by a laser, they are resistant—to a point—to finger-prints, dust, smudges, and scratches. However, surface contaminants and scratches can cause data errors. On a video player, the effect of data errors ranges from minor video artifacts to frame skipping to complete unplaya-bility. Therefore, it is a good idea to take care of your discs. In general, treat them the same way as you would a CD.

Your player cannot be harmed by a scratched or dirty disc, unless there are globs of nasty substances on it that might actually hit the lens. Still, it is best to keep your discs clean, which also will keep the inside of your player clean. Never attempt to play a cracked disc because it could shatter and damage the player. It probably does not hurt to leave a disc in the player (even if it is paused and still spinning), but leaving the player run-ning unattended for long periods of time is not advisable.

In general, there is no need to clean the lens on your player, since the air moved by the rotating disc keeps it clean. However, if you commonly use a lens cleaning disc in your CD player, you may want to do the same with your DVD player. It is best to only use a cleaning disc designed for DVD players because there are minor differences in lens positioning.

There is no need for periodic alignment of the pickup head. Sometimes the laser can drift out of alignment, especially after rough handling of the player, but this is not a regular maintenance item.

Handling and Storage

Handle discs only at the hub or outer edge. Do not touch the shiny surface with your popcorn-greasy fingers. Store the disc in a protective case when not in use. Do not bend the disc when taking it out of the case, and be careful not to scratch the disc when placing it in the case or in the player tray. Make certain that the disc is seated properly in the player tray before you close it.

Keep discs away from radiators/heaters, hot equipment surfaces, direct sunlight (near a window or in a car during hot weather), pets, small chil-dren, and other destructive forces. Magnetic fields have no effect on DVDs. The DVD specification recommends that discs be stored at a temperature between 20 and 50°C (4 and 122°F) with less than 15°C (59°F) variation per hour at a relative humidity of 5 to 90 percent.

Coloring the outside edge of a DVD with a green marker (or any other color) makes no difference in video or audio quality. Data is read based on pit interference at one-quarter of the laser wavelength, a distance of less than 165 nanometers. A bit of dye that on average is more than 3 million times farther away is not going to affect anything.

Cleaning and Repairing DVDs

If you notice problems when playing a disc, you may be able to correct them with a simple cleaning. Do not use strong cleaners, abrasives, solvents, or acids. With a soft, lint-free cloth, wipe gently in only a radial direction (in a straight line between the hub and the rim). Since the data is arranged circularly on the disc, the microscratches you create when cleaning the disc will cross more error correction blocks and be less likely to cause unrecoverable errors. Do not use canned or compressed air, which can be very cold from rapid expansion and may stress the disc thermally.

For stubborn dirt or gummy adhesive, use water, water with mild soap, or isopropyl alcohol. As a last resort, try peanut oil. Let it sit for about a minute before wiping it off. Commercial products are available to clean discs, and they provide some protection from dust, fingerprints, and scratches. Cleaning products labeled for use on CDs work as well as those that say they are for DVDs.

If you continue to have problems after cleaning the disc, you may need to attempt to repair one or more scratches. Sometimes even hairline scratches can cause errors if they just happen to cover an entire ECC block. Examine the disc, keeping in mind that the laser reads from the bottom. There are essentially two methods of repairing scratches: (1) fill or coat the scratch with an optical material or (2) polish down the scratch. Many commercial products do one or both of these, or you may wish to buy polishing compounds or toothpaste and do it yourself. The trick is to polish out the scratch without causing new ones. A mess of small polishing scratches can cause more damage than a big scratch. As with cleaning, polish only in the radial direction.

Libraries, rental shops, and other venues that need to clean many discs may want to invest in a commercial polishing machine that can restore a disc to pristine condition after an amazing amount of abuse. Keep in mind that the data layer on a DVD is only half as deep as on a CD, so a DVD can only be repolished about half as many times.

To Buy or Not to Buy

This section is intended to help someone who is thinking of buying DVD or someone who is selling DVD. Selling applies to much more than retail salespeople. Husbands may need some help selling their wives on the idea, or you may even need help convincing yourself that you absolutely cannot live another day without a DVD player. Accordingly, this section is written for the person buying a DVD player. If you are doing the selling, adjust as needed.

One caveat: DVD is like the proverbial camel in the tent. In order to make the most of the picture quality and widescreen format, you need a large TV or even a widescreen TV. And to best reproduce DVD's surround sound, you need a Dolby Digital sound system with five speakers and a subwoofer. If you already have this gear, you are in great shape. In fact, according to the Consumer Electronics Manufacturers Association, 12 million of you are in great shape on the audio side and in pretty good shape on the video side. If you can easily afford the necessary equipment for turning your living room into a theater, you probably do not need much convincing to add a DVD player. If you cannot easily afford it, perhaps you should downplay that part of the deal and think about it later when your pocketbook recovers from the DVD player. (And, in any case, it is beyond the intent of this book to hawk widescreen TVs and surround sound systems.)

DVD may seem like an extravagance, but when you come right down to it, how much does anyone really need a TV, a stereo system, and a VCR or two? Some people probably would be happier and more productive without such indulgences. Therefore, ask yourself how important it is to be entertained (or informed) and how important the quality of that entertainment (or information) is. The quiz at the end of this chapter may help.

Extol the Virtues

When you begin adding up all its features, DVD presents a very compelling face (see the beginning of Chapter 3 for more details):

- *Better video quality than anything else available.* On a good TV, the picture from DVD is truly impressive. On a widescreen TV, it will knock your socks off. Even on an average TV, however, you get a crisp picture with true colors and perfect freeze frame.

- *Exceptional multichannel surround sound.* The spatial definition of 5.1-channel Dolby Digital audio must be experienced to be believed.

However, if you only have a Dolby Surround or Dolby Pro Logic system, all DVD players will provide 4-channel surround sound. Even with only a two-speaker stereo system, DVD sound quality is in the same league as audio CD.

- *Goodies and special editions.* Movie studios are spending a surprising amount of time and money to enhance DVD movies with extra features such as running commentary, behind-the-scenes footage, storyboards, script pages, production stills, director's cut versions, alternate endings, screen tests, outtakes, theatrical trailers, interviews with cast and crew, and even Internet links (if you have the right kind of hybrid DVD player or DVD-Video computer). This extra material is usually only available on DVD, not on tape.

- *Quick access.* Do you ever want to jump straight to a favorite scene? Or perhaps you cannot remember exactly how that line you want to quote to your friends went? With a VCR, it is hardly worth the bother of fast forward and rewind. On a DVD, however, you can jump quickly to a chapter or scan at high speed through the disc. And, with most movies, if you know the time code of the scene you want, just enter it into the remote control to jump directly there in less than a second or two.

- *Think of the children.* DVD should have a certain appeal to parents. The parental control features are nice, and the discs can be played over and over without wear. The eventual low prices also should be a plus. Moreover, publishers are taking advantage of the interactive features to develop education and "edutainment" products on DVD-Video for home learning.

- *Compatibility with CD.* DVD players work with existing audio CDs. Most DVD players have digital audio outputs, a feature usually found only on deluxe-model CD players.

- *Compatibility with the future of audio.* Many people in the music industry are very excited about high-quality multichannel audio. Imagine listening to an orchestra as if you were the conductor or even the bassoon player, being able to pinpoint the locations of the instruments all around you. Or picture a carefully recorded jazz session that recreates the echoes and ambiance of a cozy nightclub. Some amazing things will be done with audio recordings, and DVD will be the premiere medium (in audio-only format or along with video). There are those who object that hardly anyone will go to that much effort to record and mix all six channels, but people said the same thing when stereo came along; consider how many TV shows and movies now use surround sound routinely.

- *Compatibility with the future of TV.* In a way, DVD is halfway between standard TV and high-definition TV (HDTV). It does not have the resolution of HDTV, but it is a digital format with widescreen capability. As good as DVD looks on a regular TV, it will look even better on an HDTV. If you are interested in this, make sure you get a DVD player with component video outputs. Or better yet, get one with digital video output when they eventually appear.

- *Long-term compatibility.* Even when a new generation of HDVD players appears the players will still play current DVD discs.

Beware of Bamboozling

A good way to undermine the case for DVD is to get carried away and overstate its capabilities or even unwittingly misrepresent its features. Make sure you read "DVD Myths" in Chapter 3. Here are a few other points to keep in mind:

- *The audio of DVD-Video is not always better than CD.* While it is true that DVD can include PCM audio at higher resolution than CD, most movies and other video on DVD use Dolby Digital audio, which is compressed to a much lower data rate. Tests have shown that there is little difference between CD audio and Dolby Digital audio to the average listener, but a discerning ear often can pick out minor deficiencies in Dolby Digital. DVD does have the advantage of discrete surround sound.

- *Not every disc will use all the features.* Just because the DVD-Video system allows different aspect ratios, multiple audio tracks, subtitles, parental control, seamless branching, chapter jumps, production stills, and so on does not mean that these features will be available on every disc. In fact, until producers become accustomed to the capabilities of DVD and production techniques adapt to accommodate additional needs, many DVDs will use few or none of these options.

- *DVD is not HDTV.* DVD is standard-definition TV (SDTV). The resolution of DVD is less than half that of HDTV (345,000 pixels versus 921,000 pixels or 2 million pixels). DVD players produce interlaced video, which is inferior to HDTV's progressive-scan video (which effectively doubles the resolution). While it is true that DVD will look better on an HDTV set than almost any other consumer video format, it is not correct to classify DVD as high-definition. It is more accurate to present it as a bridge to HDTV.

- *DVD is not better simply because it is digital.* MiniDisc and Video CD are digital, but many people are unhappy with the fidelity of these formats. The output of first-generation players is not digital—it is analog. Standard consumer displays have only analog inputs. Even component video output, although it provides a very clean signal, is not digital.

- *Laserdisc is not inferior because it is analog.* Laserdisc video is analog, but laserdiscs have both analog and digital sound tracks. Many newer laserdiscs also include an AC-3 (Dolby Digital) sound track that is essentially the same as that of DVD. Laserdisc video quality is only slightly below that of DVD.

DVD Is to Videotape What CD Is to Cassette Tape

A major disadvantage of standard DVD players is that they cannot record, but neither can CD players, and that has not stopped customers from buying over 700 million of them. Of course, it is more common to record from television than from radio, but most people have both a tape deck and a CD player, yet they think nothing of it. The same frame of mind should be cultivated for DVD—it is a companion to the VCR, not a replacement for it. Unfortunately, the copy protection features of DVD can make it harder to copy from DVD to videotape than it is to copy from CD to audiotape.

Think of It as a CD Player That Plays Movies

If your CD player is getting old and cranky or you are thinking of donating it to someone less fortunate than you and replacing it with a new model, consider a DVD player. As the price of DVD players drops, this makes ever more sense. Eventually, for less than a hundred dollars more than what you would pay for a decent CD player, you can get a DVD player that is a decent CD player and a great movie player to boot.

The Computer Connection

Just as many computers can play audio CDs in their CD drives, many new computers can play DVD-Videos in their DVD-ROM drives. Many DVDs are enhanced for use in PCs. Even after you watch a movie on your settop player, you may want to pop the disc in your DVD computer to see what else is on it.

On the Other Hand

To be fair, perhaps you do not need a DVD player. If you are happy with the quality of your VCR or laserdisc player, and if the additional features do not kindle your desire, then DVD may not be in your cards. DVD players will continue to get cheaper, and the selection of movies and other programs will continue to grow. You can always wait and reevaluate.

Also consider that DVD may not become all that it has been made out to be. DVD must compete with digital satellite, digital cable, and eventually HDTV broadcasts and even Internet videocasting.

DVD is not perfect (see Chapter 7). The first players were not even on the shelves before the designers were at work on the next generation. It is inevitable that new and improved versions of DVD will follow. At some point, an improved version of DVD designed for HDTV will be introduced. These new players will be compatible with existing discs, but new discs with new features may not work—or may only partially work—on older players. Even if there are no major changes, the quality of players and discs will improve steadily. The early bird may get the worm, but the biggest worms do not come out until evening.

The DVD-Video Buying Decision Quiz

This quiz can help you decide whether or not to get a DVD-Video player. Each answer will lead you to the next appropriate question. As you progress, you will accrue points. If you have strong feelings one way or the other, feel free to increase or decrease the suggested number of points. If you are a laserdisc owner, you may not wish to add points for features your laserdisc player already has. Your score at the end will give you an idea of how important the features and advantages of DVD are to you.

This is not a scientifically certified quiz, and some of the questions may not be especially valid, but the mere process of thinking about each question should help you evaluate various aspects of DVD technology.

1. Is audio/visual entertainment important to you?
    ❑ No   *Go to question 1a.*
    ❑ Yes   *Go to question 2.*

1a. Is audio/visual education or information important to you?

   ❏ No   *DVD probably is not for you. Stick with what you have (or don't have). Your score is 0. Go to the end.*

   ❏ Yes   *Go to question 2.*

2. Do you value high-quality video?

   ❏ No   *Go to question 3.*

   ❏ Yes   *Go to question 2a.*

2a. Do you have a widescreen TV?

   ❏ No   *Go to question 2b.*

   ❏ Yes   *DVD is practically a necessity. Add 10 points. Go to question 3.*

2b. Are you thinking of getting a widescreen TV?

   ❏ No   *Go to question 2c.*

   ❏ Yes   *DVD would be a perfect match for your widescreen TV. Get a model with component video inputs for best picture. Add 8 points. Go to question 3.*

2c. Do you have a 25-inch or larger TV?

   ❏ No   *Go to question 2d.*

   ❏ Yes   *DVD will look better on your TV than videotape or laserdisc. Add 3 points. Go to question 2e.*

2d. Would you consider getting a larger TV?

   ❏ No   *Are you sure high-quality video is important to you? If so, add 1 point. Go to question 2e.*

   ❏ Yes   *DVD will look better on your new TV than videotape or laserdisc. Get at least a 27-inch TV with s-video or component video inputs for best picture. Add 2 points. Go to question 3.*

2e. Do you want to see movies in their original wide-screen aspect ratio?

   ❏ No   *Some DVDs include a full-screen version and a wide-screen version. Add 1 point. Go to question 2f.*

   ❏ Maybe *Some DVD movies give you a choice of full-screen or wide-screen (letterbox). Add 1 point. Go to question 2f.*

   ❏ Yes   *Go to question 2f.*

2f. Do you dislike letterbox format? (black bars at the top and bottom)?

   ❏ No   *Most DVD movies can be watched in widescreen letterbox format. Add 2 points. Go to question 3.*

   ❏ Yes   *Go to question 2g.*

2g. You ought to get a widescreen TV.

   ❏ No   *Some DVD movies include both full-screen and widescreen versions, but many are widescreen only. If you really hate letterbox, subtract 2 or 3 points. Go to question 3.*

   ❏ OK   *Good choice. You should get a DVD player to match it. Add 6 points (not 10, since you had to be talked into it). Go to question 3.*

3. Do you enjoy high-quality audio?

   ❏ No   *Go to question 4.*

   ❏ Yes   *Go to question 3a.*

3a. Do you have a high-quality audio system?

   ❏ No   *Go to question 3b.*

   ❏ Yes   *DVD can provide better-than-CD audio (20 or 24 bits at 48 or 96 kHz; 192 kHz on DVD-Audio) for exceptionally pure sound reproduction. Add 4 points. Go to question 3b.*

3b. Do you have a Dolby Digital (AC-3) audio processor or receiver?

   ❏ No   *Go to question 3c.*

   ❏ Yes   *Almost all DVD movies have Dolby Digital soundtracks, so DVD will make the most of your investment. Add 8 points. Go to question 3h.*

3c. Would you consider upgrading to a Dolby Digital system?

   ❏ No   *Go to question 3d.*

   ❏ Yes   *Good choice, since this is the wave of the future. Most DVD movies have Dolby Digital soundtracks, so DVD is a perfect match for your new audio system. Add 7 points. Go to question 4.*

3d. Do you have a Dolby Surround or Dolby Pro Logic sound system?

   ❏ No   *Go to question 3e.*

   ❏ Yes   *DVD will automatically create Dolby Surround output from multichannel Dolby Digital soundtracks. Most DVD movies will sound very good on your system. Add 3 points. Go to question 4.*

3e. Would you consider getting a surround sound processor and speakers?

   ❏ No   *Go to question 3f.*

   ❏ Yes   *This is cheaper than Dolby Digital, but will still bring your TV room much closer to a theater. Almost all DVD movies have surround audio, so DVD is a perfect match for your new surround system. Add 3 points. Go to question 4.*

3f. Do you have a stereo system with external inputs?

❑ No   *Go to question 3g.*

❑ Yes   *You can hook a DVD player to your stereo for CD-quality stereo sound. Add 2 points. Go to question 4.*

3g. Do you have a stereo TV with external audio inputs?

❑ No   *If high-quality audio is important to you, you ought to get a better sound system. Add 1 point. Go to question 4.*

❑ Yes   *You can connect a DVD player to your TV for CD-quality stereo sound. Add 1 point. Go to question 4.*

3h. Do you have a DTS audio processor or receiver?

❑ No   *Go to question 4.*

❑ Yes   *Some DVD movies include DTS soundtracks. Add 5 points. Go to question 4.*

4. Do you have a laserdisc player?

❑ No   *Go to question 5.*

❑ Yes   *Go to question 4a.*

4a. Do you hate changing disc sides in the middle of a movie?

❑ No   *Go to question 4b.*

❑ Yes   *DVD movies almost always fit on one side. Add 2 to 4 points, depending on how much you hate flipping discs. Go to question 4b.*

4b. Does the whirring noise of your player bother you?

❑ No   *Go to question 4c.*

❑ Yes   *Most DVD players are as quiet as CD players. Add 2 to 4 points depending on how much the noise bothers you. Go to question 4c.*

4c. Do you think laserdisc prices are too high?

❑ No   *Go to question 5.*

❑ Yes   *DVDs are only a little more expensive than videotapes. Add 3 points. Go to question 5.*

5. Do you buy prerecorded videotapes?

❑ No   *Go to question 6.*

❑ Yes   *DVDs are better for collecting. DVDs are more versatile and often contain goodies such as extra footage and audio commentary. DVDs are easier to store and will never wear out. Most DVDs cost a little more than videotapes, but some are cheaper. Often, DVDs are available for sale when videotapes are still available for rental only. Add 4 points. Go to question 5a.*

5a. Do you buy used videotapes?
- ❏ No   *Go to question 6.*
- ❏ Yes   *As long as it is not badly scratched, a used disc will look exactly as good as when it was new. Add 3 points. Go to question 6.*

6. Do you rent videotapes?
- ❏ No   *Go to question 7.*
- ❏ Yes   *DVDs will not have a degraded picture from being rented over and over (although they may have occasional glitches caused by dirt or scratches). DVDs don't need to be rewound. Add 1 point. Go to question 7.*

7. Do you want to choose from a large selection of movies on video?
- ❏ No   *Go to question 8.*
- ❏ Yes   *It will take a while for DVD to match the selection of videotape or laserdisc. Subtract 2 or 3 points depending on how old this book is. Go to question 8.*

8. Do you want to watch a lot of old or rare movies on video?
- ❏ No   *Go to question 9.*
- ❏ Yes   *It will take time for certain movies to appear on DVD, if ever. Subtract 2 to 4 points depending on how rare your tastes are and how limited the current DVD selection is. Go to question 9.*

9. Do you use your VCR for recording?
- ❏ No   *Go to question 10.*
- ❏ Yes   *Go to question 9a.*

9a. Do you want to replace your VCR with a DVD player?
- ❏ No   *Good, since you will probably still need to use your VCR to record. Subtract 1 point. Go to question 10.*
- ❏ Yes   *Recordable DVD players are just appearing, are expensive, and the discs are not compatible with other players and recorders. Subtract 8 points. Go to question 10.*

10. Do you enjoy performance audio and video (concerts, music videos, opera, ballet, etc.)?
- ❏ No   *Go to question 11.*
- ❏ Yes   *DVD gives you the best picture and excellent audio, often in multichannel surround. Add 2 to 4 points, depending on how much of a fan you are. Go to question 11.*

11. Do you watch foreign films?
    - ❑ No   *Go to question 12.*
    - ❑ Yes   *DVD's multilingual features allow a single version to serve multiple foreign markets, making foreign films more available. Add 3 points. Go to question 11a.*

11a. Do you watch foreign tapes or discs that must be imported?
    - ❑ No   *Go to question 11b.*
    - ❑ Yes   *DVD's regional codes may prevent you from watching discs released in other countries. Subtract 2 to 5 points, depending on the number of imported discs you would want to watch. (If you watch many imported discs, consider buying a region-free player or a player from the same country as the discs.) Go to question 11b.*

11b. Do you prefer subtitles?
    - ❑ No   *Subtitles can be turned off. Add 1 point. Go to question 11c.*
    - ❑ Yes   *Films on DVD can have subtitles in many languages. Subtitles can be turned on or off. Add 2 points. Go to question 11c.*

11c. Do you prefer the original language soundtrack?
    - ❑ No   *If you favor a dubbed soundtrack, DVD movies may let you switch between up to eight languages. Add 2 points. Go to question 12.*
    - ❑ Yes   *Some foreign films on DVD have a dubbed main soundtrack, but let you switch to the original dialog. Add 2 points. Go to question 12.*

12. Would you like to watch a mature film but have your children (or your parents) watch the edited version?
    - ❑ No   *Go to question 13.*
    - ❑ Yes   *Multiple versions of a movie can be put on a single DVD. The parental control feature allows you to set a password to prevent viewing unedited versions. Add 2 points. Go to question 13.*

13. Have you ever erased a tape (accidentally or as a result of a magnetic field) or had one become warped from heat?
    - ❑ No   *Go to question 14.*
    - ❑ Yes   *Optical discs can not be erased, are unaffected by magnetic fields, and have much higher heat tolerance than tapes. Add 2 points. Go to question 14.*

14. Does your VCR ever eat tapes?
    - ❏ No   *Consider yourself lucky. Go to question 15.*
    - ❏ Yes   *DVD players are mechanically simpler and much less likely to damage discs. Add 1 point. Go to question 15.*

15. Do you like to freeze video or step through it a frame at a time?
    - ❏ No   *Go to question 16.*
    - ❏ Yes   *Most DVD players give you perfect digital still picture and let you step forward or back a frame at a time. And they don't suddenly start playing if you leave them on pause too long. Add 2 points. Go to question 16.*

16. Do you wish your VCR could instantly jump to any spot on a tape?
    - ❏ No   *Go to question 17.*
    - ❏ Yes   *DVD players can search to any chapter or timecode in about a second. Add 4 points. Go to question 17.*

17. Do you long for more control of what you watch (change viewpoints, choose endings, edit a music video your way)?
    - ❏ No   *Go to question 18.*
    - ❏ Yes   *On properly prepared discs, the multiple camera angles, seamless branching, and interactivity of DVD can make you an active viewer. Add 3 points. Go to question 18.*

18. Would you like to play interactive movies or multimedia games without buying a multimedia computer?
    - ❏ No   *Go to question 19.*
    - ❏ Yes   *Some games have editions for home DVD-Video players. DVD is perfect for interactive movies. Add 3 points. Go to question 19.*

19. Must you be one of the first to have the latest, greatest gadget?
    - ❏ No   *Go to the end.*
    - ❏ Yes   *DVD is extremely cool and is guaranteed to impress your friends. Add 5 points, unless your friends already have DVD players.*

Pencils Down.
Wasn't that fun? Add up your points and see where you stand.

45–65	You are starving for DVD. Get it now.
35–44	You will definitely enjoy DVD. Start saving up.
15–34	You may need some more convincing, or you may want to wait until DVD is more established.
Below 15	You probably do not need DVD. Or you will have to be convinced by less practical excuses.

10

DVD in Business and Education

Introduction

The effectiveness of video has long been recognized for instruction and learning, and the advantages of DVD-Video are sufficient in many cases to warrant the complete replacement of existing video equipment with DVD equipment. DVD is also a natural fit for business marketing and communications.

The Appeal of DVD

DVD has many advantages over other media, including videotape, print, CD-ROM, and the Internet. Even in the case of CD-ROM multimedia, the capacity of DVD to carry large amounts of realistic, full-screen video makes it more compelling, more effective, and more entertaining.

- *Low cost.* Production and replication costs of DVD-Video and DVD-ROM will quickly drop below that of videotape and CD-ROM, especially when cost calculations take the larger capacity into account. Corporate and government databases currently filling dozens or hundreds of CD-ROMs can be put on fewer DVD-ROMs, with 1 DVD-18 taking the place of 25 CD-ROMs. Businesses spending millions of dollars on videotapes can reduce the cost of duplication and inventory by a factor of four or more once DVD players become widespread and disc production costs drop. Just as cheap CD players can now be found for less than $75 and CD-ROM drives can be had for less than $50, the economy of scale eventually will drive the price of DVD hardware to similar levels. At this point, it becomes cost-effective to equip entire groups of recipients with "bare bones" players or with DVD-ROM upgrades for their computers, simply to reap the benefits of the medium.

- *Simple, inexpensive, reliable distribution.* Five-inch discs are easier and cheaper to mail than tapes or books. Optical discs are not susceptible to damage from magnetic fields, x-rays, or even cosmic rays, which can damage tapes or magnetic discs in transit. One DVD is easier to store than videotapes, multiple CDs, or multiple audiotapes. Production is quicker, logistics are simpler, and inventory is streamlined.

- *Ubiquity.* An apparent drawback to DVD is the lack of an installed base of players. Regardless of how long it takes DVD-Video players to begin showing up in place of VCRs, however, many businesses will soon have at least one DVD-equipped computer. As DVD-ROM overtakes CD-ROM, it will greatly amplify the audience for DVD-based material. By the end of 2000, there were over 40 million DVD players and DVD computers in the United States alone. Various forecasts indicate that between 2002 and 2010 the number of installations capable of playing DVD-Video will exceed the number of VCRs. This will occur in the business world much sooner than in the home.

- *High capacity.* A double-sided, dual-layer DVD-Video disc can hold over 8 hours of video and over 28 hours if compressed at videotape quality. Eight hours of video would require two bulky and expensive videocassettes or would take more than 25 hours to transmit over the Internet with a high-speed T1 connection. Many hours of video and hundreds of gigabytes of data can be sent anywhere in the world by slipping a few discs into an overnight express mailer.

- *Self-contained ease of use.* DVD-ROM programs obviously can include integrated instruction. The features of DVD-Video are also sufficient to provide instruction, tutorials, and pop-up help. A disc can start with a menu of programs, how-to sections, and background information. Rather than being tied to a linear taped presentation, the viewer can select appropriate material, instantly repeat any piece, or jump from section to section. Unlike previous commercial media such as videotape, audiotape, and laserdisc, DVD-Video discs need no ancillary material for training or user education—everything explaining how to use the disc can be put on the disc itself.

- *Portability.* Video presentations no longer require a VCR. A portable DVD-Video player—the size of a portable CD player—can be slipped into a briefcase and hooked up to any television or video monitor. One-on-one or small-group presentations can be done using a portable player with an integrated LCD video screen or a laptop DVD computer. Large-audience presentations can be done with a video projector and a portable player or laptop computer. Someday there will be portable video projectors with built-in DVD players. Notebook computers with DVD-ROM drives and audio/video decoding capabilities can be used for both DVD-Video and DVD-ROM multimedia presentations. Beyond presentations, portable players and laptops can be used for training and learning in any location. DVD players can even be rented in airports for trips.

- *Desktop production.* Desktop video editing and recordable DVD are doing to the video industry what desktop publishing and laser printers did to the print industry. Video production can be done from beginning to end with inexpensive desktop equipment. A complete setup will be remarkably affordable within several years of the introduction of DVD: A digital video camera (under $2000) can be plugged into a digital video editing computer (under $5000), and the final product can be assembled and recorded onto a DVD-R disc with DVD authoring software and recording hardware (less than $3000). This constitutes an entire video production studio on a desktop for under $10,000.

■ *Mixed media.* DVD bridges the gap between many different information sources. A single disc can contain all the information normally provided by such disparate sources as videotapes, newspapers, computer databases, audiotapes, printed directories, and information kiosks. Training videos can be accompanied with printable manuals, product demonstrations can include spec sheets and order forms, databases can include Internet links for updated information, product catalogs can include video demonstrations, and so on. DVD is the perfect medium for this because it provides high-quality video (unlike CD-ROM); searchable, dynamic text (unlike paper); hours and hours of random-access audio (unlike tapes); and a rapidly growing base of devices to read it all.

The Appeal of DVD-Video

The conveniences of DVD-Video—which in the home are enjoyable but not essential—are translated in the office into efficiency and effectiveness. DVD-Video also brings computers into the picture. Unlike in the home, where the value of being able to play a video disc on a computer is questionable, the usefulness of computers doing double duty as video players is clearly apparent.

The natural inclination when working with computers is to take advantage of their additional features such as keyboard entry, graphic interactivity, and so on, but in many cases this is counterproductive and inefficient. Simpler may be better.

The integration of DVD-Video features into computer multimedia is not straightforward. Most authoring environments and delivery systems still do not widely support DVD-Video content (see Chapter 11 for a few details). Even after the wrinkles are smoothed out and the learning curve is eased, developing a computer-based multimedia project may be more difficult and expensive than developing a similar project using only DVD-Video.

There are certain advantages to standard DVD-Video over DVD-ROM based multimedia:

- *Easier development at a lower cost.* For simple titles, fewer programming and design requirements exist. Creating a set of menu screens and related video can be done with low-end, low-cost DVD-Video authoring packages.

- *Easier for the customer.* The limited interface is simple to learn and is usually accessible using a remote control. Hooking a player to a TV is much simpler than getting a multimedia computer to work.

- *Familiar interface.* Menus and remote controls are similar.

- *Larger audience and no cross-platform worries.* Macintosh computers, Windows computers, workstations, DVD players, and even video game consoles can all play standard DVD-Video discs.

Certain kinds of programs lend themselves better to DVD-Video, including programs with large amounts of video, programs intended for users who may not be comfortable with computers, programs with still pictures accompanied by extensive audio, and so on. Here are a few examples of material appropriate for DVD-Video:

- Employee orientation
- Company press kits
- Corporate reports or newsletters
- Employee sensitivity training
- Emergency response training or information systems
- Product demonstrations
- Product catalogs
- Information kiosks, including product/service searches, traveler's aid, and way-finding
- Product training
- Video tours or video brochures
- Video "billboards"
- Video greeting/holiday "cards"
- Testing, including licenses and professional certification
- Video portfolios (ads, promo spots, demo reels)
- Trade show demo discs
- Point-of-sale displays

- Ambient video and music
- Video "business cards"
- Video "yearbooks"
- Lecture support resources
- Repair and maintenance manuals
- Medical informed consent information
- Patient information systems, home health special needs instructions
- Language translation assistance

Of course, DVD-Video also has its disadvantages when compared with other media (the disadvantages of DVD-Video compared with DVD-ROM are covered in the next section). Following are a few examples of material that may not be appropriate for DVD-Video. Of course, the best of both formats can be combined on a hybrid disc.

- Documentation (Not searchable, not easily read on TV.)
- Video games (Require dynamic interface with quick response.)
- Databases (DVD-Video is not well suited for large amounts of text.)
- Productivity applications (Word processing, checkbook balancing, and so on cannot be supported by DVD players.)
- Network applications (DVD data rate is higher than what the average 10-Mbps network can support.)
- Text entry
- Constantly changing databases

The Appeal of DVD-ROM

DVD-ROM has its own advantages over DVD-Video. Because a DVD-ROM can contain any sort of computer data and software, the possibilities are practically endless. DVD-ROM is appealing because of its increased capacity. There is little question that it will supersede CD-ROM as the medium of choice for computers. Some of the advantages of DVD-ROM over DVD-Video include

- *Flexibility.* Any application or content can be used. The only limitation is the target computer platform.
- *Compatibility.* Existing software can be put onto DVD-ROM with little or no change.

- *Familiar development tools.* There is no need to switch from a programming language or multimedia authoring package already in use. And most will support the improved audio/video capabilities of DVD in future versions.

- *Memory.* Unlike DVD players, computers can store information such as preferences, scores, updates, and annotations.

- *Interface.* Computers allow keyboard entry and point-and-click graphic interface.

- *Connectivity.* Computers can be connected to networks, the Internet, hard-disk drives, and so on.

Additional advantages and applications of DVD-ROM are covered in Chapter 11.

Sales and Marketing

DVD-Video can be an excellent sales and presentation tool. Imagine a complete sales presentation system contained in a portable DVD-Video player. For example, home sales presentations could be enhanced greatly by professional video supplements provided on DVD. The presenter simply would plug a portable player into the customer's TV (or put a disc in the customer's player). Unlike videotape, which must be watched straight through, a DVD can contain different segments for different scenarios, answers to common questions, and so on. The need to train sales and marketing representatives up front can be reduced by having them rely on carefully prepared presentations that can be called up as needed.

Advertising representatives can make presentations to clients using a portable DVD player rather than lugging around a computer and a videotape deck. Hundreds of 30-second ad spots, along with DVD-Video based presentation slideshows, can be put on a single disc. If the client already has a DVD player, the rep needs nothing more than a disc and some breath mints. Alternatively, the DVD-Video disc can be integrated with a laptop computer. A custom PowerPoint presentation can include ad spots that play instantly in full-screen mode from the disc.

A point-of-information station or a trade show video presentation is vastly improved with DVD. The disc can be set to loop forever—customers will not be lost because of a black screen after the tape runs out or while it rewinds. If someone is working the exhibit, he or she can respond to customer questions by quickly jumping to appropriate sections of the disc.

Product catalogs with thousands of photographs and video vignettes can be put on a single disc for a fraction of the cost of printed catalogs. Of course, the video catalog cannot be read at the kitchen table—at least not until thin DVD "videopad" players become available. Environmental resource waste from printed catalogs is becoming a big concern. Environmentally conscious companies can replace tons of paper with polycarbonate discs. By producing discs that connect back to the company via the Internet, the life of the discs can be extended with updated prices, product information, new promotions, and other supplements.

With the large capacity of DVD, companies can band together to put multiple catalogs and product videos on a single disc. This type of cooperative disc lowers the cost barrier for small companies and provides more for the recipient to choose from. It also lets companies take advantage of relationships between services and products to do cross-promotions.

A DVD can contain literally hundreds of hours of audio, any part of which can be accessed in seconds, making it the perfect vehicle for instructional audio programs.

Communications

Companies spend billions of dollars a year producing printed information, much of which requires unwieldy indexes and other reference material merely to make it accessible. And much of it becomes out of date in a very short time. Companies are learning to use CD-ROM and the Internet, but for very large publications or those which benefit from a graphic or video ingredient, DVD-ROM and DVD-Video provide an extremely cost-effective means of distribution coupled with improved access to the content.

DVD is a high-impact business-to-business communications tool. It can be used as a standalone, "no instructions needed" communication device, or it can back up an in-person presentation. The message can include high-quality corporate videos, advertising clips, interviews with company officers, video introductions, dynamic video press releases, visual instruction manuals, and documentaries of corporate events such as new office openings and seminars.

Companies can send free DVD discs in the mail to targeted audiences. DVDs make this economically feasible, unlike tapes. Unlike VHS tapes, which people often ignore because they do not want to have to sit through the whole thing, video on DVDs can be broken into small, easily digested segments that the viewer can select from on-screen menus. This will revolutionize the way businesses can communicate with their customers and with other businesses. And unlike CD-ROMs, the discs can be played in both settop players and computers. Traveling executives can be tempted to pop a free DVD into their DVD laptop computer while flying or when sitting in a hotel room.

Effective video-supported presentation demands on-the-fly, instant, context-sensitive access to any point in the footage. Linear videotapes are woefully inadequate to the task. By supporting multiple language tracks, a single DVD can even lower geographic and cultural barriers. Whether distributed to individuals for playback on PCs, shown to groups via settop players or portable PCs connected to video projectors, or mounted in a network server for remote viewing, DVD helps get the message to every recipient.

Training and Business Education

Once sufficient numbers of DVD players and DVD-equipped computers are established in businesses, there is no question that DVD will become a leading delivery format for business training. The Internet has certain advantages such as low cost and timeliness, but the demand for high-quality multimedia far exceeds the capabilities of the Internet for the near future. DVD is better suited to deliver such multimedia and can be integrated easily with the Internet to provide the best of both worlds.

DVD-Video is strictly defined and fully supported by DVD players and by DVD-Video navigation software on most DVD-equipped computers. As with all authoring systems, the ease of development is inversely proportional to the flexibility of the tools (Figure 10.1). As the parameters of the system are constrained, the complexity of the task is reduced. Since DVD-Video is quite constrained, it is relatively easy to develop for it, given the proper tools. Those considering video training programs should decide if the features of DVD-Video meet their requirements. A very simple product containing mostly menus, pictures, and movies may be developed in less time and for less money than with a complex computer authoring system.

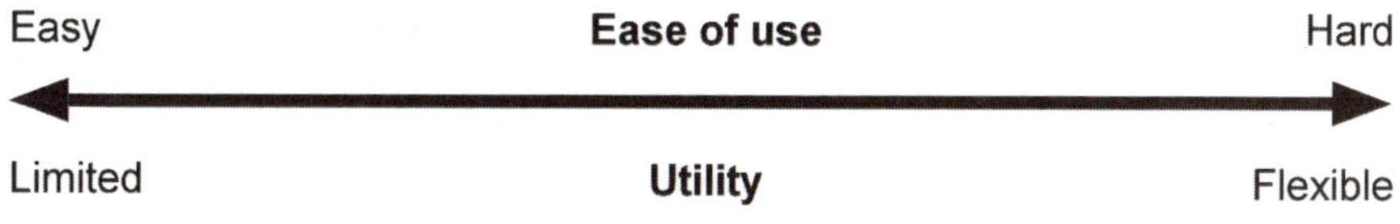

Figure 10.1
The authoring environment spectrum: utility versus ease of use

Easy	Ease of use	Hard
Limited	Utility	Flexible

DVD-Video works so well for certain corporate education and training applications that the content will sell the hardware. Player purchases will be an insignificant part of the deal. With player prices under $200, companies spending millions of dollars on video-based education and training programs should not think twice about equipping their employees or laboratories with players. Many companies that have specialized systems such as study kiosks or CD-i players should evaluate switching to DVD players.

DVD-ROM, on the other hand, covers the entire spectrum from custom-programmed software to fill-in-the blank lesson templates. Practically any authoring or software development system can produce material to be delivered on DVD-ROM. The main advantage of DVD-ROM over other media is space, data transfer speed, and cost per byte. A significant advantage of DVD-ROM over CD-ROM is responsiveness. DVD-ROM access times and transfer rates are much better than the current CD-ROM platform, which consists primarily of 2X to 6X drives and often requires painstaking optimization in order to achieve acceptable levels of performance. DVD-ROM drive transfer rates are similar to hard disks. Although access rates are slower, in most cases DVD-ROM's higher capacity and lower cost more than compensate.

Industrial Applications

DVD discs are being used increasingly in specialized applications. Custom discs can be produced on DVD-R. These discs are usually unique and proprietary—they are not designed to be for sale. Simple installations can use inexpensive, off-the-shelf players. More demanding installations may use professional DVD players, which are more reliable and can be connected to specialized hardware such as multiplayer controllers, video synchronizers, video walls, touch screens, custom input devices, lighting controllers, robotic controllers, and so on. Cheap kiosks can be made from a PC motherboard, a DVD-ROM drive, and an input device such as a trackball or touch screen.

A few examples of industrial applications include

- Kiosks and public learning stations
- Point-of-purchase displays
- Museums
- Video walls and public exhibits

- Vocational skills training
- Corporate presentations and communications
- House video in a store, bar, or dance club
- Theme park and amusement park exhibits
- Closed-circuit television
- Video simulation and video-based training
- Tourism video on buses, trains, and boats
- Hotel video channels
- Media literacy

Classroom Education

Laserdiscs were a success in education almost since day one. Teachers quickly saw the advantage of rapid access to thousands of pictures and high-impact motion video sequences. They began investing in laserdisc players and discs after seeing the effectiveness of laserdisc-based instruction in the classroom. By 1998, twenty years after their debut, over 250,000 laserdisc players were found in schools in the United States. Many are still in use, and most are enhanced with laser barcode technology to provide quick and easy access to video by scanning a barcode printed in a textbook or on a student worksheet. Computer multimedia has begun to replace laserdisc in the classroom, but the ease of popping in a laserdisc and pressing play or scanning a few barcodes may never be matched by computers with their complicated cables and software setups and their daunting troubleshooting requirements.

DVD is poised to take over from laserdisc, but in a different way. DVD-Video players will only slowly trickle into schools. DVD computers, on the other hand, will proliferate rapidly. These computers will be able to run existing CD-ROM programs and new DVD-ROM programs and also play DVD-Video discs.

Educational publishers have been slow to embrace DVD-Video. They do not see a large market, and many of them were hurt by the mass flocking of teachers to the Internet as the new source of free educational technology. Educational video titles require a large amount of work to develop and must be designed to meet curriculum standards. Until more discs intended specifically for education are produced, other titles will help fill the void.

Documentaries, historical dramas, newsreel archives, and even popular movies, TV shows, and "edutainment" programs can be adapted and repurposed for use in the classroom. The particular advantage of DVD for this application is random access. Teachers can quickly jump to any desired video segment and can skip over sections during playback. If a computer is used to play the disc, it can be programmed to show particular sections in a specific order.

Eventually, as DVD becomes a mainstream vehicle for delivering video, a large base of educational content will build up. It will be able to take advantage of features of DVD-Video such as random access to hundreds of video segments, subpicture overlays to enhance video presentation, on-screen quizzes, multiple-scenario presentations, adaptation to learner needs, multiple languages, tailored commentary tracks, and much more.

DVD on Computers

Introduction

The wall dividing home entertainment from computers used to be solid and unbreached. In recent years, it began to crumble, and then along came DVD, which knocked an enormous hole in it. DVD enriches both sides by bringing them closer together. Lamentably, each side brings to the other a new set of problems and old, bad habits.

Just as CD-ROM is used for purposes as diverse as databases, multimedia, software distribution, backups, document archiving, publication, photographic libraries, and so on, DVD-ROM is used for all of that and more.[1] It's when DVD's audio and video features enter the picture that things become more confusing.

Understanding how DVD relates to computers can become very complicated. On the one hand is DVD-ROM, which can be thought of as nothing more than an improved version of CD-ROM. On the other hand is DVD-Video, which can stand completely on its own or be tightly integrated with computers. The details get very tangled up in between the two hands. This chapter can help sort things out and keep all the fingers in the right place.

NOTE: *Many technical details of the information in this chapter are explained in Chapter 5, "Disc and Data Details," and Chapter 6, "Application Details: DVD-Video and DVD-Audio."*

DVD-Video Sets the Standard

Ironically, the MPEG-2 video and Dolby Digital audio features of DVD-Video were originally selected and implemented for home video players, but because of the influence of DVD, they are becoming the de facto video and audio standards on computers as well. Most computer display adapters now include additional circuitry to handle motion compensation, IDCT, and other processor-intensive features of MPEG digital video decoding, while

---

[1]This chapter uses the term DVD-ROM to refer to discs or content intended specifically for use on computers, as opposed to DVD-Video or DVD-Audio content. Of course, DVD-Video and DVD-Audio are applications of DVD-ROM, and audio and video are merely specific kinds of data, but no simple or unambiguous way exists for referring to computer content in particular.

many audio card manufacturers include multichannel features and digital audio output for Dolby Digital and DTS. Companies that specialize in simulating multispeaker surround sound from two computer speakers are jumping on the DVD bandwagon and ensuring that their systems work with Dolby Digital audio tracks. Operating systems developers such as Microsoft and Apple have also built support for MPEG-2 and Dolby Digital into their core multimedia layers.

Multimedia: Out of the Frying Pan . . .

Although the primary use of CD-ROM is for applications and business data, the most high-profile use of CD-ROM has been for multimedia, which is computer software that combines text, graphics, audio, motion video, and more. To battle-weary multimedia CD-ROM developers, DVD either seems like a breath of fresh air or more of the same old been there, done that. Multimedia CD-ROM products are expensive and difficult to produce. A huge disparity of computer capabilities exists among customers as well as a frustrating inconsistency of support for audio and video playback, especially on the popular Windows platform. Hundreds of multimedia CD-ROM titles are produced each year, but only a handful of them make money, in part because customers have been burned too many times by CD-ROMs that don't work on their computers.

On one hand, DVD-ROM is a new area for multimedia development without the stigma of CD-ROM and with the cachet of digital movies and surround-sound capabilities. Given the lessons learned from CD-ROM, DVD-ROM could be an opportunity to finally get it right: an ubiquitously easy-to-use, high-performance multimedia format. On the other hand, DVD-ROM has inherited many of the problems of CD-ROM on top of its own new set of problems.

Microsoft and Intel have created standard platforms and *application programming interfaces* (APIs) for DVD on PCs, but *original equipment manufacturers* (OEMs) and third-party decoder developers only half-heartedly support them. Organizations such as the Interactive Multimedia Association, the Software Publishers Association, and the OpenDVD Consortium promoted guidelines and standards in the early days of DVD, but each of these groups has disappeared. The DVD Association has assembled work groups to produce recommended practices in hopes of ensuring compatibility and interoperability between hardware and software. It remains to be seen, however, if anyone bothers to stay within the ropes during the rush for DVD gold.

A Slow Start

Within a year or two after DVD was introduced, there were about 10 times as many DVD computers as set-top DVD players. In some countries where DVD-Video was slower in getting stared, DVD computers outnumbered DVD players by 100 times. Most people expected this ratio to persist or to grow larger on the computer side, but in fact just the opposite happened.

DVD-Video sprinted to a huge success, while the DVD-ROM industry limped along slowly. By the end of 2000, DVD computers outnumbered set-top players in the U.S. by only five to seven times. Just as copy protection and a lack of titles delayed the introduction of DVD players, compatibility problems, a lack of titles, and fitful customer demand delayed DVD-ROM in the computer industry. DVD-ROM has been slow to take off for a number of reasons:

- *No pressing need*: Unlike CD-ROM, which was a quantum leap beyond floppy disks, DVD doesn't offer enough to sway many publishers. Most multimedia titles and software applications fit on one or two CDs. It's simply not worth it for publishers to move some of their titles to DVD, especially because they know that their CDs will work in DVD-ROM drives. It's hard enough to make money in the CD multimedia business without worrying about DVD.

- *No killer application*: The video and audio quality are better, but many game and animation designers focus on 3D engines instead of canned video. A number of games appeared on DVD, but many of them were merely multi-CD versions packed onto a single DVD, or slightly modified versions that used higher data rates with PC codecs for slightly better video quality. Even using MPEG-2 DVD video for transition scenes has its drawbacks, because the high-quality video shows the flaws in the graphics used for the rest of the game. Video and audio can be used for other purposes, such as video help files, but it takes more work to create the video than many developers are willing to put in.

- *Lack of a consistent development target*: Microsoft was slow in coming up with a unified solution, and in the meantime, decoder makers each created their own proprietary and incompatible solution. Even those that used the MCI standard were inconsistent in their implementation. Once Microsoft released DirectShow in 1998, things should have improved quickly, but decoder vendors and OEMs did not wholeheartedly support it. Two years after DirectShow was released, about 70 percent of the installed base of DVD PCs worked with DirectShow. No developer wanted to have to write 10 versions of their code to support proprietary decoder implementations in the remaining 30 percent. Apple was even slower to provide QuickTime support for DVD. They had a player to play movies, but by the end of 2000, they still had not produced a platform for developers to directly access DVD-Video features.

■ *No interest in bundling*: It's almost impossible to get a deal to bundle DVD-ROMs with new PCs. OEM margins are tight, and OEMs don't get much differentiation from their competitors by bundling DVD titles. Even though most early CD-ROM bundles were low-quality titles that customers never used more than once, the bundling deals helped jump-start the CD-ROM publishing industry. This never happened for DVD.

■ *Slow adoption of hardware*: Because of lack of titles, the percentage of new PCs with DVD drives has climbed slowly, especially because of competition from CD-RW drives. With the increasing popularity of build-to-order systems, customers often choose CD-RW, given the option of a recordable CD drive or a non-recordable DVD drive.

In spite of stumbling at the starting gate, DVD will still inexorably infiltrate the PC marketplace. Eventually, it will be almost impossible to buy a PC without a DVD-ROM drive (or a writable DVD variation), and the size of the installed base will steadily lure more and more content developers.

DVD-ROM for Computers

The simplest implementation of DVD on a computer is a DVD-ROM drive for reading computer data files. Support for the audio, video, and navigation features of DVD-Video is provided as an additional software or hardware layer, independent of the drive (see "DVD-Video for Computers," which follows).

A DVD-ROM drive (sometimes referred to as a DVD logical unit) is supported by the operating system as a random-access, block-oriented input device. It can be considered simply a large, read-only data storage device. The writable DVD variations are random-access, block-oriented input/output devices, although DVD-RAM is the best suited for random access and multiple rewrites.

Features

A single-speed DVD-ROM drive transfers data at 11.08 Mbps (1.321 MB/s),[2] which is just over nine times the 1.229 Mbps (150 KB/s) data rate of a single-speed CD-ROM drive (see Table 11.1). Although 20X and faster CD-ROM drives achieve a higher data transfer rate than 1X DVD-ROM drives, they push the limits of CD technology and begin to suffer from spin stability problems and read errors. A single-speed DVD-ROM drive spins only a little faster than a single-speed CD-ROM drive, but it reads data much faster. A double-speed DVD-ROM drive transfers data at 22.16 Mbps (2.642 MB/s), equivalent to an 18X CD-ROM drive. Most drives include read-ahead RAM buffers to enable burst data rates of over 100 Mbps (12 MB/s) as long as the connection between the drive and the computer is fast enough. By 2000, most DVD-ROM drives ran at 4X to 16X speeds. DVD-ROM average seek times are around 100 ms, with average access times of about 150 ms.

As with CD-ROM drives, most DVD-ROM drives include stereo output for reproducing analog audio from CDs. For internal drives, the analog audio signals are usually mixed into the computer's own audio circuitry. The computer can also read the digital audio from the CD directly. For external drives, the output can be connected to a pair of headphones or speakers. The audio feature is only for audio CDs, not DVDs. Almost no DVD-ROM drives include the decoders necessary to directly play audio or video from a DVD-Video or DVD-Audio disc. Because of cost, complexity, and copy protection issues, the audio/video decoding and playback systems are in the computer rather than in the drive.

Compatibility

All DVD-ROM drives are able to read CDs. Most drives can read variations of the CD format including CD-DA, CD-ROM, CD-ROM XA, CD-i, Video

---

[2]The oft-used figure of 1.385 refers to millions of bytes per second, not megabytes per second.

TABLE 11.1	DVD Drive Speed	Data Rate	Equivalent CD Rate	Actual CD Speed of Drive
DVD Drive Speeds	1X	11.08 Mbps (1.32 MB/s)	9X	8X-18X
	2X	22.16 Mbps (2.64 MB/s)	18X	20X-24X
	4X	44.32 Mbps (5.28 MB/s)	36X	24X-32X
	5X	55.40 Mbps (6.60 MB/s)	45X	24X-32X
	6X	66.48 Mbps (7.93 MB/s)	54X	24X-32X
	8X	88.64 Mbps (10.57 MB/s)	72X	32X-40X
	10X	110.80 Mbps (13.21 MB/s)	90X	32X-40X
	16X	177.28 Mbps (21.13 MB/s)	144X	32X-40X
	20X	221.6 Mbps (26.42 MB/s)	180X	40X+
	24X	265.92 Mbps (31.70 MB/s)	216X	40X+
	32X	354.56 Mbps (42.27 MB/s)	288X	40X+

CD, and Enhanced CD. In the case of Video CD and CD-i, specific hardware or software may be required to make use of the data after it's read from the disc. Many first-generation DVD-ROM drives, such as those from Pioneer and Toshiba, are not able to read CD-Rs, but the most recent drives have no problem reading CD-R discs. Refer to Chapter 6 and Chapter 8 for more details on DVD-ROM compatibility.

Interface

DVD-ROM drives are available with E-IDE/ATAPI or SCSI-2. The SFF 8090i (Mt. Fuji) standard extends the ATAPI and SCSI command sets to provide support for reading from DVD media, including physical format information, copyright information, regional management, decryption authentication, *burst-cutting area* (BCA), disc identification, and manufacturing information (see Table 11.2). Because the DVD extensions use the existing ATA and SCSI protocols, existing CD-ROM drivers can usually be used with DVD-ROM drives.

TABLE 11.2

DVD ATAPI/SCSI
Interface
Command
Information

Physical	
DVD book type and version	DVD-ROM, DVD-RAM, DVD-R, DVD-RW, DVD+RW; 0.9, 1.0, 2.0 etc.
Disc size	120 mm or 80 mm
Minimum read rate	2.52 Mbps, 5.04 Mbps, or 10.08 Mbps
Number of layers	One or two
Track path	PTP or OTP
Layer type	Read/write
Linear density	0.267 μm/bit, 0.293 μm/bit 0.409 to 0.435 μm/bit, 0.280 to 0.291 μm/bit, or 0.353 μm/bit
Track density	0.74 μm/track, 0.80 μm/track, 0.615 μm/track
BCA present	Yes or no
Copyright	
Protection system	None, CSS/CPPM, or CPRM
Region management	Eight bits, one for each region, set = playable, cleared = not playable
Authentication	
Disc key	2,048-byte authentication key
Media key block	24,576-byte key block (CPRM)
BCA (optional)	
BCA information	12 to 188 bytes
Manufacturing	
Manufacturing info	2,048 bytes of manufacturing information from lead-in area

Disk Format and I/O Drivers

DVD is designed around the UDF file system but includes an ISO 9660 bridge format for backward-compatibility (refer to Chapter 5 for details.) Of course, at the lowest level, a DVD disc is simply a block-oriented data storage medium and can be formatted for almost any file system: Apple HFS,

Windows FAT16, and so on. However, because UDF provides explicit cross-system support, standardization via UDF is strongly recommended. Technically, the UDF bridge format is required on all DVD discs by the DVD file system specification books.

Windows and DOS   DVD drives are supported for basic data-reading operations in any Microsoft operating system with CD-ROM extensions and ATAPI/SCSI drivers. Even DOS and Windows 3.x can usually read data from DVD drives. Only newer versions of Windows support DVD-Video (see "DVD-Video for Computers" later in this chapter).

Volume size and file size limitations are quickly revealed by DVD (see Table 11.3). Sometimes large files or files stored on a disc above four gigabytes are not readable. Limitations on the size of files or volumes that are accessible depend partly on the operating system, partly on the file system used to read DVD volumes (ISO 9660 or UDF), and partly on the underlying native file system (FAT16, FAT32, NTFS, HFS, and so on). The standard Windows autorun.inf file or Apple QuickTime autostart feature can be used to create DVD products that automatically begin execution when they are inserted into a drive.

TABLE 11.3

Operating System File and Volume Size Limitations

OS	Native File System*	Maximum File Size	Maximum Volume Size	File System Used to Read DVD
Windows 95	FAT16	Two gigabytes (2×2^{30})	Two gigabytes (2×2^{30})	ISO 9660
Windows NT 4.0	NTFS	Two terabytes (2×2^{40})	16 exabytes (16×2^{50})	ISO 9660
Windows 98, Millennium	FAT32	Four gigabytes (4×2^{30})	Eight terabytes (8×2^{40})	UDF
Windows 2000	NTFS	Two terabytes (2×2^{40})	16 exabytes (16×2^{50})	UDF
Mac OS older than 8.1	HFS	Two gigabytes (2×2^{30})		ISO 9660
Mac OS 8.1	HFS Plus	Two gigabytes (2×2^{30})	Two terabytes (2×2^{40})	UDF
Mac OS 9, X	HFS Plus	Two terabytes (2×2^{40})	Two terabytes (2×2^{40})	UDF

*Some later releases of operating systems can use newer file systems. For example, Windows 95 OSR supports FAT32.

Mac OS   Almost any Macintosh computer can support DVD-ROM, and older Macs can connect to external DVD drives using the built-in SCSI port. DVD-ROM volumes with the UDF bridge format will be recognized by the existing ISO 9660 file system extension. Full support for UDF discs also has been added with a file system extension in Mac OS 8.1. Support for reading and writing DVD-RAM, using UDF 1.5, was added in Mac OS 8.6.

DVD-Video for Computers

From a certain point of view, DVD movies on a computer seem almost an oxymoron. Why take an audio-visual experience originally designed for maximum sensory impact on a large screen with a theatrical sound system and reproduce it on a small computer screen with midget computer speakers? The answer is that computers are steadily becoming more like entertainment systems, even to the point that some home computers are designed to be installed in the living room. Not all computers are going to sprout remote controls and channel tuners, but there is much more to DVD-Video than movies. This includes educational videos, business presentations, training films, product demos, computer multimedia, DVD/Internet combinations, and more.

Computer hardware and software companies are eagerly anticipating the convergence of PCs and TVs, or at least the development of TVs to the point that they have PC-like features and PC-like operating systems, or the development of PCs to the point that they can be sold as replacements for TVs. This has motivated the hardware and software companies to cooperate amazingly well with the entertainment and consumer electronics industries. Computer companies also recognize that the issues of protecting artistic ownership rights in a digital environment must be dealt with sooner or later. It might as well be sooner, and DVD has become the first battleground. Earlier skirmishes occurred with DAT and MiniDisc but had little to do with computers. Since then, the importance of entertainment to the computer industry has increased significantly. TVs and PCs won't merge overnight, and in a certain sense, they will always be differentiated. However, their commingling will continue with DVD squarely in the center. The sooner the legal, political, and business issues are put to rest, the sooner the technical details can be tackled.

In the short term, DVD-Video for computers is important both for customer perception and as a source of content. Many customers do not understand the difference between DVD-ROM and DVD-Video. Making sure that

DVD in all its variations is supported by a DVD computer goes a long way toward alleviating customer confusion and dissatisfaction.

The DVD-Video Computer

Not just any computer can play a DVD-Video disc. A DVD drive is required, of course, but a drive alone is insufficient for the first three classifications listed in Table 11.5, and possibly the fourth and fifth. Specific hardware or software is needed. For video playback using only software, the computer must be fast enough and have enough display horsepower to keep up with full-motion, full-screen video. Otherwise, specialized hardware must be added to take on the task of decoding and playing the audio and video.

The drive speed and video memory have very little to do with playback quality. DVD-Video is designed for 1X playback, so a faster drive gains nothing more than possibly smoother scanning and faster searching. Higher speeds only make a difference when reading computer data, such as when playing a multimedia game or when using a database. Decoder software design, processor speed, and decoding acceleration in the graphics card are the keys to video quality.

Hardware DVD-Video Playback

In order to play standard DVD-Videos on any computer sold before 1997, and on those sold since without a DVD-ROM drive, a hardware upgrade kit is required. This generally includes a DVD-ROM drive and a DVD/MPEG-2 decoder card (sometimes combined with a display card). The reason a new display adapter is required is that the data bus of most computers is not fast enough to carry 30 full frames of video each second. Video data is much larger after it's decoded. For example, a 640×480 display at 16 bits per pixel requires a data rate of 147 Mbps to maintain 30 frames/sec. This is over 15 times the maximum 9.8-Mbps data rate of compressed DVD video. A 1024×768 display at 24 bits/pixel, even at a slower movie frame rate of 24 frames per second, requires over 450 Mbps. When the display card is coupled to the decoder, the data can be routed directly to video RAM, completely avoiding the bottleneck of the bus. In most cases, both Intel (Windows) and Motorola/PowerPC (Mac OS) computers require a PCI bus to handle even the compressed data.

A common option for hardware decoders is to decode the audio in software, because audio decoding requires much less processor time. Another approach is to use the minimum amount of specialized hardware. Certain MPEG decoding tasks such as motion compensation, *inverse discrete cosine transform* (IDCT), *inverse variable length coding* (IVLC), and even subpicture decoding can be performed by additional circuitry on a video graphics chip. This improves the performance of software decoders and frees up the central processor and memory bus to perform the rest of the decoding as well as other simultaneous tasks. This is called *hardware decode acceleration*, *hardware motion comp*, or *hardware assist*. Some card markers also call it *hardware decode*, even though they don't do all the decoding in hardware.

Most new graphics cards take this approach. These cards provide hardware-accelerated DVD-Video playback by dedicating a small portion of the graphics circuitry to handle MPEG decoding tasks. This usually involves giving over a few thousand gates on a chip with hundreds of thousands of gates and essentially provides a hardware speedup of about 35 percent for free. Microsoft defined the DirectX VA (Video Acceleration) API to standardize the hardware acceleration of software decoding in Windows 2000. The API was extensively improved and incorporated into DirectX 8 in late 2000 for general use in Windows operating systems.

Software DVD-Video Playback

Fast computers with the latest graphics hardware can play DVD-Video discs using only software. With the advent of 500-MHz Pentium III systems with decode acceleration circuitry in the graphics card, hardware DVD decoding is no longer necessary. The minimum system is about a 350-MHz Pentium II or a 300-MHz G3, although these are not always fast enough to keep from dropping frames (see Table 11.4).

Because MPEG video uses YC_bC_r colorspace (as opposed to RGB) with 4:2:0 sampling and interlaced scanning, a graphics card with native support for these formats is advantageous. All modern graphics cards provide hardware colorspace conversion (YC_bC_r to RGB) and videoport extensions. PCI graphics cards can usually keep up with the load, but *Advanced Graphics Port* (AGP) graphics cards provide a better DVD-Video performance.

Of course, nothing's ever this cut and dried. Some DVD-Video discs contain only MPEG-1 video and audio, which is much less demanding than MPEG-2 video and Dolby Digital audio. In this case, a 133-MHz 486 or any PowerPC is generally sufficient to play the disc. In other words, any computer that can play a Video CD should also be able to play DVD-Video that contains MPEG-1 data instead of MPEG-2. By the end of 1996, over 15 mil-

TABLE 11.4

Target Platform for
Software DVD-
Video Playback

	Windows	Mac OS
CPU	Pentium II	G3
Minimum speed	350 MHz	300 MHz
Recommended speed	400 MHz	350 MHz
Memory	16 to 32 MB SDRAM	16 MB
Video bus	AGP or PCI	AGP or PCI
Minimum video memory	2 MB	2 MB
Bus	PCI or PCI/ISA	PCI
Drive controller	ATAPI (SFF 8090) IDE/SCSI, bus master DMA, and 12 Mbps min.	Built-in SCSI (SFF 8090) or IDE
Audio	AC '97 chipset, 48 kHz	Standard built-in
Software	Audio/video decoder, DVD-Video navigator	Apple DVD Video player

lion computers in the U.S. were capable of playing MPEG-1 video. DVD-Video player software is usually bundled with a DVD-ROM computer or a DVD-ROM upgrade kit.

DVD-Video Drivers

Decoder driver software is required for video and audio, along with specialized DVD I/O, such as for CSS authentication. A complication is that the multimedia content of DVD is streaming data. Unlike traditional computer data, which can be read into memory in its entirety or in small chunks, audio and video data is continuous and time-critical. In the case of DVD, a constant stream of data must be fed from the DVD-ROM; split into separate streams; fed to decryptors, decoders, and other processors; combined with graphics and other computer-generated video; possibly combined with computer-generated audio; and then fed to the video display and audio hardware. This data stream, which can be monstrous when decompressed, may pass through the CPU and the bus many times on its trip from the disc to the display and the speakers. In order to make this process as efficient as possible, optimized low-level operating support is essential. Some operating systems such as Apple's MacOS with QuickTime already support streaming data. Others, such as Microsoft Windows, only support streaming data properly in the newer versions that include Direct-Show.

Microsoft Windows Architecture Microsoft did not use its aging MCI interface to support DVD-Video, but they instead moved to the DirectShow architecture and streaming class drivers based on the *Win32 Driver Model* (WDM). DirectShow is the streaming multimedia component of DirectX, the core audio/video technology set for Windows. Direct-Show provides standardized compatibility and an abstraction from the underlying decoders. A developer who writes a DVD application using the DirectShow API does not need to worry whether the system uses a hardware decoder or a software decoder because the interface is the same. DirectShow is essentially a virtual machine that handles hardware/software decoding, WDM drivers, demultiplexing, DVD navigation, hardware acceleration, video overlay, colorkey, VPE, CSS, APS, regional management, and so on.

The WDM streaming class driver interconnects device drivers to handle multiple data streams. The driver handles issues such as synchronization, *direct memory access* (DMA), internal and external bus access, and separate bus paths (such as sending the video display data over a specialized channel to avoid overwhelming the system bus). Microsoft has implemented a generic design that enables hardware manufacturers to develop a single minidriver to interface their component to the WDM streaming class driver. For example, an MPEG-2 video decoder card maker, a Dolby Digital audio decoder card vendor, or an IEEE 1394/FireWire interface card developer each have to write only one minidriver that conforms to a single driver model in order to operate with each other (see Figure 11.1). Software decoders are implemented as DirectShow filters, which are code components designed according to the DirectShow COM API.

Full support for DVD-Video is available in Windows 95, Windows 98, Windows 2000, Windows Millennium, and newer operating systems. DVD-Video support is also built in to DirectShow, which can be installed on Windows 95 and is included in newer versions of Windows. Microsoft DirectShow is not available for Windows NT 4.0, so no OS-level support exists for DVD-Video. Some third-party applications are also available for playing DVD-Video in Windows NT 4.0.

Windows does not include an MPEG-2 or Dolby Digital decoder. These must be provided by third parties. Almost all Windows PCs with a DVD-ROM drive include a full DVD player and decoder package. Early on, many

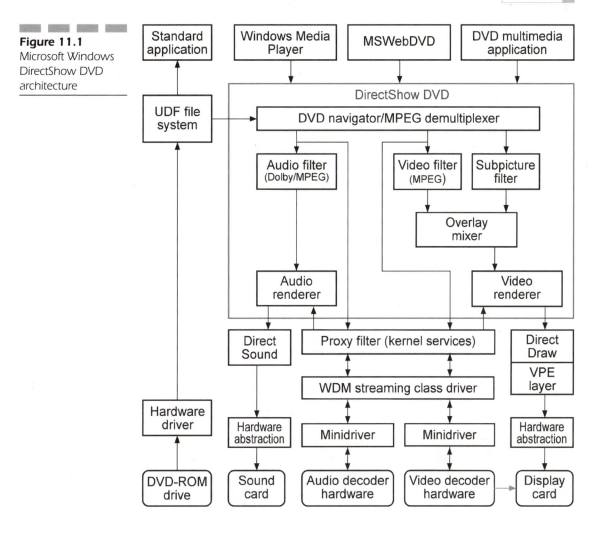

Figure 11.1
Microsoft Windows
DirectShow DVD
architecture

of these decoders were not compatible with DirectShow, but all now support DirectShow. Support for hardware decoders in DirectShow is provided with WDM device drivers (see Figure 11.1).

DVD-Video playback and navigation is supported by DirectShow versions 5.2 and later. The DirectDraw API also supports video display. Because full frame-rate decoded video has too high a bandwidth to be passed back and forth on most current implementations of the PCI bus, it often must travel on a separate path to the display adapter. Intel has designed the AGP architecture to support this. Microsoft's DirectDraw *hardware abstraction layer* (HAL) and *video port extension* (VPE) provide hardware-independent support for AGP and similar parallel bus or parallel memory access architectures.

Regionalization is supported by the Windows operating system, which gets the region from the drive. The decoder driver may also include its own regional code. If the disc does not contain code that matches the code in the decoder driver and the code stored by the OS, the system will refuse to play it.

The Windows operating system also supports copy protection. Windows and DirectShow do not provide a decryption module, but act as an agent for facilitating the exchange of authentication keys between the DVD-ROM drive and the hardware or software decryption module. Video display adapter cards are required to implement Macrovision in order to receive decoded video from the system.

NOTE: *For details of regional management and copy protection, see the sections later in this chapter and in Chapter 4.*

Older versions of Windows can't play movies from DVD-Video discs or play MPEG-2 video without support from add-on hardware and software. DVD upgrade kits intended for use with DVD-Video must include separate navigation software for playing movies and other DVD-Video titles. For additional DVD-Video functionality, such as playing MPEG-2-encoded video and Dolby Digital-encoded audio, they also must include MCI or Direct-Show drivers. The MCI command set was mostly standardized in 1997, but MCI support began disappearing after that. By 2000, most older MCI DVD applications did not work on many DVD PCs.

Apple Macintosh Architecture Apple was very slow to support DVD, which was surprising given the synergy between DVD and Apple's artistic-minded customers. According to Apple, two-thirds of all multimedia CD-ROM titles are developed on Mac OS computers, including some that are designed to be used only on the Windows platform. Unfortunately, during the period when Apple was concentrating on rebuilding market share and stock value, it largely ignored DVD. Only in 1999 when iMacs were released with DVD-ROM drives and high-end G4s were released with DVD-RAM drives did Apple begin to again pay attention to the technology that so suitably matched its user profile.

Early developers of DVD hardware and software for Macintoshes went out of business. Apple finally developed its own player software for Mac OS 9 that at first relied on hardware cards, and then later was implemented in software with hardware decode acceleration in the graphics card. Software decoding in some Mac G4s was accelerated by the AltiVec Velocity Engine portion of the CPU.

The stream-oriented architecture of QuickTime works well for media types such as MPEG in Mac OS as well as on Microsoft Windows computers. However, the interactive features of DVD-Video don't fit so easily into the QuickTime design. MPEG-1 software for playback was provided by the QuickTime MPEG extension, which was released in February of 1997. Apple and other hardware and software vendors developed codecs for MPEG-2 video, Dolby Digital audio, and other DVD-Video streams to work with QuickTime, but as of 2000 and the arrival of QuickTime 4.0, still no standardized support existed for DVD-Video or MPEG-2 playback in QuickTime.

DVD-Video Details

Hardware decoders can be built on specialized cards that use a VGA *pass-through* system for analog overlay. The pass-through system takes the output of the VGA display adapter and keys the video into pixels in a certain color range. *Hardware overlay* is also used for both hardware and software decoders, where the video is sent to the display adapter via a special bus, the VPE. The display adapter then overlays the video into the display using a color key. The key color is often magenta, which may flash now and then. The overlay approach is primarily designed to compensate for video hardware that is not powerful enough to deal with memory-mapped video at 30 frames per second. As video cards become more powerful and system busses become faster, the video port and overlay approach will give way to memory-mapped surfaces. The advantage of this is that the video can then be manipulated like other graphics surfaces, with geometric deformations, surface mapping, and so on. For example, MPEG video can be wrapped onto a spinning object or reflected onto a simulated window.

Because computer pixels are square, while DVD pixels are rectangular, the video must be scaled to the proper aspect ratio. The exact number of pixels is not as important as the displayed picture aspect ratio. Optimal square-pixel resolutions without scaling the video are 720 $\times$ 540 (no horizontal scaling) or 640 $\times$ 480 for NTSC with no vertical scaling or 768 $\times$ 576 for PAL with no vertical scaling. For full-screen viewing, the entire 1.33 screen should be used (800 $\times$ 600, 1024 $\times$ 768, and so on). On modern graphics cards with high-quality scaling hardware, video that is scaled up to fit the entire screen should look very close to windowed video in quality. Optimal square-pixel resolutions for anamorphic video without scaling the video are 720 × 405 (no horizontal scaling) or 854 × 480 for NTSC (no vertical scaling) or 1024 × 576 (no vertical scaling) for PAL.

Video often looks better if the display refresh rate is adjusted to match the source frame rate of material on the disc: 60 Hz for NTSC interlaced video (2 × 30), 72 Hz for film-sourced video (3 × 24), 75 Hz for PAL interlaced video (3 × 25). This avoids temporal aliasing such as tearing and jerkiness. A good sweet spot, still out of reach of most monitors, is 120 Hz (4 × 30 and 5 × 24).

DVD By Any Other Name: Application Types

The confusing part about DVD and computers is where the DVD-Video ends and the computer begins. The only constant is the DVD-ROM drive itself. Some computer upgrade kits include a drive and an entire DVD player on a card. Other upgrade kits or DVD-Video-enabled computers include complete DVD audio and video decoding hardware. Others include only small amounts of extra graphics circuitry to accelerate DVD-Video playback. Some systems rely entirely on software. Still other systems contain only a DVD-ROM drive and are incapable of playing video or audio. It's clear that a text database on DVD-ROM, for example, has nothing to do with the DVD-Video standard. It's also mostly clear that a blockbuster Hollywood movie that can be viewed on a computer monitor relies heavily on the DVD-Video standard. But what if the disc containing the blockbuster movie includes additional computer-only bonuses such as a screen saver or a computer game based on the movie?

The following classifications of DVD products in Table 11.5 may help clarify the variety of options. Note that the classifications cover both DVD-ROM and DVD-Video.

Pure DVD-Video

A pure DVD-Video disc is designed entirely for use on a standard DVD-Video player. It can also be played on a DVD-Video-enabled computer, but

	Classification	Content	Ice Cream Analogy	Level
TABLE 11.5 DVD Product Classifications for Computers	Pure	DVD-Video only	Vanilla	2
	Bonus	DVD-Video plus computer supplements	Sprinkles	2,1
	Augmented	DVD-Video plus computer enhancements or framework	Carmel swirl	2,1
	Split	Independent DVD-Video and computer versions	Neapolitan	2,1
	Multimedia	DVD-ROM for computer, any format of audio and video	Sundae	0
	Data	DVD-ROM for computer data or applications only	Frozen yogurt	0

nothing is different (other than the tiny computer screen and the chair that's not as comfortable as the living room couch). In this case, the computer serves merely as an expensive DVD-Video player. The computer's DVD-Video navigation software is all that is needed or used.

The target audience for this type of product is movie and video viewers. Note that the flexibility of the DVD-Video format enables a pure DVD-Video to contain still images, text screens, additional footage, supplementary audio, and other enhancements generally associated with special editions. Examples of pure DVD-Videos include movies, product demo videos, and instructional videos.

Computer Bonus DVD-Video

Just as Enhanced CDs contain both music to be played on a CD audio player and additional goodies to be used on a computer, bonus DVD-Video products contain extra material designed for a computer. The disc can be played normally on a DVD-Video player, while the video part of the disc can also be played with the computer's DVD-Video navigation software. The primary audience is similar to that of pure DVD-Video. The difference is that the disc also contains computer software or computer data that the producer thinks will make the disc more appealing: movie scripts, searchable databases of supplemental information, still pictures in computer format, instruction manuals in electronic form, short video clips in computer video format, screen savers, icons, computer games, Internet Web links, and so on.

Computer-augmented DVD-Video

An augmented DVD product can be played on a DVD-Video player, but it is amplified and changed when played on a computer. The viewer might have more control over the sequence of the video, or the video might respond to the actions of the viewer. Interludes in the video may offer a game or puzzle, or perhaps the video is subsumed entirely by a computer-based game or environment in which the video plays a supplemental or reward role. In this case, the computer might present video segments in a different order or generate additional graphics and create a composite image.

For example, a murder mystery could be played and enjoyed in a straightforward way on a DVD-Video player and perhaps even have multiple endings. But when the same disc is put into a computer, the viewer can choose to become a character in the drama. If the viewer chooses to play the role of the detective, the viewer can interrogate characters, search for records, collect and manipulate objects, answer questions, and become an interactive participant controlling the environment and outcome of the game.

Split DVD-Video/DVD-ROM

This type of disc can be used in a player or in a computer, but no overlap exists between the two. The computer version is usually related to the video player version, but no content is shared. It's basically a pure DVD-Video product on one part of the disc and a multimedia DVD-ROM on the other part (see the following paragraph).

Multimedia DVD-ROM

A multimedia DVD-ROM product cannot be played in a DVD-Video player. It's intended for use in a computer only, like a traditional multimedia CD-ROM. The disc may actually contain audio and video in the same MPEG and Dolby Digital format as a DVD-Video disc, but it can be played back only on a computer. The disc may also contain video at higher resolutions and different aspect ratios than those supported by the DVD-Video format.

The multimedia DVD-ROM is the most visible application of DVD technology on computers. Current CD-ROM multimedia producers will gravitate toward DVD-ROM depending on their needs and the changing platforms of their customers. Multimedia programs requiring more than one CD are early candidates. As soon as the installed base of DVD-ROM drives in a given market reaches critical mass, associated multimedia publishers can switch their existing content from CD-ROM to DVD-ROM. They may want to take advantage of the increased capacity and higher minimum data rate of DVD-ROM to include more content and to improve the quality of audio and video.

The multimedia DVD-ROM could even become the primary application of DVD at home, if major computer companies such as Microsoft, Compaq, and Intel succeed with their plans of capturing the home market with a computer-based, multimedia, HDTV entertainment system.

Data Storage DVD-ROM

With all the audio and video hype swirling around DVD-ROM, it's easy to overlook it as a simple data storage solution. DVD-ROM can hold the same data as hard disks, floppy disks, and CD-ROMs, but in vastly greater quantity. Text documents, graphical documents, databases, catalogs, software applications, and so on can all be stored on DVD-ROM—the bigger, the better. Most CD-ROMs are not multimedia products, and multidisc sets are common.

DVD Production Levels

DVD development can also be classified according to the way the content is produced and used and what it demands from the PC. Three classifications exist for DVD development: level 0, level 1, and level 2.

Level 0   Level 0 discs contain data and software only; they do not require MPEG-2 decoding. The content of a level 0 disc has computer data, perhaps even multimedia, but no dependence on MPEG-2 video. Traditional computer software development tools are used as well. No special features are needed in the computer to play back level 0 discs, other than a DVD drive. Obviously, level 0 discs don't play in standard set-top DVD players.

Level 1   Level 1 uses only MPEG-2 video files with no DVD-Video requirements. A level 2 disc contains only ISO MPEG-2 program stream files. It does not use anything from the DVD-Video (or DVD-Audio) specification. The advantage of such a disc is that the DVD authoring process is not needed. Once the video is encoded as an MPEG-2 file, it can be played back directly by the application software. This makes testing and development easier, especially because meaningful filenames can be used (such as intro.mpg and bouncingball.mpg), rather than cryptic DVD-Video filenames (such as vts_01_01.vob).

The drawbacks are that level 1 discs don't play in set-top players. Also, because many PC DVD decoders are designed specifically for DVD playback and not for general MPEG-2 playback, compatibility problems occur with older computers, especially those using hardware decoders. Jumping around within files is also tricky, because it requires the proper byte offsets to be predetermined at development time or that a slow timecode search be used, which demands that the decoder/playback software scan through the file to find the right position.

It's possible to produce video files with a DVD authoring system and play the resulting .vob files as MPEG-2 files. This works reasonably well, as long as the video is sequential. If interleaving is used for multiple camera angles or multistory branching, each interleaved block will be played in sequence, resulting in fragmented scenes.

Level 1 discs require a computer with an MPEG-2 decoder that has the capability to play individual MPEG-2 files. Level 1 development is supported by both Microsoft DirectShow and Apple QuickTime.

Level 2 Level 2 discs rely fully on the DVD-Video specification. The video and audio content is created with a DVD authoring system, so the disc can play in a set-top DVD player. The computer uses the DVD-Video presentation and navigation structures to present the content. Instead of specifying a filename to play a particular video segment, the application specifies a title and chapter, or a title and timecode. Because the features of DVD-Video are used, with time maps and other navigation information, access is very fast and accurate. Features such as angles, subtitles, multiple audio tracks, and seamless branching can be used, as long as they are authored on the disc. Care must be taken during authoring to make sure that all entry points on the disc are accessible. Either a chapter point must be at each place on the disc that the application would want to jump to or the video must be authored as a one-sequential-PGC title so that timecode searches can be used.

Level 2 discs require a computer with an MPEG-2 decoder and DVD navigator software with an API to provide application control over DVD playback. Level 2 development is supported by Microsoft DirectShow. As of late 2000, Apple QuickTime does not support level 2.

WebDVD

Too many pundits and predictors would have us believe that television and movies will be streaming over the Internet any day now. The truth is, streaming video on demand is only for those who aren't very demanding. It's predicted that there will be five to seven million broadband users in U.S. by the end of 2001. That's about only five percent of the total number of households. And how broad is broadband? Half of those five to seven million will be using DSL modems, which literally require 24 hours to download a DVD movie. The other half will have cable modems, which might be a bit faster, as long as no one else in the neighborhood is bogging down the shared network with their own movie downloads. The inescapable fact is that broadband Internet simply is not developing at a rate that will make the delivery of DVD-quality video feasible any time soon. It will eventually happen, but not before 2010 for the typical household. The development of true broadband will happen even slower in countries with less-developed Internet infrastructures.

The idea of TV shows and movies delivered to living rooms over the Internet is appealing, but it's far from reality. Consider that in 1978, a laserdisc connected to a computer could provide interactive, full-screen, full-motion video. Admittedly, the video was often on a separate screen, but it was a good start for interactive multimedia. Since then, many supposed advances in technology have actually taken us in the wrong direction from that promising beginning. When QuickTime appeared on the scene in the late '80s, many people were so impressed that digital video from a CD-ROM worked at all that they didn't seem to notice that it filled only a fraction of the screen at less than 10 frames per second. The technology slowly improved, especially with the introduction of MPEG-1 and Video CD, but the quality of video on a computer was still below that of TV. Eventually the Internet arrived, which was painstakingly slow to send full-screen graphics, let alone full-screen video. The newest player in this strange back-and-forth game is DVD. Finally, a mainstream technology exists with a fast-enough data rate and a high-enough capacity for high-quality video and audio to be completely integrated into a computer. But now that we're used to the instant gratification of e-mail and Internet search engines, a static, physical medium such as DVD seems somewhat limiting. So why can't we have the best of both worlds?

Unlike the Internet, with its slow data rates, tiny, fuzzy video, long download times, and lost data packets, DVD has very high data rates for high-quality video and audio, guaranteed availability (as long as the right disc is in the drive), and fast access to enormous amounts of data. DVDs can be distributed for free in magazines, at store displays, or in the mail. A DVD-9 sent via express mail can deliver in 24 hours, anywhere in the world, eight gigabytes of content that would literally take two weeks to download on a 56k modem. Until true broadband Internet capabilities are available and affordable to everyone who wants them, one of the best ways a Web site can provide a high-impact experience is to depend on a local DVD to supplement HTML pages with hours of video and days of audio.

WebDVD, also known as connected DVD, Internet DVD, online DVD, enhanced DVD, and many other names, is the simple but powerful concept of enhancing DVD-Video or DVD-ROM with Internet technology. From the reverse point of view, it can be seen as enhancing Web pages with video, audio, and data from a DVD. WebDVD combines the best of DVD with the best of the Internet. It's the marriage of two of today's hottest technologies, which complement each other beautifully. It's not a new idea; CD-ROMs have been linked to the Internet for years, but the character of DVD elevates it to a new level. DVD makes PCs true audio/video machines, with better-than-TV video and better-than-CD audio; and the Internet links them together, creating links between creators and consumers and communities.

Until the Internet becomes a true broadband highway for average home viewers, DVD has an interesting role to play. When a DVD disc is placed in the local drive of a computer or an Internet-connected set-top box, it can supply the quality video that the Web sorely lacks. An educational Web site can play detailed instructional videos from the local DVD drive. A movie fan site can pull up favorite clips. A director, presenting a live on-line chat, can pull up behind-the-scenes footage hidden on the disc inserted in each viewer's drive, delivering it simultaneously to thousands of viewers with a "virtual bandwidth" that would bring any Web server to its knees. A Web-enhanced DVD movie can connect to the Internet for promotional deals, viewer discussion forums, multiplayer games, and more. A marketing disc, perhaps inserted in a magazine, can entice viewers with gorgeous product videos backed up by Web pages for on-the-spot purchases. The possibilities are endless, as long as the customer already has the disc or can be easily supplied with one.

When a DVD is designed to connect to the Internet, it becomes a different kind of product. Even after the disc leaves the hands of the developer, it can be improved or refreshed. What would otherwise be a static medium becomes dynamic and renewable, extending the life of the product. WebDVD titles can take advantage of the timeliness of the Internet with automatic software updates, supplemental content as new information becomes available, updated links to related Web pages, current news and events, special offers of associated services and products, and even communities of customers who participate in online chats or discussion groups and provide feedback about the content of the disc. For example, the discographies included on many DVDs are often out of date by the time you watch the film. A WebDVD discography is never out of date.

Web content, such as HTML pages viewed in a Web browser, can be greatly enhanced with embedded audio and video assets, including MPEG-2 video and multichannel Dolby Digital audio, played from local DVD storage. Conversely, DVD-Video titles can be enhanced with supplements and dynamic information from the Internet. Such DVD-Video titles will play normally on standalone DVD players, but when played on a DVD-enabled PC or a Web-enabled set-top DVD player, they will be enriched with additions delivered via the Web. Here are some examples of such additions:

- *A movie with built-in Web links or Web wrappers.* Hundreds of Hollywood movies have been enhanced for use in a computer. These discs play normally in a DVD-Video player, but when put into the DVD-ROM drive in a PC, much more becomes available, such as links to information about the movie, an active screenplay that can play associated scenes from the movie, multimedia annotations, interactive games, special product discounts for Web purchases, and much more. Some movie enhancements are intriguing, some are frivolous, and some are downright lame, but none have yet come close to taking advantage of the power of a PC to take the movie experience to a different level.

- *A custom interface for the PC-based use of a DVD.* HTML pages on the disc or on a remote Web site can control the DVD, giving instant access to various sections of the disc, controlling angles, changing audio, and more. This kind of custom interface can be quick and simple, and much easier to use than the on-disc menus that must be accessed by a standard DVD remote control or through the DVD player software interface. Hundreds of HTML development tools are at hand to quickly and easily create specialized interfaces in thousands of different styles.

- *A content pool for dynamic front ends.* Advertising companies and corporate video development companies can use WebDVD for sales and marketing presentations. Rather than carrying VCRs along with laptops for presentations, presenters can show everything from the laptop. Hundreds or thousands of video and audio clips can be collected on the DVD: product demos, customer testimonials, tutorials, sales presentations, and so on. Custom HTML pages can be created to organize and present selected clips and other information from the disc. A new DVD can be shipped to offices and sales reps every few months or once a year. In the interim, different creative teams can put together new PowerPoint presentations and custom HTML pages. The resulting presentations, perfectly tailored to specific needs or specific clients, can be quickly e-mailed or downloaded anywhere in the world, even though they include full-screen video. The high-bandwidth content pulled from the DVD doesn't need to be downloaded. Urgent deadlines can be easily met with new pages that can be dashed out over e-mail, as long as the necessary video or audio is on the disc.

■ *HTML-based reference material on DVD.* Corporations, government agencies, and other institutions can use DVD to make massive databases of text, pictures, graphics, audio, and video available. Because the interface is done with HTML, large Web sites can be easily moved onto a DVD. New content can be put on a Web site between DVD releases, and it can be integrated into the existing body of HTML content for the next version of the disc. The DVD-based content is available in situations where no Internet connection exists, such as when traveling or for presentations.

■ *Commentaries, annotations, and analyses.* Anyone who loves (or hates) a movie enough that they want to share their opinion with the rest of the world can use WebDVD to create their own personal tour of the DVD. They can put their analysis on a Web site with clickable text or buttons that play associated clips. Visitors to the Web site only need to put the proper disc into their DVD-ROM drive. Such a WebDVD site could even include a custom audio track, streamed over the Internet and synchronized to the DVD, supplying a running commentary (either the serious kind or the *Mystery Science Theater 3000* kind). Because the person viewing the Web page owns their copy of the disc, there should be no copyright worries.

A student of Shakespeare can create a Web site that enhances any standard DVD version of one of the bard's plays. As the video runs, notes can appear on the side explaining the historical context, what is happening, what the actors are saying, and so on. An alternative design might give the viewer the option to stop the video at any point and bring up background information. Web-based annotation works well with many kinds of DVDs, such as documentaries, historical dramas, or even TV shows. A WebDVD page that explained all the in-jokes and references in episodes of *The Simpsons* could be a huge hit. Teacher's lectures, educational programs, and all manner of instructional video can be enhanced with pop-up annotations and ancillary information to enrich the learning process.

■ *Customized advertising and commerce.* A DVD-based department store catalog could offer an interactive showcase of all the merchandise, complete with audio and video. Because the disc is connected to the store via the Internet, the consumer can get current prices, check availability, order merchandise, communicate with a shopping consultant, pay bills, and so on. The advantage of the DVD component is that the customer does not have to wait for slow downloads of pictures and even slower downloads of low-quality video. If the DVD is sent free in the mail, it creates a push to get the customer to visit the Web site.

■ *Sharing high-bandwidth content.* Until high-speed Internet connections are available between research groups and corporate divisions, WebDVD is an excellent tool for sharing large collections of data such as 3D visualizations, rendered graphics, and video simulations. The video and data can be put on a recordable DVD and shipped anywhere in the world. Recipients need only put the disc in their DVD PC and connect to a Web page that provides playback control, indexes, notes, and even collaborative discussions.

WebDVD Applications

Obviously, WebDVD can be implemented easily on a computer. Traditional set-top DVD-Video players can be enhanced with WebDVD capabilities, just as Internet appliances can be enhanced with DVD drives. As the digital convergence continues, other devices such as satellite receivers and cable boxes will implement both Web and DVD features.

Still, most WebDVD development centers on Web browsers, specifically Microsoft Internet Explorer, which provides the best development environment for DVD. In addition, WebDVD applications can be created using PowerPoint, Macromedia Director, ToolBook, Visual Basic, and other multimedia tools. When references are made in this chapter to Web pages controlling playback, keep in mind that the same thing can often be done in other environments as well.

WebDVD applications can be built on existing standards and specifications such as the ATSC's *DTV Application Software Environment* (DASE), the *Advanced TV Enhancement Forum* (ATVEF), the *World Wide Web Consortium* (W3C), and, of course, the DVD Forum's DVD-Video specification.

The continuum of WebDVD possibilities runs from a simple startup menu to a sophisticated application that displays video combined with synchronized content from the Internet. Some examples include the following:

- A menu that appears on the screen when the disc is inserted. The simplest "surf or show" menu would have two options: connect to a Web site or play the disc.

- Hyperlink buttons that appear in player interfaces or over motion video when associated Web content is available

- HTML information that appears in a separate window, synchronized to the video

- An HTML-based API to control DVD playback in an embedded window or in full-screen mode

- DVD-enhanced Web pages, where high-quality DVD content is played from the local drive in place of low-quality streaming Web content

- Web-updated DVD pages, where new or time-sensitive information on the DVD is superseded by content from the Web

- A custom application with a window playing DVD-Video, a Web-based sidebar synchronized with the video, and a chat window to talk to people viewing the same content

Tools and standards for WebDVD are still rudimentary. A fully developed standard for WebDVD might include some the following features:

- A way to associate URLs with a volume or with portions of a volume, such as titles and chapters. This could include a strategy for falling back to HTML files stored on the DVD in cases where an Internet connection is not available.

- A standard way for an HTML-based front end to be activated in place of the standard DVD-Video playback action. A WebDVD disc inserted into a standard movie player would play normally, but when inserted into a WebDVD player, an HTML interface, possibly loaded from a remote server over the Internet, would take charge. If desired, this HTML interface could start the video in full-screen mode, giving the user the option to watch it normally or to switch to a more interactive viewing mode.

- A way to recognize that a specific disc is in the DVD-ROM drive. A remote Web page should be able to query the system for a unique identifier to verify that the proper disc is available.

- Drive independence. An HTML page used by thousands of computers from a Web server can't know which device or drive letter to use to access the DVD. A DVD URL, something like DVD:title/chapter, could provide this.

■ Platform recognition. In order to work across a variety of WebDVD devices, a DVD-enhanced Web page may need to know which platform it's running on, because different platforms may have different APIs for controlling DVD-Video playback. In addition, it may be useful to know if the device is connected to a standard TV so that graphics, font sizes, and layout can be adjusted for the lower-resolution, interlaced display.

■ A way to synchronize DVD content and Web content. This could take the form of EIA-746 triggers, but it has a few disadvantages, such as a lack of support for 625/50 (PAL) video and the need to embed triggers into the video early in the production process. The trigger concept could be expanded by providing synchronization data in an external file stored on the DVD or on the Internet. Other standards such as SMIL/HTML+TIME could be extended to support synchronization of DVD content.

■ A way to supercede resources on the disc with updated versions from the Internet or, conversely, to provide a resource from the disc when the Internet is not accessible or the Internet resource is unavailable. This could happen at the system level so that it is transparent to the HTML browser; there could be a way for scripts to determine if an Internet connection exists.

■ Support for conditional access to encrypted information on the disc. Content providers may want to make portions of the disc unavailable until an authentication process is performed with an Internet server. The incentive of unlocking additional video or other data can motivate viewers to visit a Web site. For example, a new video clip that can be unlocked each month will keep visitors returning frequently to a Web site. Conditional access can be implemented in various ways: free unlocking after visiting a Web site, unlocking in return for taking a survey or filling out a form, paid unlocking, timed unlocking, and so on.

Enabling WebDVD

HTML with associated scripting languages and related technologies has become the easiest and most widely supported computer development environment. More authoring systems, utilities, support services, and reference resources are available for HTML than for traditional languages such as C, Visual Basic, or Macromedia Director. Also, more people are versed in HTML than in any programming language. Millions of people, from children to grandmothers, know how to create a Web page. It's easy to add an object to play DVD-Video in a window on a Web page. Ironically, some of the most interesting applications of WebDVD don't even use an Internet connection. They simply rely on the ease of HTML authoring to create interactive DVDs for PCs.

The key is that a WebDVD platform must provide a way to integrate a window into the HTML page that shows video from the DVD. The best approach enables the video to be fully integrated into the page, with HTML objects overlaid on top of the video. Full-screen mode, where the video takes up the entire window, with HTML-based controls and information boxes that can pop up over the video, is essential for full-fledged WebDVD applications. At the other end of the spectrum, some WebDVD platforms only enable limited control of video playback in a separate window. Others play the video and simply link to URLs in a separate browser window. In most cases, DVD playback control is accomplished using a script language such as ECMAScript (JavaScript) to call methods and access properties defined by the DVD API of the platform. In some cases, declarative DVD control is also possible, using markup tags such as those defined for SMIL or HTML+TIME.

WebDVD pages can be jazzed up with any other Web technology supported by the browser, such as CSS, DHTML, XML, CGI, and so on. Even proprietary or platform-specific formats, such as 360-degree picture viewers, ASP, Flash, Shockwave, Real Networks audio, Microsoft Windows Media Audio, and so on, can be used in conjunction with the DVD playing in a window on the same HTML page. In other words, anything that works in a Web browser can be enhanced with DVD. This is a key point: the hard part of WebDVD development is not the DVD part; it's everything else that goes with it. Too many people get hung up trying to figure out what they can and can't do with WebDVD. The answer is that if you can do it in a Web browser, you can probably add DVD to it.

The golden rule of WebDVD: Anything you can do with the Internet or HTML, you can combine with DVD.

TABLE 11.6 WebDVD Needs	DVD-quality video and audio	Use existing disc or produce new content.
	PC control of video and audio	Use WebDVD features of authoring system or write scripts or generate SMIL/HTML+TIME tags.
	Synchronization of PC content	Use WebDVD features of authoring system or write scripts based on time event and list of timecode or on ATVEF trigger events.
	Unlocking of content on disc	Use simple tricks to hide the content on disc or use conditional access and encryption tools designed to work with DVD-Video content.
	Linking to URLs	Put HTML pages on disc. Minimize links from disc to Web because destinations may move or disappear.
	Multiple platforms	Write generic control scripts that don't depend on a specific WebDVD object and API.

Creating WebDVD

The preliminary step in creating a WebDVD is to determine which parts of the project are specific to DVD and which are not. Elements and processes that are not specific to WebDVD include page layout, script and application programming, customer registration, e-commerce and online sales, the general mechanics of conditional access and payment, user authentication, searches, and printing. These can usually be accomplished with tools that have no WebDVD components, and by developers who know nothing about DVD. Only a few tasks are specific to WebDVD (see Table 11.6).

The HTML content of a WebDVD can reside in essentially four places:

- On the DVD, designed to work without a connection to the Internet.
- On the DVD, designed to connect to the Internet, pulling content from the Internet or linking to other HTML pages.
- On the Web, designed to be activated by the DVD. A single link on the DVD jumps to the Web page, which then controls the disc and provides additional content and links to other Web pages.
- On the Web, intended to work with a DVD. The HTML page requires that the disc be inserted in the local drive so that the HTML page can control the disc.

Making and Accessing the Audio and Video   Video content needs to be identified or created. If the WebDVD project uses an existing disc, the only task is to identify the titles, chapters, and timecodes needed to access parts of the disc. If the project requires a new disc, the video and audio content needs to be created. You must decide whether to author the content as DVD-Video or simply put MPEG-2 or other multimedia files on the DVD-ROM. In general, authoring a DVD-Video disc is the best approach. The DVD-Video content will play in all DVD platforms, even those with no Web-DVD support, and DVD-Video-specific features such as subpictures and camera angles can also be used. DVD-Video volumes are more compatible with computers and with set-top WebDVD players than plain MPEG-2 files.

Access to the DVD-Video zone of the disc is accomplished with the Web-DVD API of the target platform, which must provide a way to jump to specific titles and chapters, and preferably to specific timecodes as well. For simple WebDVD titles intended only to show related HTML pages as the video plays, an authoring system with WebDVD support may be all you need; you can type in URLs to be associated with each chapter. For more complicated WebDVD titles, you will need to write scripts and possibly create lists or databases that link the video to the Web or the Web to the video. In either case, a title number plus a chapter number, or a title number plus a timecode (hours: minutes: seconds: frames), is the link to the video. A URL (such as http://dvddemystified.com/webdvd), possibly with a script to be executed on the page, is the link to the Web. URLs may be relative (no http:// at the front), in which case the HTML pages come from the DVD itself.

If a new disc is created, it should be authored with WebDVD in mind. *User operation controls* (UOPs) should be used sparingly. Disabling the fast-forward, next, and menu operations to lock the viewer into introductory logos, legal warnings, and ads will also prevent control by HTML pages. Scripts on the page won't be able to jump to desired locations on the disc, while the necessary user operations are blocked. Commands and GPRMs should be used judiciously. A disc that depends on certain values being placed in GPRMs may behave in unexpected ways when the HTML page that controls it skips over sections that set or check values. Video access points should be identified ahead of time. That is, every place in the video that you might want to have the HTML page jump to should be marked as a chapter during DVD-Video authoring because chapter access is extremely fast, easy, and accurate. It's possible to jump to an arbitrary point in a video sequence using timecodes, but only if the title is authored in one sequential PGC form. Most players can only jump to the I frame nearest a timecode, so the granularity of access is usually about 0.5 seconds.

Designing the Interface A common approach to a WebDVD interface design is to try to create links from the video to the Internet or PC world. For other than very simple titles, this is a dead-end approach. The main reason is that the DVD-Video specification has no provisions for jumping outside of its limited universe. A much more flexible approach is to design the disc so that the computer takes control and wraps the DVD-Video inside its own much larger universe. Menus and other features authored on the disc can still be used, but the PC can provide additional control and content. The autorun or autostart mechanism can be used to launch an HTML page when the disc is first inserted. The HTML page can take over with its own menus and windows, or it can play the video in full-screen mode to mimic normal disc playback, perhaps placing a small icon in the corner that the user can click to gain access to enhanced content. Because the PC can do a better job of implementing a user interface, providing graphical enhancement and controlling the video in a more interactive way, a PC-centric approach works better than a DVD-Video-centric approach.

Synchronization Fancy WebDVD titles may synchronize PC-based content with the video or audio playing from the disc. This can be done in a variety of ways. The most straightforward method requires that the WebDVD platform generate a time signal or a recurring time interval event. The page can repeatedly poll the current time position or can respond to the time event. A list of synchronization points can be compared to the current playback time position. When a synchronization point is passed, the page can show text, activate a link, start audio playback, or begin an animation. SMIL/HTML+TIME extensions make this much easier in browsers that support them and that are DVD-aware, such as Internet Explorer 5.5.

Alternative approaches to synchronization require more support in authoring and playback platforms. For example, the EIA-746 (TV Crossover Links) standard used by WebTV and supported by ATVEF provides a method for embedding URLs and script "triggers" into NTSC line 21 Text Mode service channel 2 (T2). Because DVD supports Closed Caption information, interactive television programs can be published on DVD after they are broadcast. A WebDVD system that reads the line 21 triggers would produce the same interactive experience as an enhanced TV system. Video producers can therefore "publish" once but deliver to multiple platforms. Here is an example of an EIA-746 trigger:

```
<http://dvddemystified.com/webdvd/>[name:Learn more about Web-
connected DVD][expires:20030801][script:DVDlink()][1F32]
```

In this example, an interactive TV system supporting EIA-746 will, after receiving this trigger, display an icon over the video indicating to the user that enhanced content is available. If the viewer clicks on the icon, the specified URL is accessed and the specified script is activated. A WebDVD system that is ATVEF-compliant would do something similar after reading the trigger from the Closed Caption component of the DVD-Video stream.

Connections Most WebDVD titles connect from the disc to the Internet, or from one Web page to another. WebDVD authors must make the commitment to maintain Web-based content and links. Along with the ability to extend the reach of a DVD comes the responsibility to keep the extensions working.

Internet links change with alarming frequency. A year after producing a WebDVD title, you may discover that half the links to other sites no longer work. The best way to deal with this is to make sure that all the links that are permanently stored on the disc point only to sites under your control. You can then use redirection mechanisms to detour the user to other sites. If the sites move or disappear, the central Web site can be updated as needed. Some WebDVD tool providers such as InterActual offer a redirection service. A simpler approach than redirection is to simply make sure that all pages that link to other sites are stored on your own Web site instead of the disc. These pages can then be updated whenever needed to maintain working connections to the rest of the Internet.

Conditional Access Hiding or locking the content on a disc is done for many reasons and in many different ways. You might want to play special hidden tracks as an incentive for customers to visit your Web site. You might want to charge customers for access to value-added content. Or you might simply have extra video clips on the disc that only make sense when played through a custom PC interface.

The simplest approach is to put video clips in titles that aren't accessible through any of the menus on the disc. Users who know how to use the title search feature of their player will be able to play the clips, but most users won't even know to look for them. For added security, extras can be authored in a separate DVD-Video volume that is stored in a subdirectory of the disc. That is, the .VOB and .IFO files that would normally be stored in the video_ts root directory are stored in a different directory, possibly a few levels down in other directories. A standard DVD player will never see the secondary volume, but a PC application can be designed to read the DVD-Video content from its non-standard location. Another option for securing access to video is to add a precommand to each restricted section to check for a special value in a GPRM. The PC will store the special value in the GPRM, possibly after a transaction or authorization process.

For robust security and conditional access, a variety of encryption and authentication technologies are available from companies such as Broad-Bridge Media, CrypKey, Greenleaf, and SpinWare.

WebDVD for Windows

Microsoft provides free basic components for WebDVD development as part of the Windows operating system. DirectShow, part of the DirectX set of core audio/video technologies, provides a standardized framework for DVD decoders and DVD applications. C++ programs can write directly to the DirectShow API, but most WebDVD developers will use the Windows Media Player or the MSWebDVD object. Each is an ActiveX control that can be added to any HTML page to create a scriptable DVD window. The controls are built on top of DirectShow, so they require that a DirectShow-compatible DVD decoder be installed in the computer. Because ActiveX support is required, Windows Media Player and MSWebDVD don't work in Netscape (unless it has since been updated to support ActiveX). Also, because of design flaws in Windows Media Player versions 5.2 to 7.x, the DVD features work only in Internet Explorer, but not in PowerPoint, Visual Basic, and other ActiveX hosts.

The scriptable DVD APIs for Windows Media Player and MSWebDVD provide access to essentially all the features of the DVD-Video format. These include title search and play, chapter search and play, timecode search and play, operation of onscreen menu buttons, audio track selection, video angle selection, subpicture selection, scan, slow, step, and so on. Various properties enable scripts to determine the status of playback, to query for features available on the disc, to respond to events, and so on. The feature set is sufficient that a complete DVD-Video player can be implemented in HTML, even though most WebDVD applications use only a few basic commands for playing video clips.

Adding DVD to a Web page using Windows Media Player or MSWebDVD is far simpler than many Web page creation tasks, but it does require scripting. For those looking for easier but possibly more limited solutions, various options are available. PCFriendly, the software used for most PC-enhanced movies, is available to developers directly from InterActual and also as part of WebDVD development tools provided by DVD-authoring system vendors. InterActual and others also make plug-ins for HTML-authoring systems to simplify the process of creating WebDVD pages. For DVD-enhanced Power-

Point presentations, PCFriendly-based add-ins are available from companies such as Zuma Digital and Daikin. Many DVD-Video authoring tools from companies such as Daikin, Sonic Solutions, and Spruce Technologies also include basic WebDVD features. Also, a number of Xtras add DVD playback to Macromedia Director projects. The sample disc that comes with this book includes demos of various WebDVD tools.

WebDVD for Macintosh

As of fall 2000, Apple QuickTime doesn't yet provide control of DVD-Video. Until this happens, it's difficult to develop WebDVD applications that work on Macs. Apple is implementing JavaScript control of the Mac OS DVD player application, which at least provides the rudimentary capability for HTML pages to control DVD playback in a separate window. InterActual's second-generation WebDVD player software supports scriptable control of the Apple DVD player, providing the first glimmer of cross-platform Web-DVD tools.

No plug-ins exist for Macintosh Web browsers, and without QuickTime support, they would have to be written to talk directly to specific hardware or software DVD players, thus limiting future compatibility.

WebDVD for All Platforms

Other computer platforms such as Linux and BeOS will slowly improve their support for DVD playback and may eventually provide DVD scripting APIs for WebDVD applications. Sophisticated DVD set-top DVD players using iDVD and Nuon technology support basic enhancements to DVD and will probably add WebDVD capabilities. New DVD-based game consoles such as Sony's PlayStation 2 and Microsoft's Xbox also have the power and connectability needed for WebDVD.

Microsoft Windows is far and away the leading WebDVD platform and is the natural focus for most WebDVD developers. The key to being able to embrace other platforms as they become worthy of WebDVD is to design HTML pages and scripts in a platform-independent manner. The best option for those interested in developing cross-platform WebDVD pages is to create a set of standard DVD playback functions, or "stubs," which are used instead of calling an ActiveX object directly. When DVD scripting features become available on other platforms, the standard playback scripts can be expanded to check which platform they are running on in order to call the proper function of the platform-specific browser plug-in. This requires that either a new disc be produced or that the WebDVD pages be stored on the Web, but, of course, this is the beauty of WebDVD. The Web pages can be updated whenever needed, long after the disc has shipped.

Here is an example of a set of standardized WebDVD functions. The first section is an example of how the function would be called from all HTML pages that control the disc. This code would never need to be changed:

```
<img src="images/scene1.gif" onClick="goChapter(1, 1);">
<img src="images/scene2.gif" onClick="goChapter(1, 2);">
<img src="images/scene3.gif" onClick="goChapter(2, 8);">
```

To begin with, a Windows-only version of the function could be used. It might look something like this:

```
function goChapter(titleNum, chapterNum){
  MediaPlayer.DVD.ChapterPlay(titleNum, chapterNum);
}
```

Later, as other platforms join the game, the function might be extended, as illustrated by the following pseudocode:

```
function goChapter(titleNum, chapterNum){
  if platform == Windows {
    MediaPlayer.DVD.ChapterPlay(titleNum, chapterNum);
  } else if platform == Mac OS {
    QuickTimeDVD.search(titleNum, chapterNum);
  } else if platform == Nuon {
    NuonDVDControl.playChapter(titleNum, chapterNum);
  }
}
```

By using the include feature of HTML scripting languages, a complete WebDVD function library could be stored online and be updated as needed without requiring any changes to the HTML pages that reference it. The Haiku group of the *DVD Association* (DVDA) is working to standardize this approach to cross-platform WebDVD scripting.

Why Not WebDVD?

Of course, WebDVD doesn't make sense in certain applications. For example, video on demand is intended to provide video without requiring a physical copy. Video or audio collections that change frequently or have recurring additions are difficult to keep up to date on distributed media.

Large libraries of video that can't fit on a single disc are generally not appropriate for WebDVD, unless DVD jukeboxes can support them.

WebDVD obviously requires DVD PCs or WebDVD players. DVD still has a ways to go before it catches up to the installed base of CD-ROM computers. However, it's possible to put short bits of DVD content on a CD-ROM for WebDVD-type applications, as long as the target computers have the power and the necessary software or hardware to play back DVD video.

Sending data over the Internet is generally perceived to be free. In spite of the costs that are hidden within access fees, subscriptions, and hundreds of charges, specific data transactions usually have no direct charges. DVDs cost money to produce. Even when replication charges or the cost of blank discs eventually drops below 50 cents, DVD can't compete with the virtual free ride of the Internet.

Why WebDVD?

As the Internet inexorably entangles itself into everything we do, Web-enhanced products will become the norm. Customers will be disappointed if the DVDs they buy don't have Web features. Home and business Internet users will become increasingly frustrated that the quality of video and audio is so inferior to their TV and VCR or DVD player. The convergent evolution of set-top devices such as cable TV boxes, game consoles, personal video recorders, and WebTV will lead to the integration of the Internet into TVs, DVD players, and every kind of information and entertainment device. The advantages of WebDVD, such as customizable and updateable content, dynamic data integration, customer-publisher communication, e-commerce, and more, make it very appealing to developers and customers alike. Today's Web-enhanced DVD titles are just the tip of the iceberg, given the unlimited potential that is unlocked by the marriage of DVD and Internet.

Besides computer geeks who do it just because they can, a surprising number of people watch movies on a PC. For example, those with limited living space, such as college students, can make a PC double as a TV. Airplane travelers with laptop computers can watch their own movies without getting a permanent kink in their neck from craning to see the projection screen from the first row. Even if you have a home theater with a standard DVD-Video player, there's no reason you can't enjoy the movie from your comfy chair, and then later pop it into your computer to see what else is available. One out of five copies of Web-enhanced movies are viewed on a PC. Keep in mind also that WebDVD goes beyond movies. It works even better for personal improvement videos, documentaries, education, business training, product info, and so on.

The fact is that eventually we will all watch movies on a computer. We just won't notice. Today a computer is one of the best solutions for progressive DVD display and for digital television. It will probably be the first solution for high-definition DVD. The trend begun by WebTV and accelerated by Tivo and ReplayTV will continue, reaching the day when every TV and set-top box is smarter and more connected than a Pentium III.

Traditional linear storytelling will always be an important part of the entertainment mix, but interactive cinema will continue to grow. DVD, and especially WebDVD, is the first technology that makes Hollywood-quality interactive cinema possible for the average producer and available for the average consumer. Until Internet bandwidth catches up, WebDVD provides *portable broadband*. This is a proven, practical, and functional way to deliver high-quality interactive video to the worldwide market.

Copy Protection Details

Copy protection has been a thorn in the side of DVD PCs, just as DVD PCs have been a thorn in the side of copy protection. But without a reasonable level of protection for copyrights, many content producers are not willing to publish on DVD.

This section discusses some of the technical details of copy protection systems as they apply to PCs. Many of the mechanisms of newer copy protection schemes, such as *content protection for prerecorded media* (CPPM) and *content protection for recordable media* (CPRM), are based on the original CSS design, so most of the details of authentication and encryption are covered under the CSS heading. Refer to Chapter 4 and Chapter 7 for general information on copy protection and ramifications of copy protection.

Content-Scrambling System (CSS)

CSS is the protection system used for prerecorded DVD-Video discs. Copyright information is stored in every sector of the disc, indicating if the sector is allowed to be copied or not.

Encryption is applied at the sector level. If a disc contains encrypted data, the corresponding decryption keys are stored in the sector headers and in the control area of the disc in the lead-in, which is not directly readable by a PC. The keys can be read only by the drive in response to certain I/O commands that are controlled by authentication procedures. When an encrypted sector is encountered, the drive and the decoder exchange a set of keys, further encrypted by bus obfuscation keys to prevent eavesdropping by other programs. This authentication process eventually produces the key used by the decryptor to decrypt the data. Until an authentication success flag is set in the drive, it does not read encrypted sectors. The drive itself does not decrypt the data; it merely participates in the two-way process of establishing an authenticated key and then sends the encrypted data to the decryptor (see Figure 11.2).

NOTE: *In the context of CSS,* scrambling *and* encryption *mean the same thing.*

All components participating in this process must be licensed to use CSS. Although regional management is technically independent of copy protection, it's included as a requirement of CSS-compliant components. See the Regional Management section at the end of this chapter for a few other details and also refer to Chapter 4.

CSS requires that the analog output of decrypted video be covered by an *analog protection system* (APS) such as Macrovision or similar. That is, display adapters with TV outputs are required by license to include Macrovision circuitry. The decoder or operating system queries the card. If the card reports that it has no TV output or that it has Macrovision-protected output, the system will send it decrypted, decoded video. Otherwise, the card will not receive video and will show a black screen when playing encrypted DVDs.

Digital bus output, such as USB or IEEE 1394 (Firewire) must be protected by a *digital protection system* (DPS). The standards for these systems are still under development, although DTCP is the forerunner (see the following section). Digital video output, such as DVI, must also be protected (see the DHCP section).

Video, audio, and subpictures are stored in MPEG packs. Each pack is one sector, so encryption can be selectively applied. About 15 percent of sectors are encrypted on a disc because 100 percent encryption can be a burden on software decoders. Encrypted content must be decrypted before the

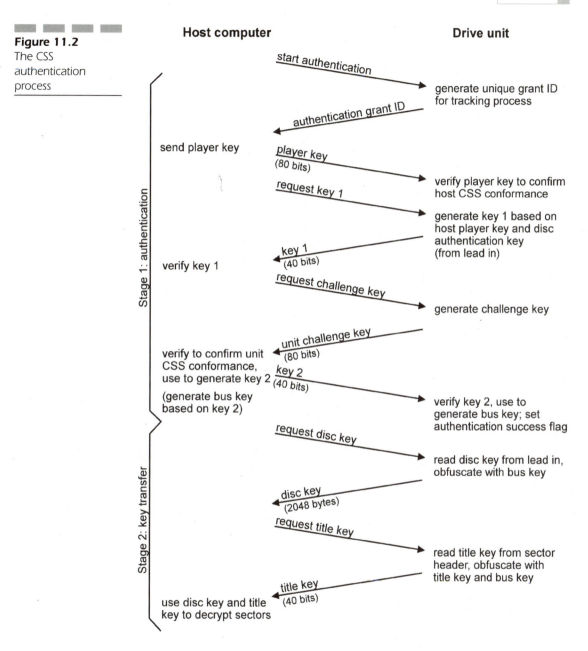

Figure 11.2
The CSS
authentication
process

MPEG video decoder, audio decoder, or subpicture decoder can process it.
The decryptor can be part of the same circuitry as the decoder, or it can be

independent. The decryptor and the decoder can be implemented in hardware or software.

The system is expected to protect the decrypted data from being copied. This is the state in which the content is considered to be most valuable. Once it has been decoded, requirements to protect the content are less stringent, especially because no mechanism protects RGB VGA output. After the video and audio is decrypted and decoded, the video is sent to the display and the audio is sent to the audio hardware.

CSS Authentication and Decryption CSS is essentially a cryptographic method for the distribution and management of cryptographic keys. The decoder or the OS (referred to as the player in this discussion) and the DVD drive engage in a handshaking protocol in which all the communication between them is encrypted. After verifying that the decryption module is registered and not compromised (which is done by using a specific hash algorithm to match the decoder key to a key in the key block on the disc), the DVD drive passes the content key, copy control information, and encrypted content to the decryption module. The decrypted content is then sent on a secure channel to the decoder. The decoded content, along with the copy control information, is communicated to the video controller where it is converted to video signals. The control information tells the video controller if an analog protection scheme must be applied prior to delivering the video signals to the display.

The authentication process works roughly as follows (refer to Figure 11.2). The player and the drive generate random numbers and send them to each other. Each encrypts the number using a CSS hash function (which turns a 40-bit number into an 80-bit number) and sends back the result, called a *challenge key*. If the recipient can decode the number using the same function, then it knows that the other device is a bona fide member of the CSS club. Once the authentication process is complete, the drive sets a flag to provide access to encrypted sectors. All sectors on the disc can then be read (by any program), but encrypted sectors must be decrypted to be of any use.

The second stage of CSS is to decrypt protected content. Each device uses challenge keys from the authentication step to generate a bus key. The bus key, which is never sent over the bus, is used to encrypt further key exchanges using a simple XOR operation. Each player has a key (or set of keys) assigned to it by the licensing authority. Each disc has a 2048-byte key block stored in the control area. The key block contains disc keys encrypted with 409 player keys. The purpose of this arrangement is that if a player key is known to be compromised, future discs can omit the encrypted key for that player, rendering the player unable to retrieve the disc key. Once a player retrieves and decrypts the disc key from its slot in the key block, it can use the disc key to decrypt the title key stored in the sector header of each encrypted sector. Each VTS on the disc uses a different title key. The decrypted title key is used to decrypt the video or audio contents of sectors that are encrypted with the CSS stream cipher.

CSS Licensing and Compliance CSS-compliant computers are required to have the following:

- A CSS-licensed DVD-ROM drive with authentication hardware
- CSS-licensed decoding hardware/software, including an authenticator and decryptor
- CSS-licensed video components
- Regional management support
- Protection of digital output
- Protection of analog video output (other than RGB)
- Prevention against copying decrypted files
- Rejection of "no copies allowed" material on a recordable disc (may be provided by the drive or system)

Any computer industry segment involved in the production or integration of hardware or software components that are affected by CSS may be required to obtain a license and abide by its restrictions, as follows:

Operating System Makers A CSS license is not officially required, but the operating system should protect decrypted data so that it cannot be copied either as files or as data streams. For example, a video T filter must not be allowed to split off a stream of decrypted video during playback. The OS must also not enable complete bit-image copies of DVD-Video discs.

The OS must support the regional management for DVD-Video discs if not handled elsewhere. The OS must provide for, or not interfere with, DPS and APS information in the video output.

Application Developers A CSS license may be required if the application directly manipulates DVD-Video files. The application is subject to the same restrictions as an OS (refer to the preceding section).

Disc Manufacturers Companies that master and replicate encrypted discs must have a CSS license in order to apply disc keys and encrypt sectors.

They are responsible for maintaining the secrecy of the keys and the encryption algorithms.

Drive Makers A CSS license is required only if the drive supports CSS authentication. If so, the drive must perform authentication handshaking and key obfuscation. The drive must also include firmware that stores a region code and enables it to be changed a limited number of times by the user.

Component Makers A CSS license is required for any hardware or software component, such as a decoder or decryption module, which directly deals with the decrypted data stream. Decrypting components are strictly licensed because they use the "secret" descrambling algorithm and the "secret" keys. The components must safeguard the algorithm, the keys, and the decrypted data. Licensed components can only be sold to other CSS-licensed integrators or OEMs.

A CSS license is also required for DVD add-in cards that incorporate DVD-Video playback features. The hardware must protect decrypted digital and analog outputs.

A CSS license is not required for graphic cards and other video display components, but other licensed components are not allowed to be connected to display components that include television video output unless analog protection is provided. If a method is devised for protecting RGB outputs, it may be required as well.

System Integrators and OEMs A CSS license is required for companies that produce or assemble DVD-Video-enabled computers. The final system must comply with all requirements, including video output protection, digital output protection, decrypted file protection, and regional management. No hardware or software can be added that enables the circumvention of any license requirements. Partly assembled subsystems must be distributed only to licensed integrators or resellers.

Retailers and Resellers A CSS license is required only for retailers who do an additional assembly or integration of CSS-related components. The restrictions are then the same as for system integrators.

Users No license is required, although users who desire to assemble their own systems might not be able to legally purchase licensed components. The goal is to make DVD-Video a plug-and-play system, with the end user unaffected by CSS entanglements.

A Recipe for Resistance The DVD Copy Protection Technical Working Group worked hard to make the licensing process as simple and unrestrictive as possible, but anything with this many rules and requirements can easily go awry. Strenuous objections on the part of computer hardware and software companies had to be overcome, but the golden glow of Hollywood movies on computers was a strong incentive to compromise.

Most of the computer industry had no say in these negotiations. This is an industry notorious for its lack of regard for regulations and artificial barriers. Computer software developers long ago gave up the fight for copy-protected software because it did little to slow down those determined to make copies. However, it sorely inconvenienced honest users who were unable to make legal backup copies or had problems simply installing or uninstalling the software they had purchased. Microsoft, for instance, loses billions of dollars to illegal software copying, yet it still makes plenty of money. The company educates users, employs sophisticated anti-counterfeiting techniques such as holograms and thermographic ink, and aggressively pursues counterfeiters and commercial pirates, but it does not usually encrypt its software applications or use other technical protection methods. Hollywood, however, is not this sanguine about its bread-and-butter assets.

A worse problem is the potentially onerous restrictions on computer retailers and computer owners. Many mom and pop computer stores assemble low-cost systems from diverse components. How happy will they be with the licensing requirements demanded of them? How many of them will simply turn to the inevitable gray market for supplies? And what about the computer owner who simply wants to upgrade to a new video card? Even a lowly computer owner who purchases a system assembled from licensed components by a licensed integrator is presumably barred from buying a new licensed component because these components can only be sold to licensed integrators. Of course, nothing can prevent end users from swapping in different video boards without APS and making copies to their hearts' content. Clearly, however, the same factors of cost and convenience that keep consumers from engaging in mass duplication of audio CDs and videotapes apply to DVD as well, with or without complex technical copy protection schemes.

The CSS algorithm and keys were supposed to be a big secret, but the algorithm was reverse engineered and all the keys were derived by computer hackers. Security experts who analyzed the CSS implementation noted that its 40-bit key length made it easily compromised through brute force attacks, especially because only 25 bits of the key were uniquely employed. Still, CSS prevents the average user from using a computer to copy a DVD movie, which is what it was intended to do.

Content Protection for Prerecorded Media (CPPM)

CPPM is the content protection system used for DVD-Audio discs. Copyright information is stored in every sector of the disc, indicating whether or not the sector is allowed to be copied. Some sectors of the disc are encrypted. Licensing and compliance rules are similar to those described in the earlier section on CSS. Also refer to the discussion of CPSA and CPPM in Chapter 4.

Each side of a CPPM-protected disc has a secret album identifier placed in the control area during replication. This is similar to the CSS disc key. A 56-bit key is applied to 64-bit chunks of data for the encryption/decryption process. The final 1,920 bytes of an MPEG pack (a sector on the disc) are encrypted. DVD-Audio data is interleaved with five 64-bit bit key conversion values used to vary the CPPM encryption after each pack.

Each disc contains a media key block stored in a file on the disc. The media key block data is logically ordered in rows and columns that are used during the authentication process to generate a decryption key from a specific set of player keys (or device keys). Each column may have up to 65,536 rows. The first generation of CPPM for DVD-Audio defines 16 device key columns, resulting in a maximum file size of 3,145,728 bytes (96 ECC blocks). Someone designing CPPM has a sense of humor because a successful media key verification step returns the value DEADBEEF in hex digits. If the device key is revoked, the media key block-processing step results in a key value of 0. As with CSS, the media key block can be updated to revoke the use of compromised player keys.

The CPPM authentication process roughly works as follows:

1. *Authentication and key sharing.* The drive and computer carry out a CSS-style authentication and key exchange process. If successful, each calculates a shared bus key.

2. *Encrypted transfer of control area data.* The computer requests that the drive read sector 2 from the control data area of the disc. The drive encrypts (obfuscates) the data using the bus key. This is also the same as in CSS.

3. *Decryption of the album identifier.* The computer decrypts sector 2 using the bus key and extracts the album identifier from bytes 80 through 87. The album identifier is then used to decrypt content on the disc. Unlike CSS, CPPM does not use an intermediate title key cryptographic step.

Content Protection for Recordable Media (CPRM)

CPRM is the protection system used for recordable DVD discs, regardless of the content being recorded. The licensing and compliance rules are similar to those described in the earlier section for CSS. Also refer to the discussion of CPSA and CPRM in Chapter 4.

The key to CPRM is that each disc contains a unique, 64-bit *media ID* permanently etched into the BCA. The first four bits are reserved, the second four bits identify the media type, the next two bytes are a manufacturer ID (assigned by the licensing authority), and the last five bytes are the unique serial number. The content stored on the disc is encrypted using a title key, which is then encrypted with a *media unique key* generated from the media ID and from the media key block. Because the media ID, which is physically part of the disc, is required to decrypt the title key in order to decrypt the content for playback, any content copied to another medium will not have the necessary keys. Neither the media key block nor the media ID need to be kept secret.

CPRM uses a media key block, similar to that used in CPPM, to store a record of keys associated with sets of device keys. Recorders and players process the media key block in a prescribed way, using their device keys, to produce a media key, which is then combined with the media ID to generate a media unique key. The authentication process uses the existing CSS mechanism to read the media key block. A cryptomeria (C2) hash function is used to verify the integrity of the media key block and the media ID.

The media key block is stored during disc manufacturing in the embossed control data area of the lead-in. Sectors 4 through 15, a 24,576-byte extent, hold the media key block pack, which is repeated 12 times.

Sectors on the disc contain MPEG-2 packs. Video packs, audio packs, and subpicture packs, referred to as AV packs, can be encrypted. Each pack contains a 56-bit *title key conversion value* that varies the encryption from pack to pack. *Real-time data information* (RDI) packs, used by DVD-VR, are not encrypted. Each RDI pack indicates the CGM and APS status of subsequent AV packs. Because this information is not protected from potential tampering, devices actually control the copying of the recorded content based on whether or not it is encrypted. Unencrypted content can be copied without restriction, whereas CPRM-encrypted content is not allowed to be copied.

The encryption process, performed by a DVD recorder when writing new protected content to a disc, works roughly as follows. The recording device reads the media key block from the disc and uses its 16 device keys to calculate the media key. It also reads the media ID from the disc. Using the two values, it calculates a media unique key. A single 56-bit title key is used for all protected video and audio data on the disc. The recording device examines the title key status field (in the VMGI_MAT of a DVD-VR disc, or in the AMGI_MAT of a DVD-AR disc) to determine if a title key is already recorded on the disc. If not, the recording device generates a random title key and records it on the disc. For each AV pack to be encrypted, the recording device uses the pack's title key conversion data (generated as the pack is formatted) to calculate a 56-bit *content key*. The content key is then used to encrypt the last 1,920 bytes of the pack.

The decryption process, performed by a CPRM-compliant DVD reader, works roughly as follows. The player reads the media key block from the disc and uses its 16 device keys to calculate the media key. The player reads the media ID from the disc and combines it with the media key to calculate the media unique key. The player reads the encrypted title key from the disc and uses the media unique key to decrypt it. Each pack on the disc contains title key conversion data, which is combined with the title key to calculate a 56-bit content key to decrypt the last 1,920 bytes of the pack.

HDCP

High-bandwidth digital content protection (HDCP) is used with digital video monitor interfaces such as DVI (refer to Chapter 4 for an overview). HDCP has three components: authentication and key exchange, encryption, and revocation. The HDCP ciphers and related circuitry are implemented in DVI digital transmitters and receivers and add approximately 10,000 gates to each device. Each transmitter and receiver has *programmable read-only memory* (PROM) to hold device keys. Receivers are integrated into display devices, and transmitters are integrated into computers. HDCP software drivers)are required in the host computer.

HDCP authentication is a cryptographic process that verifies that the display device is authorized to receive protected content. A licensed computer and a licensed display each have secret keys, supplied by the HDCP licensing authority, stored in PROM circuitry. Keys are 56 bits long. Each device has an array of 40 keys and a corresponding 40-bit binary *key-selection vector* (KSV). The computer begins the authentication process by sending its KSV and a random 64-bit value. The display then sends back its own KSV, and the computer checks that the received KSV has not been revoked.

The two devices use the exchanged data to calculate a shared value that will be equal if both devices have a valid set of keys. This shared value is then used to encrypt and decrypt protected data. A block cipher is used during the authentication process. Reauthentication occurs approximately every two seconds (128 data frames) to confirm that the link is still secure. That is, the computer checks that it has not been unplugged from an authorized device and plugged into an unauthorized device.

Content is encrypted in the computer (at the transmitter) so that devices tapped into the connection cannot make any unauthorized copies. A stream cipher is XORed with the video data to provide pixel-by-pixel content protection. Because the receiving device has calculated the same cipher key, it can decrypt the incoming stream. If the encrypted content is viewed on a display device without decryption, it appears as random noise.

If a display device is compromised and its secret keys are exposed, the licensing administrator places the device's KSV on a key-revocation list, which is carried by *system renewability messages* (SRMs). The computer updates its key-revocation list whenever it receives a new SRM, which can come from prerecorded or broadcast sources, or from a connected device.

Region Management Details

Before 2000, regional management was handled independent of the drive and most DVD-ROM drives had no built-in region code. These drives were designed according to *region playback control phase 1* (RPC1). Because RPC1 drives did not contain hardware support for region management, the software player application, the decoder, or the operating system was responsible for maintaining the region code. The region code could be set once by the user on initial use. Some older DVD decoders were preset for a specific region. Generally speaking, the user could not change a decoder's region.

To make it easier for drive manufacturers to ship drives anywhere in the world, for computer makers to allow their customers to set the proper region code for where they live, and for users to change the region code if they move to a different region, *region playback control phase 2* (RPC2) was implemented. As of January 1, 2000, all drive manufacturers are required to make only RPC2 drives. (This deadline was pushed back numerous times from earlier dates.) RPC2 drives maintain region code and region change count information in firmware. The user can change the region of the drive up to five times. After that, it can't be changed again unless the vendor or manufacturer resets the drive. RPC2 moves region management into the drive hardware where it's more secure and more easily controlled.

An RPC2 drive has four region states that can be queried and set by the host computer:

- *None*: The drive region has not been set. The computer must set the initial region before playing discs with region information. When the region is first set, the region-setting counter goes from 5 to 4.

- *Set*: The drive region has been set. The region-setting counter will decrement on each change until it reaches 1.

- *Last chance*: The region-setting counter is 1, and the drive region can be set only one more time. In order to change to a new region, it must be the same as an inserted single-region disc.

- *Permanent*: The region-setting counter is 0, and the region can no longer be changed. At this point, the drive region can be reinitialized by the vendor to the None state.

Essentials of DVD Production

Introduction

This chapter covers the basics of producing DVD-Video titles and DVD-ROM titles. It is not a detailed production guide, but rather an overview of the production processes that can serve as a guide and checklist, as well as an introduction for those who want to understand what's involved. Aspects of DVD authoring that are confusing, misunderstood, or overlooked are also covered.

General DVD Production

Creating a DVD is a tricky process. DVD-Video and DVD-Audio production demand knowledge and training, especially for highly interactive projects. Even DVD-ROM production is surprisingly more difficult than CD-ROM production. This is partly because DVD has more capabilities and more flexibility and thus more potential for mistakes, and partly because the tools for DVD haven't gone through 15 years of refinement. The tools will undoubtedly improve and the process will become easier, but in the meantime, do not assume that making a DVD is a walk in the park.

Special utility software is needed for DVD-ROM formatting and for writing DVD-Rs. Specialized production systems are required for DVD-Video and DVD-Audio. Movie studios, in conjunction with DVD hardware partners, developed proprietary production systems to create content before the introduction of DVD. At the introduction, one authoring system was commercially available, and only for SGI workstations: Scenarist DVD, created by Daikin and distributed by Sonic Solutions. A few compression and premastering systems were commercially available from companies including Daikin/Sonic Solutions, Minerva, Innovacom, and Zapex. Commercial production and compression systems ranged from $80,000 to $250,000. Following the introduction of DVD, dozens of new production systems began to be developed. By 2000, the tools ranged from barebones authoring applications priced at $75 to full-featured authoring packages that cost around $20,000, but could do far more than the original $100,000 systems.

Producing a DVD is significantly different from producing a videotape, laserdisc, audio CD, or CD-ROM. There are basically three stages: authoring (including encoding), formatting (premastering), and replication (mastering). DVD production costs are not much higher than for other media, unless the extra features of DVD-Video or DVD-Audio such as multiple sound tracks, camera angles, synchronized lyrics, and so on are employed.

Authoring costs are proportionately the most expensive part of DVD production. Video, audio, and subpictures must be encoded, menus have to be laid out and integrated with audio and video, and control information has to be created. This all has to be multiplexed into a single data stream and finally laid down in a low-level format.

In comparison, videotapes don't have authoring or formatting costs to speak of, aside from video editing, although they are more expensive to replicate. The same is true for laserdiscs, which also include premastering and mastering costs. CDs require formatting and mastering, but cost less than DVDs. Since DVD production is based mostly on the same equipment used for CD production, mastering and replication costs will eventually drop to CD levels. Double-sided or dual-layer DVDs cost slightly more to replicate since they are more difficult to produce.

Project Examples

DVD projects can take a variety of forms. The list in Table 12.1 is by no means exhaustive, but illustrates the ways DVD can go far beyond a portion of linear video transferred from tape.

CD or DVD?

For computer applications, CD and DVD do not differ greatly other than their capacity. If audio and video are involved, then CD—particularly Video CD—may be an option if video quality is not paramount. For short video that can be played on computers, DVD formatted content can be put on a CD.

For those considering moving from CD to DVD, the financial break-even point is at 2 or 3 discs. Although it may be cheaper to produce and replicate 2 CDs compared to 1 DVD, cost savings also come from simpler packaging, lower shipping and inventory costs, and so on.

DVD-5 or DVD-9 or DVD-10?

The disc type is generally mandated by the size of the content. Single-sided DVD-9 discs and double-sided DVD-10 discs are similar in capacity. Although DVD-10 discs hold slightly more and are slightly cheaper, DVD-9 is almost always a better choice, since it allows a full label and doesn't

TABLE 12.1

Project examples

Project	Example
Video and audio without menus, graphics, or options	Archived video
Video and audio with a single menu to choose selections	Home videos
Video and audio with a main menu and submenus with supplements	Simple movie, educational disc
Audio with a few menus and stills	Music albums (DVD-Video or DVD-Audio)
Multichannel audio and music videos with menus and supplemental video	Music video albums (DVD-Video or DVD-Audio)
Main menu and submenus for video format, audio language, subtitles, angles, chapters, and so on	Special edition movie
Any of the above with computer content or Web connectivity on hybrid disc	PC-enhanced movie
DVD-ROM with data or applications	Auto parts database, computer game, and application suite

require flipping. Rumors of production problems and compatibility issues are just that—rumors. Early production yield problems were solved long ago, and minor dual-layer compatibility problems affect an insignificant number of readers.

DVD-Video or DVD-ROM or Both?

A DVD-Video disc plays in any DVD system: DVD players, DVD computers, DVD game consoles, and so on. It has more limitations, making project design and production easier in some ways. In general, DVD-Video discs require no tech support. DVD-ROM has more flexibility, and it has better support for text display, data manipulation, searching, and interactivity. The best approach is to combine both, using the best qualities of each. Of course, this option requires more work. See Chapter 11 for more on computer use of DVD and enhancing DVD-Video and DVD-Audio discs with Internet technology.

DVD-Video and DVD-Audio Production

DVD-Video and DVD-Audio production require specialized equipment and software. Melding the computer world and the audio/video world, in ways that are unfamiliar to many, is a complex process. Computer graphics artists, accustomed to square pixels and an RGB palette, must learn about non-square video pixels, video-safe colors, overscan, and interlace artifacts. Video engineers, accustomed to headroom and waveforms, have to deal with YC_bC_r colorspace, sampling, compression, and bit rates. Audio engineers also must deal with issues such as multichannel mixing, apportioning the LFE channel, downmixing, sampling frequencies and word sizes, and synchronizing multiple audio tracks to video.

DVD production work is done in a variety of ways and places. Each business model is optimized for the market it serves, with attendant advantages and disadvantages. Smaller businesses may provide services for only part of the DVD production process, such as menu design, subtitles, or audio formatting, while large service houses can provide a one-stop shopping experience for an entire project. Several types of DVD production environments are available.

- *Movie / music studios or subsidiaries* develop their own products from content they create.

- *Film / video / audio post-production houses* produce DVD as a sideline to complement core business

- *Replication facilities* operate authoring services to generate more disc manufacturing business and can provide one-stop shopping.

- *DVD shops* focus primarily on encoding and authoring.

- *Specialty DVD shops* focus on market niches such as government, museums, or weddings.

- *Corporate A / V departments* develop discs for internal training or external marketing and may provide production services to other companies.

- *Individual or small-group projects* involve one person or a few people with a vision, such as independent filmmakers.

A big studio has experience creating hundreds of titles. They have a large staff with experience in what works and what doesn't. They have the financial wherewithal to stay in business, and can eat cost overruns if something goes wrong. On the downside, they operate in factory mode and are not as flexible. Most creative people don't function well within this environment. Internal titles or big customers come first, potentially sidelining smaller projects. Costs may be higher, but the quality generally matches the price.

The traditional film, video, and audio post facility has the production background and experience to make DVD work, even though DVD is much more complicated than film transfer and editing. If problems with the assets occur, such as bad tapes, or missing pieces, a good post house usually has the equipment to make repairs on the spot. Professional artists and creative people tend to be on hand, and good service is an expected way of life. The downside may be that DVD is not the prime focus of the facility.

Replication facilities that offer DVD premastering services can handle most elements of a job from input to packaged product. Disc manufacturers need titles to feed the production lines, and they will do plenty in order to keep the production lines going. This includes authoring and premastering, often at very competitive rates. Replicators can afford to use compression and authoring as a loss leader in order to make money in replication. However, the sideline authoring service may not carefully focus on quality and creativity. The user of such packaged services is usually a price conscious customer, likely to switch suppliers for the next cheaper deal.

The dedicated DVD shop tends to be a small but focused group, originally from a creative multimedia background. The personnel skill sets are well suited and dedicated to all things DVD, and good service is the norm. Often, the president is also the person that runs the machines, does the shipping, and makes the coffee. DVD-only companies are more likely to try new techniques. They offer more value for the dollar, since they have to work harder just to survive. The disadvantage is they may be undercapitalized and unable to keep up to changing conditions and technology. If a project goes bad, it could seriously affect their bottom line. Some will grow to be big players, but many will come and go.

Specialty shops are the best at what they do in a limited role, and they know enough to stay in their niche and be successful. The disadvantage is competition with all-in-one operations. Producers must establish multiple relationships with suppliers and providers of related services. Technology can change and make niche markets obsolete.

Corporate audio/video departments often evolve DVD production abilities from existing services. In some cases, the DVD side of things may be awkwardly grafted onto existing structures, or in other cases, the DVD group may be well positioned to understand the needs of corporate video and know how to apply unique features of DVD.

Sometimes the best titles spring from the creative drive of one person. This person may do it alone, finding the cheapest production tools possible, or may beg, borrow, or barter service from others. The most difficult part is finding a distributor to put the creator's labor of love into the marketplace.

Tasks and Skills

Many skills and processes are needed for DVD production. The larger the facility, the more specialized each position is. No facility is likely to have all facets down to perfection, but many have multiple areas of expertise or have relationships with other service providers. Small DVD production shops rely on outside facilities to do much of the production work.

The following list covers the essential job positions of DVD production. These could be combined into a few positions or spread across multiple people. Rare is the person who excels at all tasks of DVD production, although it's a good idea to cross-train personnel so they can help out in other areas as needed.

- A *project designer* works with clients and is responsible for layout, design, and general budget, and therefore, must understand the steps of the production process.

- A *project manager* deals with bottlenecks, should be familiar with all other production tasks, works with clients for testing and signoff and is responsible for schedule, coordination, budget details and asset management.

- A *colorist* transfers film to video, is responsible for color correction, and often works with the director or *director of photography* (DP).

- A *video editor* is responsible for assembling, editing, and preparing all the video assets for encoding; uses video editing systems, usually digital video editing software; may need to restore and clean up footage; may do compositing of video and computer graphics; should be familiar with video encoding and DVD authoring.

- An *audio editor* is responsible for assembling, mixing, editing, and preparing all the audio assets for encoding; uses audio editing systems, usually software; is often required to sweeten or clean up audio; must conform audio to match edited video, especially when multiple languages are involved; and should be familiar with audio encoding and DVD authoring.

- A *video compressionist* must understand MPEG-2 compression, preferably VBR; needs to be familiar with DVNR equipment; should be familiar with colorist and video editor jobs; have working knowledge of DVD authoring process; and must be familiar with professional videotape equipment.

- An *audio compressionist* must understand multichannel audio and principles of perceptual audio encoding; must have in-depth knowledge of details of Dolby Digital encoding, and DTS if needed; must be familiar with professional digital and analog audio recording equipment; should be familiar with audio editor job; and have working knowledge of DVD authoring process.

- A *graphic artist* is responsible for graphic design, including menus and stills; may also create subpictures; may be responsible for user interface design; may also work on disc label and package design; uses graphics packages such as Adobe Photoshop; and must understand nuances of digital video and analog video.

- A *3D animator* produces animation sequences for menus and transitions; uses graphics packages such as Adobe After Effects; must understand nuances of digital video and analog video.

- An *author* uses specialized DVD authoring applications; must have detailed knowledge of DVD specification and navigation design. Advanced authoring requires programming knowledge.

- A *tester* is responsible for QC of product; reviews encoded audio, video, and subtitles; tests navigation; and should be familiar with digital video and audio.

The Production Process

The DVD video and audio production process can be broken into basic steps, which are shown in roughly chronological order in Table 12.2 The quality control (QC) and testing steps are spread throughout most of a project. Proper project design and careful asset management are key to the successful creation of a DVD title.

Authoring refers to the process of designing and creating the content of a DVD-Video title. The more complex and interactive the disc, the more authoring is required. The authoring process is similar to other creative processes such as making a movie, writing a book, or building an electronic circuit. It may include creating an outline, designing a flowchart, writing a script, sketching storyboards, filming video, recording and mixing audio, taking photographs, creating graphics or animation, designing a user interface, laying out menus, defining and linking menu buttons, writing captions, creating subpicture graphics, encoding audio and video, and then assembling, organizing, synchronizing, and testing the material. Authoring

TABLE 12.2	Project planning	Develop schedule and milestones, storyboard video, lay out disc, design navigation, budget bits.
The DVD production process	Asset preparation	Collect, create, capture, process, edit, and encode video, audio, graphics (menus, subpictures, stills), and data. Check source assets, digitizing, and encoding.
	Authoring	Import, synchronize, and link together assets; create additional content, define and describe content; simulate and check compliance with DVD specification.
	Formatting	Multiplex and create volume image. Emulate, check compliance, run check discs, and compare checksums. Write DLT or DVD-R.
	Replication	Create master image, record glass master, mold discs, bond, print, and QC final copies.
	Packaging and distribution	Insert discs and printed material, insert source tags, shrink wrap, box, and ship.

is done on an authoring system, which may be integrated with encoding systems and a premastering system.

Authoring sometimes refers only to the specific step of arranging the assets using an authoring system. Encoding and authoring are sometimes referred to as premastering, although premastering more accurately refers to the final step of multiplexing files, formatting a UDF bridge image, and writing it to DLT in preparation for mastering and replication.

Production Decisions

During the production process, many decisions between different options need to be made. The choices made depend on the particular project and also on personal preference.

Service Bureau or Home Brewed? You must decide whether to do most or all of the work yourself, or whether to outsource it to a DVD production house. (Unless, of course, you are a DVD production house.) The advantages of producing a title yourself are that you have full control of the project and can make changes as needed. If you intend to produce a number of DVDs, you can build a base for long-term production. Service bureaus have the advantage of knowledge and experience. They own most of the expensive production equipment needed to produce a DVD. They have trained personnel and can usually produce a disc quickly if needed. They have tools for testing and may even have a collection of players and DVD PCs for testing. They usually have connections with replicators.

To find an appropriate service facility for your DVD project, check with other media producers in your industry, or look in trade magazines and features for knowledgeable DVD personnel and service bureaus. The better companies are well known and are often featured in news articles. Advertisements display the companies with marketing budgets, but often the best service providers remain so busy that they don't need to advertise for more business.

With a list of several companies in hand, do a comparative analysis between them. Request company information, including background, personnel, equipment, DVD experience, production capacity, and other available services that can help you with your project. Ask for demonstration or sample discs. Once you choose a facility, develop a good relationship and communicate regularly during projects.

VBR or CBR?   *Variable bit rate compression* (VBR) enables better quality at longer playing times, since it uses disc capacity more efficiently. Good VBR encoding is somewhat of a black art and requires an experienced compressionist. *Constant bit rate* (CBR) is cheaper and faster, since it doesn't require multiple passes. For discs without a lot of video, high data rate CBR can achieve the same quality as VBR.

NTSC or PAL?   DVDs in 525/60 (NTSC) format will play in over 95 percent of all DVD players, since PAL and SECAM players play NTSC discs. DVDs in 625/50 (PAL/SECAM) format provide better resolution, but will not play in most NTSC players.

PCM or Dolby Digital or MPEG Audio or DTS?   Dolby Digital audio format gives the widest coverage. Every DVD player, including computers, can play Dolby Digital audio tracks. Every player is also required to play PCM audio, but PCM doesn't leave much room for video. DTS audio is a secondary format that appeals to an audiophile niche. Most players and computers must be connected to a DTS decoder, since they can't natively decode DTS. MPEG audio is not widely used.

Multiregion and Multilanguage Issues

Before you even begin designing a DVD-Video disc, ask yourself what its distribution will be. Will it stay within a single geographical region, or do you want it to sell worldwide? Do you want to restrict the regions in which it will play? Unless you expect special exclusive regional distribution arrangements, there is no reason to restrict the regions of your disc. If you make an all-region disc, you are free to distribute it anywhere in the world. But if you restrict distribution to one or a few regions, you may be sorry later when you want to broaden distribution; you'll have to reformat and remaster the disc.

Also recognize that if you want to set region codes on your discs, you must also CSS encode the contents. Region code restrictions in players are required only in conjunction with CSS. It's possible to set the region flags on the disc without using CSS, but players are then not required to honor the region settings.

If you decide to use regional management, then obviously you need to decide which regions. Again, broadening the scope of the disc early may be a life saver later, when marketing plans become more ambitious. Table A.25 lists what regions apply to which countries. You must also decide the languages for which you will make menus. You must also choose which audio languages and which subtitle languages to put on the disc. Carefully check the audio and subtitle tracks to make sure they match the video you are using. In many cases, an existing audio dub or subtitle set is from a different cut of the movie, and will have to be modified or redone to match the cut you are using. Audio from film usually has to be sped up 4 percent for PAL, so make sure the audio properly matches the video.

If you cover multiple regions, or even if you only cover a single region such as Europe/Japan, you may need to consider certification. Each country has its own requirements for classifications and certification. A disc with a "children's" rating in one country may end up with a "teen" rating in another country. If you have to cut scenes to get a particular classification, you will have to deal with conforming the audio and any subpictures. Using DVD-Video parental management features usually will not work. The classification boards rate a work according to all the versions on the disc, regardless of how you have restricted access to them.

You must also decide if you want a single package (often called a SKU—*stock-keeping unit*) or multiple packages for multiple regions. In some cases, creating multiple SKUs is easier, since you can concentrate on making each version appropriate for its region, rather than trying to make a single version work everywhere. For example, you might want to make three versions of a disc for release in Europe: one for the UK and Ireland, one for Germany and Scandinavia, and one for central Europe (France, Spain, Italy, and so on). Multiple SKUs can also help with legal restrictions concerning how soon the home video version can be sold or rented after theatrical release.

Regardless of the number of regions, if you expect to use more than one language on the disc, especially for menus, keep all the graphic and video elements as language-neutral as possible. Keep text separate from graphics so it can be easily changed for other languages. Use subpictures, including forced subpictures, to put text on video, rather than permanently setting the text into the video. Keep in mind that American English is different from British English, regarding usage and spelling. Likewise there are differences between American Spanish and Castillian Spanish, between Canadian French and continental French, between Brazilian Portuguese and Portugal Portuguese, and so on. In all cases, check to make sure you have proper rights for the content. A particular work may have only been licensed for distribution in one country, or a foreign language audio dub may be restricted for use in certain countries.

Supplemental Material

One of the great appeals of DVD is that a disc can hold an incredible variety of added content to enhance the featured program. Being able to watch a movie a second time and listening to the comments of the director, writer or actors, gives viewers a deepened appreciation for the movie and for the filmmaking process. Adding more video, audio, graphic, and text information can greatly enrich a disc. Anyone producing a DVD should consider how the product can be enhanced with supplements; not just because it's neat or expected, but to truly improve the viewing experience. The following list provides some examples of added content for DVD-Video and DVD-Audio discs. Some types of added content, especially text, and information that can become out of date, may be better provided as computer-based or Web-based supplement. These are marked with *.

- Multiple languages and subtitles, menus in other languages
- Biographies: information about actors, directors, writers, and crew, describing careers, influences, and so on*
- Filmographies: information about actors, directors, writers, and crew describing previous films, acting experience, awards, and so on*
- Commentaries: audio or subtitle commentaries from people who worked on the film, people who have studied the film, people with historical information, and so on
- Song list: a chapter index to all the songs on the soundtrack
- Backgrounders: historical context, timeline, important details, related works*

- "Making of" documentary
- Production notes and photos, behind-the-scenes footage, lobby cards, memorabilia
- Live production details, such as synchronized storyboards, screenplay, or production notes*
- Trivia quiz or game
- "Easter eggs:" hidden features that are found by selecting disguised buttons on menus
- Related titles: "If you liked this disc, you should see...," lists or previews of discs with similar characteristics or with the same actors
- Trailers or ads for other titles
- Bibliographies: where to find more information on subject matter*
- Web links: Web sites for the movie, actors, studio, producer, sponsor, fans, and such*
- Excerpts for computer use: screenplay, documents, scanned photographs, sound bites, and so on for use on a computer*
- Related computer applications: screen savers, games, quizzes, digitized comic books, novelizations, and so on*
- Additional content more easily programmed or accessed through ROM on a computer
- Product catalog and ordering system to sell related goods such as other DVD titles, action figures, and such*
- Newspaper and magazine articles or reviews*

Scheduling and Asset Management

The secret to a smooth and successful project is a carefully monitored schedule. The schedule enables you to allocate resources and determine bottlenecks. Project planning software is very helpful. In many cases you will start with a rough schedule for laying out the project, then produce a final schedule based on the project plan.

Throughout the entire production process, keep track of the assets. Make a list of each asset, when it's due and when it arrived, what format the source is in, when it was digitized and encoded and by whom, what the filename is, when the client signed off on each stage of asset conversion or production, and so on. Clarify who owns the copyright for new content created during production.

Project Design

The project design document is the blueprint from which everyone will work; it is the important first step of creating a disc. Decide up front the scope and character of your title. Make a plan and try to stick to it. Authoring is a creative process, but if you continually change the project as it goes along, you will waste time and money. Determine the level of interactivity you want. The more interactive, the more complex the layout, and thus the more time you need to spend at this early stage.

Laying out the disc generally involves the following:

- *Flowchart:* Diagram each part of the title. Decide what appears when the disc is first inserted. Does the disc go straight to the video or to a menu? Diagram each menu and how it links to other menus and video segments. For a complex title, unless you create a clean design with clear relationships between the various hierarchical levels, your viewers will become lost when trying to navigate the disc. Each box on the flowchart should indicate how the viewer will move to other boxes, and what keys or menu buttons they will use to do so (see Figure 12.1).

- *Storyboard:* Draw pictures of the menus and the video segments. You may want to draw a storyboard picture for each chapter point. Make sure that transitions between menus and from menus to video are smooth.

- *Prototype:* Create rough versions of the menus in PhotoShop or PowerPoint or another graphics program. Simulate jumping from one menu to another by using layers or by opening new files. Bring in other people to test the simulation. Make sure the experience is smooth, clear, and easy.

- *UOP controls:* For every menu and video segment, determine what the user will be able to do. Start with all user operations enabled, and only disable them for a good reason. Do not needlessly frustrate your viewers by restricting their freedom. Specify user operation results.

- If you have a dual-layer disc, choose PTP or OTP (RSDL). For sequential playback across layers, you will need RSDL. In this case, start thinking about where the layer switch will be. For two versions of a movie, such as pan & scan and widescreen, you will probably use PTP layout.

The layout process serves as a preflight check to make sure you have accounted for all the pieces and that they work together. Good layout docu-

Figure 12.1
Sample flowchart

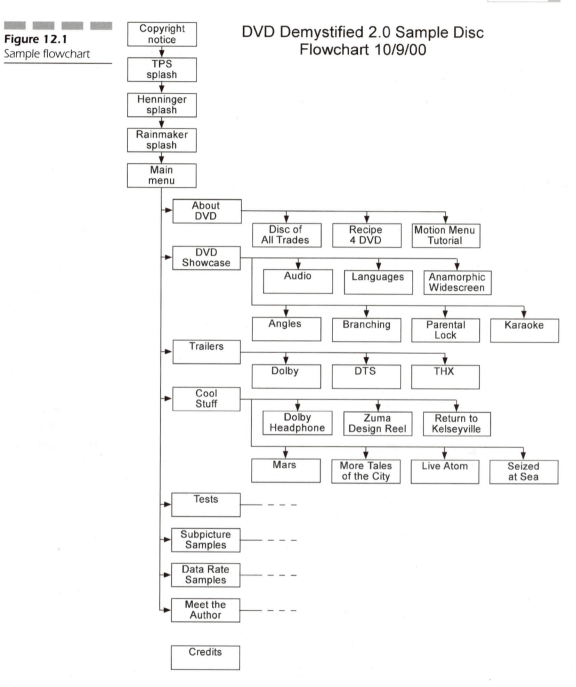

DVD Demystified 2.0 Sample Disc
Flowchart 10/9/00

ments provide a roadmap to be used by the entire team throughout the remainder of the project: a checklist for coordinating assets and schedules, a framework for menus and transitions, a guide for bit budgeting, an outline for the authoring program, a checklist for testing, and so on.

Many authoring systems attempt to provide layout and storyboarding tools in the application. Some succeed better than others. In general, at least a basic layout should be done on paper or in a separate program (or in Pioneer's DVDesigner layout tool) before jumping into the authoring program, especially since the authoring stations are often key pieces of expensive equipment that are much in demand.

Menu Design

Menus make the DVD; they set the tone and define the experience of using a disc. They are also fun to design. The possibilities are endless, so you must make many decisions up front. Do you want still menus or moving menus? If you have a moving menu, consider carefully how to design the video so that it loops smoothly from the end, back to the beginning. If a great deal of movement occurs, consider starting with a still and ending with the same still. Fade the audio down just before the end, and fade it back up at the beginning.

Do you want live transitions? That is, do you want a video or animated chunk that appears at the beginning of the main menu, or between menus and video segments? Make sure the transitions are smooth and appropriate. Even if you don't use motion transitions, jumping from one design and color scheme to a completely different one can be very jarring. Likewise, jumping from one type of audio in the menu to a completely different type of audio in the video program can be grating.

Do you want background audio in the menu? Be very careful with this. One of the most obnoxious features of DVD discs are menus with audio that is too loud and too intrusive. Drop the audio to a pleasingly low level; then cut the level in half. Loop the video and audio, playing it over and over at least twenty times, to discover just how annoying your creation can be.

Do you want multilanguage menus? If so, keep the text separate from the graphics and video so that it can be easily changed for each language version of the menu. If you are creating the original backgrounds in English, leave 30 percent more space for translations.

Will you use subpicture highlights or action buttons? Action buttons jump to a different menu, optionally with motion, when the button is highlighted. This approach lets you use full-color graphics for highlighting rather than one-color subpicture highlights. You can even do fancy motion highlighting such as flaming or spinning buttons, but it's much slower on many players. It will also cause problems on PCs, since moving a mouse over the button will usually not cause the action highlight effect. If you use action buttons, be sure to test them on more than one computer.

Do you want to idle out? That is, do you want the menu to automatically jump somewhere else after a particular amount of time? If so, choose a long timeout period. Make sure the menu doesn't time out while the viewer is still reading it.

Menu Creation Menus are produced with a still or motion video background and a subpicture highlight overlay. The actual "art" of the buttons is usually in the video background, while the highlights are used to color or surround the currently selected button (see Figure 12.2). One of the tricky parts of menu creation is implementing the subpicture highlights. There are essentially three ways to create highlights: as a separate color-coded graphic, as layers in a Photoshop file, or directly in the authoring system.

Four highlight pixel types exist (see Chapter 6 for details). The background pixel type is always transparent for the video to show through the pixel. When using separate graphics, the background pixels are usually mapped to white by the authoring system. The other three pixel types can be used for highlight effects. These are usually mapped to blue, red, and black in separate graphics. The four pixels types have one set of colors and transparencies when the button is highlighted, and a separate set of colors and transparencies when the button is activated. This means you can use the same shape for highlighting and activating, but with different colors, or you can use different shapes by having the activation pixels be transparent when the button is highlighted, and the highlight pixels be transparent when the button is activated. Since activation lasts less than a second, you may decide it's not worth creating separate graphic designs.

The upshot is that you can use three colors (with three transparency levels) for button highlighting. Unless you do some clever antialiasing or you stick with rectangular designs, your highlight graphics will look rough and jaggy.

If you use the auto action feature of DVD to create "action buttons" or "24-bit rollover buttons," you must make a separate graphic for each highlighted button. That is, if the menu has 6 buttons, then you have to make 6 graphics, each one with a different button in its highlight state. Each version is actually a different menu page that is automatically jumped to when the user changes the button selection.

When you lay out a menu, consider interbutton navigation. The user has four directional arrow keys to use in jumping from button to button. Try to

Figure 12.2
Menu subpicture
highlight overlay

Menu video graphic

Subpicture graphic

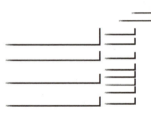

lay out the menu so that it's obvious which button will be selected when a particular arrow key is pressed. At the same time, don't forget that your disc inevitably will be played in a computer with a mouse interface, so don't rely on the user moving sequentially from button to button. In general, it's best to have button links wrap around. That is, when the user presses the down key on the bottom button, the top button should be selected. When the user presses the up arrow key on the top button, the bottom button should be selected, likewise for left and right movement. The reasoning is that it's better to have a keypress do something than nothing. The brief confusion experienced by novice users when the arrow keys wrap around is preferable to the angry frustration of users who expect arrows to wrap when they don't.

Tips and Tricks   The following tips will assist you in the process of creating a menu:

- The number one mistake DVD menu designers make is failing to differentiate between a highlighted and non-highlighted button. For example, the viewer sees a screen with two buttons; one is yellow, one is green. Which one is highlighted? If the user presses an arrow key, the yellow button turns green and the green button turns yellow. The user still can't tell which one is highlighted. This can even be confusing with more than two buttons if some of the buttons are different colors in their unhighlighted state. In many cases, it's better to put a box or a circle around a button, or have an icon or pointer appear next to it, rather than only changing the color or contrast.

- Make sure the highlighted color is different enough from the unhighlighted color so that button selections are visible. Do not rely on menu simulation in the computer, since colors can end up much closer on a TV.

▪ Each menu should prehighlight the button that the user is most likely to select. For example, if the main menu is shown before the feature, the "play" button on the menu should be preselected so the user can simply press "Enter" or "Play" to start. This is especially important since the "Play" key on most remote controls does the same thing as the "Enter" key when in a menu. Beginning DVD users will intuitively press the "Play" key when they reach the first menu.

▪ Author all menus so that the button that takes the viewer out of the menu is highlighted upon return. If a video segment or another menu jumps to a menu, the button that returns the viewer to where he or she just came from should be highlighted.

▪ Design your menus for numerical access. Most remote controls enable the user to press the 1 through 9 keys on a menu to directly choose a button. Arrange the buttons in proper numerical order. Be careful with invisible buttons. On a disc with several menus and complex navigation, consider putting a number next to each button.

▪ An introductory video sequence for the main menu will become a flaming irritant if the user has to sit through it every time they return to the menu. Use the intro for initial play only, then stifle it for later menu accesses.

▪ Motion transitions between menus can be enjoyable, but keep them short and sweet; never make transitions longer than two seconds. When creating the transitions, play them over and over and imagine yourself pressing a menu button each time the video loops. If it seems like a long time between button presses, then it is too long a time.

▪ If you have video in 16:9 anamorphic form, you will also want to make the menus work in 4:3 mode. One approach is to make all your menus in widescreen form, then use the built-in pan & scan feature to have the player crop the sides. Make sure that no button art is outside the area that will be cropped. You will need to make two sets of button highlight graphics, as explained in the subpicture section.

▪ Make sure the button highlight/selection rectangle is over the top of the button art in the background. If you put the rectangle somewhere else on the screen, the disc will still work with arrow keys, but the button will not work when a computer user clicks on it.

▪ Double check that the subpicture overlay aligns with the underlying background. If they don't match, the highlight will jump or be offset. See the graphics preparation section for details on making graphics match motion video.

■ Buttons are subpictures, and subpictures go away when the user presses fast forward or rewind. To avoid this problem with motion menus or with buttons over the top of video, disable the FF/REW user operations or refresh the subpictures every few seconds.

■ Temper creativity with good user interface design. Weigh functionality over aesthetics and experimentation. Frustrated users will head for the eject button and will never see your avant-garde designs.

Navigation Design

Navigation is one of the hardest aspects to get right. In addition to the confusion of "Title" keys, "Menu" keys, and the inconsistency of navigation on different discs, it's difficult to balance creativity and playfulness against ease of use.

When you create the flowchart, identify the effect of important menu keys at each menu. Clearly indicate where the viewer will go when they press "Top" ("Title"), "Menu," or "Return" ("GoUp"). If you have several menus, use a hierarchical design that the user will be able to intuitively understand. If you use more than three levels in your hierarchy, try combining menus to flatten the structure. Keep all selectable features and options as close to the main menu page as possible, with a minimal number of button selections. Make it as easy to navigate back up through menus as to navigate in the down direction.

Understand the difference between the "Top" ("Title") key and the "Menu" key. "Top" means "take me to the very top, to the table of contents." "Menu" means "take me to the most appropriate menu for where I am." For multi-level menu structures, the "Menu" key can be used to go up one level, or the "Return" key can accomplish this feature. A well-designed disc will program the "Return" ("GoUp") function to act somewhat like the back button on a Web browser. Each time the viewer presses "Return," they move up one level until they reach the top (see Figure 12.3). Every remote should include the three basic navigation keys: "Top" ("Title"), "Menu," and "Return." Take advantage of the intended function of each (see Figure 12.4).[1]

---

[1]Unfortunately, some player interfaces, especially on computers, leave out the Top (Title) key or the Return key. However, the incompetence of some player designers should not cause you to castrate your disc. The more discs there are that properly use the navigation keys, the sooner players will be fixed to provide them all. In the meantime, you can overcome the problem by providing on-screen buttons for navigation. See the following paragraph.

Figure 12.3
Example menu
structure

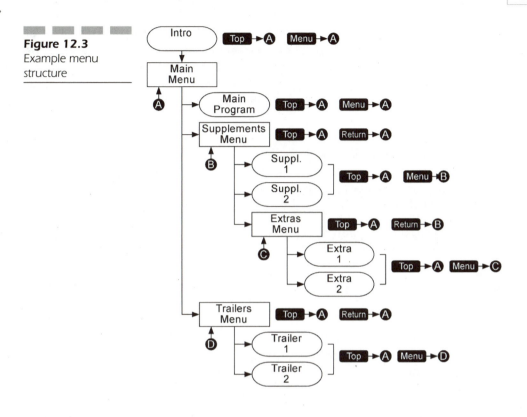

Figure 12.4
Basic archetypal
remote control

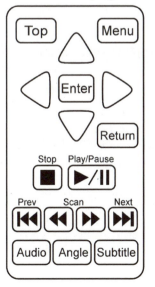

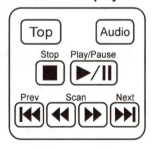

Always enable navigation with on-screen buttons as an alternative to remote control keys. Assume that the viewer has only the directional arrow keys and the "Enter" key, or assume that the viewer has only a mouse. Make sure that buttons labeled "back" or "main menu" or something similar are included to enable the viewer to navigate through all the menus without relying on any other keys.

Figure out where each chapter point will be. Anywhere the viewer might want to jump to should be a chapter point. Identify the chapter points by timecode on the storyboard or flowchart. Do this before the video is encoded, since each chapter entry point in the video must be encoded as an I-frame at a GOP header.

NOTE: *Rule of thumb: There should be no more than 10 minutes between chapters.*

Tips and Tricks   The following tips may assist your efforts to make user navigation easier and more clear.

■ If you only have one menu, make both the "Top" ("Title") key and "Menu" key go to it. Most discs do not have multiple titles on them, so the "Title" key should act like the "Menu" key.

■ Never disable the "Top" ("Title") or "Menu" keys. The viewer should always be able to leave the current video segment by pressing either of these keys.

■ Do not author the disc so that pressing the "Top" ("Title") key starts over and forces the viewer to watch logos and FBI warnings over again.

■ Make sure the "Next" and "Prev" keys always work. At minimum, they should move through chapters.

■ Consider placing a "How to use this disc" segment on every disc you make. It can explain the basics of navigation using the arrow keys, "Enter," "Top" ("Title"), "Menu," and "Return." Keep in mind that the "Top" button may be labeled "Title" or "Guide" or something else, and that the "Menu" button may be labeled "DVD Menu" or something else (see Chapter 6).

Balancing the Bit Budget

Bit budgeting is a critical step before you encode the audio and video. You must determine the data rate for each segment of the disc. If you underestimate the bit budget, your assets won't fit and you will have to re-encoded and re-author. If you overestimate the bit budget, you will waste space on your disc that could have been used for better quality encoding.

The idea is to list all the audio and video assets and determine how many of the total bits available on the disc can be allocated to each. Part of the early bit budgeting process involves balancing program length and video quality. Two axes of control exist: data rate and capacity. Within the resulting two-dimensional space, you must balance title length, picture quality, number of audio tracks, quality of audio tracks, number of camera angles, amount of additional footage for seamless branching, and other details. The data rate (in megabits per second) multiplied by the playing time (in seconds) gives the size (in megabits). At a desired level of quality, a range of data rates will determine the size of the program. If the size is too big, reduce the amount of video, reduce the data rate (and thus the quality), or move up to a bigger capacity disc. Once you have determined the disc size and the total playing time, maximize the data rate to fill the disc, which will provide the best quality within the other constraints. See Figure 12.5 to get a general idea of needed disc sizes. A DVD-5 holds 1 to 9 hours of video or 2 to 160 hours of audio. A DVD-9 holds 2 to 16 hours video or 3 to 300 hours of audio. Of course the video quality at high-end playing times is much lower than what is expected from DVD.

The easiest way to make a bit budget is to use a spreadsheet. One is included on the sample disc that comes with this book. For simple projects with only one video segment, use the DVDCalc spreadsheet on the sample disc. Some authoring programs will calculate bit budgets for you when you use their layout features.

To simplify calculations, keep track of sizes in megabits, rather than megabytes (see Table 12.3). Allow an overhead of 3 to 4 percent for control data and backup files which are added during formatting and multiplexing. This also allows a bit of breathing room to make sure everything fits on the disc.

In addition to fitting the assets into the total space on the disc, also make sure that the combined data rate of the streams in a video program do not exceed the maximum data rate (or *instantaneous bit rate*) of 10.08 Mbps.

Figure 12.5
Data rate vs. capacity

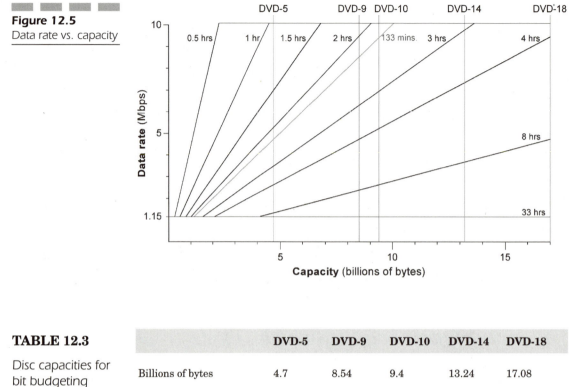

TABLE 12.3

Disc capacities for
bit budgeting

	DVD-5	DVD-9	DVD-10	DVD-14	DVD-18
Billions of bytes	4.7	8.54	9.4	13.24	17.08
Megabits	37,600	68,320	75,200	105,920	136,640
4% overhead	1,504	2,733	3,008	4,237	5,466
Adjusted capacity	36,096	65,587	72,192	101,683	131,174

Subtract the audio data rates from the maximum to get the remaining data rate for video. Lowering the maximum data rate reduces the ability of a variable bit rate encoder to allocate extra bits when needed to maintain quality.

Example A typical bit budgeting process for a disc with a single main video program, plus a few ancillary segments, goes as follows (see Table 12.4). Calculate the total bit rate and size of the main program's audio tracks and subpicture tracks. If there are additional video segments such as animated logos, motion menus, and previews, calculate their sizes (including audio and subpictures), and add them to the total size of the main program's audio and subpicture tracks. Still logos and copyright

TABLE 12.4

Sample bit budget

Element	Time	Data rate	Size
English audio track (5.1)	110 minutes	0.448 Mbps	110×60×0.448 = 2,957 Mbits
English audio track (2.0)	110 minutes	0.192 Mbps	110×60×0.192 = 1267 Mbits
Spanish audio track (5.1)	110 minutes	0.448 Mbps	110×60×0.448 = 2,957 Mbits
4 subpicture tracks	110 minutes	0.010 Mbps	4×110×60×0.01 = 264 Mbits
Total (simultaneous with movie)	—	1.128 Mbps	—
2 motion menus and 3 transitions	48 seconds	8 Mbps (incl. audio)	48×8 = 384 Mbits
Intro logos	32 seconds	8 Mbps (incl. audio)	32×8 = 192 Mbits
Movie preview	2 minutes	4.5 Mbps	2×60×4.5 = 540 Mbits
Audio for preview	2 minutes	0.192 Mbps	2×60×0.192 = 23 Mbits
Interview	3 minutes	4.5 Mbps	2×60×4.5 = 810 Mbits
Audio for interview	3 minutes	0.192 Mbps	2×60×0.192 = 35 Mbits
Total (non-movie elements)	—	—	9429 Mbits
Space remaining for movie	—	—	36,096*-9429 = 26,667 Mbits
Movie	110 minutes	Avg: 26,667/(110×60) = 4.04 Mbps Max: 10.08-1.128 = 8.95 Mbps	

*From Table 12.3.

warnings should be authored as timed stills, which have no impact on disc space. Subtract the total size of ancillary material from the disc capacity minus overhead. This gives remaining capacity for the main program. Calculate the average video bit rate for the main program by dividing the remaining capacity by the running time. Also calculate the maximum video bit rate by subtracting the combined audio and subpicture bit rate from 10.08. These numbers will be fed into the encoder when compressing the video for the main program. Note that minor elements such as subpicture tracks and motion menus are usually so small that they can often be omitted from calculations. If they were left off of the example in Table 12.4, the average data rate would come out to 4.17 Mbps, which is close enough.

An alternative bit budgeting approach is to divide the disc capacity (in bits) by the total running time of all segments (in seconds) to get the average data rate (in Mbps). Adjust up or down for quality of different programs as needed. For each program, subtract the combined data rate of its audio and subpicture streams to get video rate. It's a good idea to follow up by summing the sizes of each segment (including audio) to check that the total doesn't exceed disc capacity.

Tips and Tricks   The following list of tips will help you in balancing the bit budget:

- Don't forget motion menus and menu transitions. Don't forget to leave room for any added PC content. A few programs or HTML files are not worth worrying about, but large multimedia applications and installer packages need to be accounted for.

- Each camera angle increases the data; two camera angles double it, three angles triple it, and so on. Reduce the maximum data rate according to the number of angles (see Table 6.34).

- If you have two or more programs that contain much of the same data, consider using the seamless branching feature to combine the duplicated segments.

- Still images take up so little space that you don't need to include them in your calculations. Unless you are including more than four subpictures, you may wish to omit subpictures from the calculations, since the space they take up is negligible.

Asset Preparation

A typical project requires dozens of assets from a variety of sources. A complicated project may require thousands (see Table 12.5). Each asset must be properly prepared before it can be fed into the DVD authoring process. The following sections cover the details for preparing each type of asset.

Preparing Video Assets

The final quality of DVD video is primarily dependent on three things: the condition and quality of the source material, the visual character of the material, and the data rate allocated to the video during encoding. The better the quality of the source, the better job the encoder can do. Always use the best quality source. Insist on a digital copy whenever possible, such as D1, or Digital Betacam. Beta SP will do in a pinch, but in many ways it is inferior to DVD's capabilities.

	Type	Assets	Tasks
TABLE 12.5	Video	Source video tapes (movie, trailer, supplements)	Digitize, clean up, encode
Typical project assets		Already encoded video (logos, computer graphics)	Check
	Audio	Original language source tape	Digitize, synchronize
		Foreign language source tapes	Digitize, check length, synchronize
		Commentary or other supplemental audio sources	Digitize, synchronize
		Already encoded audio	Check synchronization
	Graphics	Menu graphics	Create/integrate graphics, create button highlights
		Supplemental graphics	Create/integrate graphics, create button highlights
		Production stills and other photos	Digitize, correct color, create button highlights
	Subtitles	Text files	Add timecodes
		Graphic files	Create timecode-filename mapping file

If the video is of poor quality or contains noise, use *digital video noise reduction* (DVNR) to improve it. Noise is random, high-entropy information, which is antithetical to MPEG-2 encoding. Efficient encoding depends on reducing redundant information, which is obscured by noise. Noise can come from grainy film, dust, scratches, video snow, tape dropouts, cross-color from video decoders, satellite impulse noise, and other sources. Complex video, including high detail and high noise levels, can be dealt with in two ways: increase the data rate, or decrease the complexity. The DVNR process compares the video across multiple frames and removes random noise. After DVNR, the video compresses better, resulting in better quality at the same data rate.

If possible, keep the production path short and digital. A frequent irony of the process is that video is edited on a computer, then output to tape in order to be encoded back into a computer file. If you use a *non-linear digital editing system* (NLE), look into options for directly outputting DVD-compliant MPEG-2 streams. Anticipate the need for aspect ratio conversion. Make sure the various sources are in the proper aspect ratio (4:3 or 16:9), especially for assets that need to be edited together.

Make sure all the video on a disc (or on a single side) uses the same video standard. Be prepared to convert from PAL to NTSC or vice versa. Good-quality conversion is difficult. Unless you will be converting hundreds of clips, it's better to find a video production house with the equipment and expertise to do it correctly. Be aware that the audio must be conformed to match (see the following section).

You may need to choose between *variable bit rate encoding* (VBR) or *constant bit rate encoding* (CBR). VBR doesn't directly improve the quality, it only means that quality can be maintained at lower average bit rates. For best quality of long-playing video, VBR is the only choice. For short video segments, CBR set at or near the same data rate of VBR will produce similar results with less work and lower cost. For VBR encoding, set the average and the maximum bit rate as determined in the bit budgeting step.

Tips and Tricks   The following tips will help you in preparing video assets:

- The display rate must be 25 frames per second for 625/50 (PAL/SECAM) video or 29.97 frames per second for 525/60 (NTSC) video, not 30 frames per second. This is especially important when using NLE systems that allow many different frame rates. Many authoring tools will reject streams with non-compliant frame rates. Note that the display frame rate can be different from the source frame rate. Twenty-four frame-per-second film can be encoded to display at 29.97 frames per second.

- Ensure that the encoder can produce DVD-compliant streams, with proper GOP size, display rates, and so on. Also look for an encoder that can do *segment-based re-encoding*, which enables you to re-encode small segments as needed, rather than reprocessing hours of video to correct a single, small problem.

- If you are working with film transferred to video, make sure the encoder can do inverse telecine.

- Chapters and programs must start at a sequence header or GOP header, and the GOP must be closed, not open. Each cell must also start at a sequence header. If you have very precise points for chapter marks, make sure they are encoded properly, otherwise the authoring software may move the chapter point to the nearest I frame.

- In general, the easiest way to put still video on a disc is to author each still as a menu.

- Static video, where the picture doesn't change over time, such as FBI warnings, logos, and so on, should be authored as a timed still. Not only does this look better, with no movement or variation when played back, it can save space on the disc.

- Make sure the video does not have burned-in subtitles. Use the DVD subpicture feature instead, particularly for multilingual discs.

- If you want to pack a large amount of video on a disc, don't try to encode to MPEG-2 under 2 Mbps. Use MPEG-1 between 1 and 1.86 Mbps; it will look better than MPEG-2 at the same data rates. You could also use "half D1" resolution (352×480) at rates of about 1.5 to 3 Mbps, which can look surprisingly good. Be aware that although half D1 video is mandatory in the DVD-Video specification, it will cause problems on a few players. Before encoding the video, use heavy *digital video noise reduction* (DVNR) and liberal application of blurring filters in video processing programs. These steps reduce the high-frequency detail in the video which makes MPEG encoding more efficient. Also try reducing the color depth from 24 bits to something between 16 bits and 23 bits. Do not use a dither process which increases high-frequency detail. This can help with encoding efficiency and can reduce posterization artifacts.

- If video quality is a top consideration, create or request your content in progressive format if at all possible: shoot on film or use a progressive digital video camera. If your video assets come from an interlaced source, consider converting to progressive form before encoding. Depending on the nature of the video, progressive conversion may cause artifacts that unacceptably degrade the video. For some types of interlaced content, however, a high-quality progressive conversion will result in a DVD that looks significantly better when played on progressive DVD players and computers. See Chapter 3 for more on progressive video.

Preparing Animation and Composited Video

Video produced on a computer, or camera-source video that is composited with computer graphics, has its own set of issues. The most important thing to keep in mind is avoiding too much detail, especially vertical detail, which wreaks havoc on interlaced TVs. Avoid thin horizontal lines; use the anti-aliasing and motion blur features of your animation or editing software whenever possible. Use a low-pass or blur filter, but don't overdo it because too much blurring reduces resolution on progressive-scan players and computers.

For 525-line (NTSC) video, render animations at 720×540 and scale down to 720×480 before encoding. For 625-line (PAL) video, render animations at 768×576 and scale down to 720×576 before encoding, or if your software supports it, render with D1 or DV pixel geometry.

If at all possible, do not use the "print to video" or similar TV video rendering feature of your software, which will almost always result in interlaced output. Instead, render at 24 fps progressive or 30 fps progressive, then encode for 29.97 fps display. This requires an encoder that can handle progressive video input files and perform "synthetic telecine" to add flags to the MPEG-2 stream for 2-3 pulldown in the player (see the following graphics preparation section for more details).

Preparing Audio Assets

Contrary to most expectations, audio is usually just as complex as video, and often more complex. You must deal with multiple tracks, synchronization, audio levels across the disc, and so on. As with video, the better quality the source, the better the results after encoding. Audio assets usually come on *modular digital multitrack* (MDM) formats such as Tascam DA-88, or you may receive PCM (WAV) files on a CD-R, or already-encoded Dolby Digital or DTS files. With multitrack tape source, make sure the track-to-channel assignments are correct. Dolby has specified that the sequence is L, R, C, LFE, Ls, Rs, (Lt, Rt), but not all production houses follow this standard.

Capture your audio mixes at the highest sample rate and word size possible. If someone else is doing the audio mixing, request that they do the same. As recording and mastering studios acquire the new equipment needed to work with high-resolution audio, 96 kHz 24 bit audio will become commonplace. For DVD-Audio or for audio-oriented DVD-Video, it's usually not a good idea to add video to a 96 kHz 24 bit PCM track due to little remaining headroom.

When converting from NTSC to PAL, or transferring from film to PAL, the audio must be conformed to match the changed video length with a 4 percent speedup, and preferably a pitch shift to restore proper pitch. If the audio is from a CD, it must be upsampled from 44.1 kHz to 48 kHz for DVD-Video. Many audio files produced on PCs are also at 44.1 kHz. DVD-Audio supports 44.1 kHz sampling, unless a Dolby Digital version is being added for compatibility, in which case the audio must be converted to 48 kHz before encoding.

Do all necessary processing before encoding: clean up and sweeten the audio, re-mix and equalize, adjust phase, adjust speed and pitch, upsample or downsample, and so on. Audio levels can be adjusted before encoding, or for Dolby Digital, they can be adjusted during encoding by using the dialog normalization feature. If the mix was done for the theater, re-equalize for the home environment. Establish consistent audio levels and equalization throughout all audio on the disc, including menus and supplements. This is especially important with multiple audio tracks from various sources, since the viewer will be able to jump at will between them, and will be bothered by inconsistencies in level or harmonic content. Resist the temptation of locking out the audio key on the remote control, which will only annoy your viewers. Instead, get all sources in PCM format so they can be matched and harmonized before being encoded.

Tips and Tricks   The following tips will assist you in preparing the audio assets:

- Make sure timecode is included with the audio so that it can be synchronized with the video.
- It's almost always a good idea to mute the audio for a second or so at the beginning of each segment. This ensures that the player (or receiver) doesn't clip the audio as it begins the decoding process.
- If you are doing 5.1-channel Dolby Digital encoding, or using multichannel DVD-Audio PCM tracks, test the downmix.
- If you do your own encoding, request the *Dolby Digital Professional Encoding Manual* from Dolby Labs.

The Zen of Subwoofers   If you understand the proper use of the LFE channel, you belong to an elite minority. As Dolby Labs has pointed out over and over, apparently with little success, LFE does not equal subwoofer. All 5 channels of Dolby Digital are full range which means they are just as capable of carrying bass as the LFE channel. Most theaters have full-range speakers. They don't need separate subwoofers since most of the speakers have a built in subwoofer. In the home, it's the responsibility of the receiver or the subwoofer crossover to allocate bass frequencies to the subwoofer speaker. This is not the responsibility of the DVD player (unless it has a built-in 5.1-channel decoder) or of the engineer mixing the audio in the studio.

The LFE channel is heavily overused in most audio mixes. A case in point is a certain famous movie released on laserdisc with Dolby Digital 5.1 tracks in which the booming bass footsteps of large dinosaurs were placed exclusively in the LFE channel. When downmixed for two-channel audio systems, the bass was discarded by the decoder, leaving mincing, tiptoeing dinosaurs; this was not the fault of downmixing or the decoder. It could have been fixed by a separate 2-channel mix, but it also could have been fixed with a proper 5.1 mix or a proper 5.0 mix. Had the footsteps been left in the main 5 channels, which are all full range, the bass effects would have come through properly on all home systems.

Dolby Digital decoders ignore the LFE channel when downmixing. This is because the typical stereo or 4-channel surround system doesn't have a subwoofer. Again, this does not mean that the LFE channel represents the subwoofer, only that the extra "oomph" intended to be supplied in the LFE channel is inappropriate and could muddy the audio. Dolby Digital decoders all have built-in bass management, which directs low-frequency signals from all six channels to the subwoofer, if one exists. Therefore, the LFE channel should be reserved only for added emphasis to very low frequency effects such as explosions and jets.[2]

As illustrated by the tiptoeing dinosaurs, moving low-frequency audio to the LFE channel does not make the soundtrack better in the home environment. This mistaken approach is what has caused complaints about the LFE channel being omitted in downmixing. Nothing important in a film soundtrack, including low-frequency audio, should ever be moved out of the main 5 channels. Isolating low-frequency audio in the LFE channel does not remove any sort of burden from the center and main speakers since the built-in bass management already does this.

It's possible to mix all "subwoofer sounds" in Dolby Digital 5.0 without an LFE channel at all. Some theatrical releases are mixed this way. Such a mix will play the same in 5.1-channel home theaters as in cinemas, since the decoder and receiver automatically route the bass to the speakers that can best reproduce it. The .1 channel is present only for an extra kick that the audio engineer might decide should only appear in full 5.1 audio systems.

Slipping Synchronization   An occasional problem with DVD is lack of proper synchronization between the audio and the video (see Chapter 7 for more detail). Players are partly to blame for the problem, but things can be done during production to keep synchronization tight.

Check the timecodes on the audio to make sure they are correct. A handy trick for testing audio sync, especially when no timecodes exist and it has to be aligned by hand, is to bump the audio track forward several frames, check playback, bump it back several frames from the original position, and check playback again. If moving it one direction had little or no effect, while

---

[2]Roger Dressler, Technical Director for Dolby, has remarked that the LFE channel should be renamed to the "explosions and special effects" channel in order to avoid confusion.

moving it in the other direction had a very big effect, chances are that the original position was not optimal.[3] Make sure the master print was not incorrectly dubbed before being transferred to video.

If sync steadily worsens, check to see if non-drop timecode is being used on the audio master. The difference between 29.97 and 30 fps will slowly move the audio out of sync with the video. Also make sure the film chain in the telecine machine is run at the proper video speed. If sync suddenly changes, suspect a slip during reel change in the telecine process or an edit in the NLE system independent of the audio.

Preparing Subpictures

Subpictures are used for two things: graphic overlays (usually subtitles) and menu highlighting. Menus are dealt with in a following section. The possible uses for subpicture overlays are endless. They can expand and alter existing video, emphasize or de-emphasize regions on the screen, cover over or censor, tint the video a different color, add a logo bug, and so on.

A handy feature of subpictures is that they can be forced to appear under program control. This is especially useful for multilanguage discs since the video can be kept clean of all subtitles, but they can be made to appear in the appropriate language during normal playback of the disc.

Subpictures can be used to create limited animation overlaying the video. A subpicture rate of 15 per second is the practical limit for many players, although anything faster than 2 per second will cause problems on some players. Lowering the maximum data rate for the other streams can help avoid problems.

Effects such as fade, wipe, and crawl can be performed by the player. A good authoring system will be able to add the subpicture display commands needed to create these effects.

Tips and Tricks   The following tips will assist you in preparing subpictures:

■ Subpictures cannot cross chapter or program points.

■ The size of a subpicture frame is limited, but because subpictures are RLE compressed, it's difficult to determine the size ahead of time. It's easiest to let the authoring software flag subpictures that are too big.

---

[3]When using this trick, keep in mind that synchronization problems in audio delayed behind the video are less perceptible than in audio that precedes the video.

Subpictures for Anamorphic Video Subpictures accompanying anamorphic video are not cropped or scaled by the player. This means that separate subpictures must be created for each allowed display mode (widescreen, letterbox, and pan & scan). The basic steps are as follows:

1. Start out with a 16:9 overlay in square-pixel 854×480 resolution for NTSC, 1024×576 for PAL.

2. For widescreen mode, resize the subpicture to 720×480 for NTSC or 720×576 for PAL. This squeezes in the horizontal direction to make an anamorphic version of the subpicture.

3. For letterbox mode, add transparent (white) pixels to the top and bottom to make the image 854×640 for NTSC or 1024×768 for PAL. You can do this in PhotoShop by changing the canvas size. This accounts for the letterbox mattes that are added by the player. Resize the image to 720×480 for NTSC or 720×576 for PAL.

4. For pan & scan mode, crop the sides of the picture to make it 720×540 for NTSC or 768×576 for PAL. You can do this in PhotoShop by changing the canvas size. Then resize the image to 720×480 for NTSC or 720×576 for PAL.

Subtitles Subtitling, particularly creating foreign language subtitles, is a tricky job best given to a subtitling house. Provide the subtitling service with a master copy that has clear audio and *burnt-in timecode* (BITC) that exactly matches the DVD video master. This is critical for frame-accurate subtitling work; timing is very important.

You may wish to request that the subtitle source be provided in text form rather than graphic form since text is much easier to change, but text subtitles must then be turned into graphics.

If you are creating your own subtitles, consider the following:

- Don't use serif fonts. Sans serif fonts are easier to read on a video screen, especially since the serifs tend to create single-pixel horizontal lines that produce interlaced twittering.

- Use a drop shadow or an outline, either black or translucent (using the subpicture transparency feature).

- Each new subtitle should appear at the scene change, not at the dialog start. This provides more on-screen time, and studies show that it reduces reading fatigue, even though it's annoying to hearing users who have turned on the subtitles to accompany the audio.

- Center the subtitles. If there is more than one line, left-justify the lines within the centered position (or right-justify the lines for right-to-left writing systems). Justifying the lines makes them easier to scan.

- Try to limit titles to two lines. Keep the first line shorter, which covers less video and clues in the eye that there's another line.

- If there is more than one speaker, indicate speaker change with a hyphen at the beginning of the line.

- Display subtitles for no less than 1 second and no longer than 7 seconds.

- Keep in mind that jumping into the middle of a scene, or using fast forward or rewind, will not refresh a subtitle. Subtitles with long periods will take a long time to appear. To avoid this problem, you may wish to repeat the same subtitle at intervals of 5 to 10 seconds.

- Proofread! With thousands of subtititles, especially in foreign languages, it's easy for misspellings to occur.

Preparing Graphics

Graphics are generally used for menus. They are also used for slideshow sequences such as photo galleries and information pages, although these are usually implemented as menus. Graphics can be tricky, since the traditional tools for creating computer graphics are not always well suited for television graphics.

Colors   Graphics should be created in 24-bit mode. Choose RGB colorspace (or YUV colorspace, if available) rather than CMYK or other print-oriented colorspace. The file format (TIFF, BMP, and so on) doesn't really matter, as long as it's one that your authoring system can import. Avoid JPEG file format, since JPEG compression can cause artifacts that may be compounded by MPEG compression. If you must use JPEG, choose the highest quality level.

Use NTSC-safe colors. Overlay satured colors, especially bright reds and yellows, will bleed on an NTSC television. Keep the saturation of all colors below 90 percent. Another way to accomplish the same thing is to keep RGB values below 230. Some graphics programs have an NTSC color filter that will tone down colors that are out of the NTSC-safe gamut. Use the filter on your graphics until you get a feel for the colors. Certain color combinations don't mesh in NTSC colorspace. For example, light blue next to peach will usually bloom badly. The only way to avoid problems like this is to check the graphics on an NTSC monitor.

Image Dimensions   The pixel density of the image doesn't matter. Whether you use 72 dpi or 96 dpi or even 300 dpi, what matters is the final pixel dimensions of the output. Pixel geometry does matter. Most computer graphics programs use square pixels. DVD is based on the ITU-R BT.601 formats that do not use square pixels. In some cases you don't need to worry about the difference, but in other cases, such as when the graphics have squares or circles, you need to account for pixel aspect ratio changes or else you will get rectangles and ovals. There's a 9 percent vertical distortion for NTSC, and a 9 percent horizontal distortion for PAL.[4]

To account for pixel geometry differences, create 525-line (NTSC) video graphics at 720×540 and then scale them down to 720×480 before encoding. Create 625-line (PAL) video graphics at 768×576 and scale them down to 720×576 before encoding. Do not constrain proportions when scaling (uncheck the link icon in Photoshop). The resulting picture will look slightly squashed—this is normal. Some authoring software will automatically scale the video down from larger sizes. If so (and if the scaling algorithm produces good quality results), this will save a step. Try to avoid scaling up from 640×480, since this will make the picture fuzzy.

For 16:9 anamorphic video, create 525-line (NTSC) video graphics at 854×480 or 960×540, then scale to 720×480. The 960×540 size is useful if you plan on cropping the graphics to 4:3. Create 625-line (PAL) video graphics at 1024×576, then scale to 720×576.

If your graphics production software gives you the option of using video pixel aspect ratios, consider using it, but keep in mind that images on the computer screen may appear distorted. Choose DV, which matches DVD. If DV is not an option, then choose D1, but be aware that you may have to crop 6 lines to match DVD picture dimensions.[5] Some graphics software will also let you set the picture aspect ratio. Choose 4:3 (1.33) or 16:9 (1.78).

One way to avoid all this is to simply create the graphics at native DVD resolution. If you're working with text or with images that look ok when

---

[4]You may be wondering why the distortion isn't 12.5 percent for NTSC (540/480) and 6.7 percent for PAL (768/720). Technically, these are the correct differences in raw pixel dimensions, but the pixel aspect ratios for MPEG-2 video also take into account the horizontal overscan ratio. So the visual distortion is 9 percent (see Table 6.22).

[5]DV and D1 video formats use the same pixel aspect ratio, but D1 uses 486 lines for 525-line (NTSC) video, while DV uses 480.

slightly distorted, then don't bother with resizing. This has the advantage of avoiding artifacts caused by downscaling.

If you need to match graphics to video, such as for menu transitions or motion menu subpicture overlays, then you will usually need to slightly alter the scaling process above for 525-line (NTSC) video. The exact details depend on your video source and your MPEG-2 video encoder. Video from D1 tape has pixel dimensions of 720×486. To conform to MPEG-2 dimensions of 720×480, some encoders scale and some encoders crop the extra six lines; however, no consistent approach exists. Some crop 3 from the top and 3 from the bottom, others crop 6 from the bottom, some crop 2 and 4, and so on.[6] You will need to do some research or testing of your encoder to find out what it does. Then, instead of scaling to 720×480, scale to 720×486 and then crop the picture the same way your video encoder does. This will ensure that everything lines up properly when transitioning from graphics to video or vice versa. Another approach is to render graphics as one-second stills at the end of the motion video. After encoding, you can extract the first I-frame of the still sequence and use it for the menu graphic. Since it was encoded along with the video, it will match perfectly.

Safe Areas   Most televisions employ overscan, which covers the edges of the picture. Overscan can hide as much as 10 percent of the picture. Two common "safe areas" are used to account for this. The *action-safe* area defines a 5 percent boundary as a guideline for the area where action or important video content should be kept. The *title-safe* area defines a 10 percent boundary as a guideline for the area in which text and other vital information should be kept (see Figure 12.6). The DVD Demystified sample disc includes safe area templates. Some programs have built in video safe templates. In After Effects, for example, press the apostrophe key to turn them on and off. Most modern TVs don't have more than 5 percent overscan, but older or cheaper models may approach 10 percent. Computers, LCD or plasma screens, and many video projectors have little or no overscan.

Video Artifacts   Because there are one billion interlaced televisions in the world that your DVD video might be displayed on, you must consider interlace artifacts. The fundamental problem is that since only every other horizontal scan line is displayed at a time, thin horizontal lines disappear

---

[6]Cropping an odd number of lines (1, 3, or 5) is a bad thing for video, since it changes field dominance and can cause interlacing artifacts. For stills, field dominance is irrelevant.

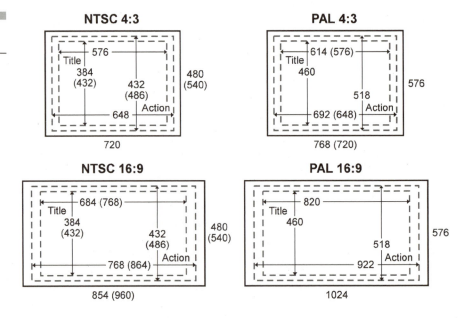

and reappear 60 times a second. This interlaced *twitter* effect (also called *flicker*) can be especially bad with computer-generated graphics. In addition to interlace artifacts, chroma crawl and color crosstalk from composite video signals make thin lines or sharp color transitions problematic, despite their orientation.

To reduce these effects, use the antialiasing and feathering features of your graphics software whenever possible. Make sure that all horizontal lines are at least two pixels thick. To be even more careful, keep the thickness of horizontal lines at multiples of two pixels.[7] Use gradual color transitions instead of sharp contrasts between dark and light colors. To adapt existing graphics, or as a final pass before encoding, apply a blur filter to the entire image or to offending areas. For example, a Gaussian blur of 0.5 to 1 has little visual effect but helps to reduce video artifacts. Also, because blurring reduces high-frequency detail, the MPEG compression will be more efficient, possibly resulting in higher quality.

---

[7]This advice only applies when you are working at native NTSC video resolution of 480 lines. If you create graphics at higher resolutions, the number of pixels will change when the graphic is scaled down. This doesn't apply to 625-line (PAL) video, since it's generally scaled horizontally, not vertically.

Text is also affected by interlacing and other video artifacts. Small point sizes produce single-pixel lines, which are bad, especially the horizontal lines created by serifs. Small text sizes are also difficult to read from standard viewing distances. The smallest object that normal human vision can discern subtends 1 minute of arc on the retina. Studies have shown that for legibility the height of a lower case character must subtend at least 9 or 10 minutes of arc; more as the viewer moves off axis. ANSI standards recommend 20 to 22 minutes of arc, with a minimum of 16 minutes. (See Chapter 9 for more information on viewing distances and resolution.) Angular measurements are independent of screen size. This means, for example, that at a viewing distance of 5 feet, 21 minutes of arc equates to 0.37 inches, while at a viewing distance of 15 feet, 21 minutes of arc results in 1.10 inches. As a rule, stick with 24-point or larger sizes. 14-point, bold text should be the absolute minimum. Use antialiased text whenever possible.

Tips and Tricks Here are a few tips that will help you in preparing graphics. Table 12.6 also provides some guidelines to follow.

- Proofread all graphics before importing them into the authoring system. Don't rely on the graphic artist.
- If you work with clients, have them sign off on every graphic and menu page before beginning authoring.

Putting It All Together (Authoring)

Once all the assets are prepared and proofed, bring them into the authoring system and knit them all together. A variety of authoring paradigms are available. Some systems stick close to the DVD specification, while others try to hide terminology and details. Most provide an onscreen flowchart, timeline, or combination of both. Large production environments may benefit from networked authoring workstations, where each station focuses on a specific task (encoding, menu creation, asset integration, simulation, formatting, and so on) and the project data flows from station to station over a high-speed network.

At the authoring stage, specify basic parameters of the volume such as video format and disc size. Import audio and video, synchronize them, and add chapter points. Then, apportion titles and title sets. Create menus by importing graphics and subpictures, link buttons together for directional highlighting, and link buttons to video segments or to other menus. Those

TABLE 12.6		525-line video (NTSC)	625-line video (PAL/SECAM)
Video graphics checklist	Pixel geometry	Create at 720×540, resize to 720×480. (For 16:9 anamorphic, create at 854×480 or 960×540, resize to 720×480.) Or create at 720×480 if vertical distortion is not a problem.	Create at 768×576, resize to 720×576. (For 16:9 anamorphic, create at 1024×576, resize to 720×576.) Or create at 720×576 if horizontal distortion is not a problem.
	Colors	No bright reds or yellows. Keep saturation below 90 percent or RGB values below 230.	Not an issue.
	Safe areas	Keep text and important detail about 70 pixels from the sides, and about 50 pixels from the top and bottom.	Keep text and important detail about 70 pixels from the sides, and about 56 pixels from the top and bottom.
	Small detail	Avoid single-pixel lines, especially horizontal. Use antialising and blurring. Use feathering and color gradations instead of sharp transitions.	
	Text	14 points minimum. Avoid serifs. Antialias.	

are the essential authoring tasks. More complex projects require much more detailed work.

Once everything is imported and connected, simulate the final product. Good authoring systems can show menus, video, and audio on the fly, giving a reasonable facsimile of how the disc will look in a player. Take advantage of the simulation feature to test navigation and overall layout before going any farther.

The authoring system is responsible for verifying compliance with the DVD specification. The better the authoring system, the better a job it will do of warning you when you have violated the spec. Better authoring systems will also provide information or warnings from the "reality spec"— they will warn you about things that are allowed by the spec but may cause problems on certain players. Some authoring systems also implement behind-the-scenes workarounds to avoid known player deficiencies.

Region coding and CSS or CPPM flags can be added in the authoring system. In general, the actual encryption is done by the replicator. This is the best option since the replicator will have already executed a license and have been given sets of disc keys and media key blocks.

Tips and Tricks The following tips will help you with the authoring process:

- Make sure you have more hard disk storage space than you think you could possibly need; DVD assets take up huge amounts of space. The first time you have to suspend a project because of a bottleneck and start work on a different project, you will either be driven mad by the amount of time it takes to offload a project to tape, or you will be glad you have the disk space needed to work on another project on the same station.

- If you must do a layer change in the middle of the video, look for slow-moving pictures, a still, or a fade to black. The secret to a smooth transition is low data rate, which gives more time to the DVD player to refocus on the second layer before the buffer underflows. At the encoding stage, make the data rate as low as possible for a few seconds going into the layer change, as well as a few seconds coming out of the layer change.

Formatting and Output

Once the authoring process is complete and everything works in simulation, it's time to create the final disc image. The authoring system will multiplex all the assets together, adding navigation and control information according to the DVD specification, and it will write out a special set of .VOB, .AOB, .IFO, and .BUP files in a VIDEO_TS directory (for DVD-Video or DVD-Audio discs with video) or an AUDIO_TS directory (for DVD-Audio-only discs). You will often end up going through this step more than once after finding a problem during emulation or when testing with a DVD-R test disc.

The data files can be written to a hard drive. This is useful for emulation testing, where you use a computer software player to test for problems that don't show up when simulating the disc in the authoring system. The data files can be written to DVD-R (or other writable DVD media) using the UDF bridge format. This creates a fully functional DVD disc that can be tested on standard players and other computers.

Once everything is working perfectly, the output can be directed to *digital linear tape* (DLT). The disc image is written as an ANSI tape file, along with a DDP control file that includes additional information to be used by the replicator. All DVD replicators accept compact DLT type III or type IV in DDP 2.0 format. Some replicators will also accept DVD-Rs (which they usually turn around and write to DLT). Discs to be protected with CSS or CPPM must be submitted on DLT, since DVD-Rs have only 2048-byte sectors without extra header information. A dual layer disc takes two DLTs. Only one layer can be written per tape.

Testing and Quality Control

Testing is the most important factor in the success of a project. If you work with clients, quality control is vital to your reputation. Testing should never be left to the end; it must be an ongoing process from the beginning.

The first testing stage is source asset checking. Never assume that your supplier will get it right. QC all incoming assets, at least until you have verified which suppliers you can rely on to always get it right. Although it seems like a redundancy and drain on time and resources, it will save you money and time in the long run. Screen all incoming assets for quality and accuracy. View every video tape and encoded video file. Audition all audio tapes and audio files. View every graphic and subpicture file, checking for spelling mistakes and missing elements. If you can't check foreign language subtitles, make sure the supplier guarantees accuracy and takes responsibility for costs associated with reauthoring the disc in case of subtitle errors. Check timecodes and synchronization between audio and video and between subtitles and video.

Review video and audio during and after compression. Check encoded audio against the master to spot sync, level, and equalization problems. Check all text before it goes to the graphic artist. Provide text to the graphic artist as text files that they copy and paste (or otherwise import) into the graphics application. Do not let them type text—they always misspell. If you work with clients, have the client proof and sign off on all the prepared assets (video, audio, and graphics) before authoring. Check navigation, layout, and general functionality with the simulation features of the authoring system. Don't worry about audio and video quality checking at this point, since simulation is not accurate enough.

After outputting formatted files, do emulation testing with computer software players. At this point, check whether the computer has reliable playback, and QC video and audio, especially synchronization. Since the computer displays the video in progressive mode, this is a good time to check for unanticipated effects of progressive display. Also be sure to check menu functionality, since the mouse-based user interface changes the way things work. Emulation testing can catch errors before you spend the time and expense of writing a DVD-R for further testing.

After emulation testing, output to DVD-R for testing on standard players and on other PCs. Test as many players as you can, and test with as many of the popular PC hardware and software decoders as you can. However, keep in mind that DVD-Rs don't play properly in all players, and can even cause spurious errors. Reserve some of this *matrix testing* time for replicated check discs. Consider sending the DVD-R to a verification service or running it through verification software or hardware. If the video has Closed Captions, connect a player to a TV with a Closed Caption decoder and make sure they appear correctly.

When you are sure everything works, send the DLT off to the replicator and request check discs. Check discs are a short replication run from the stamping master. A check disc is essentially the same as a final mass-produced disc. If you approve the check disc, the replicator can use the existing master for the full replication run. Most replicators provide one round of check discs for free as part of their replication service. However, they may charge a fee if you find errors and have to send in a new DLT for another set of check discs. Unless you have thoroughly tested with DVD-Rs and are comfortable with the potential of having thousands of shiny coasters at your disposal, don't skip the check disc step. If you are making a dual-layer disc, check discs are indispensable since you can't fully test with a DVD-R. You can break the disc into smaller parts to check on DVD-R, but you can't test layer changes and full navigation.

Tips and Tricks   The following tips may help your testing and quality control be more affective and efficient:

- Test on a cheap TV. Most of the production process is done with high-end, high-quality video equipment, but many of your customers will watch your work on a cheap, old television. Make sure that details of video and graphics, menu highlights, and so on are not lost on the average viewer.

- It's almost impossible to test all the permutations of a highly interactive disc; try to break things down into testable subsections.

- If you have computer software that can mount a formatted DVD image file from DLT or a hard drive, mount the volume for more thorough emulation. This will find file system errors that normal emulation will not find.

- If you put computer applications or large amounts of computer data on the disc, use a checksum program to verify the accuracy of the original data against DVD-Rs, DLTs, check discs, and final discs. A file-compare program such as windiff is also useful for verifying filenames and data integrity. Be careful when testing DVD-Rs, since some DVD-ROM drives have problems reading DVD-Rs and will report errors that will not be present on a check disc or the final replicated copies.

■ If you put Macintosh files on the disc, make sure that the icons display properly and that the resource forks are still intact (use ResEdit or similar utility). Check that the filenames are correct when played on a Windows PC.

Table 12.7 is also provided to help you test QC issues on your project.

NOTE: *Anyone serious about producing DVD-Video must invest in three things: an expensive, high-quality monitor; a cheap TV; and a DVD PC.*

Replication, Duplication, and Distribution

At this point, things are mostly out of your hands. Your DLT, representing weeks or months or years of your life, is in the hands of someone else. Most replicators have excellent processes in place and will take good care of your project.

The replicator will use the DLT to cut a glass master and make stamping masters. Some replicators first verify the DLT formatting and check for spec compliance. The replicator will send you check discs made from the master. Once you approve the check disc, then thousands or millions of copies of your disc will be churned out. See Chapter 5 for technical details of mastering and replication.

If you want to serialize the discs or put other custom information in the BCA, a small number of replicators have the necessary equipment. If you need to make less than a hundred discs, you may want to use DVD-R duplication instead. You can burn DVDs by hand, buy a DVD-R bulk-duplication system, or have a service bureau duplicate the discs.

Most replicators can do *source tagging*: inserting *electronic article surveillance* (EAS) labels into packaging. The two most popular source tagging technologies are acousto-magnetic tags from Sensormatic and RF tags from Checkpoint Systems.

TABLE 12.7	☐ Intro sequence	Does the disc do what's intended when inserted? Are the intros annoyingly longer than you realized?
Testing checklist	☐ Menus	Check arrow movement in all four directions from each button. Test that Top (Title), Menu, and Return functions work. Check any motion loops and idle outs.
	☐ User operations	Which functions work? Which are locked out? Are they supposed to be locked out?
	☐ Video	Verify quality. Check chapter search and Previous/Next functions. Check any angles, branching, and layer switch. Test on interlaced and progressive equipment.
	☐ Audio	Make sure languages match the language name displayed by the player. Test multichannel playback and stereo downmix. Carefully look for audio sync problems.
	☐ Subpictures	Check that subtitles appear at proper times. Make sure languages match the language name displayed by the player.
	☐ Closed Captions	Verify that Closed Captions can be decoded by a TV.
	☐ Parental control	Verify that disc or section lockout works. Try the disc at each parental level, and also with no parental level set.
	☐ DVD-ROM filenames	Compare ISO 9660 and UDF filenames. Make sure long filenames were not truncated and that special characters are intact.
	☐ DVD-ROM data integrity	Verify readability of all files. (If errors with a DVD-R, try a different drive.)

Disc Labeling Many options for disc labeling are available, such as silkscreen printing, offset printing, pit art, and holograms. The area of a disc available for label printing normally extends from the 46 mm to 116 mm diameter points (the data area runs from 48 mm to 116 mm). For special applications, the print area can start at 34 mm, just past the clamping area. The BCA area is at 44.6 mm to 47 mm, so discs that use the BCA must start the label at about 48 mm. A DVD-10, DVD-14, or DVD-18, with data on both sides, has a narrow ring from only 39.5 to 43.5 mm for printing.

Silkscreening applies layers of colored ink to the label side of the disc. If more than two colors are used, or if the disc has large ink coverage areas, care must be taken that the disc does not warp as the ink dries. Offset web printing causes fewer warping problems from ink shrinkage than screen printing, but it is more expensive.

Pit art is produced by creating a stamper that embosses a graphic design onto the back substrate. The black print of the source graphic will appear as a mirrored surface, while the white areas of the graphic will appear as a frosted surface. The pit art area extends from 40 mm diameter to 118 mm. The source graphic is a one-bit monochrome image, usually 2048×2048 pixels.

Package Design   Packaging is the spokesperson for your disc. Make it reflect the contents of the disc and grab the attention of the customer. Even with Internet shopping, the disc jacket influences the buyer's impression of the work.

Design with the collector in mind. Part of the reason people buy DVDs rather than renting them or taping them (or waiting for the network broadcast) is because they enjoy collecting. Use quality artwork on the package. Produce an informative and enjoyable insert. Consider including little goodies in the package that relate to the disc.

Explain what's on the disc. Do not assume that the customer is familiar with the contents. Include a list of titles, chapters, extras, and so on. Use standard text and icons to indicate the format of the contents. Clearly identify number and type of audio tracks, aspect ratio, disc region, video format, running time, and so on (see Figure 12.7). Sometimes it's a good idea to

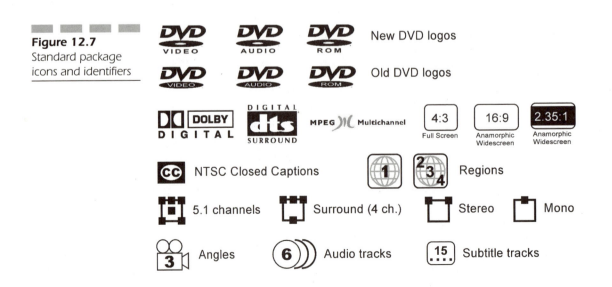

Figure 12.7
Standard package icons and identifiers

delay package design until the end, since details of aspect ratio, sound-tracks, extras, and so on may change during production.

Tips and Tricks The following advice will help you to avoid problems during the replication, duplication and distribution processes:

■ If you need a rush job, schedule replication early, since large orders (especially things such as Star Wars discs or Microsoft operating systems) can sometimes swallow all available replication capacity. Establishing a relationship with one or two replicators will help when you have special demands.

■ If you want to use CSS, CPPM, or Macrovision make sure the replicator is appropriately licensed.

Production Maxims

For your edification and enjoyment, here is a small collection of maxims, reminders, and rules of thumb—many of them learned the hard way—collected from various DVD authors:

■ Players display time (and sometimes chapters) only for one_sequential_PGC titles.

  ▪ Corollary: Timecode search, particularly on computers, only works with one_sequential_PGC titles.

  ▪ Corollary: DVD players that use barcodes require timecode, and thus one_sequential_PGC titles.

■ Playback between PGCs is non-seamless. Cell commands are executed in a non-seamless manner.

■ Audio will mute during a non-seamless break.

■ A layer change (on an RSDL disc) does not have to be at a chapter or program boundary, only at a cell boundary.

■ All titles in a title set must have the same video aspect ratio (4:3 or 16:9) and audio format.

■ Programs are usually the same as chapters (PTTs), but programs that are not chapters behave in special ways. To be safe, always put a chapter mark on every program.

■ Next/previous commands can only be used in the title domain. The return (go up) command can be used in all domains.

- Entry points (chapters, programs, and cells) must start at a GOP header and must be at least 0.4 seconds apart.

  - Corollary: An I frame is not necessarily a GOP header. To create an entry point, you must force a GOP header, not just an I frame.

- Subpictures cannot cross an entry point (chapters, programs, or cell).

- GPRMs are reset to zero on title search or stop.

- Cell commands are executed after the cell is presented.

- If you want to play continuously from chapter to chapter, do not put a next program command at the end of the program, since the player will interrupt playback to execute the command. There is no need to add any commands, since playback will automatically go to the next chapter.

- Never assume that anyone, even the client, knows what the DLT contains. After authoring and writing to a DLT, fill out the replicator's project sheet personally to avoid inaccuracies.

- Any amount of QC on a replicated check disc is worth more than the cost of paying for a disc run that failed for the simplest problem.

DVD-ROM Production

Many details of producing a DVD for computers are covered in Chapter 11. This section goes over a few of the basics. Also see the testing section earlier in this chapter.

File Systems and Filenames

DVD-ROM discs are specified to use the UDF bridge format, which includes ISO 9660 and Micro UDF, but a DVD can be formatted with any file system: full UDF, ISO 9660 only, Windows NTFS, Apple HFS, and so on. ISO 9660 Level 1 limits filenames to 8 characters plus a 3-character extension. Directory names are limited to 8 characters with no extension. ISO 9660 Level 3 limits filenames to 30 characters (not counting the period between filename and extension), although it's possible to store up to 221 characters as the file identifier. Directory names are limited to 31 characters, but 221 are possible. UDF supports filenames of up to 255 characters using the Unicode character set, which supports characters from almost every language. However, an operating system has no guarantee that it will be able to correctly display a given Unicode character.

To provide better compatibility with operating systems such as Windows 95, DOS, and Unix that are unable to read the UDF file system on DVD volumes, Microsoft recommends that DVD image formatting software, including DVD-Video and DVD-Audio authoring systems, implement the Joliet extensions to allow long filenames within the ISO 9660 file system. Long filename support is especially important to publishers putting legacy software onto DVD-ROM volumes.

The Joliet extensions to ISO 9660 provide for filenames longer than 30 characters and for the Unicode character set in file and directory names. The use of Joliet extensions does not violate the DVD specification or the Micro UDF specification. Standardized directory names of VIDEO_TS and AUDIO_TS and the standardized names of files within these directories follow the 8.3 filename limitations, but these limitations do not apply outside of the video zone and audio zone.

Joliet applies only to the ISO 9660 section of the UDF bridge format and is independent of UDF, which has its own provisions for long file names. Windows 98 and Windows 2000 use UDF and do not read the ISO 9660 portion of UDF bridge DVDs. The intent behind using Joliet extensions with DVD volumes is to make them behave more like today's CD-ROMs when used with an OS that recognizes Joliet but does not recognize UDF.

The UDF file system also supports Macintosh resource fork, including file creator and file type information. The authoring system or DVD-ROM formatting software must be able to recognize Macintosh file information in order to preserve it for UDF formatting. On a Windows-based authoring system, software that supports Macintosh files, such as *PC MacLAN* from Miramar Systems, is needed.

It's possible to make a bootable DVD-ROM. The Micro UDF spec specifically forbids boot records, but most computers don't look for UDF boot records, they look for ISO 9660 boot records according to the El Torito specification. Very few authoring systems and DVD-ROM formatting utilities can create a bootable disc. If you need this capability you may even have to produce a DVD-ROM image and modify it by hand to create the boot sector before writing the image disc or DLT.

Bit Budgeting

Bit budgeting on DVD-ROM is a breeze compared to DVD-Video or DVD-Audio. You just add up all the file sizes and see if they fit on a disc; however, a few potential pitfalls might arise. Don't forget the difference between billions of bytes and gigabytes. The operating system will report file sizes in gigabytes, but DVD capacities are usually given in billions of bytes (see Chapter 1.) Don't forget to leave some room for the directory entries.

When adding up file sizes, use the actual file size, not the "size on disk" value, since the space taken on the hard disk is dependent on the block size. But likewise, space taken up by each file on a DVD can be more than the actual file size. For a very tight bit budget, adjust the size of each file by the DVD sector size: divide the file size in bytes by 2048, round up, then multiply by 2048. This will give the sector-adjusted file size, which represents the true amount of space the file occupies on the DVD.

Creating a DVD-9 is not as simple as it might seem. You need two DLTs, but simply dividing your project in half and creating two disc images will not work. A DVD-9 has a single UDF bridge directory section located on the first layer that references files on both layers. If you use formatting software to create two images, you will get two volumes, each with its own directory section. The first layer will work fine, but the second layer will be inaccessible. Use a DVD authoring application or a DVD-ROM formatting utility that knows how to create DVD-9 images, preferably one that lets you specify where the layer break occurs.

AV File Formats

When creating a multimedia DVD for playback on PCs, you have essentially three choices for the AV format: DVD-Video files (.vob), MPEG files (.mpg), or computer-oriented media files (.avi, .mov, .mp3, and so on). The advantages of using DVD-Video content are that the disc will also play in standard DVD players, the features of DVD-Video such as angles, seamless branching, fast chapter access, and subpictures are available, and the video will be more compatible with PC decoders, which are oriented more toward DVD than MPEG-2. The advantages of using MPEG-2 program stream files are that you don't have to use a DVD-Video authoring system; you can easily create and test playback. You can also make each video clip a file with a meaningful name rather than trying to find the right access point in the middle of a .vob file.

The advantages of using other formats are that the development tools are simple, widely available, and often inexpensive. Since the data rate of most DVD drives is at least equivalent to an 18x CD-ROM drive, high data rates can be used with codes such as Cinepak, Sorensen, and Windows Media Video (MPEG-4) to provide surprisingly good video. It may also be much easier to create discs that work on both Windows and Macintosh computers by using tools such as Macromedia *Director*, Click2Learn *Toolbook*, or even HTML authoring applications combined with multimedia playback plugins.

Hybrid Discs

If you are adding DVD-ROM content to a DVD-Video or DVD-Audio disc (or adding DVD-Video or DVD-Audio content to a DVD-ROM disc), be mindful of a few things. Data in the DVD-Audio and DVD-Video zones must come first on the disc and must be contiguous. If you have a set of VOB and IFO files in a VIDEO_TS directory or a set of AOB and IFO files in an AUDIO_TS directory and you feed the files into a DVD-ROM formatting application, it must recognize the DVD-Video or DVD-Audio files and physically place them on the disc in the proper way. The IFO files include references to sector offsets, which requires that the IFO and AOB/VOB files be in proper sequential order on the disc. If not, the disc will fail to play on many DVD players. It will play fine in software players on PCs, since they ignore physical placement of data on the disc. If you add more than a few DVD-ROM files to a DVD-Video or DVD-Audio disc, put them in a subdirectory. Too many extra files in the root directory will cause some DVD players to fail.

The Future of DVD

Introduction

In the fall of 2000, as this book was being finished, DVD was on the verge of turning four years old. It had succeeded beyond many expectations but also failed to meet other expectations. Given the wide range of forecasts and markets, this is no surprise. This chapter examines past and present forecasts, along with possibilities for DVD over the next decade or so. At the very least, it should prove entertaining and informative to reread this chapter four or five years hence.

Even though DVD PCs were five to seven times as popular as home DVD players in 2000, about 10,000 movie titles were available, compared to only 200 computer software titles. Over 100 million DVDs discs were shipped in 1999, with expectations of shipping 230 million or more in 2000, potentially surpassing VHS revenues. Before the end of 2000, 10 percent of U.S. homes had a DVD player. Although DVD has a long way to go before it reaches the 98 percent penetration of color TV and the 94 percent penetration of VCRs, now that it has passed the inflection point of 10 percent, its growth will presumably continue to accelerate faster than any previous consumer electronics entertainment technology.

The success of DVD-Audio is assured only by the success of DVD-Video. If DVD-Audio had to stand on its own, it would most likely fall flat on its face. Instead, however, it rides the coattails of DVD-Video. By 2002 or thereabouts, most DVD players will play both DVD-Audio and DVD-Video discs, and the distinctions between formats will largely disappear. DVD-Audio discs will simply be a higher-fidelity variation of DVD.

Although the growth of the DVD PC market has been slower than generally expected, it will inevitably displace the CD-ROM PC market. CD-ROM media will continue to be widely used, however, but CD-ROM drives will soon suffer the fate of single-density floppy drives (remember those?).

The argument about whether DVD will succeed or not has been settled, but now the question is how well will it succeed? How long will it take prerecorded DVDs to outsell VHS tapes? When will DVD recorders replace VCRs? Will the DVD-VR and DVD-AR formats succeed? How will hard-disk-based *personal video recorders* (PVRs) factor in? When will the next generation of high-density DVD appear? How much will it hold? What will the high-definition video format be like? Will we ever need another audio format?

The variables involved in predicting the road of DVD are extremely complex, but this has not deterred many people from making plentiful predictions from the sensible to the outrageous.

The Prediction Gallery

The DVD format was announced in December of 1995. In the 11 months between then and the appearance of the first DVD players for sale in Japan, the only people making money from DVD were journalists and analysts. Considering the disparity between their forecasts, some of them were making more money than they deserved. It has been interesting to look back and see who had the most accurate auguries. Here is an interesting potpourri of prognostications, all made in 1996:

Philips: 25 million DVD-ROM drives worldwide by the year 2000 (10 percent of projected 250 million optical drives).

Pioneer: 500,000 DVD-ROM drives sold in 1997, 54 million sold in 2000.

Toshiba: 120 million DVD-ROM drives in 2000 (80 percent penetration of 150 million PCs). Toshiba projects that it will no longer make CD-ROM drives in the year 2000.

International Data Corporation (IDC): 10 million DVD-ROM drives sold in 1997, 70 million sold in 2000 (surpassing CD-ROM), and 118 million sold in 2001. Over 13 percent of all software will be available on DVD-ROM in 1998. DVD recordable drives will make up more than 90 percent of the combined CD/DVD recordable market in 2001, with DVD drive units installed in 95 percent of computer systems with fixed storage. IDC later revised its figures to 3.7 million in 1997, 10.8 million in 1998, 36.5 million in 1999, and 95 million in the year 2000.

AMI: An installed base of seven million DVD-ROM drives by 2000.

Intel: 70 million DVD-ROM drives by 1999 (sales will surpass CD-ROM drives in 1998).

InfoTech: 1.1 million DVD-Video players in 1997, with 10 million by 2000. 1.2 million DVD-ROM drives in 1997, with 39 million by 2000. 250,000 interactive (video game or cable TV set-top) DVD players in 1997, with six million by 2000. InfoTech revised its DVD-Video predictions in January of 1997 to a slightly lower 820,000 units in the first year, with an installed base of 80 million by 2005.

Dataquest: One-year sales of 15 million DVD-Video players in 2000, with a DVD optical drive market of $35 million by 1996 and $4.1 billion by 2000.

Paul Kagan Associates: 12 million DVD-Video players in the U.S. after five years. Total DVD-Video business of $6 billion per year by the year 2000.

Simba: Annual DVD software sales of up to $35 million by 1997 and $100 million by 1999.

Freeman Associates: DVD-ROM drive sales of 89 million by 2001, accompanied by zero sales of CD-ROM drives.

Electronic Industry Association of Japan (EIAJ): A combined U.S. and Japanese DVD-Video and DVD-ROM market of $23.6 billion by 2005, with a DVD-ROM drive in about 80 percent of all new PCs.

Microsoft: As many as 100 million PC-based digital TV sets by 2005 (predicted in March 1997).

Intel: By the year 2000, all PCs shipped will be DTV receivers (predicted in March 1997).

It's enlightening to compare the DVD projections with the success rate of previous consumer products (see Figure 13.1).

Figure 13.1
Consumer electronics success rates

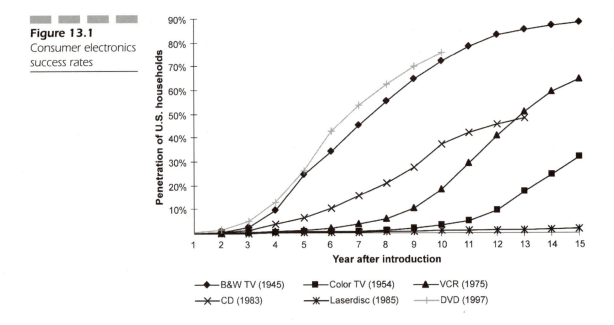

The reality was that 349,000 DVD players were sold to dealers in the U.S. in 1997. First-year sales of DVD players were more than twice that of VCRs from 1975 to 1977 and more than 12 times the first-year sales of CD players in 1983. Just over one million DVD players were sold in 1998. Over four million DVD players were sold in 1999 for a U.S. total of about 5.5 million at the end of 1999. DVD reached 16 percent penetration of home theater systems in 1999. In October 2000, the 10-millionth U.S. DVD player was shipped. In three and half years since its U.S. introduction, DVD players achieved the mark that VCRs took eight years to reach, and CD players took eight years to match. Sales were on track for around nine million player sales (worth roughly $1.9 billion) in 2000 alone, to reach an expected grand total in the U.S. of over 14 million DVD players. About 40 million discs shipped in the U.S. in 1998 and about 100 million discs shipped in 1999, with expectations of over 230 million title shipments in 2000, representing more than $4 billion in retail revenue. In Europe, 270 thousand players shipped to dealers in 1998, 1.6 million in 1999, and approximately four million in 2000, for a total of just under six million. Twenty million DVD-Video discs shipped in Europe in 1999, with expected shipments of 50 million in 2000. Two hundred and forty thousand DVD players were sold in Japan in 1999, and 388 thousand were sold in 1999. Sony sold over two million PlayStation 2 consoles since its March 2000 introduction, essentially tripling the number of DVD players in Japan. The worldwide total of DVD players at the end of 2000 was expected to be around 26 million.

For comparison, an installed base of about 700 million CD players and about 240 million CD-ROM drives existed worldwide at the end of 2000. Forty-seven million new CD players were expected to be sold in 2000, equaling sales of $4.5 billion. Total audio shipment revenues in 1999 passed the $8 billion mark for the first time since 1995. Audio system revenues in 1999 were $2.1 billion. By the end of 2000, about 600 million VCRs and 1.3 billion televisions were owned worldwide. Over 93 percent of American households had at least one VCR. Nearly 23 million new VCRs were sold in the U.S. in 1999, compared to just under 24 million in 2000. American consumers spent $9 billion to rent about four billion videotapes in 2000. They spent $500 million renting DVDs. The pay-per-view market had grown to $1 billion (after 15 years). Over 40 million TVs were expected to be sold in 2000. More than 90,000 digital TVs were sold in 1999, followed by expected sales of 600,000 in 2000. Home satellite dish installations were expected to climb from 3.6 million in 1999 to 4.1 million in 2000, for revenues of about $992 million. About 40 million Video CD players were in homes, mostly in Asia, by the end of 2000. Sales of digital cameras hit two million units in 1999.

Total consumer electronics sales reached $81 billion in 1999 and is expected to hit $85 billion in 2000. Total PC sales will hit a record $16.8 billion in 2000 with expected sales of 16.8 million units. PC software accounted for $4.5 million in sales in 1999 and will surpass $5.2 billion in 2000, a 16 percent rise. Of the $6.9 billion in revenues that the interactive entertainment industry generated in 1999, $1.3 billion came from PC games.

Here's a sampling of expectations at the beginning of the new millennium. The following forecasts were made in 2000:

DVD Forum: $16 billion in DVD-ROM/RAM drive sales in 2001, $4 billion in DVD audio players, and $8 billion in DVD video players/recorders.

Technicolor: The number of PCs equipped with DVD-ROM drives will reach more than 130 million units worldwide by 2001.

Jon Peddie and Associates: DVD player sales in the U.S. in 2004 will be nearly 20 million units.

IRMA: Thirteen million players will ship in the U.S in 2000, with 31 million sold in 2003. Two hundred and eighteen million DVD-Video titles will ship in 2000, with 360 million units in 2001. DVD-ROM and DVD-Audio will ship a combined 70 million discs in 2001.

Screen Digest: Western Europe will have 5.4 million DVD players by the end of 2000, in 3.5 percent of TV households. The total will rise to 47 million by 2003, or about 30 percent of TV households. DVD spending will rise from $450 million in title purchases and $18 million in rentals ($470 million total) in 1999 to $4.9 billion in sales and $1 billion in rentals ($5.8 billion total) in 2003. VHS spending will fall from $6.1 billion in 1999 to $3 billion in 2003.

Baskerville Communications: Worldwide spending on DVDs will surpass spending on videos by 2003. By 2010, 625 million DVD players will be in homes worldwide (55 percent of TV households), with spending on DVDs at $64.6 billion versus $2.5 billion for VHS. Also by 2010, China will have the highest number of homes with DVD players at 127 million.

Sony: Ten million PlayStation 2 units will be shipped worldwide by the end of March 2001: four million in Japan, three million in the U.S., and three million in Europe.

InfoTech: Approximately 25 million DVD video recorders will be sold by 2005.

Disk/Trend: DVD-ROM drive sales will pass CD drive sales in 2001, with 60.3 million DVD units sold versus 56.8 million CD units. By 2002, consumers will buy 92.8 million DVD drives, compared to 30.3 million CD drives.

Hewlett Packard: More than 140 million DVD-ROM drives and DVD video players will be in use by the end of 2001.

Understanding and Solutions: The attach rate of DVD-ROM PCs in Western Europe will hit nine percent in 2000, 20 percent in 2001, 37 percent in 2002, and 58 percent in 2003.

Strategy Analytics: Global DVD hardware sales (players and drives) will reach 46 million units in 2000, including 21 million in the U.S. and 17 million in Europe. The average price of a DVD player in 2001 will fall to $200 in the US and $270 in Europe. Fourteen percent of U.S. homes and five percent of European homes will own at least one TV-based DVD player by the end of 2000. By 2002, 58 percent of U.S. homes will own at least one DVD device (a player, game console, or PC). DVD PCs, which accounted for 75 percent of the installed base at the beginning of 2000, will fall to 59 percent by 2002 as TV-based DVD becomes more widespread. Video titles accounted for over 90 percent of the software market in 1999. By 2005, their share will have fallen to 43 percent, while DVD-ROM will account for 28 percent and games formats will total 24 percent. Worldwide DVD-Video disc shipments in 2000 will reach nearly 400 million units, growing to 2.3 billion by 2005, worth $44 billion. Including other DVD formats (games, ROM, and audio), shipments will reach 3.7 billion by 2005, worth nearly $100 billion.

Dataquest: DVD drives outsold CD-RW in 1999 (16.2 million versus 12.5 million), but CD-RW will take the lead in 2000 with 28.7 million CD-RW units shipped compared with 22.6 million DVD drives. By 2002, DVD will overtake CD-RW, and by 2004 a predicted 105 million DVD drives will be shipped versus 28 million CD-RW drives.

Paul Kagan Associates: Hard-disk-based PVR sales in the U.S. will reach one million in 2000, 12 million in 2004, and 40 million by 2008.

Forrester Research: Fourteen million PVRs will be in homes by 2004. About 80 percent of U.S. homes could have PVRs by 2010.

New Generations of DVD

The development of DVD was planned to proceed in stages. DVD-Video was released first as a consumer entertainment product, closely followed by DVD-ROM for computers. Record-once DVD-R appeared in the second half of 1997, with erasable DVD-RAM coming in the middle of 1998. DVD-Audio was originally expected to appear along with DVD-Video. The specification for DVD-Audio, minus copy protection, was ready long before the players, which showed up in small numbers in Japan at the end of 1999 and began to appear in the rest of the world in the second half of 2000. DVD-RW followed a schedule similar to DVD-Audio. A no-show for three years running, DVD+RW should finally make an appearance in 2001.

Early ambitions to make all versions of DVD physically compatible were derailed by late additions to DVD-RAM, which made it incompatible with first-generation DVD-ROM drives. Little effort was expended to make existing players and drives compatible with DVD-RAM until the DVD Multi initiative of 2000, which will still take a few years to become established.

Also planned from early on, but in more nebulous terms, was an improved high-density DVD format for HDTV with even more capacious data storage. Deep thinkers are undoubtedly envisioning third and fourth generations of DVD for digital film, three-dimensional video, and the always-growing computer storage demands. When the next incarnation of DVD technology is developed, using smaller wavelength lasers, the same problem that DVD readers had with CD-R discs will occur; the dyes used in the DVD-R polymer recording material may not properly reflect other wavelengths of light. The solution will probably be three-laser designs: one for CD and CD-R, one for DVD and DVD-R, and one for HD-DVD and HD-DVD-R. Or perhaps DVD-Rs can be made with improved material that works with lasers of both wavelengths.

The Death of VCRs

DVD video recorder technology is all over the map, but once DVD video recorders duck below the magic price point of $500, they will begin their inevitable replacement of VCRs. Some projections expect that DVD video recorders will outsell VCRs by 2005. VCR revenues in the U.S. peaked in 1996, and prerecorded VHS sales peaked in 1999. VCR sales growth in 2000 was less than five percent above 1999, compared to over 100 percent sales growth of DVD players. Once recordable DVD technology begins to take hold, consumer demand for compatibility between formats will hopefully push manufacturers toward better cooperation.

The displacement of cassette tapes by CDs can be compared to the potential displacement of VCRs by DVD recorders. It's a reasonable analogy, although widespread, easy-to-use DVD recorders will become available much sooner than recordable CD players, which still haven't caught on for general consumer use. It's interesting to note that it took until 2000, 17 years after the introduction of CD audio, for prerecorded cassette tapes to begin disappearing completely from music store shelves. If this historical perspective is applied to DVD, we might expect VHS tapes to begin fading away before 2014.

Other technologies, such as hard-disk-based video recorders, will compete with DVD recorders to become the VCR of the future, but the final outcome will be a melding of both technologies: digital recorders with hard disks for time shifting and DVD drives for playing movies and making long-term copies of recordings from the hard disk.

Eventually, broadband Internet connections will reach the point where DVD-quality video can be streamed to living rooms and offices on demand. At about the same time, high-definition video will finally reach mainstream status, which will require another bump in connected bandwidth. Once again, DVD (or more specifically HD-DVD) will fill the need for the delivery, storage, and archiving of high-bandwidth, high-quality content. Techniques such as downloading video overnight to a hard drive will help overcome bandwidth deficiencies, but even when high-definition video or digital cinema are delivered in real time over the Internet, there will still be a need for making permanent physical copies.

DVD Players, Take 2

Manufacturers will continue to improve DVD players in the endless race to best the competition. Future players will include digital connections for video, audio, and direct bitstream output; video processing circuitry to improve picture quality; progressive-scan and resolution-enhancement circuitry to make discs with interlaced video look better on digital TVs and HDTVs; improved interactivity; and other features such as smooth reverse play, artifact suppression, and simulated surround sound for headphones. Future generations of DVD players will also include phone or Internet connections for WebDVD applications. Such connections will also make pay-per-view discs possible.

Player models from the first few years lack digital video connectors, primarily because copy protection standards and digital interchange protocols are still under development. At some point, however, DVD players will benefit from the new digital interconnection standards developed for consumer electronics and computers. Once the standards are finalized and the copy protection systems are settled on, new DVD players will convert playback information into the proper format. This should be trivial for encoded MPEG audio/video and encoded Dolby Digital or DTS audio, but it may be more complicated for subpictures. Obviously, the menus and random-access features of DVD will not be available if the outgoing digital bitstream is recorded for later playback, but the primary purpose of a digital connection is to provide a pristine signal for the decoder in the digital television or the digital media center equipment.

The IEEE 1394 external serial bus system is the best candidate for digital connections. It uses a foolproof connector adopted from video game consoles. Apple Computer's original design for the product was called FireWire, a name that stuck even after the IEEE adopted it as standard number 1394. Its first major support was from Sony with its digital video cameras, although Sony chose to call it the DV connector, and later the iLink connector. Over a hundred companies now support the IEEE 1394/FireWire/DV system in their products. Initial implementations support data rates of 100 to 400 Mbps, and new versions can support data rates over 2 Gbps. At these speeds, even uncompressed digital video signals can be accommodated.

Intel's *Universal Serial Bus* (USB) standard is intended primarily for computer peripherals such as mice and keyboards. Standards are inevitably stretched beyond their original purview, however, and newer versions of USB are capable of carrying compressed audio and video information.

The advantages of digital connections are as follows:

- **A single cable to carry audio, video, and even power** (Sony's iLink connector does not carry power, however). DVD players can fairly bristle with a ludicrous number of connectors to cover all possible scenarios: two audio connectors for stereo sound, one or more coax digital audio connectors, one or more optical digital audio connectors, six audio connectors for 5.1-channel sound (or seven connectors for 6.1), a composite video connector, an s-video connector, and three connectors for component video. Of course, a digital connector is yet one more, but in theory it will be the last new connector ever needed, since it will be able to accommodate all future signals.

- **Daisy chaining.** IEEE 1394 enables dozens of devices to talk to each other with only one cable connected between each of them. Other star configurations can also be used (see Figure 13.2).

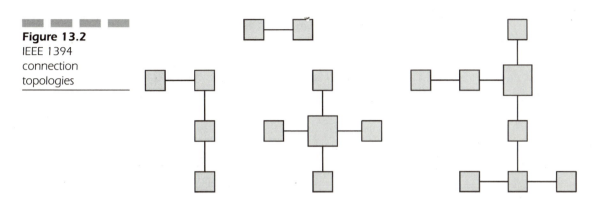

Figure 13.2
IEEE 1394
connection
topologies

- **Little or no electronic interference.** Because the signals are digital, they are practically immune to degradation from external sources.

Many new computers have both USB and IEEE 1394 interfaces. In some cases, the digital audio and video from the computer's DVD-ROM drive will be available to external digital devices, possibly even DVD recorders.

Hybrid Systems

The DVD format will be combined with other products and systems in intriguingly diverse ways. DVD-ROMs will be integrated with the Internet, home DVD players will add Internet connections, and home Internet-TV appliances may add DVD-Video players or DVD-ROM drives. New pay-per-view systems may be built around DVD players. Discs will appear for free in the mail to be viewed or purchased by having the player make a quick phone call to authorize the charge to an account or credit card. Both home and commercial video game systems will include DVD-Video and DVD-ROM units as well. Public information kiosks will also incorporate DVD technology. Car navigation systems will take advantage of DVD storage capacity for maps and graphics and might even pipe DVD movies to a monitor in the back seat. As MPEG video and Dolby Digital audio become more common, the digital decoder may move into the television or stereo fields, making the player an inexpensive transport that, along with other devices such as a digital satellite receiver, digital cable box, digital TV tuner, or video phone connection, would simply feed compressed signals to the video display. Video edition applications on PCs, or even video-editing features of cameras and set-top boxes, will enable users to record digital pictures and camcorder footage to be shared with others.

The living room of the cost-conscious consumer may be the primary setting for splicing DVD into other formats. These possibilities are explored further in the section entitled "The Changing Face of Home Entertainment," which follows.

HD-DVD

Before DVD players had even hit the streets, DVD developers began talking about and demonstrating new high-density or high-definition versions. HD-DVD, as it's commonly called, is expected to be a reality in the year 2002 or 2003[1]. Given the typical gap between expectations and their fulfillment, products based on HD-DVD will more likely appear around the year 2005.

The essential advancement is a blue laser, which has a smaller wavelength than the visible red lasers of DVD and can therefore read smaller pits. This laser is projected to yield an increase in data density three to ten times greater than DVD.

A new format of this type will obviously require new players. HD-DVD discs will not be playable in older players, but the new players will undoubtedly play older discs. They will probably even make the video from older discs look better with improved circuitry, new digital signal processing, and progressive-scan output. It will also be possible to make hybrid discs with a standard DVD data layer on one side and a high-definition data layer on the other side, so that the discs will work in all DVD players.

At a certain level, DVD must remain competitive with HDTV. This can be accomplished in the short run with progressive-scan players, but in the long run it will require true HD video. If someone can watch a recent blockbuster movie in high-definition on broadcast HDTV or digital cable, they'll be less interested in renting the DVD if it only plays in 480-line, interlaced, standard definition.

The *Consumer Electronics Association* (CEA) projects that the first 10 million DTV units will be sold by 2003, the next 10 million by 2005, and 10.8 million will be sold in 2006 alone.

Speculation has arisen that a double-headed player reading both sides of the disc at the same time could double the data rate for applications such

---

[1]HD-DVD probably would have been shortened to HDVD, but the term has already been appropriated for high-density volumetric display.

as HDTV. This is currently impossible since the layer 0 track spirals go in opposite directions. The DVD standard would have to be amended to enable reverse spirals on the second side. This still might not get off the ground, since large buffers or precise alignments in both the radial and angular dimensions would be required to keep both sides in sync.

A possible DVD-36 disc that would have four data layers per side has also been proposed. The data density would remain the same as in standard DVDs, but the extra layers would double the capacity. Existing players could read two of the layers, while newer players could read all four.

Beyond blue lasers lie even more interesting formats. In 1999, Constellation 3D Inc. began touting *fluorescent multilayer disc* (FMD) technology. Here data is stored in pits on multiple layers of fluorescent material. When a laser excites the material, it emits light that can be read more easily than the reflected coherent light of a laser. In theory, hundreds of data layers can be used to create very high-capacity volumetric storage media. In 2000, Constellation 3D Inc. claimed that enhanced 25GB DVD players using FMD technology would be available in mid-2001. A more likely scenario is that FMD might be used for the third generation of DVD, after blue-laser DVD.

Other future possibilities include diamond-based x-ray lasers, charged particle beam recorders, near-field scanning microscopy, 3D holographic recording, and any number of creative new technologies for manipulating matter in strange ways to store vast amounts of data.

No Laserdisc-Sized DVD

Inevitably, rumors have circulated about DVD technology being applied to the larger 30-cm (12-inch) form factor of laserdiscs. Variations of laserdisc technology have been used for high-definition television, such as the HDLD format designed for the Japanese HiVision system, but it's improbable that such a format will be adopted for commercial digital video, given the disadvantages of larger discs. CD and DVD production lines can be adapted to higher-density formats as long as the physical diameter does not change.

Large discs require larger molds, which are more difficult to inject molten polycarbonate into, and which can be more difficult to separate from the pressed discs, especially with the tiny pit sizes used by DVD. Large discs are also more subject to warping, and they create more stress and slippage at the bonding layer, which can damage the reflective data layer. Large discs require larger players and stronger motors (making battery-powered portable players all but impossible) and have more noise and stability problems because of the higher velocity at the outer edge.

None of these, however, is the real reason a 12-inch digital disc will never exist for general consumer use. The most important factor is the computer. Twelve-cm CDs and DVDs are the right size for computers. The drives fit in existing bays and in laptops. The computer side of the scale weighs ever more heavily on traditional consumer electronics products.

Twelve-inch discs are dinosaurs. The development of new high-density optical media is oriented toward the more practical 12-cm (five-inch) format. Because of the insatiable demand for higher capacities, it's also not likely that a smaller format will become a mainstream product, although smaller eight-cm (three-inch) discs will flourish in products such as camcorders.

DVD for Computer Multimedia

During the development and introduction of DVD, many people were confused by the difference between DVD-Video and DVD-ROM. As computers evolve from data processors to media processors, this distinction will become mostly irrelevant. DVD movies already play equally well in a movie player or a computer.

Current computers are powerful enough to handle DVD decoding in software, and adventurous users have learned that they provide the most flexibility and best quality for home theater setups. New video display hardware for computers can enhance the picture with deinterlacing and can even accommodate widescreen video displays. Computer display subsystems are on the verge of being able to integrate graphics and video at 30 frames/second or higher, without requiring complex sideports and video busses. Digital I/O connections such as IEEE 1394/FireWire are becoming common, enabling the computer to capture, process, combine, and generate high-quality digital video and audio.

Of course, HDTV will come along and raise the stakes, possibly once again requiring specialized video hardware for a year or so while CPUs and software catch up. The game of leapfrog between hardware and software has been played out many times before; they will no doubt continue hopping past each other throughout their future development.

Faster Sooner

Unlike CD-ROM drives, which took years to slowly move from 1X speeds to 2X speeds, then 3X and beyond, DVD-ROM drives hit 2X in less than a year and 16X in less than four years. DVD drives will continue to speed up, coming close to the performance level of average hard drives.

Promising technology from Zen Research, Inc. splits the laser light to illuminate multiple tracks simultaneously and read the data in parallel. The result of this multibeam process is a faster data rate at disc rotation speeds about 20 percent slower, resulting in better performance and more reliable reading.

The Death of CD-ROM

As evidenced in the preceding "Prediction Gallery," the range of dates given for the disappearance of CD-ROM would make a fine office pool, but all agree that CD-ROM will become obsolete. This will not happen overnight, to be sure, since CD-ROM had a 100 million-unit lead on DVD-ROM when it was introduced, but it's conceivable that few CD-ROM drives will be made after the year 2002. It's also a pretty sure bet that the last production lines will have changed over before 2005.

The insatiable demand for storage space will propel DVD-ROM into the computer mainstream. Because DVD-ROM drives can read CDs, the only real barriers are support and price. DVD-ROMs will be supported in upcoming operating system upgrades, as well as by hardware upgrade kits. Prices will also drop to CD-ROM levels because of economies of scale and because most of the mechanisms and components are similar to existing CD technology.

Just as Microsoft has pushed the acceptance of new operating systems by ensuring that desirable new software is only available on the new platform, and just as software publishers have driven the acceptance of CD-ROM drives by making products available only on CD, the establishment of DVD-ROM drives will be hastened by the desire of software publishers for a bigger vessel in which to distribute their products.

The Interregnum of CD-RW

Rewritable CD (CD-RW), finalized at the end of 1996 and introduced shortly after DVD-ROM, had a tough row to hoe. It finally brought the Holy Grail of erasability to the basic CD format, but only by being incompatible with all existing CD drives and players. Had the makers of DVD-RAM been aggressive and seized the opportunity, they could have taken over the writable market, especially if they had introduced DVD-RAM drives that could write to CD-R discs. Instead, DVD-RAM drives remained high-priced specialty items, while CD-RW drives sold better than expected. This was not because CD-RW itself was a success, but because CD-R media became astoundingly cheap. For every CD-RW disc sold in 2000, about 100 CD-R discs were sold. In 1999, 1.7 billion CD-Rs were produced worldwide to meet a demand between 1.3 and 1.4 billion, while CD-RW demand was just over 15 million worldwide. People bought CD-RW drives so they could record CD-Rs, specifically music CD-Rs that would play in existing CD audio players. In 2000, CD-RW drives outsold DVD drives 28.7 million to 22.6 million. Once writable DVD drives that can also write to CD-R discs are available for under $400, they will begin encroaching on the CD-R/RW drive market.

Standards, Anyone?

Computer standards often develop like meandering cowpaths that later become so well traveled that they end up being paved into meandering roads. DVD offers both promise and hardship: a new format that can be done well or a new jungle through which too many crooked paths can be blazed.

Take, for instance, DVD's predecessor, MPEG-1 video on CD, which was a standards nightmare. Each hardware and software implementation of MPEG-1 was slightly different, with companies designing their own different MCI interfaces and data formats. The market for MPEG-1 video hardware decoding, which at first looked so promising as an alternative to quarter-screen, low-resolution video codecs, took a nosedive before it ever got off the ground. Software publishers discovered that their movies might play fine on one card, but not on another. The OpenMPEG Consortium was created in an attempt to resolve the problems but did not receive sufficient industry support and was never able to accomplish much.

Some of the members of the OpenMPEG Consortium, now sadder and wiser, formed the OpenDVD Consortium with more pragmatic goals. No attempt was made to define a standard, and instead, the organization provided a common meeting ground for computer DVD developers and served as a clearinghouse for information. Unfortunately, the OpenDVD Consortium barely got off the ground before it faded away.

The *Interactive Multimedia Association* (IMA), with slightly loftier goals, established numerous *technical working groups* (TWGs) in December of 1996 to deal with issues such as standardized architectures, cross-platform interactive media formats, hybrid Internet-DVD applications, data interchange formats for DVD authoring and production, technical safeguards, and compliance testing. The IMA was later absorbed by the *Software Pub-*

lishers Association (SPA), which lost interest in DVD as it went through its own metamorphoses and name changes. The *DVD Association* (DVDA), established in 1999, also formed working groups for standardizing DVD applications and production processes. It remains to be seen what kind of success they will have.

Microsoft's DirectShow provides the broadest standard for DVD application development, in spite of poor support from computer makers and decoder software developers. InterActual's DVD application development software, which builds on top of DirectShow but also supports other proprietary decoders along with other platforms such as Apple Macintoshes and Nuon-based players, may become the de facto standard.

Mr. Computer Goes to Hollywood

As digital technology steadily improves, the computer will become as much a media machine as a computation machine. Early distinctions between DVD-Video and DVD-ROM will become less important as new computers fully incorporate high-fidelity audio and video, resulting in thoroughly amazing applications as the interactive power of Silicon Valley and the sensory impact of Hollywood coalesce on desktop computers.

The Changing Face of Home Entertainment

The year 1996 presaged the digital revolution of home entertainment. *Direct broadcast digital satellite* (DBS) systems were introduced and purchased by the millions. DVD was unveiled, though it did not manage to appear anywhere but in Japan before the end of the year. In December of 1996, the FCC approved the U.S. DTV standard[2] that ushered in 1997, the Year of Waiting for HDTV.[3]

DBS and DVD each have one foot in the past and one foot in the future. They use digital signal recording and compression methods to squeeze the

---

[2]DTV is a set of broadcast standards for digital HDTV transmission in the U.S. Other countries may develop different HDTV standards, such as DVB in Europe.

[3]Followed by 1998, the Year No One Can Afford HDTV. Followed by 1999, the Year No One Paid Much Attention to HDTV. Followed by 2000, the Year of Wondering If the DTV Format Will Be Changed.

most quality into their limited transmission and storage capacities, but they convert the signal to analog format for display on conventional televisions. As digital televisions slowly push their creaky predecessors out of living rooms and offices, the pure digital signal of DVD and similar technologies will be ready and waiting for them. Video from existing discs and players will look better than ever on widescreen digital televisions. New-generation DVD players will enhance existing discs by generating progressive-scan video and by supplying it via digital connectors, leaving behind the flicker and flutter of interlaced video and enabling all television personalities everywhere to wear pinstriped fabrics and sit in front of miniblinds.

It will take many years for digital television to establish even a foothold, yet this period will also witness the computerization of television. This does not mean that every television will sprout a floppy disk drive, a mouse, and system crash error messages. It does mean, however, that as consumer electronics companies and home computer makers endeavor to make their products more suitable to the basic entertainment needs, communication needs, and productivity needs of home users, features from computers such as onscreen menus, pointers, multiple windows, and even an Internet connection will become commonplace. DVD is the perfect format for this environment by being able to carry content for both the television, the computer, and all the variations in between.

In the meantime, the traditional television is getting better with each passing year. The term "big-screen TV" applies to ever-larger sizes. Thirty-five-inch sets are now the price of 25-inch sets a few years ago, and of 19-inch sets a few years before that. Video projectors are coming close to the magic $2000 price point, where they'll compete with direct-view and rear-projection big-screen TVs. Widescreen TVs have been available for a while, but they have not caught the attention of most American buyers. Widescreen popularity in Europe and in Japan, where over 60 percent of new television purchases are widescreen, is helping to drive prices down, and the introduction of widescreen H/DTV will help even further. The biggest impediment has been a lack of movies and other video sources in widescreen format. DVD is filling this hole with its native support for wide aspect ratio displays.

Convergence will happen with systems other than the computer and TV. The prime candidates are cable set-top boxes, satellite receivers, video game consoles, Internet TV boxes, telephones, cellular phones, hand-held computers, and, of course, DVD players (see Figure 13.3). Engineers are designing new boxes that combine the features of digital satellite receivers and DVD players, since they share many of the same MPEG-2 digital video components. Digital cable boxes will combine the features of cable TV tuners and DVD players and will top it off with an Internet connection. Hard-disk based recorders are selling well and will add DVD recording capabilities once prices are low enough. Program listings from the Internet can be shown on the TV screen, for instance, so that shows can be marked for recording with the click of a remote control, and the system automatically compensates for delays caused by ball games or newsbreaks. Video game consoles include DVD drives for game play, giving them the capability to double as DVD movie players.

In short, because of the flexibility and interconnectability of inexpensive digital electronics, all the independent pieces we have grown accustomed to can be tossed in a big hat, shaken around, and pulled back out as new, mixed-and-matched systems with more features at lower prices. Customers will benefit from, and soon come to expect, personalization, freedom of location, and freedom of time. Our electronic devices will enable us to access anything, anytime, in the way we want.

At the end of 1999, digital video products generated 11 percent of the net consumer electronics market revenue. The CEA expects this number to double by 2003.

Even movie theaters are being changed by digital technology. Digital cinema trials around the world in 2000 were surprisingly better than almost anyone expected. The digital movie files were delivered on DVD-ROM discs but will eventually be sent over Internet connections. International release dates have been steadily getting closer to U.S. release dates, as the old model of movie distribution has been forced to change by Internet advertising that hits all audiences simultaneously. Movies that used to take six months to creep into theaters around the world now make a worldwide release in only one or two months.

DVD in the Classroom

Unlike the home entertainment market, where laserdiscs never gained more than a tiny videophile niche, the education market adopted laserdisc technology early on and in a big way. Over 200,000 laserdisc players have been sold to schools in the U.S., almost one for every building. So, the big question is, will DVD replace laserdisc? The answer is no, not directly.

Most of the appeal of DVD is lost in a classroom environment. Lessons usually never last more than 45 minutes, so there's no pressing need for hours of uninterrupted video. Surround-sound audio is pointless, especially since most laserdiscs and VCRs use the cheap speakers in the television. Laserdisc video quality is already as good or better than most of the class-

Chapter 13

Figure 13.3
Media convergence
in the digital age

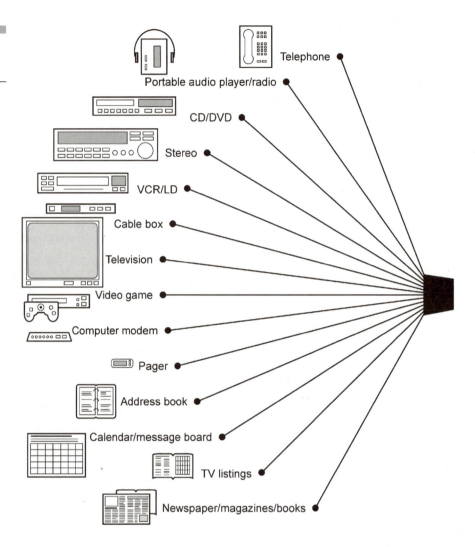

room televisions it's shown on. Multilingual audio tracks and subtitles are nice, but it's already possible to put four audio tracks on a laserdisc by using both channels of analog audio and both channels of digital audio. DVD's menus and interactivity are an improvement over passive laserdiscs, but most educators would prefer to make the jump to computers and gain significantly more power and flexibility.

Laserdisc players maintain advantages over DVD players, however. Any still picture or movie can be accessed almost instantly by entering a five-digit frame number on the remote control or, easier yet, by scanning a

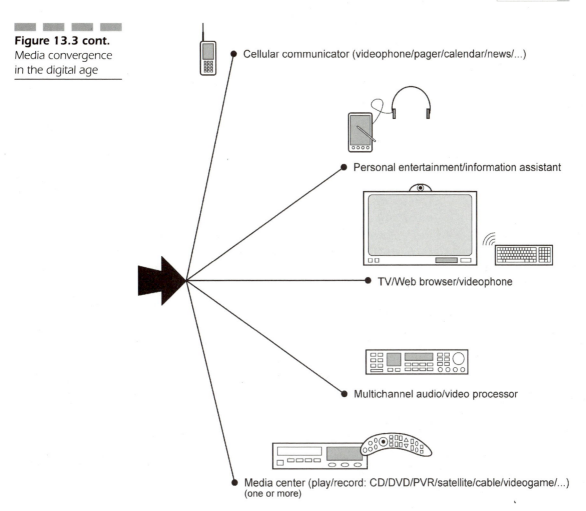

Figure 13.3 cont.
Media convergence
in the digital age

Cellular communicator (videophone/pager/calendar/news/...)

Personal entertainment/information assistant

TV/Web browser/videophone

Multichannel audio/video processor

Media center (play/record: CD/DVD/PVR/satellite/cable/videogame/...)
(one or more)

printed bar code. The biggest advantage is established content. Over 3,000 educational laserdiscs have been developed over the past 20 years or so. Many of these are still just as effective and useful as they were when they were first developed. It's improbable that such a wealth of visual resources will ever be developed for DVD-Video, especially since educational publishers will wait for schools to buy players, while schools wait for the publishers to make DVD-Video products.

Educational media publishers have already been spread thin by too many new formats, many of which seem as much a step backward as forward. Videotape appeared shortly after laserdisc, and though it could record, it had poor picture quality, no decent still-frame capability, and no random access. The subsequent "advance" was digitized video, using QuickTime and Video for Windows, playing from a CD. It was flexible and could be integrated with computers, but it lost vital detail with its quarter-screen fuzzy picture and required a TV converter box or a video projector for full-class presentation. Publishers were next asked to support Video CD, which provided full-screen digital video but still lacked quality and had almost no installed base of players. Then came the Internet, which was highly interactive and timely, but essentially precluded motion video. It's little wonder that gun-shy publishers are not eager to embrace yet another newfangled medium.

The failure of DVD-Video to unseat laserdisc in the classroom hardly means that laserdisc's days are unnumbered, however. Beginning in 1996, educational sales of players and discs dropped almost as quickly as in the home market. Competition from computers, and especially the Internet, is finally proving too much for the reliable but no longer glamorous workhorse, which will slowly decline over the next decade. The end of the heyday of laserdiscs signals the decline of all stand-alone audio/visual players for training and education.

DVD will succeed in the educational environment, but via computers rather than DVD-Video players. Long before educational DVD discs are plentiful enough, new computers being purchased by schools will be able to play DVD-Video discs as well as educational DVD-ROM software. Ironically, this promises to provide the market that educational publishers need in order to embrace DVD-Video. Cleverly designed products will be able to work in DVD computers at school and in DVD-Video players at home. Home education, both formal and informal, will drive the demand for this kind of product.

In the long run, multimedia PCs will replace classroom TVs, VCRs, overhead projectors, and laserdiscs, just as these have replaced filmstrip projectors and 16mm movies. Most lightweight content will come from the Internet, but the necessary graphics, audio, and video will be provided by DVD. Inasmuch as technology adoption in schools typically lags behind business and home use by several years, the process may take some time. This is unfortunate, since the case for computer and media integration in education is more compelling than almost anywhere else. Technology is not a magic elixir that will cure the ills of the education system, but it is a very powerful tool that when wielded properly can be truly effective. DVD-Video as currently implemented does not seem to be well suited as an educational tool, but computers combined with DVD in both -ROM and -Video form promise to substantially advance the state of the art by providing the kind of high-capacity knowledge bases and high-impact sensory environments that foster more effective learning.

The Far Horizon

In the very long term, the Internet will merge with cable TV, broadcast TV, radio, telephones, satellites, and eventually even newspapers and magazines. In other words, the Internet will take over the communications world. News, movies, music, advertising, education, games, financial transactions, e-mail, and most other forms of information will be delivered via this giant network. Internet bandwidth, currently lagging far behind the load being demanded of it, will eventually catch up, as did other systems such as intercontinental telephone networks and communications satellites. Discrete media such as DVD will then be relegated to niches such as software backups, archiving, time-shift "taping," and collector's editions of movies. Why go to a software store to buy a DVD-ROM or to a movie rental store to rent a DVD when you can have it delivered right to your computer or your TV or even your digital video recorder? In the intervening years, however, DVD in all its permutations and generations promises to be the definitive medium for both computers and home entertainment.

APPENDIX A

Quick Reference

Figure A.1
DVD-Video
conversion formulas

Formula		Example
Time (hours) =	$\dfrac{\text{space (G bytes)} \times 8000}{\text{data rate (Mbps)} \times 3600}$	$\dfrac{4.7}{5.5}$ x 2.22= 1.9 hours
Data rate (Mbps) =	$\dfrac{\text{space (G bytes)} \times 8000}{\text{time (hours)} \times 3600}$	$\dfrac{8.5}{3}$ x 2.22= 6.3 Mbps
Space (G bytes) =	$\dfrac{\text{data rate (Mbps)} \times \text{time (hours)} \times 3600}{8000}$	5 x 2 x 0.45 = 4.5 G bytes

For gigabytes (instead of billions of bytes) replace 8000 with 8590 (or replace 2.22 with 2.39, and 0.45 with 0.42)

Figure A.2
Data rate versus
playing time

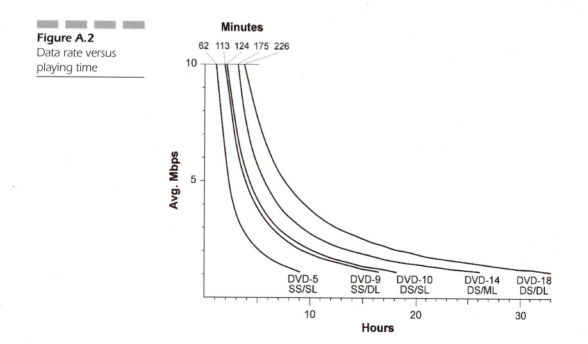

Figure A.3
Data rate versus
capacity

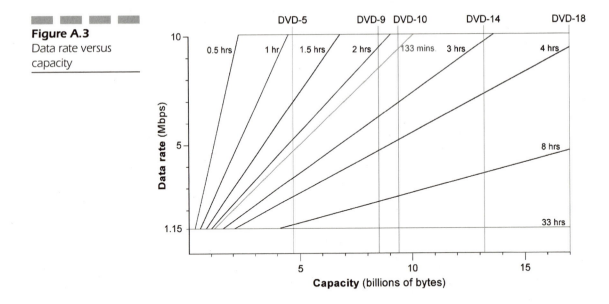

TABLE A.1

Meanings of Prefixes

Abbreviation	SI Prefix	IEC Prefix	IEC Abbr.	Common Use	Computer Use	Difference
k or K	kilo	kibi	Ki	$1000\ (10^3)$ [k]	$1024\ (2^{10})$ [K]	2.4%
M	mega	mebi	Mi	$1{,}000{,}000\ (10^6)$	$1{,}048{,}576\ (2^{20})$	4.9%
G	giga	gibi	Gi	$1{,}000{,}000{,}000\ (10^9)$	$1{,}073{,}741{,}824\ (2^{30})$	7.4%
T	tera	tebi	Ti	$1{,}000{,}000{,}000{,}000\ (10^{12})$	$1{,}099{,}511{,}627{,}776\ (2^{40})$	10%

TABLE A.2

DVD Capacities

Type	Sides/layers[a]	Billions of bytes[b]	Gigabytes[b]	Typical hours[c]	Min. to max. hours[d]	Typical audio hours[e]	Min. to max. audio hours[f]
12-cm size							
DVD-ROM (DVD-5)	SS/SL	4.7	4.37	2.25	1-9	4.5	1.7-63
DVD-ROM (DVD-9)	SS/DL	8.54	7.95	4	1.9-16.5	8.3	3.1-296
DVD-ROM (DVD-10)	DS/SL	9.4	8.75	4.5	2.1-18.1	9.1	3.4-326
DVD-ROM (DVD-14)	DS/ML	13.24	12.33	6.25	2.9-25.5	12.8	4.8-459
DVD-ROM (DVD-18)	DS/DL	17.08	15.91	8	3.8-33	16.5	6.2-593
DVD-R 1.0	SS/SL	3.95	3.67	1.75	0.9-7.6	3.8	1.4-137
DVD-R(G) 2.0	SS/SL	4.7	4.37	2.25	1-9	4.5	1.7-163
DVD-R(G) 2.0	DS/SL	9.4	8.75	4.5	2.1-18.1	9.1	3.4-326
DVD-R(A) 2.0	SS/SL	4.7	4.37	2.25	1-9	4.5	1.7-163
DVD-RAM 1.0	SS/SL	2.58	2.4	1.25	0.6-4.9	2.5	0.9-89
DVD-RAM 1.0	DS/SL	5.16	4.8	2.5	1.1-9.9	5	1.9-179
DVD-RAM 2.0	SS/SL	4.7	4.37	2.25	1-9	4.5	1.7-163
DVD-RAM 2.0	DS/SL	9.4	8.75	4.5	2.1-18.1	9.1	3.4-326
DVD-RW 1.0	SS/SL	4.7	4.37	2.25	1-9	4.5	1.7-163
DVD-RW 1.0	DS/SL	9.4	8.75	4.5	2.1-18.1	9.1	3.4-326
DVD+RW 2.0	SS/SL	4.7	4.37	2.25	1-9	4.5	1.7-163
DVD+RW 2.0	DS/SL	9.4	8.75	4.5	2.1-18.1	9.1	3.4-326
CD-ROM[g]	SS/SL	0.682	0.635	0.25[g]	0.2-1.3	0.7	0.2-23
DDCD-ROM	SS/SL	1.36	1.28	0.5[g]	0.3-2.6	1.3	0.5-47

continues

TABLE A.2 cont.

DVD Capacities

Type	Sides/layers[a]	Billions of bytes[b]	Gigabytes[b]	Typical hours[c]	Min. to max. hours[d]	Typical audio hours[e]	Min. to max. audio hours[f]
8-cm size							
DVD-ROM	SS/SL	1.46	1.36	0.75	0.3–2.8	1.4	0.5–50
DVD-ROM	SS/DL	2.65	2.47	1.25	0.6–5.1	2.6	1–92
DVD-ROM	DS/SL	2.92	2.72	1.5	0.6–5.6	2.8	1.1–101
DVD-ROM	DS/ML	4.12	3.83	2	0.9–7.9	4	1.5–143
DVD-ROM	DS/DL	5.31	4.95	2.5	1.2–10.2	5.1	1.9–184
DVD-RAM 2.0	SS/SL	1.46	1.36	0.75	0.3–2.8	1.4	0.5–50
DVD-RAM 2.0	DS/SL	2.92	2.72	1.5	0.6–5.6	2.8	1.1–101
CD-ROM[g]	SS/SL	0.194	0.18	0.07[h]	0–0.3	0.2	0.1–6
DDCD-ROM	SS/SL	0.388	0.36	0.14[h]	0.1–0.7	0.4	0.1–13

[a]Writable DVDs have only one layer per side. DVD-14 (and the corresponding eight-cm size) has one layer on one side and two layers on the other.

[b]Reference capacities in billions of bytes (10^9) and gigabytes (2^{30}). Actual capacities can be slightly larger if the track pitch is reduced.

[c]Approximate video playback time, given an average data rate of 4.7 Mbps. Actual playing times can be much longer or shorter (see next column).

[d]Minimum video playback time at the highest data rate of 10.08 Mbps. Maximum playback time at the MPEG-1 data rate of 1.15 Mbps.

[e]Typical audio-only playback time at the two-channel MLP audio rate of 96 kHz and 24 bits (2.3 Mbps).

[f]Minimum audio-only playback time at the highest single-stream PCM audio rate of 6.144 Mbps. Maximum audio-only playback time at the lowest Dolby Digital or MPEG-2 data rate of 64 kbps.

[g]Mode 1, 74 minutes (333,000 sectors) or 21 minutes (94,500 sectors). Audio/video times are for comparison only.

[h]Assuming that the data from the CD is transferred at a typical DVD video data rate, about four times faster than a single-speed CD-ROM drive.

TABLE A.3

Playing Times for Various Data Rates

Video (Average)	Data Rate (Mbps) (Audio Tracks & Format)	Playing Time per Disc, minutes (hours)					
		Total[a]	DVD-5	DVD-9	DVD-10	DVD-14	DVD-18
3.5	1.344 (3 DD5.1)	4.88	128 (2.1)	233 (3.8)	256 (4.2)	361 (6)	466 (7.7)
3.5	0.896 (2 DD5.1)	4.44	141 (2.3)	256 (4.2)	282 (4.7)	397 (6.6)	513 (8.5)
3.5	0.448 (1 DD5.1)	3.99	157 (2.6)	285 (4.7)	314 (5.2)	442 (7.3)	571 (9.5)
3.5	3.072 (8 DD5.1)	6.61	94 (1.5)	172 (2.8)	189 (3.1)	266 (4.4)	344 (5.7)
3.5	1.536 (1 48/16 PCM)	5.08	123 (2)	224 (3.7)	246 (4.1)	347 (5.7)	448 (7.4)
8.7	1.344 (3 DD5.1)	10.08	62 (1)	112 (1.8)	124 (2)	175 (2.9)	225 (3.7)
9.6	0.448 (1 DD5.1)	10.08	62 (1)	112 (1.8)	124 (2)	175 (2.9)	225 (3.7)
7.0	0.896 (2 DD5.1)	7.94	78 (1.3)	143 (2.3)	157 (2.6)	222 (3.7)	286 (4.7)
6.0	0.896 (2 DD5.1)	6.94	90 (1.5)	164 (2.7)	180 (3)	254 (4.2)	328 (5.4)
6.0	0.384 (2 DD2.0)	6.42	97 (1.6)	177 (2.9)	195 (3.2)	274 (4.5)	354 (5.9)
5.0	0.896 (2 DD5.1)	5.94	105 (1.7)	191 (3.1)	211 (3.5)	297 (4.9)	383 (6.3)
4.0	0.896 (2 DD5.1)	4.94	126 (2.1)	230 (3.8)	253 (4.2)	357 (5.9)	461 (7.6)
3.0	0.896 (2 DD5.1)	3.94	159 (2.6)	289 (4.8)	318 (5.3)	448 (7.4)	578 (9.6)
2.0	0.192 (1 DD2.0)	2.23	280 (4.6)	510 (8.5)	561 (9.3)	790 (13.1)	1020 (17)
1.86[b]	0.192 (1 DD2.0)	2.09	299 (4.9)	544 (9)	599 (9.9)	843 (14)	1088 (18.1)
1.5[b]	0.192 (1 DD2.0)	1.73	361 (6)	657 (10.9)	723 (12)	1019 (16.9)	1314 (21.9)
1.15[c]	0.224 (1 Layer II)	1.41	443 (7.3)	805 (13.4)	886 (14.7)	1248 (20.8)	1610 (26.8)
1.15[b]	0.064 (1 DD1.0)	1.25	499 (8.3)	908 (15.1)	999 (16.6)	1407 (23.4)	1816 (30.2)
1.0[b]	0.064 (1 DD1.0)	1.10	567 (9.4)	1031 (17.1)	1135 (18.9)	1599 (26.6)	2062 (34.3)
0.7[d]	0.064 (1 MP3)	0.80	779 (12.9)	1416 (23.6)	1558 (25.9)	2195 (36.5)	2832 (47.2)

Note: DD = Dolby Digital

[a]Total data rate includes four subpicture streams (0.04 Mbps)

[b]MPEG-1 video

[c]Video and audio rates equivalent to Video CD

[d]MPEG-4 video. It will not play on a standard DVD-Video player.

TABLE A.4

Approximate Audio Playing Times at Various Data Rates

| | | Playing Time per Disc (hours) | | | | | | | | | |
| | | No Video | | | | | +4 Mbps Video (Av.) | | | | |
Format	kbps	DVD-5	DVD-9	DVD-10	DVD-14	DVD-18	DVD-5	DVD-9	DVD-10	DVD-14	DVD-18
DD 1.0	64	163.1	296.5	326	459.7	593	2.5	4.6	5.1	7.2	9.3
DD 2.0	192	54.3	98.8	108.6	153.2	197.6	2.4	4.5	4.9	7	9
DD 5.1	384	27.1	49.4	54.3	76.6	98.8	2.3	4.3	4.7	6.7	8.6
DD 5.1 max	448	23.3	42.3	46.5	65.6	84.7	2.3	4.2	4.6	6.6	8.5
2 DD 5.1	768	13.5	24.7	27.1	38.3	49.4	2.1	3.9	4.3	6.1	7.9
2 DD 5.1 max	896	11.6	21.1	23.2	32.8	42.3	2.1	3.8	4.2	6	7.7
MPEG 7.1 max	912	11.4	20.8	22.8	32.2	41.6	2.1	3.8	4.2	5.9	7.7
3 DD 5.1	1152	9	16.4	18.1	25.5	32.9	2	3.6	4	5.7	7.3
3 DD 5.1 max	1344	7.7	14.1	15.5	21.8	28.2	1.9	3.5	3.9	5.5	7.1
PCM 48/16 stereo	1536	6.7	12.3	13.5	19.1	24.7	1.8	3.4	3.7	5.3	6.8
PCM 48/20 stereo	1920	5.4	9.8	10.8	15.3	19.7	1.7	3.2	3.5	4.9	6.4
8 DD 5.1	3072	3.3	6.1	6.7	9.5	12.3	1.4	2.6	2.9	4.1	5.3
PCM 96/20 stereo	3840	2.7	4.9	5.4	7.6	9.8	1.3	2.4	2.6	3.7	4.8

Note: DD = Dolby Digital

TABLE A.5

Stream Data Rates

	Minimum (kbps)	Typical (kbps)	Maximum (kbps)
MPEG-2 video	1500[a]	3500	9800
MPEG-1 video	900[a]	1150	1856
PCM (DVD-Video)	768	1536	6144
MLP/PCM (DVD-Audio)	n/a	6900	9600
Dolby Digital	64	384	448
MPEG-1 audio	64	192	384
MPEG-2 audio	64	384	912
Subpicture	n/a	10	3360

[a]Not an absolute limit but a practical limit below which video quality is too poor.

TABLE A.6

Physical Characteristics of DVD

Thickness	1.2 mm (+0.30/−0.06) (two bonded substrates)
Substrate thickness	0.6 mm (+0.043/−0.030)
Spacing layer thickness	55 μm (±15)
Mass	13 to 20 g (12-cm disc) or 6 to 9 g (8-cm disc)
Diameter	120 or 80 mm (±0.30)
Spindle hole diameter	15 mm (+0.15/−0.00)
Clamping area diameter	22 to 33 mm
Inner guardband diameter	33 to 44 mm
Burst cutting area diameter	44.6 mm (+0.0/−0.8) to 47 (±0.10) mm
Lead-in diameter	45.2 to 48 mm (+0.0/−0.4)
Data diameter	48 mm (+0.0/−0.4) to 116 mm (12-cm disc) or 76 mm (8-cm disc)
Lead-out diameter	Data + 2 mm (70 mm min. to 117 mm max. or 77 mm max.)
Outer guardband diameter	117 to 120 mm or 77 to 80 mm
Radial runout (disc)	<0.3 mm, peak to peak
Radial runout (tracks)	<100 μm, peak to peak

continues

TABLE A.6 cont.

Physical Characteristics of DVD

Index of refraction	1.55 ($\pm$0.10)
Birefringence	0.10 μm max.
Reflectivity	45 to 85% (SL), 18 to 30% (DL)[a]
Readout wavelength	650 or 635 nm (640 $\pm$15) (red laser)
Polarization	Circular
Numerical aperture	0.60 ($\pm$0.01) (objective lens)
Beam diameter	1.0 mm ($\pm$0.2)
Optical spot diameter	0.58 to 0.60 μm
Refractive index	1.55 ($\pm$0.10)
Tilt margin (radial)	$\pm$0.8°
Track spiral (outer layer)	Clockwise
Track spiral (inner layer)	Clockwise or counterclockwise
Track pitch	0.74 μm ($\pm$0.01 avg., 0.03 max.)
Pit length	0.400 to 1.866 μm (SL), 0.440 to 2.054 μm (DL) (3T to 14T)
Data bit length (avg.)	0.2667 μm (SL), 0.2934 μm (DL)
Channel bit length (avg.)	0.1333 ($\pm$0.0014) μm (SL), 0.1467 ($\pm$0.0015) μm (DL)
Jitter	<8% of channel bit clock period
Correctable burst error	6.0 mm (SL), 6.5 mm (DL)
Maximum local defects	100 μm (air bubble), 300 μm (black spot), no more than six defects between 30 and 300 μm in an 80-mm scanning distance
Rotation	Counterclockwise to readout surface
Rotational velocity[b]	570 to 1630 rpm (574 to 1528 rpm in data area)
Scanning velocity[b]	3.49 m/s (SL), 3.84 m/s (DL) ($\pm$0.03)
Storage Temperature	-20 to 50°C (-4 to 112°F), $\leq$ 15°C/h variation (59°F/h)
Storage humidity	-5 to 90% relative, 1 to 30 g/m^3 absolute, $\leq$10%/h variation
Operating temperature	-25 to 70°C (-13 to 158°F), $\leq$50°C sudden change (122°F)
Operating humidity	-3% to 95% relative, 0.5 to 60 g/m^3 absolute, $\leq$30% sudden change

[a]SL = single layer; DL = dual layer.

[b]Reference value for a single-speed drive.

TABLE A.7

DVD and CD
Characteristics
Comparison

	DVD	CD
Thickness	1.2 mm (2 × 0.6)	1.2 mm
Mass (12 cm)	13 to 20 g	14 to 33 g
Diameter	120 or 80 mm	120 or 80 mm
Spindle hole diameter	15 mm	15 mm
Lead-in diameter	45.2 to 48 mm	46 to 50 mm
Data diameter (12 cm)	48 to 116 mm	50 to 116 mm
Data diameter (8 cm)	48 to 76 mm	50 to 76 mm
Lead-out diameter	70 to 117 mm	76 to 117 mm
Outer guardband dia. (12 cm)	117 to 120 mm	117 to 120 mm
Outer guardband dia. (8 cm)	77 to 80 mm	77 to 80 mm
Reflectivity (full)	45% to 85%	70% min.
Readout wavelength	650 or 635 nm	780 nm
Numerical aperture	0.60	0.38 to 0.45
Focus depth	0.47 μm	1 (± 2 μm)
Track pitch	0.74 μm	1.6 μm (1.1 μm[a])
Pit length	0.400 to 1.866 μm (SL), 0.440 to 2.054 μm (DL)[b]	0.833 to 3.054 μm (1.2 m/s), 0.972 to 3.560 μm (1.4 m/s); [0.623 to 2.284 μm[a] (0.90 m/s)]
Pit width	0.3 μm	0.6 μm
Pit depth	0.16 μm	0.11 μm
Data bit length	0.2667 μm (SL), 0.2934 μm (DL)	0.6 μm (1.2 m/s), 0.7 μm (1.4 m/s)
Channel bit length	0.1333 μm (SL), 0.1467 μm (DL)	0.3 μm
Modulation	8/16	8/14 (8/17 w/merge bits)
Error correction	RS-PC	CIRC (CIRC7[a])
Error correction overhead	13%	23%/34%[c]
Bit error rate	10^{-15}	10^{-14}

continues

TABLE A.7 cont.

DVD and CD Characteristics Comparison

	DVD	CD
Correctable error (1 layer)	6 mm (SL), 6.5 mm (DL)	2.5 mm
Speed (rotational)[d]	570 to 1600 rpm	200 to 500 rpm
Speed (scanning)[d]	3.49 m/s (SL), 3.84 m/s (DL)	1.2 to 1.4 m/s (0.90 m/s[a])
Channel data rate[d]	26.15625 Mbps	4.3218 Mbps (8.6436 Mbps[a])
User data rate[d]	11.08 Mbps	1.41 Mbps/1.23 Mbps[c]
User data:channel data	2048:4836 bytes	2352:7203/2048:7203[c]
Format overhead	136 percent	206 percent/252 percent[c]
Capacity	1.4 to 8.0 GB per side	0.783/0.635 GB[c]

[a]Double-density CD

[b]SL = single layer, DL = dual layer

[c]CD-DA / CD-ROM Mode 1.

[d]Reference value for a single-speed drive.

TABLE A.8

Comparison of
MMCD, SD, and
DVD

	MMCD	SD	DVD
Diameter	120 mm	120 mm	120 mm
Thickness	1.2 mm	2×0.6 mm	1.2 mm
Sides	1	1 or 2	1 or 2
Layers	1 or 2	1 or 2	1 or 2
Data area (diameter)	46 to 116 mm	48 to 116 mm	48 to 116 mm
Min. pit length	0.451 μm	0.400 μm	0.400 μm
Track pitch	0.84 μm	0.74 μm	0.74 μm
Scanning velocity	4.0 m/s	3.27 m/s	3.49 m/s
Laser wavelength	635 nm	650 nm	650 or 635 nm
Numerical aperture	0.52	0.60	0.60
Modulation	8/16	8/15	8/16
Channel data rate	26.6 Mbps	24.54 Mbps	26.16 Mbps
Max. User data rate	11.2 Mbps	10.08 Mbps	11.08 Mbps
Avg. User data rate	3.7 Mbps	4.7 Mbps	4.7 Mbps
Capacity (single layer)	3.7 G bytes	5.0 G bytes	4.7 G bytes
Capacity (dual layer)	7.4 G bytes	9.0 G bytes	8.54 G bytes
Sector size	2048 bytes	2048 bytes	2048 bytes
Error correction	CIRC+	RS-PC	RS-PC
Stated playing time	135 minutes	140 minutes	133 minutes
Video encoding	MPEG-2 VBR	MPEG-2 VBR	MPEG-2 VBR
Audio encoding	MPEG-2 Layer II	AC-3	AC-3, MPEG-2, PCM, etc.

TABLE A.9

Data Storage
Characteristics of
DVD

Modulation	8/16 (EFMPlus)
Sector size (user data)	2048 bytes
Logical sector size (data unit 1)	2064 bytes (2048 + 12 header + 4 EDC)
Recording sector size (data unit 2)	2366 bytes (2064 + 302 ECC)
Unmodulated physical sector (data unit 3)	2418 bytes (2366 + 52 sync)
Physical sector size	4836 (2418 × 2 modulation)
Error correction	Reed-Solomon product code (208,192,17) × (182,172,11)
Error correction overhead	15% (13% of recording sector: 308/2366)
ECC block size	16 sectors (32,678 bytes user data, 37,856 bytes total)
Format overhead	16% (37,856/32,678)
Maximum random error	280 in eight ECC blocks
Channel data rate[a]	26.16 Mbps
User data rate[a]	11.08 Mbps
Capacity (per side, 12 cm)	4.37 to 7.95 GB (4.70 to 8.54 billion bytes)
Capacity (per side, eight cm)	1.36 to 2.48 GB (1.46 to 2.66 billion bytes)

[a]Reference value for a single-speed drive.

TABLE A.10

DVD-Video Physical
Units

Unit	Maximum
Video title set (VTS)	99 per disc
Video object set (VOBS)	99 per VTS
Video object (VOB)	32767 per VOBS
Cell	255 per VOB
Video object unit (VOBU)	
Pack (PCK)	2048 bytes
Packet (PKT)	

TABLE A.11

DVD-Video
Logical Units

Unit	Maximum
Title	99 per disc
Parental block (PB)	
Program chain (PGC)	999 per title, 16 per parental block
Part of title (PTT)	999 per title, 99 per one-sequential-PGC title
Program (PG)	99 per PGC
Angle block (AB)	
Interleave block (ILVB)	
Interleave unit (ILVU)	
Cell pointer	255 per PGC

TABLE A.12

DVD-Video format

Multiplexed data rate	Up to 10.08 Mbps
Video data	One stream
Video data rate	Up to 9.8 Mbps (typical avg. 3.5)
TV system	525/60 (NTSC) or 625/50 (PAL)
Video coding	MPEG-2 MP@ML/SP@ML VBR/CBR or MPEG-1 VBR/CBR
Coded frame rate	24 fps[a] (film), 29.97 fps[b] (525/60), 25 fps[b] (625/50)
Display frame rate	29.97 fps[b] (525/60), 25 fps[b] (625/50)
MPEG-2 resolution	$720 \times 480, 704 \times 480, 352 \times 480$ (525/60); $720 \times 576, 704 \times 576, 352 \times 576$ (625/50)
MPEG-1 resolution	352×240 (525/60); 352×288 (625/50)
MPEG-2 GOP max.	36 fields (525/60), 30 fields (625/50)
MPEG-1 GOP max.	18 frames (525/60), 15 frames (625/50)
Aspect ratio	4:3 or 16:9 anamorphic[c]
Pixel aspect ratio	Refer to Table 6.22

[a]Progressive (decoder performs 2-3 or 2-2 pulldown)

[b]Interlaced (59.94 fields per second or 50 fields per second)

[c]Anamorphic only allowed for 720 and 704 resolutions.

TABLE A.13	Sample frequency	48 kHz
Dolby Digital Audio Details	Sample size	Up to 24 bits
	Bit rate	64 to 448 kbps; 384 or 448 kbps typical
	Channels (front/rear)[a]	1/0, 2/0, 3/0, 2/1, 2/2, 3/1, 3/2, 1+1/0 (dual mono)
	Karaoke modes	L/R, M, V1, V2

[a]LFE channel can be added to all variations.

TABLE A.14	Sample frequency	48 kHz only
MPEG Audio Details	Sample size	Up to 20 bits
	MPEG-1	Layer II only
	MPEG-1 bit rate	64 to 192 kbps (mono), 64 to 384 kbps (stereo)
	MPEG-2	BC (matrix) mode only
	MPEG-2 bit rate[a]	64 to 912 kbps
	Extension streams[b]	5.1-channel, 7.1-channel
	Channels (front/rear)[c]	1/0, 2/0, 2/1, 2/2, 3/0, 3/1, 3/2, 5/2 (no dual channel or multilingual)
	Karaoke channels	L, R, A1, A2, G
	Emphasis	None
	Prediction	Not allowed

[a]MPEG-1 Layer II stream + extension stream(s)

[b]AAC (unmatrix, NBC) not allowed

[c]LFE channel can be added to all variations.

TABLE A.15

PCM Audio Details

Sample frequency	48 or 96 kHz
Sample size	16, 20, or 24 bits
Channels	1, 2, 3, 4, 5, 6, 7, or 8
Karaoke channels	L, R, V1, V2, G

TABLE A.16

DTS Audio Details

Sample frequency	48 kHz
Sample size	Up to 24 bits
Bit rate	64 to 1536 kbps; 768 or 1536 kbps typical
Channels (front/rear)[a]	1/0, 2/0, 3/0, 2/1, 2/2, 3/2, 3/3 (no multilingual)
Karaoke modes	L/R, M, V1, V2

[a]LFE channel can be added to all variations.

TABLE A.17

Subpicture Details

Data	0 to 32 streams
Data rate	Up to 3.36 Mbps
Unit size	53,220 bytes (up to 32,000 bytes of control data)
Coding	RLE (max. 1440 bits/line)
Resolution[a]	Up to 720 × 478 (525/60) or 720 × 573 (625/50)
Bits per pixel	Two (defining one of four types)
Pixel types	Background, foreground, emphasis-1, emphasis-2
Colors[a]	Four of 16 (from four-bit palette,[b] one per type)
Contrasts[a]	Four of 16 (from four-bit palette,[b] one per type)

[a]Area, content, color, and contrast can be changed for each field.

[b]Color palette and contrast can be changed every PGC.

	No.	Description	Access	Values	Default Value
TABLE A.18 *Player System Parameters (SPRMs)*	0	Preferred menu language	Read-only	Two lowercase ASCII letters (ISO 639)	Player-specific
	1	Audio stream number	Read/write	0 to 7 or 15 (none)	15 (Fh)
	2	Subpicture stream number and on/off state	Read/write	b0–b5: 0 to 31 or 62 (none) or 63 (forced subpicture) b6: display flag (0 = do not display)	62 (3Eh)
	3	Angle number	Read/write	1 to 9	1
	4	Title number in volume	Read/write	1 to 99	1
	5	Title number in VTS	Read/write	1 to 99	1
	6	PGC number	Read/write	1 to 32,767	Undefined
	7	Part of title number	Read/write	1 to 99	1
	8	Highlighted button number	Read/write	1 to 36	1
	9	Navigation timer	Read-only[a]	0 to 65,536 (seconds)	0
	10	PGC jump for navigation timer	Read-only[a]	1 to 32,767 (PGC in current title)	Undefined
	11	Karaoke audio mixing mode	Read/write	b2: mix ch2 to ch1 (0 = do not mix) b3: mix ch3 to ch1 b4: mix ch4 to ch1 b10: mix ch2 to ch0 b11: mix ch3 to ch0 b12: mix ch4 to ch0	0
	12	Parental management country code	Read-only	Two uppercase ASCII letters (ISO 3166) or 65,535 (none)	Player-specific
	13	Parental level	Read/write	1 to 8 or 15 (none)	Player-specific

continues

No.	Description	Access	Values	Default Value
14	Video preference and current mode	Read-only	b10–b11: preferred display aspect ratio 0 (00b): 4:3 2 (01b): not specified 3 (10b): reserved 4 (11b): 16:9 b8–b9: current video output mode 0 (00b): normal (4:3) or wide (16:9) 1 (01b): pan-scan (4:3) 2 (10b): letterbox (4:3) 3 (11b): reserved	Player-specific
15	Player audio capabilities	Read-only	b2: SDDS karaoke (0 = cannot play) b3: DTS karaoke b4: MPEG karaoke b6: Dolby Digital karaoke b7: PCM karaoke b10: SDDS b11: DTS b12: MPEG b14: Dolby Digital	Player-specific
16	Preferred audio language	Read-only	Two lowercase ASCII letters (ISO 639) or 65,535 (none)	65,535 (FFFFh)
17	Preferred audio language extension	Read-only	0 = Not specified 1 = Normal audio 2 = Audio for visually impaired 3 = Director comments 4 = Alternate director comments	0
18	Preferred subpicture language	Read-only	Two lowercase ASCII letters (ISO 639) or 65,535 (none)	65,535 (FFFFh)

continues

TABLE A.18 cont.	No.	Description	Access	Values	Default Value
Player System Parameters (SPRMs)	19	Preferred subpicture language extension		0 = Not specified 1 = Normal subtitles 2 = Large subtitles 3 = Subtitles for children 5 = Normal captions 6 = Large captions 7 = Captions for children 9 = Forced subtitles 13 = Director comments 14 = Large director comments 15 = Director comments for children	0
	20	Player region code (mask)	Read-only	One bit set for corresponding region (00000001 = region 1, 00000010 = region 2, etc.)	Player-specific
	21	Reserved			
	22	Reserved			
	23	Reserved for extended playback mode			

ᵃBits within the word are referred to as b0 (low order bit) through b15 (high order bit).

TABLE A.19

Video resolution

Format		VHS (1.33)	VHS (1.78)	VHS (2.35)	LD (1.33)	LD (1.78)	LD (1.85)	LD (2.35)	VCD (1.33)	VCD (1.78)	VCD (2.35)
NTSC	TVL	250	250	250	425	425	425	425	264	264	264
	H pixels	333	333	333	567	567	567	567	352	352	352
	V pixels	480	360	272	480	360	346	272	240	180	136
	Total pixels	159,840	119,880	90,576	272,160	204,120	196,182	154,224	84,480	63,360	47,872
PAL	TVL	240	240	240	450	450	450	450	264	264	264
	H pixels	320	320	320	600	600	600	600	352	352	352
	V pixels	576	432	327	576	432	415	327	288	216	163
	Total pixels	184,320	138,240	104,640	345,600	259,200	249,000	196,200	101,376	76,032	57,376

Format		DVD (1.33/1.78)	DVD (1.85)	DVD (2.35)	DTV3 (1.33)	DTV3 (1.78)	DTV3 (2.35)	DTV4 (1.78)	DTV4 (2.35)
NTSC	TVL	540/405	540/405	540/405	720	720	720	1,080	1,080
	H pixels	720	720	720	1,280	1,280	1,280	1,920	1,920
	V pixels	480	461	363	960	720	545	1080	817
	Total pixels	345,600	331,920	261,360	1,228,800	921,600	697,600	2,073,600	1,568,640
PAL	TVL	540	540	720					
	H pixels	720	720	720					
	V pixels	576	554	436					
	Total pixels	414720	398,880	313,920					

continues

TABLE A.19 cont.

Video resolution

1. DTV is neither PAL nor NTSC. The values are placed in the NTSC rows for convenience.

2. Wide aspect ratios (1.78 and 2.35) for VHS, LD, and VCD assume a letterboxed picture. For comparison, letterboxed 1.66 aspect ratio resolution is about seven percent higher than 1.78. Letterbox is also assumed for DVD and DTV at a 2.35 aspect ratio. DVD's native aspect ratio is 1.33; it uses anamorphic mode for 1.78. DTV's native aspect ratio is 1.78.

3. The very rare 1.78 anamorphic LD has the same pixel count as 1.33 LD. Anamorphic LD letterboxed to 2.35 has almost the same pixel count as 1.78 LD (567×363). The mostly non-existent 1.78 anamorphic VHS has the same pixel count as 1.33 VHS. Anamorphic VHS letterboxed to 2.35 has almost the same pixel count as 1.78 VHS (333×363). No commercial 2.35 anamorphic format exists and no corresponding stretch mode exists on widescreen TVs.

4. TVL is lines of horizontal resolution per picture height. For analog formats, the customary value is used; for digital formats, the value is derived from the actual horizontal pixel count adjusted for the aspect ratio. DVD's horizontal resolution is lower for 1.78 because the pixels are wider. Pixels for VHS and LD are approximations based on TVL and scan lines.

5. Resolutions refer to the medium, not the display. If a DVD player performs automatic letterboxing on a 1.85 movie (stored in 1.78), the displayed vertical resolution on a standard 1.33 TV is the same as from a letterboxed LD (360 lines).

TABLE A.20

Resolution Comparison of Different Video Formats

Format	VCD (1.78)	VCD (1.33)	VHS (1.78)	VHS (1.33)	LD (1.78)	LD (1.33)	DVD (1.78/1.33)	DTV3 (1.78)	DTV4 (1.78)
Horizontal pixels	352	352	333	333	567	567	720	1280	1,920
Vertical pixels	180	240	360	480	360	480	480	720	1,080
Total pixels	63,360	84,480	119,880	159,840	204,120	272,160	345,600	921,600	2,073,600
x VCD (16:9)		*4:3*	1.89	2.52	3.22	4.30	5.45	14.55	32.73
x VCD (4:3)			*1.42*	1.89	*2.42*	3.22	4.09	*10.91*	*24.55*
x VHS (16:9)				*4:3*	1.70	2.27	2.88	7.69	17.30
x VHS (4:3)					*1.28*	1.70	2.16	*5.77*	*12.97*
x LD (16:9)						*4:3*	1.69	4.51	10.16
x LD (4:3)							1.27	*3.39*	*7.62*
x DVD (16:9/4:3)								2.67	6.00
x DTV3 (16:9)									2.25

Note: 16:9 aspect ratios for VHS, LD, and VCD are letterboxed in a 4:3 picture. Comparisons between different aspect ratios are not as meaningful. These are shown in *italics*. Comparisons at 1.85 or 2.35 aspect ratios are essentially the same as at 1.78 (16:9).

TABLE A.21	Native data	Native rate (kbps)	Compression	Compressed Rate (kbps)*	Ratio	Percent
Compression Ratios	720 × 480 × 12 bits × 24 fps	99,533	MPEG-2	3,500	28:1	96
	720 × 480 × 12 bits × 24 fps	99,533	MPEG-2	6,000	17:1	94
	720 × 576 × 12 bits × 24 fps	119,439	MPEG-2	3,500	34:1	97
	720 × 576 × 12 bits × 24 fps	119,439	MPEG-2	6,000	20:1	95
	720 × 480 × 12 bits × 30 fps	124,416	MPEG-2	3,500	36:1	97
	720 × 480 × 12 bits × 30 fps	124,416	MPEG-2	6,000	21:1	95
	352 × 240 × 12 fps × 24 bits	24,330	MPEG-1	1,150	21:1	95
	352 × 288 × 12 fps × 24 bits	29,196	MPEG-1	1,150	25:1	96
	352 × 240 × 12 fps × 30 bits	30,413	MPEG-1	1,150	26:1	96
	2 ch × 48 kHz × 16 bits	1,536	Dolby Digital 2.0	192	8:1	87
	6 ch × 48 kHz × 16 bits	4,608	Dolby Digital 5.1	384	12:1	92
	6 ch × 48 kHz × 16 bits	4,608	Dolby Digital 5.1	448	10:1	90
	6 ch × 48 kHz × 16 bits	4,608	DTS 5.1	768	6:1	83
	6 ch × 48 kHz × 16 bits	4,608	DTS 5.1	1,536	3:1	67
	6 ch × 96 kHz × 20 bits	11,520	MLP	5,400	2:1	53
	6 ch × 96 kHz × 24 bits	13,824	MLP	7,600	2:1	45

*MPEG-2 and MLP compressed data rates are an average of a typical variable bit rate

TABLE A.22

Player and Media Compatibility

Disc	DVD-Video Player	DVD-ROM Drive	DVD/LD Player	LD Player	CD Player	CD-ROM Drive	Video CD Player
DVD-Video	yes	depends[1]	yes	no	no	no	no
DVD-ROM[2]	no	yes	no	no	no	no	no
LD	no	no	yes	yes	no	no	no
Audio CD	yes	yes	yes	yes[3]	yes	yes	yes
CD-ROM[4]	no	yes	no	no	no	yes	no
CD-R[5]	few[6]	some[6]	few[6]	yes[3]	yes	yes	yes
CD-RW[5]	yes	yes	yes	no	yes	some[7]	no
CDV	part[8]	part[8]	usually[3]	usually[3]	part[8]	part[8]	part[8]
Video CD	some[9]	depends[1]	some[9]	no	no	depends[1]	yes
Photo CD	no	depends[6,10]	no	no	no	depends[10]	no
CD-i	no	depends[11]	no	no	no	depends[11]	no

[1]Computer requires hardware or software to decode and display audio/video.

[2]DVD-ROM containing data other than standard DVD-Video files.

[3]Most newer LD players can play audio from a CD and both audio and video from a CDV.

[4]CD-ROM containing data other than standard CD digital audio.

[5]CD-R/RW containing CD digital audio data.

[6]DVD units require an additional laser tuned for CD-R readout wavelength.

[7]Only MultiRead CD-ROM drives can read CD-RW discs.

[8]CD digital audio part of disc only (no video).

[9]Not all DVD players can play Video CDs.

[10]Computer requires software to read and display Photo CD graphic files.

[11]Computer requires hardware or emulation software to run CD-i programs.

TABLE A.23

Compatibility of Writable DVD Formats

	DVD Unit	DVD-R(G) Unit	DVD-R(A) Unit	DVD-RW Unit	DVD-RAM Unit	DVD+RW Unit
DVD-ROM disc	Reads	Reads	Reads	Reads	Reads	Reads
DVD-R(G) disc	Usually reads	Reads, writes	Reads, does not write	Reads, often writes	Reads	Reads
DVD-R(A) disc	Usually reads	Reads, does not write	Reads, writes	Reads, does not write	Reads	Reads
DVD-RW disc	Usually reads	Reads	Reads	Reads, writes	Usually reads	Usually reads
DVD-RAM disc	Rarely reads	Does not read	Does not read	Does not read	Reads, writes	Does not read
DVD+RW disc	Usually reads	Usually reads	Usually reads	Usually reads	Usually reads	Reads, writes

TABLE A.24

ISO 639 Language
Codes

Language	Code	Hex	Dec
Abkhazian	ab	6162	24930
Afar	aa	6161	24929
Afrikaans	af	6166	24934
Albanian	sq	7371	29553
Amharic	am	616D	24941
Arabic	ar	6172	24946
Armenian	hy	6879	26745
Assamese	as	6173	24947
Avestan[1]	ae	6165	24933
Aymara	ay	6179	24953
Azerbaijani	az	617A	24954
Bashkir	ba	6261	25185
Basque	eu	6575	25973
Bengali; Bangla	bn	626E	25198
Bhutani	dz	647A	25722
Bihari	bh	6268	25192
Bislama	bi	6269	25193
Bosnian[1]	bs	6273	25203
Breton	br	6272	25202
Bulgarian	bg	6267	25191
Burmese	my	6D79	28025
Byelorussian	be	6265	25189
Cambodian	km	6B6D	27501
Catalan	ca	6361	25441
Chamorro[1]	ch	6368	25448
Chechen[1]	ce	6365	25445
Chichewa; Nyanja[1]	ny	6E79	28281

continues

TABLE A.24 cont.	Language	Code	Hex	Dec
ISO 639 Language Codes	Chinese	zh	7A68	31336
	Church Slavic[1]	cu	6375	25461
	Chuvash[1]	cv	6376	25462
	Cornish[1]	kw	6B77	27511
	Corsican	co	636F	25455
	Croatian	hr	6872	26738
	Czech	cs	6373	25459
	Danish	da	6461	25697
	Dutch	nl	6E6C	28268
	English	en	656E	25966
	Esperanto	eo	656F	25967
	Estonian	et	6574	25972
	Faeroese	fo	666F	26223
	Fiji	fj	666A	26218
	Finnish	fi	6669	26217
	French	fr	6672	26226
	Frisian	fy	6679	26233
	Galician	gl	676C	26476
	Georgian	ka	6B61	27489
	German	de	6465	25701
	Greek	el	656C	25964
	Greenlandic	kl	6B6C	27500
	Guarani	gn	676E	26478
	Gujarati	gu	6775	26485
	Hausa	ha	6861	26721
	Hebrew[2]	iw	6977	26999
	Herero[1]	hz	687A	26746

continues

Language	Code	Hex	Dec
Hindi	hi	6869	26729
Hiri Motu[1]	ho	686F	26735
Hungarian	hu	6875	26741
Icelandic	is	6973	26995
Indonesian[3]	in	696E	26990
Interlingua	ia	6961	26977
Interlingue	ie	6965	26981
Inuktitut[1]	iu	6975	26997
Inupiak	ik	696B	26987
Irish	ga	6761	26465
Italian	it	6974	26996
Japanese	ja	6A61	27233
Javanese	jw	6A77	27255
Kannada	kn	6B6E	27502
Kashmiri	ks	6B73	27507
Kazakh	kk	6B6B	27499
Kikuyu[1]	ki	6B69	27497
Kinyarwanda	rw	7277	29303
Kirghiz	ky	6B79	27513
Kirundi	rn	726E	29294
Komi[1]	kv	6B76	27510
Korean	ko	6B6F	27503
Kuanyama[1]	kj	6B6A	27498
Kurdish	ku	6B75	27509
Laothian	lo	6C6F	27759
Latin	la	6C61	27745
Latvian, Lettish	lv	6C76	27766

continues

TABLE A.24 cont.

ISO 639 Language Codes

Language	Code	Hex	Dec
Letzeburgesch[1]	lb	6C62	27746
Lingala	ln	6C6E	27758
Lithuanian	lt	6C74	27764
Macedonian	mk	6D6B	28011
Malagasy	mg	6D67	28007
Malay	ms	6D73	28019
Malayalam	ml	6D6C	28012
Maltese	mt	6D74	28020
Manx[1]	gv	6776	26486
Maori	mi	6D69	28009
Marathi	mr	6D72	28018
Marshall[1]	mh	6D68	28008
Moldavian	mo	6D6F	28015
Mongolian	mn	6D6E	28014
Nauru	na	6E61	28257
Navajo[1]	nv	6E76	28278
Ndebele, North[1]	nd	6E64	28260
Ndebele, South[1]	nr	6E72	28274
Ndonga[1]	ng	6E67	28263
Nepali	ne	6E65	28261
Northern Sami[1]	se	7365	29541
Norwegian	no	6E6F	28271
Norwegian Bokmål[1]	nb	6E62	28258
Norwegian Nynorsk[1]	nn	6E6E	28270
Occitan, Provençal	oc	6F63	28515
Oriya	or	6F72	28530
Ossetian; Ossetic[1]	os	6F73	28531

continues

TABLE A.24 cont.

ISO 639 Language Codes

Language	Code	Hex	Dec
Oromo (Afan)	om	6F6D	28525
Pali[1]	pi	7069	28777
Pashto, Pushto	ps	7073	28787
Persian	fa	6661	26209
Polish	pl	706C	28780
Portuguese	pt	7074	28788
Punjabi	pa	7061	28769
Quechua	qu	7175	29045
Rhaeto-Romance	rm	726D	29293
Romanian	ro	726F	29295
Russian	ru	7275	29301
Samoan	sm	736D	29549
Sangro	sg	7367	29543
Sanskrit	sa	7361	29537
Sardinian[1]	sc	7363	29539
Scots Gaelic	gd	6764	26468
Serbian	sr	7372	29554
Serbo-Croatian[4]	sh	7368	29544
Sesotho	st	7374	29556
Setswana	tn	746E	29806
Shona	sn	736E	29550
Sindhi	sd	7364	29540
Singhalese	si	7369	29545
Siswati	ss	7373	29555
Slovak	sk	736B	29547
Slovenian	sl	736C	29548
Somali	so	736F	29551

continues

Language	Code	Hex	Dec
TABLE A.24 cont. ISO 639 Language Codes			
Spanish	es	6573	25971
Sundanese	su	7375	29557
Swahili	sw	7377	29559
Swedish	sv	7376	29558
Tagalog	tl	746C	29804
Tahitian[1]	ty	7479	29817
Tajik	tg	7467	29799
Tamil	ta	7461	29793
Tatar	tt	7474	29812
Telugu	te	7465	29797
Thai	th	7468	29800
Tibetan	bo	626F	25199
Tigrinya	ti	7469	29801
Tonga	to	746F	29807
Tsonga	ts	7473	29811
Turkish	tr	7472	29810
Turkmen	tk	746B	29803
Twi	tw	7477	29815
Uighur[1]	ug	7567	30055
Ukrainian	uk	756B	30059
Urdu	ur	7572	30066
Uzbek	uz	757A	30074
Vietnamese	vi	7669	30313
Volapuk	vo	766F	30319
Welsh	cy	6379	25465
Wolof	wo	776F	30575
Xhosa	xh	7868	30824

continues

Language	Code	Hex	Dec
Yiddish[5]	ji	6A69	27241
Yoruba	yo	796F	31087
Zhuang[1]	za	7A61	31329
Zulu	zu	7A75	31349

Note: The DVD specification refers to ISO 639:1988, which has since been updated. Because the normative reference is to the 1988 version, it is recommended that old codes be used in disc production. It is recommended that players recognize old codes and new codes.

[1]Added after original publication.

[2]Hebrew was changed from iw to he after the original publication.

[3]Indonesian was changed from in to id after the original publication.

[4]Serbo-Croatian was deprecated after the original publication in favor of Bosnian (bs), Croatian (hr), and Serbian (sr).

[5]Yiddish was changed from ji to yi after the original publication.

TABLE A.25

ISO 3116 Country
Codes and DVD
Regions

Country	ISO codes			DVD region
Afghanistan	AF	AFG	004	5
Albania	AL	ALB	008	2
Algeria	DZ	DZA	012	5
American Samoa	AS	ASM	016	1
Andorra	AD	AND	020	2
Angola	AO	AGO	024	5
Anguilla	AI	AIA	660	4
Antarctica	AQ	ATA	010	?
Antigua and Barbuda	AG	ATG	028	4
Argentina	AR	ARG	032	4
Armenia	AM	ARM	051	5
Aruba	AW	ABW	533	4
Australia	AU	AUS	036	4
Austria	AT	AUT	040	2
Azerbaijan	AZ	AZE	031	5
Bahamas	BS	BHS	044	4
Bahrain	BH	BHR	048	2
Bangladesh	BD	BGD	050	5
Barbados	BB	BRB	052	4
Belarus	BY	BLR	112	5
Belgium	BE	BEL	056	2
Belize	BZ	BLZ	084	4
Benin	BJ	BEN	204	5
Bermuda	BM	BMU	060	1
Bhutan	BT	BTN	064	5
Bolivia	BO	BOL	068	4
Bosnia and Herzegovina	BA	BIH	070	2

continues

Country	ISO codes			DVD region
Botswana	BW	BWA	072	5
Bouvet Island	BV	BVT	074	?
Brazil	BR	BRA	076	4
British Indian Ocean Territory	IO	IOT	086	5
Brunei Darussalam	BN	BRN	096	3
Bulgaria	BG	BGR	100	2
Burkina Faso	BF	BFA	854	5
Burundi	BI	BDI	108	5
Cambodia	KH	KHM	116	3
Cameroon	CM	CMR	120	5
Canada	CA	CAN	124	1
Cape Verde	CV	CPV	132	5
Cayman Islands	KY	CYM	136	4
Central African Republic	CF	CAF	140	5
Chad	TD	TCD	148	5
Chile	CL	CHL	152	4
China	CN	CHN	156	6
Christmas Island	CX	CXR	162	4
Cocos (Keeling) Islands	CC	CCK	166	4
Colombia	CO	COL	170	4
Comoros	KM	COM	174	5
Congo	CG	COG	178	5
Congo, the Democratic Republic of the	CD	COD	180	5
Cook Islands	CK	COK	184	4
Costa Rica	CR	CRI	188	4
Cote D'Ivoire	CI	CIV	384	5

continue

TABLE A.25 cont.	Country		ISO codes		DVD region
ISO 3116 Country Codes and DVD Regions	Croatia (Hrvatska)	HR	HRV	191	2
	Cuba	CU	CUB	192	4
	Cyprus	CY	CYP	196	2
	Czech Republic	CZ	CZE	203	2
	Denmark	DK	DNK	208	2
	Djibouti	DJ	DJI	262	5
	Dominica	DM	DMA	212	4
	Dominican Republic	DO	DOM	214	4
	East Timor	TP	TMP	626	3
	Ecuador	EC	ECU	218	4
	Egypt	EG	EGY	818	2
	El Salvador	SV	SLV	222	4
	Equatorial Guinea	GQ	GNQ	226	5
	Eritrea	ER	ERI	232	5
	Estonia	EE	EST	233	5
	Ethiopia	ET	ETH	231	5
	Falkland Islands (Malvinas)	FK	FLK	238	4
	Faroe Islands	FO	FRO	234	2
	Fiji	FJ	FJI	242	4
	Finland	FI	FIN	246	2
	France	FR	FRA	250	2
	French Guiana	GF	GUF	254	4
	French Polynesia	PF	PYF	258	4
	French Southern Territories	TF	ATF	260	?
	Gabon	GA	GAB	266	5
	Gambia	GM	GMB	270	5
	Georgia	GE	GEO	268	?

continues

Country	ISO codes			DVD region
Germany	DE	DEU	276	2
Ghana	GH	GHA	288	5
Gibraltar	GI	GIB	292	2
Greece	GR	GRC	300	2
Greenland	GL	GRL	304	2
Grenada	GD	GRD	308	4
Guadeloupe	GP	GLP	312	4
Guam	GU	GUM	316	4
Guatemala	GT	GTM	320	4
Guinea	GN	GIN	324	5
Guinea-Bissau	GW	GNB	624	5
Guyana	GY	GUY	328	4
Haiti	HT	HTI	332	4
Heard and McDonald Islands	HM	HMD	334	4
Holy City (Vatican City State)	VA	VAT	336	2
Honduras	HN	HND	340	4
Hong Kong	HK	HKG	344	3
Hungary	HU	HUN	348	2
Iceland	IS	ISL	352	2
India	IN	IND	356	5
Indonesia	ID	IDN	360	3
Iran, Islamic Republic of	IR	IRN	364	2
Iraq	IQ	IRQ	368	2
Ireland	IE	IRL	372	2
Israel	IL	ISR	376	2
Italy	IT	ITA	380	2

continues

TABLE A.25 cont.	Country		ISO codes		DVD region
ISO 3116 Country Codes and DVD Regions	Jamaica	JM	JAM	388	4
	Japan	JP	JPN	392	2
	Jordan	JO	JOR	400	2
	Kazakhstan	KZ	KAZ	398	5
	Kenya	KE	KEN	404	5
	Kiribati	KI	KIR	296	4
	Korea, Democratic People's Republic of	KP	PRK	408	5
	Korea, Republic of	KR	KOR	410	3
	Kuwait	KW	KWT	414	2
	Kyrgyzstan	KG	KGZ	417	5
	Lao People's Democratic Republic	LA	LAO	418	3
	Latvia	LV	LVA	428	5
	Lebanon	LB	LBN	422	2
	Lesotho	LS	LSO	426	2
	Liberia	LR	LBR	430	5
	Libyan Arab Jamahiriya	LY	LBY	434	5
	Liechtenstein	LI	LIE	438	2
	Lithuania	LT	LTU	440	5
	Luxembourg	LU	LUX	442	2
	Macau	MO	MAC	446	3
	Macedonia, the Former Yugoslav Republic of	MK	MKD	807	2
	Madagascar	MG	MDG	450	5
	Malawi	MW	MWI	454	5
	Malaysia	MY	MYS	458	3
	Maldives	MV	MDV	462	5

continues

Country	ISO codes			DVD region
Mali	ML	MLI	466	5
Malta	MT	MLT	470	2
Marshall Islands	MH	MHL	584	4
Martinique	MQ	MTQ	474	4
Mauritania	MR	MRT	478	5
Mauritius	MU	MUS	480	5
Mayotte	YT	MYT	175	5
Mexico	MX	MEX	484	4
Micronesia, Federated States of	FM	FSM	583	4
Moldova, Republic of	MD	MDA	498	5
Monaco	MC	MCO	492	2
Mongolia	MN	MNG	496	5
Montserrat	MS	MSR	500	4
Morocco	MA	MAR	504	5
Mozambique	MZ	MOZ	508	5
Myanmar	MM	MMR	104	3
Namibia	NA	NAM	516	5
Nauru	NR	NRU	520	4
Nepal	NP	NPL	524	5
Netherlands	NL	NLD	528	2
Netherlands Antilles	AN	ANT	530	4
New Caledonia	NC	NCL	540	4
New Zealand	NZ	NZL	554	4
Nicaragua	NI	NIC	558	4
Niger	NE	NER	562	5
Nigeria	NG	NGA	566	5
Niue	NU	NIU	570	4

continues

TABLE A.25 cont.

ISO 3116 Country Codes and DVD Regions

Country	ISO codes			DVD region
Norfolk Island	NF	NFK	574	4
Northern Mariana Islands	MP	MNP	580	4
Norway	NO	NOR	578	2
Oman	OM	OMN	512	2
Pakistan	PK	PAK	586	5
Palau	PW	PLW	585	4
Panama	PA	PAN	591	4
Papua New Guinea	PG	PNG	598	4
Paraguay	PY	PRY	600	4
Peru	PE	PER	604	4
Philippines	PH	PHL	608	3
Pitcairn	PN	PCN	612	4
Poland	PL	POL	616	2
Portugal	PT	PRT	620	2
Puerto Rico	PR	PRI	630	1
Qatar	QA	QAT	634	2
Reunion	RE	REU	638	5
Romania	RO	ROM	642	2
Russian Federation	RU	RUS	643	5
Rwanda	RW	RWA	646	5
Saint Kitts and Nevis	KN	KNA	659	4
Saint Lucia	LC	LCA	662	4
Saint Vincent and the Grenadines	VC	VCT	670	4
Samoa	WS	WSM	882	4
San Marino	SM	SMR	674	2
Sao Tome and Principe	ST	STP	678	5

continues

TABLE A.25 cont.

ISO 3116 Country Codes and DVD Regions

Country	ISO codes			DVD region
Saudi Arabia	SA	SAU	682	2
Senegal	SN	SEN	686	5
Seychelles	SC	SYC	690	5
Sierra Leone	SL	SLE	694	5
Singapore	SG	SGP	702	3
Slovakia (Slovak Republic)	SK	SVK	703	2
Slovenia	SI	SVN	705	2
Solomon Islands	SB	SLB	090	4
Somalia	SO	SOM	706	5
South Africa	ZA	ZAF	710	2
South Georgia and the South Sandwich Islands	GS	SGS	239	4
Spain	ES	ESP	724	2
Sri Lanka	LK	LKA	144	5
St. Helena	SH	SHN	654	5
St. Pierre and Miquelon	PM	SPM	666	1
Sudan	SD	SDN	736	5
Suriname	SR	SUR	740	4
Svalbard and Jan Mayen Islands	SJ	SJM	744	2
Swaziland	SZ	SWZ	748	2
Sweden	SE	SWE	752	2
Switzerland	CH	CHE	756	2
Syrian Arab Republic	SY	SYR	760	2
Taiwan	TW	TWN	158	3
Tajikistan	TJ	TJK	762	5
Tanzania, United Republic of	TZ	TZA	834	5

continues

TABLE A.25 cont.

ISO 3116 Country Codes and DVD Regions

Country	ISO codes			DVD region
Thailand	TH	THA	764	3
Togo	TG	TGO	768	5
Tokelau	TK	TKL	772	4
Tonga	TO	TON	776	4
Trinidad and Tobago	TT	TTO	780	4
Tunisia	TN	TUN	788	5
Turkey	TR	TUR	792	2
Turkmenistan	TM	TKM	795	5
Turks and Caicos Islands	TC	TCA	796	4
Tuvalu	TV	TUV	798	4
Uganda	UG	UGA	800	5
Ukraine	UA	UKR	804	5
United Arab Emirates	AE	ARE	784	2
United Kingdom	GB	GBR	826	2
United States	US	USA	840	1
United States' Minor Outlying Islands	UM	UMI	581	1
Uruguay	UY	URY	858	4
Uzbekistan	UZ	UZB	860	5
Vanuatu	VU	VUT	548	4
Venezuela	VE	VEN	862	4
Vietnam	VN	VNM	704	3
Virgin Islands (British)	VG	VGB	092	4
Virgin Islands (U.S.)	VI	VIR	850	1
Wallis and Futuna Islands	WF	WLF	876	4
Western Sahara	EH	ESH	732	5
Yemen	YE	YEM	887	2
Yugoslavia	YU	YUG	891	2
Zambia	ZM	ZMB	894	5
Zimbabwe	ZW	ZWE	716	5

APPENDIX B

Standards Related to DVD

DVD is based on or has borrowed from dozens of standards developed over the years by many organizations. Most of the standards in this appendix are listed as normative references in the DVD format specification books.

Physical Format Standards

Disc format
—*ECMA 267: 120 mm DVD - Read-Only Disc* (DVD-ROM part 1)
—*ECMA 268: 80 mm DVD - Read-Only Disc* (DVD-ROM part 1)
—*ECMA-279: 80 mm (1.23 Gbytes per side) and 120 mm (3.95 Gbytes per side) DVD-Recordable Disc* (DVD-R 1.0)
—*ECMA-272: 120 mm DVD Rewritable Disc* (DVD-RAM)
—*ECMA-273: Case for 120 mm DVD-RAM Discs*
—*ECMA-274: Data Interchange on 120 mm Optical Disc using +RW Format - Capacity: 3.0 Gbytes and 6.0 Gbytes* (DVD+RW 1.0)

Device interface:
—*SFF 8090 ATAPI / SCSI* (Mt. Fuji; *INF8090i*)

Physical connection:
—*ANSI X3.131-1994: Information Systems-Small Computer Systems Interface-2* (SCSI-2)
—*ANSI X3.277-1996: Information Technology-SCSI-3 Fast-20*
—*ANSI X3.221-1994: Information Systems-AT Attachment Interface for Disk Drives* (EIDE/ATA)
—*ANSI X3.279-1996: Information Technology-AT Attachment Interface with Extensions* (ATA-2)
—*IEC 60958: Digital audio interface.*
—*IEC 60856: Prerecorded optical reflective videodisc system (PAL).*
—*IEC 60857: Prerecorded optical reflective videodisc system (NTSC).*
—*IEEE 1394-1995 IEEE Standard for a High Performance Serial Bus* (FireWire)

System Standards

File system:
—*OSTA Universal Disc Format Specification: 1996* (Appendix 6.9) "OSTA UDF Compliant Domain" of *ISO / IEC 13346:1995 Volume*

and file structure of write-once and rewritable media using non-sequential recording for information interchange. (Note: *ECMA 167, 2d edition, 1994,* is equivalent to *ISO/IEC 13346:1995.*)

—*ISO 9660:1988 Information processing-Volume and file structure of CD-ROM for information interchange* (Note: Equivalent to *ECMA 119, 2d edition, 1987.*)

—*ECMA TR/71* (UDF Bridge)

—*Joliet CD-ROM Recording Specification, ISO 9660:1988 Extensions for Unicode* (Microsoft)

MPEG-2 system:

—*ISO/IEC 13818-3:1998 Information technology-Generic coding of moving pictures and associated audio information: Systems* (ITU-T H.222.0) (program streams only, no transport streams)

CD:

—*IEC 60908 (1987-09) Compact disc digital audio system* (Red Book)

CD-ROM:

—*ISO/IEC 10149:1995 Information technology-Data interchange on read-only 120 mm optical data disks* (CD-ROM) (Yellow Book) (Note: Equivalent to *ECMA 130, 2nd Edition, June 1996*)

—Philips/Sony Orange Book part-II Recordable Compact Disc System

—Philips/Sony Orange Book part-III Recordable Compact Disc System

—*IEC 61104: Compact Disc Video System, 12 cm* (CDV Single).

Video Standards

MPEG-1 video

—*ISO/IEC 11172-2:1993 Information technology-Coding of moving pictures and associated audio for digital storage media at up to about 1.5 Mbit/s-Part 2: Video*

MPEG-2 video

—*ISO/IEC 13818-2:1996 Information technology-Generic coding of moving pictures and associated audio information: Video* (ITU-T H.262)

Source video:

—*ITU-R BT.601-5 Studio encoding parameters of digital television for standard 4:3 and widescreen 16:9 aspect ratios*

NTSC video:
>—*SMPTE 170M-1994 Television-Composite Analog Video Signal-NTSC for Studio Applications*
>—*ITU-R BT.470-4 Television Systems*

PAL video:
>—*ITU-R BT.470-4 Television Systems*

Additional video signals:
>—*ETS 300 294 Edition 2:1995-12 Television Systems; 625-Line Television: Wide Screen Signaling* (WSS)
>—*ITU-R BT.1119-1 Widescreen signaling for broadcasting. Signaling for widescreen and other enhanced television parameters*
>—*IEC 61880 (1998-01) Video systems (525/60) - Video and accompanied data using the vertical blanking interval - Analogue interface* (CGMS-A; NTSC line 20; PAL/SECAM/YUV line 21)
>—*EIA/IS 702 Copy Generation Management System (Analog).* (CGMS-A; NTSC line 21; YUV line 21)
>—*ETS 300294* (PAL/SECAM CGMS-A)
>—*EIAJ CPX-1204* (NTSC widescreen signaling and CGMS-A)
>—*ITU-R BT.1119-1 Widescreen signaling for broadcasting. Signaling for widescreen and other enhanced television parameters* (PAL CGMS-A)
>—*EIA-608 Recommended Practice For Line 21 Data Service* (NTSC Closed Captions)
>—*EIA-746 Transport Of Internet Uniform Resource Locator (URL) Information Using Text-2 (T-2) Service* (TV links; ATVEF triggers)
>—*ETS 300 294 Edition 2:1995-12* (Film/camera mode)

Audio Standards

Dolby Digital audio (AC-3)
>—*ATSC A/52 1995*

MPEG-1 audio
>—*ISO/IEC 11172-3:1993 Information technology-Coding of moving pictures and associated audio for digital storage media at up to about 1,5 Mbit/s-Part 3: Audio*

MPEG-2 audio:
>—*ISO/IEC 13818-3:1995 Information technology-Generic coding of moving pictures and associated audio information-Part 3: Audio*

Digital audio interface:
—*IEC 60958 (1989-02) Digital audio interface* (Type II-Consumer, "SP/DIF")
—*IEC 60958-2 (1994-07) Digital audio interface-Part 2: Software information delivery mode*
—*IEC 61937-1 Interfaces For Non-Linear PCM Encoded Audio Bitstreams Applying IEC 60958 - Part 1: Non-Linear PCM Encoded Audio Bitstreams For Consumer Applications* (also *ATSC A/52 Annex B: AC-3 Data Stream in IEC 958 Interface*)
—*EIAJ CP-340* (optical digital audio; "Toslink")

Recording codes:
—*ISO 3901:1986 Documentation-International Standard Recording Code (ISRC)*

Other Standards

Language codes
—*ISO 639:1988 Code for the representation of names of languages* (see Table A.24)

Country codes:
—*ISO 3166:1993 Codes for the representation of names of countries* (see Table A.25)

Text information:
—*ISO/IEC 646:1991 Information technology-ISO 7-bit coded character set for information interchange*
—*ISO 8859-1:1987 Information processing-8-bit single-byte coded graphic character sets-Part 1: Latin alphabet No. 1*
—*ISO 8859-2:1987 Information processing-8-bit single-byte coded graphic character sets-Part 2: Latin alphabet No. 2*
—*ISO/IEC 2022:1994 Information technology-Character code structure and extension techniques*
—*JIS, Shift-JIS*, and others

Digital A/V interface:
—*IEC 61883 Standard for Digital Interface for Consumer Electronic Audio/Video Equipment* (transport protocol for IEEE 1394)
—*1394 Trade Association Audio/Video Control Digital Interface Command Set (AV/C)* (control protocol for IEEE 1394).

APPENDIX C

References and Information Sources

For an up-to-date list of references and information sources, plus lists of companies serving the DVD industry, visit dvddemystified.com and the DVD FAQ (dvddemystified.com/dvdfaq.html).

Recommended References

Benson, K. Blair. *Television Engineering Handbook: Featuring HDTV Systems* (revised ed.). McGraw-Hill, 1992. ISBN: 007004788X.

Dunn, Julian. *Sample Clock Jitter and Real-Time Audio over the IEEE1394 High-Performance Serial Bus*. Preprint 4920, 106th AES Convention, Munich, May 1999.

Dunn, Julian, and Ian Dennis. *The Diagnosis and Solution of Jitter-Related Problems in Digital Audio*. Preprint 3868, 96th AES Convention, Amsterdam. February 1994.

Haskell, Barry G., Atul Puri, and Arun N. Netravali. *Digital Video: An Introduction to MPEG-2*. Chapman & Hall, 1996. ISBN: 0412084112.

Jack, Keith. *Video Demystified* (2d ed.). Hightext Publications, 1996. ISBN: 187870723X.

Mitchell, Joan L., William B. Pennebaker, and Chad E. Fogg. *MPEG Video: Compression Standard*. Chapman & Hall, 1996. ISBN: 0412087715.

Negroponte, Nicholas and Marty Asher. *Being Digital*. Vintage Books, 1996. ISBN: 0679762906.

Pohlmann, Ken C. *Principles of Digital Audio* (3d ed.). McGraw-Hill, 1995. ISBN: 0070504695.

Poynton, Charles A. *Digital Video and HDTV: Pixels, Pictures, and Perception*. John Wiley & Sons, 2001. ISBN: 0471384895.

——. *A Technical Introduction to Digital Video*. John Wiley & Sons, 1996. ISBN: 047112253X.

Solari, Stephen J. *Digital Video and Audio Compression*. McGraw-Hill, 1997. ISBN: 0070595380.

Watkinson, John. *The Art of Digital Audio* (2d ed.) Butterworth-Heinemann, 1994. ISBN: 0240513207.

——. *Compression in Video and Audio*. Focal Press, 1995. ISBN: 0240513940.

——. *An Introduction to Digital Audio*. Focal Press, 1994. ISBN: 0240513789.

DVD Information and Licensing

General DVD Information
DVD Forum

www.dvdforum.org

Shiba Shimizu Bldg. 5F

2-3-11, Shibadaimon,

Minato-ku, Tokyo 105-0012

Japan

+81-3-5777-2881, fax +81-3-5777-2882

DVD Specification and Logo
DVD Format/Logo Licensing Corporation (DVD FLLC)

www.dvdfllc.co.jp

Shiba Shimizu Bldg. 5F

2-3-11, Shibadaimon,

Minato-ku, Tokyo 105-0012

Japan

+81-3-5777-2881, fax +81-3-5777-2882

Patent Licensing (DVD: Hitachi/Matsushita/Mitsubishi/Time Warner/Toshiba/Victor Pool)
Toshiba Corporation

DVD Business Promotion and Support

1-1 Shibaura 1-chome,

Minato-ku, Tokyo 105-01

Japan

+81-3-3457-2473, fax +81-3-5444-9430

Patent Licensing (DVD: Philips/Pioneer/Sony Pool)
Philips Standards and Licensing

www.licensing.philips.com

Licensing Support

Building SFF-8

P.O. Box 80002

5600 JB Eindhoven

The Netherlands

Fax +31-40-2732113

Patent Licensing (DVD)
Thomson Multimedia

Director Licensing

46 Quai Alphonse Le Gallo

92648 Boulogne Cedex

France

33 1 4186 5284, fax 33 1 4186 5637

Patent Licensing (Optical Disc)
Discovision Associates

2355 Main Street, Suite 200

Irvine, CA 92614

949-660-5000, fax 949-660-1801

Patent Licensing (MPEG)
MPEG LA, LLC

www.mpegla.com

250 Steele Street, Suite 300

Denver, Colorado 80206

303-331-1880, fax 303-331-1879

Patent Licensing (Dolby Digital and MLP Audio)

Dolby Laboratories Licensing Corporation

www.dolby.com

100 Potrero Avenue

San Francisco, CA 94103-4813

415-558-0200, fax 415-863-1373

Patent Licensing (CD and DVD Packaging)

Business Development Europe (BDE) (inside EU)

International Standards & Licensing (IS&L) (outside EU)

Copy Protection Licensing

Macrovision Corporation

www.macrovision.com

1341 Orleans Drive

Sunnyvale, California 94089

408-743-8600, fax 408-743-8610

Copy Protection Licensing

DVD Copy Control Association (CCA)

Digital Transmission Licensing Administrator (DTLA)

4C Entity, LLC

225 B Cochrane Circle

Morgan Hill, CA 95037

408-776-2014, fax 408-779-9291

Newsletters and Industry Analyses

Adams Media Research

Market research

tomadams@ix.netcom.com

15B West Carmel Valley Rd.

Carmel Valley, CA 93924

408-659-3070, fax 408-659-4330

The CD-Info Company (CDIC)
Industry directories, newsletters, and other publications
www.cd-info.com, info@cd-info.com
4800 Whitesburg Drive a30-283
Huntsville, AL 35802-1600
205-650-0406, fax 205-882-7393

Cahners In-stat Group
www.instat.com
275 Washington St.
Newton, MA 02458
617-630-3900

Centris
www.centris.com
Santa Monica Studios
1817 Stanford
Santa Monica, CA 90404
877-723-6874, fax 310-264-8776

Computer Economics
www.computereconomics.com
5841 Edison Place
Carlsbad, CA 92008
800-326-8100, fax 760-431-1126

Corbell Publishing
www.corbell.com
4676 Admiralty Way, Suite 300
Marina del Rey, California 90292
310-574-5337, fax 310-574-5383

Dataquest

Market research

www.dataquest.com

251 River Oaks Parkway

San Jose, CA 95134-1913

408-468-8000, fax 408-954-1780

Ernst & Young

www.ey.com

Home Recording Rights Coalition

www.hrrc.org

P.O. Box 14267

Washington, DC 20044

800-282-8273

InfoTech

Market research

www.infotechresearch.com

Box 150, Skyline Dr.

Woodstock, VT 05091-0150

802-763-2097, fax 802-763-2098

International Data Corporation (IDC)

Market research

www.idcresearch.com

5 Speen Street

Framingham, MA 01701

508-872-8200, fax 508-935-4015

Jon Peddie Associates (JPA)

www.jpa.com

100 Shoreline Hwy, Bldg. A, 2nd Floor

Mill Valley, CA 94941

415-331-6800, fax 415-331-6211

Knowledge Industry Publications, Inc. (KIPI)

Newsletters, magazines, conferences

www.kipinet.com, 800-800-5474

701 Westchester Avenue

White Plains, NY 10604

914-328-9157, fax 914-328-9093

Market Vision

Market research

www.webcom.com/newmedia, mktvis@cruzio.com

326 Pacheco Avenue, Suite 200

Santa Cruz, CA 95062

408-426-4400, fax 408-426-4411

Paul Kagan Associates

Market research

126 Clock Tower Place

Carmel, CA 93923-8734

408-624-1536

SIMBA Information Inc.

Market research, newsletters

www.simbanet.com, info@simbanet.com

11 River Bend Drive South

P.O. 4234

Wilton, CT 06907

203-358-0234, fax 203-358-5824

Strategy Analytics

17-21 Napier Road

Luton, Bedfordshire

LU1 1RF

United Kingdom

+44 (0)1582 405678, fax: +44 (0)1582 454828

Magazines

Digital Video Magazine
www.dv.com
411 Borel Ave., Suite 100
San Mateo, CA 94402
415-358-9500, 888-776-7002, fax 415-358-8891

DVD Report
www.kipinet.com/dvd
Knowledge Industry Publications
Suite 101W
701 Westchester Avenue
White Plains, NY 10604
800-800-5474, fax 914-328-9093

EMedia Professional (formerly CD-ROM Professional)
www.onlineinc.com/emedia
649 Massachusetts Ave., Suite 4
Cambridge, MA 02139
617-492-0268, fax 617-492-3159

Medialine News (formerly Replication News)
Miller Freeman PSN, Inc.
2 Park Avenue, Suite 1820
New York, NY 10016
415-905-2200, fax 415-905-2239

One to One
Miller Freeman Entertainment Group
8 Montague Close, London Bridge
London SE1 9UR
UK
+44-171-620-3636

Standards Organizations

Audio Engineering Society (AES)/AES Standards Committee (AESSC)

www.aes.org

60 E. 42nd St.

New York, NY 10165-2520

212-661-8528, fax 212-682-0477

American National Standards Institute (ANSI)

www.ansi.org

11 West 42nd Street

New York, NY 10036

212-642-4900, fax 212-398-0023

Commission Internationale de l'Éclairage/International Commission on
Illumination (CIE)

ciecb@ping.at

IE Central Bureau, Kegelgasse 27

A-1030 Vienna, Austria

43 (01) 714 31 87/0 , fax 43 (01) 713 0838/18

Deutsches Institut für Normung/German Institute for Standardization
(DIN)

www.din.de, postmaster@din.de

Burggrafenstrasse 6, D-10787

Berlin, Germany

49 30 26 01-0, fax 49 30 26 01 12 31

European Telecommunications Standards Institute (ETSI)

www.etsi.fr

Route des Lucioles, F-06921

Sophia Antipolis, Cedex, France

33 4 92 94 42 00, fax 33 4 93 65 47 16

European Broadcasting Union (EBU)

www.ebu.ch

European Computer Manufacturers Association (ECMA)

www.ecma.ch, helpdesk@ecma.ch

114 Rue de Rhône,

CH-1204 Genève 20, Switzerland

41 22 735 3634

International Electrotechnical Commission (IEC)

www.iec.ch

3 rue de Varembé, Case postale 131

1211 Genève 20, Switzerland

41 22 919 02 11, fax 41 22 919 03 00

International Organization for Standardization (ISO)

www.iso.ch, central@iso.ch

1 rue de Varembé, Case postale 56

CH-1211 Genève 20, Switzerland

41 22 749 01 11, fax 41 22 733 34 30

International Telecommunication Union (ITU)

www.itu.int, sales@itu.int

Sales Service

Place de Nations

CH-1211 Genève 20, Switzerland

41 22 730 6141 (English), 41 22 730 6142 (French), 41 22 730 6143
  (Spanish), fax 41 22 730 5194

National Committee for Information Technology Standards (NCITS)

(Formerly the Accredited Standards Committee X3, Information
  Technology)

www.ncits.org

Optical Storage Technology Association (OSTA)

www.osta.org

311 E. Carrillo St.

Santa Barbara, CA 93101

805-962-1541

Society of Motion Picture & Television Engineers (SMPTE)

www.smpte.org, smpte@smpte.org

595 W. Hartsdale Ave.

White Plains, NY 10607-1824

914-761-1100, fax 914-761-3115

Other Related Organizations

Acoustic Renaissance for Audio (ARA)

www.meridian.co.uk/ara, ara@meridian.co.uk, negishi@gcds.canon.co.jp

Business Software Alliance (BSA)

www.bsa.org

1150 18th Street N.W., Suite 700

Washington, DC 20036

202-872-5500, fax 202-872-5501

Computer and Business Equipment Manufacturer's Association (CBEMA)

1250 Eye St., Suite 200

Washington, DC 20005

202-737-8888, fax 202-638-4922

Consumer Electronics Association (CEA)

CEA, a sector of the EIA, represents U.S. manufacturers of audio, video, consumer information, accessories, mobile electronics, and multimedia products.

www.ce.org

2500 Wilson Blvd.

Arlington, VA 22201-3834

703-907-7600, fax 703-907-7675

The DVD Association

www.dvda.org

Electronic Industries Association (EIA)

A 72-year-old trade association representing all facets of electronics manufacturing.

www.eia.org

2500 Wilson Boulevard

Arlington, VA 22201-3834

703-907-7600, fax 703-907-7601

Information Technology Industry Council (ITI)

www.itic.org

Motion Picture Association of America (MPAA)

The MPAA serves as the advocate of the American motion picture, home video, and television production industries.

www.mpaa.org

Recording Industry Association Of America (RIAA)

www.riaa.com

1330 Connecticut Avenue NW, Suite 300

Washington, DC 20036

202-775-0101

SFF (Small Form Factor) Committee

250-1752@mcimail.com

14426 Black Walnut Ct.

Saratoga, CA 95070

408-867-6630x303, fax 408-867-2115

Video Software Dealers Association (VSDA)

www.vsda.org

16530 Ventura Blvd., Suite 400

Encino, CA 91436

818-385-1500, fax 818-385-0567

GLOSSARY

1080i   1080 lines of interlaced video (540 lines per field). This usually refers to a 1920 × 1080 resolution in a 1.78 aspect ratio.

1080p   1080 lines of progressive video (1080 lines per frame). This usually refers to a 1920 × 1080 resolution in a 1.78 aspect ratio.

2-2 pulldown   The process of transferring 24-frame-per-second film to video by repeating each film frame as two video fields. (See Chapter 3, "DVD Technology Primer," for details.) When 24-fps film is converted via a 2-2 pulldown to 25-fps 625/50 (PAL) video, the film runs four percent faster than normal.

2-3 pulldown   The process of converting 24-frame-per-second film to video by repeating one film frame as three fields, and then the next film frame as two fields. (See Chapter 3 for details.)

3-2 pulldown   An uncommon variation of 2-3 pulldown, where the first film frame is repeated for three fields instead of two. Most people mean 2-3 pulldown when they say 3-2 pulldown.

4:1:1   The component digital video format with one C_b sample and one C_r sample for every four Y samples. This uses 4:1 horizontal downsampling with no vertical downsampling. Chroma is sampled on every line, but only for every four luma pixels (one pixel in a 1 × 4 grid). This amounts to a subsampling of chroma by a factor of two compared to luma (and by a factor of four for a single C_b or C_r component). DVD uses 4:2:0 sampling, not 4:1:1 sampling.

4:2:0   The component digital video format used by DVD, with one C_b sample and one C_r sample for every four Y samples (one pixel in a 2 × 2 grid). This uses 2:1 horizontal downsampling and 2:1 vertical downsampling. C_b and C_r are sampled on every other line, in between the scan lines, with one set of chroma samples for each two luma samples on a line. This amounts to a subsampling of chroma by a factor of two, compared to luma (and by a factor of four for a single C_b or C_r component).

4:2:2   The component digital video format commonly used for studio recordings, with one C_b sample and one C_r sample for every two Y samples (one pixel in a 1 × 2 grid). This uses 2:1 horizontal downsampling with no vertical downsampling. This allocates the same number of samples to the chroma signal as to the luma signal. The input to MPEG-2 encoders used for DVD is typically in 4:2:2 format, but the video is subsampled to 4:2:0 before being encoded and stored.

4:4:4   A component digital video format for high-end studio recordings, where Y, Cb, and Cr are sampled equally.

480i   480 lines of interlaced video (240 lines per field). This usually refers to 720 × 480 (or 704 × 480) resolution.

480p   480 lines of progressive video (480 lines per frame). 480p60 refers to 60 frames per second, 480p30 refers to 30 frames per second, and 480p24 refers to 24 frames per second (film source). This usually refers to 720 × 480 (or 704 × 480) resolution.

4C   The four-company entity consisting of IBM, Intel, Matsushita, and Toshiba.

525/60   The scanning system of 525 lines per frame and 60 interlaced fields (30 frames) per second. This is used by the NTSC television standard.

5C   The five-company entity that consists of IBM, Intel, Matsushita, Toshiba, and Sony.

625/50   The scanning system of 625 lines per frame and 50 interlaced fields (25 frames) per second. This is used by PAL and SECAM television standards.

720p   720 lines of progressive video (720 lines per frame). This offers a higher definition than standard DVD (480i or 480p). 720p60 refers to 60 frames per second, 720p30 refers to 30 frames per second, and 720p24 refers to 24 frames per second (film source). This usually refers to a 1280 x 720 resolution in a 1.78 aspect ratio.

8/16 modulation   The form of modulation block code used by DVD to store channel data on the disc. *See* modulation.

AAC   Advanced audio coder. An audio-encoding standard for MPEG-2 that is not backward-compatible with MPEG-1 audio.

AC   Alternating current. An electric current that regularly reverses direction. It has been adopted as a video term for a signal of non-zero frequency. Compare this to DC.

AC-3   The former name of the Dolby Digital audio-coding system, which is still technically referred to as AC-3 in standards documents. AC-3 is the successor to Dolby's AC-1 and AC-2 audio coding techniques.

access time   The time it takes for a drive to access a data track and begin transferring data. In an optical jukebox, the time it takes to locate a specific disk, insert it in an optical drive, and begin transferring data to the host system.

ActiveMovie   The former name for Microsoft's DirectShow technology.

ADPCM   Adaptive differential pulse code modulation. A compression technique that encodes the difference between one sample and the next. Variations are lossy and lossless.

AES   The Audio Engineering Society.

AES/EBU   A digital audio signal transmission standard for professional use, defined by the Audio Engineering Society and the European Broadcasting Union. Sony/Philips digital interface (S/P DIF) is the consumer adaptation of this standard.

AGC   Automatic gain control. A circuit designed to boost the amplitude of a signal to provide adequate levels for recording. *See* Macrovision.

aliasing   A distortion (artifact) in the reproduction of digital audio or video that results when the signal frequency is more than twice the sampling frequency. The resolution is insufficient to distinguish between alternate reconstructions of the waveform, thus admitting additional noise that was not present in the original signal.

AMGM_VOBS   The Video Object Set for Audio Manager Menu.

analog   A signal of (theoretically) infinitely variable levels. Compare this to digital.

angle   In DVD-Video, this is a specific view of a scene, usually recorded from a certain camera angle. Different angles can be chosen while viewing the scene.

ANSI   American National Standards Institute (see Appendix C, "References and Information Sources").

AOTT_AOBS   Audio Object Set for Audio-Only Title.

apocryphal   Of questionable authorship or authenticity; erroneous or fictitious. The author of this book is fond of saying that the oft-cited 133-minute limit of DVD-Video is apocryphal.

application format   A specification for storing information in a particular way to enable a particular use.

artifact   An unnatural effect not present in the original video or audio, produced by an external agent or action. Artifacts can be caused by many factors, including digital compression, film-to-video transfer, transmission errors, data readout errors, electrical interference, analog signal noise, and analog signal crosstalk. Most artifacts attributed to the digital compression of DVD are in fact from other sources. Digital compression artifacts always occur in the same place and in the same way. Possible MPEG artifacts are mosquitoes, blocking, and video noise.

aspect ratio   The width-to-height ratio of an image. A 4:3 aspect ratio means the horizontal size is a third wider than the vertical size. The standard television ratio is 4:3 (or 1.33:1). The widescreen DVD and HTDV aspect ratio is 16:9 (or 1.78:1). Common film aspect ratios are 1.85:1 and 2.35:1. Aspect ratios normalized to a height of one are often abbreviated by leaving off the :1.

ASV (Audio Still Video)   A still picture on a DVD-Audio disc.

ASVOBS   Audio Still Video Object Set.

ATAPI   Advanced Technology Attachment (ATA) Packet Interface. An interface between a computer and its internal peripherals such as DVD-ROM drives. ATAPI provides the command set for controlling devices connected via an IDE interface. ATAPI is part of the Enhanced IDE (E-IDE) interface, also known as ATA-2. ATAPI was extended for use in DVD-ROM drives by the SFF 8090 specification.

ATSC   The Advanced Television Systems Committee. In 1978, the Federal Communications Commission (FCC) empaneled the Advisory Committee on Advanced Television Service (ACATS) as an investigatory and advisory committee to develop information that would assist the FCC in establishing an advanced broadcast television (ATV) standard for the U.S. This committee created a subcommittee, the ATSC, to explore the need for and to coordinate development of the documentation of Advanced Television Systems. In 1993, the ATSC recommended that efforts be limited to a digital television system (DTV), and in September 1995 issued its recommendation for a DTVstandard, which was approved with the exclusion of compression format constraints (picture resolution, frame rate, and frame sequence).

ATV   Advanced television with significantly better video and audio than standard TV. Sometimes used interchangeably with HDTV, but more accurately encompasses any improved television system, including those beyond HDTV. ATV is also sometimes used interchangeably with the final recommended standard of the ATSC, which is more correctly called DTV.

authoring   For DVD-Video, authoring refers to the process of designing, creating, collecting, formatting, and encoding material. For DVD-ROM, authoring usually refers to using a specialized program to produce multimedia software.

autoplay (or automatic playback)   A feature of DVD players that automatically begins playback of a disc if so encoded.

bandwidth   Strictly speaking, this is the range of frequencies (or the difference between the highest and the lowest frequency) carried by a circuit or signal. Loosely speaking, this is the amount of information carried in a signal. Technically, bandwidth does not apply to digital information; the term data rate is more accurate.

BCA   Burst cutting area. A circular section near the center of a DVD disc where ID codes and manufacturing information can be inscribed in barcode format (refer to Figure 5.4).

birefringence   An optical phenomenon where light is transmitted at slightly different speeds depending on the angle of incidence. Also refers to light scattering due to different refractions created by impurities, defects, or stresses within the media substrate.

bit A binary digit. The smallest representation of digital data: zero/one, off/on, no/yes. Eight bits make one byte.

bitmap An image made of a two-dimensional grid of pixels. Each frame of digital video can be considered a bitmap, although some color information is usually shared by more than one pixel.

bit rate The volume of data measured in bits over time. Equivalent to data rate.

bits per pixel The number of bits used to represent the color or intensity of each pixel in a bitmap. One bit enables only two values (black and white), two bits enable four values, and so on. Bits per pixel is also referred to as color depth or bit depth.

bitstream Digital data, usually encoded, that is designed to be processed sequentially and continuously.

bitstream recorder A device capable of recording a stream of digital data, but not necessarily capable of processing the data.

BLER (Block error rate) A measure of the average number of raw channel errors when reading or writing a disc.

block In video encoding, an 8×8 matrix of pixels or DCT values representing a small chunk of luma or chroma. In DVD MPEG-2 video, a macroblock is made up of six blocks: four luma and two chroma.

blocking A term referring to the occasional blocky appearance of compressed video (an artifact). Blocking is caused when the compression ratio is high enough that the averaging of pixels in 8×8 blocks becomes visible.

Blue Book The document that specifies the CD Extra interactive music CD format. The original CDV specification was also in a blue book. *See* Enhanced CD.

Book A The document specifying the DVD physical format (DVD-ROM). Finalized in August 1996.

Book B The document specifying the DVD-Video format. Mostly finalized in August 1996.

Book C The document specifying the DVD-Audio format.

Book D The document specifying the DVD record-once format (DVD-R). Finalized in August 1997.

Book E The document specifying the rewritable DVD format (DVD-RAM). Finalized in August 1997.

B picture (or B frame) One of three picture types used in MPEG video. B pictures are bidirectionally predicted, based on both previous and following pictures. B pictures usually use the least number of bits and they do not propagate coding errors because they are not used as a reference by other pictures.

bps   Bits per second. A data rate unit.

brightness   Defined by the CIE as the attribute of a visual sensation according to which area appears to emit more or less light. Loosely, it is the intensity of an image or pixel, independent of color, that is, its value along the axis from black to white.

buffer   A temporary storage space in the memory of a device that helps smooth data flow.

burst   A short segment of the color subcarrier in a composite signal that is inserted to help the composite video decoder regenerate the color subcarrier.

B-Y, R-Y   The general term for color-difference video signals carrying blue and red color information where the brightness (Y) has been subtracted from the blue and red RGB signals to create B-Y and R-Y color-difference signals. Refer to Chapter 3, "DVD Technology Primer."

byte   A unit of data or data storage space consisting of eight bits, commonly representing a single character. Digital data storage is usually measured in bytes, kilobytes, megabytes, and so on.

caption   A textual representation of the audio information in a video program. Captions are usually intended for the hearing impaired and therefore include additional text to identify the person speaking, off-screen sounds, and so on.

CAV   Constant angular velocity. Refers to rotating disc systems in which the rotation speed is kept constant, where the pickup head travels over a longer surface as it moves away from the center of the disc. The advantage of CAV is that the same amount of information is provided in one rotation of the disc. Contrast with CLV and ZCLV.

C_b, C_r   The components of digital color-difference video signals carrying blue and red color information, where the brightness (Y) has been subtracted from the blue and red RGB signals to create B-Y and R-Y color-difference signals (refer to Chapter 3, "DVD Technology Primer").

CBEMA   Computer and Business Equipment Manufacturers Association (Refer to Appendix C, "References and Information Sources.")

CBR   Constant bit rate. Data compressed into a stream with a fixed data rate. The amount of compression (such as quantization) is varied to match the allocated data rate, but as a result, quality may suffer during high-compression periods. In other words, the data rate is held constant, while quality is allowed to vary. Compare this to VBR.

CCI   Copy control information. Information specifying if the content is allowed to be copied.

CCIR Rec. 601   A standard for digital video. The CCIR changed its name to ITU-R, and the standard is now properly called ITU-R BT.601.

CD   Short for compact disc, an optical disc storage format developed by Philips and Sony.

CD-DA   Compact disc digital audio. The original music CD format, storing audio information as digital PCM data. Defined by the Red Book standard.

CD+G   Compact disc plus graphics. A CD variation that embeds graphical data in with the audio data, allowing video pictures to be displayed periodically as music is played. Primarily used for karaoke.

CD-i   Compact disc interactive. An extension of the CD format designed around a set-top computer that connects to a TV to provide interactive home entertainment, including digital audio and video, video games, and software applications. Defined by the Green Book standard.

CD-Plus   A type of Enhanced CD format using stamped multisession technology.

CD-R   An extension of the CD format that enables data to be recorded once on a disc by using dye-sublimation technology. It is defined by the Orange Book standard.

CD-ROM   Compact disc read-only memory. An extension of the Compact disc digital audio (CD-DA) format that enables computer data to be stored in digital format. Defined by the Yellow Book standard.

CD-ROM XA   CD-ROM extended architecture. A hybrid CD that enables interleaved audio and video.

CDV   A combination of laserdisc and CD that places a section of CD-format audio on the beginning of the disc and a section of laserdisc-format video on the remainder of the disc.

cell   In DVD-Video, a unit of video with a duration that is anywhere from a fraction of a second to several hours long. Cells enable the video to be grouped for sharing content among titles, interleaving for multiple angles, and so on.

CEA   The Consumer Electronics Association. A subsidiary of the Electronics Industry Association (EIA). (Refer to Appendix C, "References and Information Sources.")

CGMS   The Copy Guard Management System. A method of preventing copies or controlling the number of sequential copies allowed. CGMS/A is added to an analog signal (such as line 21 of NTSC). CGMS/D is added to a digital signal, such as IEEE 1394.

challenge key   Data used in the authentication key exchange process between a DVD-ROM drive and a host computer, where one side determines if the other side contains the necessary authorized keys and algorithms for passing encrypted (scrambled) data.

channel   A part of an audio track. Typically, one channel is allocated for each loudspeaker.

channel bit   The bits stored on the disc after being modulated.

channel data   The bits physically recorded on an optical disc after error-correction encoding and modulation. Because of the extra information and processing, channel data is larger than the user data contained within it.

chapter   In DVD-Video, a division of a title. Technically, it is called a part of title (PTT).

chroma (C′)   The nonlinear color component of a video signal, independent of the luma. It is identified by the symbol C′ (where ′ indicates nonlinearity), but it is usually written as C because it's never linear in practice.

chroma subsampling   Reducing the color resolution by taking fewer color samples than luminance samples. (*See* 4:1:1 and 4:2:0.)

chrominance (C)   The color component (hue and saturation) of light, independent of luminance. Technically, chrominance refers to the linear component of video, as opposed to the transformed nonlinear chroma component.

CIE   Commission Internationale de l'Éclairage/International Commission on Illumination. (Refer to Appendix C, "References and Information Sources.")

CIF   The common intermediate format, which is a video resolution of 352×288.

CIRC   Cross-interleaved Reed Solomon code. An error-correction coding method that overlaps small frames of data.

clamping area   The area near the inner hole of a disc where the drive grips the disc in order to spin it.

closed captions   Textual video overlays that are not normally visible, as opposed to open captions, which are a permanent part of the picture. Captions are usually a textual representation of the spoken audio. In the U.S., the official NTSC Closed Caption standard requires that all TVs larger than 13 inches include circuitry to decode and display caption information stored on line 21 of the video signal. DVD-Video can provide closed caption data, but the subpicture format is preferred for its versatility.

CLUT   Color lookup table. An index that maps a limited range color values to a full range of values such as RGB or YUV.

CLV Constant linear velocity. This refers to a rotating disc system in which the head moves over the disc surface at a constant velocity, requiring that the motor vary the rotation speed as the head travels in and out. The further the head is from the center of the disc, the slower the rotation. The advantage of CLV is that data density remains constant, optimizing the use of the surface area. Contrast this with CAV and ZCLV.

CMI Content management information. This is general information about copy protection and the allowed use of protected content. CMI includes CCI.

codec Coder/decoder. The circuitry or computer software that encodes and decodes a signal.

colorburst *See* burst.

color depth The number of levels of color (usually including luma and chroma) that can be represented by a pixel. It is generally expressed as a number of bits or a number of colors. The color depth of MPEG video in DVD is 24 bits, although the chroma component is shared across four pixels (averaging 12 actual bits per pixel).

color difference A pair of video signals that contain the color components minus the brightness component, usually B-Y and R-Y (G-Y is not used, since it generally carries less information). The color-difference signals for a black-and-white picture are zero. The advantage of color-difference signals is that the color component can be reduced more than the brightness (luma) component without being visually perceptible.

colorist Someone who operates a telecine machine to transfer film to video. Part of the process involves correcting the video color to match the film.

combo drive A DVD-ROM drive capable of reading and writing CD-R and CD-RW media. It may also refer to a DVD-R, DVD-RW, or DVD+RW drive with the same capability. *See* RAMbo.

component video A video system containing three separate color component signals, either red/green/blue (RGB) or chroma/color difference (YC_bC_r, YP_bP_r, YUV), in analog or digital form. The MPEG-2 encoding system used by DVD is based on color-difference component digital video. Very few televisions have component video inputs.

composite video An analog video signal in which the luma and chroma components are combined (by frequency multiplexing), along with sync and burst. This is also called CVBS. Most televisions and VCRs have composite video connectors, which are usually colored yellow.

compression The process of removing redundancies in digital data to reduce the amount that must be stored or transmitted. Lossless compression removes only enough redundancy so that the original data can be recreated exactly as it was. Lossy compression sacrifices additional data to achieve greater compression.

constant data rate or **constant bit rate**   *See* CBR.

contrast   The range of brightness between the darkest and lightest elements of an image.

control area   A part of the lead-in area on a DVD containing one ECC block (16 sectors) repeated 192 times. The repeated ECC block holds information about the disc.

CPPM   Content Protection for Prerecorded Media. Copy protection for DVD-Audio.

CPRM   Content Protection for Recordable Media. Copy protection for writable DVD formats.

CPSA   Content Protection System Architecture. An overall copy protection design for DVD.

CPTWG   Copy Protection Technical Working Group. The industry body responsible for developing or approving DVD copy protection systems.

CPU   Central processing unit. The integrated circuit chip that forms the brain of a computer or other electronic device. DVD-Video players contain rudimentary CPUs to provide general control and interactive features.

crop   To trim and remove a section of the video picture in order to make it conform to a different shape. Cropping is used in the pan and scan process, but not in the letterbox process.

CVBS   Composite video baseband signal. This is a standard single-wire video, mixing luma and chroma signals together.

DAC   Digital-to-analog converter. Circuitry that converts digital data (such as audio or video) to analog data.

DAE   Digital audio extraction. Reading digital audio data directly from a CD audio disc.

DAT   Digital audio tape. A magnetic audio tape format that uses PCM to store digitized audio or digital data.

data area   The physical area of a DVD disc between the lead in and the lead out (or middle area) that contains the stored data content of the disc.

data rate   The volume of data measured over time. The rate at which digital information can be conveyed. This is usually expressed as bits per second with notations of kbps (thousand/sec), Mbps (million/sec), and Gbps (billion/sec). Digital audio date rate is generally computed as the number of samples per second times the bit size of the sample. For example, the data rate of uncompressed 16-bit, 48-kHz, two-channel audio is 1536 kbps. The digital video bit rate is generally computed as the number of bits per pixel times the number of pixels per line times the number of lines per frame times the number of frames per second. For example, the data rate of a DVD movie before compression is usually $12 \times 720 \times 480 \times 24 = 99.5$ Mbps. Compression reduces the data rate. Digital data rate is sometimes inaccurately equated with bandwidth.

dB *See* decibel.

DBS Digital broadcast satellite. The general term for 18-inch digital satellite systems.

DC Direct current. The electrical current flowing in one direction only. Adopted in the video world to refer to a signal with zero frequency. Compare this to AC.

DCC Digital compact cassette. A digital audio tape format based on the popular compact cassette that was abandoned by Philips in 1996.

DCT Discrete cosine transform. An invertible, discrete, orthogonal transformation. Got that? A mathematical process used in MPEG video encoding to transform blocks of pixel values into blocks of spatial frequency values with lower-frequency components organized into the upper-left corner, allowing the high-frequency components in the lower-right corner to be discounted or discarded. DCT also stands for digital component technology, a videotape format.

DDWG Digital Display Working Group. *See* DVI.

decibel (dB) A unit of measurement expressing ratios using logarithmic scales related to human aural or visual perception. Many different measurements are based on a reference point of 0 dB, such as a standard level of sound or power.

decimation A form of subsampling that discards existing samples (pixels, in the case of spatial decimation, or pictures, in the case of temporal decimation). The resulting information is reduced in size but may suffer from aliasing.

decode To reverse the transformation process of an encoding method. Decoding processes are usually deterministic.

decoder 1) A circuit that decodes compressed audio or video, taking an encoded input stream and producing output such as audio or video. DVD players use the decoders to recreate information that was compressed by systems such as MPEG-2 and Dolby Digital; 2) A circuit that converts composite video to component video or matrixed audio to multiple channels.

delta picture (or **delta frame**)   A video picture based on the changes from the picture before (or after) it. MPEG P pictures and B pictures are examples. Contrast this with key picture.

deterministic   A process or model in which the outcome does not depend upon chance, and a given input always produces the same output. Audio and video decoding processes are mostly deterministic.

digital   Expressed in digits. A set of discrete numeric values, as used by a computer. Analog information can be digitized by sampling.

digital signal processor (DSP)   A digital circuit that can be programmed to perform digital data manipulation tasks such as decoding or audio effects.

digital video noise reduction (DVNR)   Digitally removing noise from video by comparing frames in sequence to spot temporal aberrations.

digitize   To convert analog information to digital information by sampling.

DIN   Deutsches Institut für Normung/German Institute for Standardization (Refer to Appendix C, "References and Information Sources.")

directory   The part of a disc that indicates which files are stored on the disc and where they are located.

DirectShow   A software standard developed by Microsoft for the playback of digital video and audio in the Windows operating system. This has replaced the older MCI and Video for Windows software.

disc key   A value used to encrypt and decrypt (scramble) a title key on DVD-Video discs.

disc menu   The main menu of a DVD-Video disc from which titles are selected. This is also called the system menu or title selection menu.

discrete cosine transform   *See* DCT.

discrete surround sound   Audio in which each channel is stored and transmitted separate from and independent of other channels. Multiple independent channels, directed to loudspeakers in front of and behind the listener, enable precise control of the soundfield in order to generate localized sounds and simulate moving sound sources.

display rate   The number of times per second the image in a video system is refreshed. Progressive scan systems such as film or HDTV change the image once per frame. Interlace scan systems such as standard television change the image twice per frame, with two fields in each frame. Film has a frame rate of 24 fps, but each frame is shown twice by the projector for a display rate of 48 fps. 525/60 (NTSC) television has a rate of 29.97 frames per second (59.94 fields per second). 625/50 (PAL/SECAM) television has a rate of 25 frames per second (50 fields per second).

Divx   Digital Video Express. A short-lived pay-per-viewing-period variation of DVD.

DLT   Digital linear tape. A digital archive standard using half-inch tapes, commonly used for submitting a premastered DVD disc image to a replication service.

Dolby Digital   A perceptual coding system for audio, developed by Dolby Laboratories and accepted as an international standard. Dolby Digital is the most common means of encoding audio for DVD-Video and is the mandatory audio compression system for 525/60 (NTSC) discs.

Dolby Pro Logic   The technique (or the circuit that applies the technique) of extracting surround audio channels from a matrix-encoded audio signal. Dolby Pro Logic is a decoding technique only, but it is often mistakenly used to refer to Dolby Surround audio encoding.

Dolby Surround   The standard for matrix encoding surround-sound channels in a stereo signal by applying a set of defined mathematical functions when combining center and surround channels with left and right channels. The center and surround channels can then be extracted by a decoder such as a Dolby Pro Logic circuit that applies the inverse of the mathematical functions. A Dolby Surround decoder extracts surround channels, while a Dolby Pro Logic decoder uses additional processing to create a center channel. The process is essentially independent of the recording or transmission format. Both Dolby Digital and MPEG audio compression systems are compatible with Dolby Surround audio.

downmix   To convert a multichannel audio track into a two-channel stereo track by combining the channels with the Dolby Surround process. All DVD players are required to provide downmixed audio output from Dolby Digital audio tracks.

downsampling   *See* subsampling.

DRC   *See* dynamic range compression.

driver   A software component that enables an application to communicate with a hardware device.

DSD   Direct Stream Digital. An uncompressed audio bitstream coding method developed by Sony. It is used as an alternative to PCM.

DSI   Data search information. Navigation and search information contained in the DVD-Video data stream. DSI and PCI together make up an overhead of about one Mbps.

DSP   Digital signal processor (or processing).

DSVCD   Double Super Video Compact Disc. A long-playing variation of SVCD.

DTS   Digital Theater Sound. A perceptual audio-coding system developed for theaters. A competitor to Dolby Digital and an optional audio track format for DVD-Video and DVD-Audio.

DTS-ES   A version of DTS decoding that is compatible with 6.1-channel Dolby Surround EX. DTS-ES Discrete is a variation of DTS encoding and decoding that carries a discrete rear center channel instead of a matrixed channel.

DTV   Digital television. In general, any system that encodes video and audio in digital form. In specific, the Digital Television System proposed by the ATSC or the digital TV standard proposed by the Digital TV Team founded by Microsoft, Intel, and Compaq.

duplication   The reproduction of media. This generally refers to producing discs in small quantities, as opposed to large-scale replication.

DV   Digital Video. This usually refers to the digital videocassette standard developed by Sony and JVC.

DVB   Digital video broadcast. A European standard for broadcast, cable, and digital satellite video transmission.

DVC   Digital video cassette. The early name for DV.

DVCAM   Sony's proprietary version of DV.

DVCD   Double Video Compact Disc. A long-playing (100-minute) variation of VCD.

DVCPro   Matsushita's proprietary version of DV.

DVD   An acronym that officially stands for nothing but is often expanded as Digital Video Disc or Digital Versatile Disc. The audio/video/data storage system based on 12- and 8-cm optical discs.

DVD-Audio (DVD-A)   The audio-only format of DVD that primarily uses PCM audio with MLP encoding, along with an optional subset of DVD-Video features.

DVD-R   A version of DVD on which data can be recorded once. It uses dye sublimation recording technology.

DVD-RAM   A version of DVD on which data can be recorded more than once. It uses phase-change recording technology.

DVD-ROM   The base format of DVD-ROM stands for read-only memory, referring to the fact that standard DVD-ROM and DVD-Video discs can't be recorded on. A DVD-ROM can store essentially any form of digital data.

DVD-Video (DVD-V)   A standard for storing and reproducing audio and video on DVD-ROM discs, based on MPEG video, Dolby Digital and MPEG audio, and other proprietary data formats.

DVI (Digital Visual Interface)   The digital video interface standard developed by the Digital Display Working Group (DDWG). A replacement for analog VGA monitor interface.

DVNR   *See* digital video noise reduction.

DVS   Descriptive video services that provide a narration for blind or sight-impaired viewers.

dye polymer   The chemical used in DVD-R and CD-R media that darkens when heated by a high-power laser.

dye-sublimation   An optical disc recording technology that uses a high-powered laser to burn readable marks into a layer of organic dye. Other recording formats include magneto-optical and phase-change.

dynamic range   The difference between the loudest and softest sound in an audio signal. The dynamic range of digital audio is determined by the sample size. Increasing the sample size does not allow louder sounds; it increases the resolution of the signal, thus allowing softer sounds to be separated from the noise floor (and allowing more amplification with less distortion). Therefore, the dynamic range refers to the difference between the maximum level of distortion-free signal and the minimum limit reproducible by the equipment.

dynamic range compression   A technique of reducing the range between loud and soft sounds in order to make dialog more audible, especially when listening at low volume levels. It is used in the down-mix process of multichannel Dolby Digital sound tracks.

EBU   European Broadcasting Union. Refer to Appendix C, "References and Information Sources."

ECC   *See* error-correction code.

ECD   Error-detection and correction code. *See* error-correction code.

ECMA   European Computer Manufacturers Association. (See Appendix C, "References and Information Sources.")

EDC   A short error-detection code applied at the end of a DVD sector.

edge enhancement   When films are transferred to video in preparation for DVD encoding, they are commonly run through digital processes that attempt to clean up the picture. These processes include noise reduction (DVNR) and image enhancement. Enhancement increases the contrast (similar to the effect of the sharpen or unsharp mask filters in Photoshop), but it can tend to overdo areas of transition between light and dark or different colors. This causes a chiseled look or a ringing effect like the haloes you see around streetlights when driving in the rain. Video noise reduction is a good thing when done well, because it can remove scratches, spots, and other defects from the original film. Enhancement, which is rarely done well, is a bad thing. The video may look sharper and clearer to the casual observer, but fine tonal details of the original picture are altered and lost.

EDS   Enhanced data services. Additional information in the NTSC line such as a time signal.

EDTV   Enhanced-definition television. A system that uses existing transmission equipment to send an enhanced signal that looks the same on existing receivers, but it carries additional information to improve the picture quality on new enhanced receivers. PALPlus is an example of EDTV. Contrast this with HDTV and IDTV.

EFM   Eight-to-14 modulation. A modulation method used by CD. The 8/16 modulation used by DVD is sometimes called EFM plus.

EIA   Electronics Industry Association. Refer to Appendix C, "References and Information Sources."

E-IDE   Enhanced Integrated Drive Electronics. These are extensions to the IDE standard that provide faster data transfers and enable access to larger drives, including CD-ROM and tape drives, using ATAPI. E-IDE was adopted as a standard by ANSI in 1994. ANSI calls it Advanced Technology Attachment-2 (ATA-2) or Fast ATA.

elementary stream   A general term for a coded bitstream such as audio or video. Elementary streams are made up of packs of packets.

emulate   To test the function of a DVD disc on a computer after formatting a complete disc image.

encode   To transform data for storage or transmission, usually in such a way that redundancies are eliminated or complexity is reduced. Most compression is based on one or more encoding methods. Data such as audio or video is encoded for efficient storage or transmission and is decoded for access or display.

encoder   1) A circuit or program that encodes (and thereby compresses) audio or video; 2) A circuit that converts component digital video to composite analog video. DVD players include TV encoders to generate standard television signals from decoded video and audio; 3) A circuit that converts multichannel audio to two-channel matrixed audio.

Enhanced CD   A music CD that has additional computer software and can be played in a music player or read by a computer. Also called CD Extra, CD Plus, hybrid CD, interactive music CD, mixed-mode CD, pre-gap CD, or track-zero CD.

entropy coding   Variable-length, lossless coding of a digital signal to reduce redundancy. MPEG-2, DTS, and Dolby Digital apply entropy coding after the quantization step. MLP also uses entropy coding.

EQ   Equalization of audio.

error-correction code   Additional information added to data to enable errors to be detected and possibly corrected. Refer to Chapter 3, "DVD Technology Primer."

ETSI   European Telecommunications Standards Institute. Refer to Appendix C, "References and Information Sources."

father   The metal master disc formed by electroplating the glass master. The father disc is used to make mother discs from which multiple stampers (sons) can be made.

field   A set of alternating scan lines in an interlaced video picture. A frame is made of a top (odd) field and a bottom (even) field.

file   A collection of data stored on a disc, usually in groups of sectors.

file system   A defined way of storing files, directories, and information about such files and directories on a data storage device.

filter   1) To reduce the amount of information in a signal. 2) A circuit or process that reduces the amount of information in a signal. Analog filtering usually removes certain frequencies. Digital filtering (when not emulating analog filtering) usually averages together multiple adjacent pixels, lines, or frames to create a single new pixel, line, or frame. This generally causes a loss of detail, especially with complex images or rapid motion. *See* letterbox filter. Compare this to interpolate.

FireWire   A standard for the transmission of digital data between external peripherals, including consumer audio and video devices. The official name is IEEE 1394, based on the original FireWire design by Apple Computer.

fixed rate   Information flow at a constant volume over time. *See* CBR.

forced display   A feature of DVD-Video that enables subpictures to be displayed even if the player's subpicture display mode is turned off. It is also designed to show subtitles in a scene where the language is different from the native language of the film.

formatting   1) Creating a disc image. 2) Preparing storage media for recording.

fps   Frames per second. A measure of the rate at which pictures are shown to create a motion video image. In NTSC and PAL video, each frame is made up of two interlaced fields.

fragile watermark   A watermark designed to be destroyed by any form of copying or encoding other than a bit-for-bit digital copy. The absence of the watermark indicates that a copy has been made.

frame   The piece of a video signal containing the spatial detail of one complete image, or the entire set of scan lines. In an interlaced system, a frame contains two fields.

frame doubler   A video processor that increases the frame rate (display rate) in order to create a smoother-looking video display. Compare this to line doubler.

frame rate   The frequency of discrete images. This is usually measured in frames per second (fps). Film has a rate of 24 frames per second, but it usually must be adjusted to match the display rate of a video system.

frequency   The number of repetitions of a phenomenon in a given amount of time. The number of complete cycles of a periodic process occurring per unit time.

G   Giga. An SI prefix for denominations of one billion (10^9).

G byte   One billion (10^9) bytes. Not to be confused with GB or gigabyte (2^{30} bytes).

Galaxy Group   The group of companies proposing the Galaxy watermarking format (IBM/NEC, Hitachi/Pioneer/Sony).

GB   Gigabyte.

Gbps   Gigabits/second. Billions (10^9) of bits per second.

gigabyte   1,073,741,824 (2^{30}) bytes. See the end of Chapter 1, "Introduction," for more information.

GOP   Group of pictures. In MPEG video, one or more I pictures followed by P and B pictures. A GOP is the atomic unit of MPEG video access. GOPs are limited in DVD-Video to 18 frames for 525/60 and 15 frames for 625/50.

gray market   Dealers and distributors who sell equipment without proper authorization from the manufacturer.

Green Book   The document developed in 1987 by Philips and Sony as an extension to CD-ROM XA for the CD-i system.

HAVi   A consumer electronics industry standard for interoperability between digital audio and video devices connected via a network in the consumer's home.

HDCD   High-definition Compatible Digital. A proprietary method of enhancing audio on CDs.

HDTV   High-definition television. A video format with a resolution approximately twice that of conventional television in both the horizontal and vertical dimensions, and a picture aspect ratio of 16:9. Used loosely to refer to the U.S. DTV System. Contrast this with EDTV and IDTV.

H/DTV   High-definition/digital television. A combination of acronyms that refers to both HDTV and DTV systems.

hertz   *See* Hz.

hexadecimal   Representation of numbers using base 16.

HFS   Hierarchical file system. A file system used by Apple Computer's Mac OS operating system.

High Sierra   The original file system standard developed for CD-ROM, later modified and adopted as ISO 9660.

horizontal resolution   *See* lines of horizontal resolution.

HQ-VCD   High-Quality Video Compact Disc. Developed by the Video CD Consortium (Philips, Sony, Matsushita, and JVC) as a successor to VCD. It has evolved into SVCD.

HRRA   Home Recording Rights Association.

HSF   *See* High Sierra.

HTML   Hypertext markup language. This is a tagging specification, based on the standard generalized markup language (SGML), for formatting text to be transmitted over the Internet and displayed by client software.

hue   The color of light or a pixel. The property of color determined by the dominant wavelength of light.

Huffman coding   A lossless compression technique of assigning variable-length codes to a known set of values. The values occurring the most frequently are assigned the shortest codes. MPEG uses a variation of Huffman coding with fixed code tables, often called variable-length coding (VLC).

Hz   Hertz. A unit of frequency measurement that determines the number of cycles (repetitions) per second.

I picture (or I frame)   In MPEG video, this is an intra picture that is encoded independent from other pictures (see intraframe). Transform coding (DCT, quantization, and VLC) is used with no motion compensation, resulting in only moderate compression. I pictures provide a reference point for dependent P pictures and B pictures and enable random access into the compressed video stream.

i.Link   Trademarked Sony name for IEEE 1394.

IDE   Integrated Drive Electronics. An internal bus or standard electronic interface between a computer and internal block storage devices. IDE was adopted as a standard by ANSI in November 1990. ANSI calls it Advanced Technology Attachment (ATA). *See* E-IDE and ATAPI.

IDTV   Improved-definition television. A television receiver that improves the apparent quality of the picture from a standard video signal by using techniques such as frame doubling, line doubling, and digital signal processing.

IEC   International Electrotechnical Commission. Refer to Appendix C, "References and Information Sources."

IED   ID error correction. An error-detection code applied to each sector ID on a DVD disc.

IEEE   Institute of Electrical and Electronics Engineers, an electronics standards body.

IEEE 1394   A standard for the transmission of digital data between external peripherals, including consumer audio and video devices. Also known as FireWire.

IFE   In-flight entertainment.

I-MPEG   Intraframe MPEG. An unofficial variation of MPEG video encoding that uses only intraframe compression. I-MPEG is used by DV equipment.

interframe   Something that occurs between multiple frames of video. Interframe compression takes temporal redundancy into account. Contrast this with intraframe.

interlace   A video scanning system in which alternating lines are transmitted, so that half a picture is displayed each time the scanning beam moves down the screen. An interlaced frame is made of two fields. Refer to Chapter 3, "DVD Technology Primer.")

interleave   To arrange data in alternating chunks so that selected parts can be extracted while other parts are skipped over, or so that each chunk carries a piece of a different data stream.

interpolate   To increase the pixels, scan lines, or pictures when scaling an image or a video stream by averaging together adjacent pixels, lines, or frames to create additional inserted pixels or frames. This generally causes a softening of still images and a blurriness of motion images because no new information is created. Compare this to filter.

intraframe   Something that occurs within a single frame of video. Intraframe compression does not reduce temporal redundancy but enables each frame to be independently manipulated or accessed. *See* I picture. Compare this to interframe.

inverse telecine   The reverse of 2-3 pulldown, where the frames that were duplicated to create 60-fields/second video from 24-frames/second film source are removed. MPEG-2 video encoders usually apply an inverse telecine process to convert 60-fields/second video into 24-

frames/second encoded video. The encoder adds information enabling the decoder to recreate the 60-fields/second display rate.

ISO International Organization for Standardization. Refer to Appendix C, "References and Information Sources."

ISO 9660 The international standard for the file system used by CD-ROM. ISO 9660 allows filenames of only eight characters plus a three-character extension.

ISRC International Standard Recording Code.

ITU International Telecommunication Union. Refer to Appendix C, "References and Information Sources."

ITU-R BT.601 The international standard specifying the format of digital component video. Currently at version 5 (identified as 601-5).

Java A programming language with specific features designed for use with the Internet and HTML.

JCIC Joint Committee on Intersociety Coordination.

JEC Joint Engineering Committee of EIA and NCTA.

jewel box The plastic clamshell case that holds a CD or DVD.

jitter A temporal variation in a signal from an ideal reference clock. Many kinds of jitter can occur, including sample jitter, channel jitter, and interface jitter. Refer to Chapter 3, "DVD Technology Primer."

JPEG Joint Photographic Experts Group. The international committee that created its namesake standard for compressing still images.

k Kilo. An SI prefix for denominations of one thousand (10^3). Also used, in capital form, for 1,024 bytes of computer data (see kilobyte).

k byte One thousand (10^3) bytes. Not to be confused with KB or kilobyte (2^{10} bytes). Note the small "k."

karaoke Literally empty orchestra. The social sensation from Japan where sufficiently inebriated people embarrass themselves in public by singing along to a music track. Karaoke was largely responsible for the success of laserdisc in Japan, thus supporting it elsewhere.

KB Kilobyte.

kbps Kilobits/second. Thousands (10^3) of bits per second.

key picture (or key frame) A video picture containing the entire content of the image (intraframe encoding), rather than the difference between it and another image (interframe encoding). MPEG I pictures are key pictures. Contrast this with delta picture.

kHz Kilohertz. A unit of frequency measurement. It is one thousand cycles (repetitions) per second or 1,000 hertz.

kilobyte   1,024 (2^{10}) bytes. Refer to Chapter 1, "Introduction," for more information.

land   The raised area of an optical disc.

laserdisc   A 12-inch (or 8-inch) optical disc that holds analog video (using an FM signal) and both analog and digital (PCM) audio. Laserdisc was a precursor to DVD.

layer   The plane of a DVD disc where information is recorded in a pattern of microscopic pits. Each substrate of a disc can contain one or two layers. The first layer, closest to the readout surface, is layer 0; the second is layer 1.

lead in   The physical area that is 1.2 mm or wider preceding the data area on a disc. The lead in contains sync sectors and control data including disc keys and other information.

lead out   On a single-layer disc or PTP dual-layer disc, this is the physical area 1.0 mm or wider toward the outside of the disc following the data area. On an OTP dual-layer disc, this is the physical area 1.2 mm or wider at the inside of the disc following the recorded data area (which is read from the outside toward the inside on the second layer).

legacy   A term used to describe a hybrid disc that can be played in both a DVD player and a CD player.

letterbox   The process or form of video where black horizontal mattes are added to the top and bottom of the display area in order to create a frame in which to display video using an aspect ratio different than that of the display. The letterbox method preserves the entire video picture, as opposed to pan and scan. DVD-Video players can automatically letterbox an anamorphic widescreen picture for display on a standard 4:3 TV.

letterbox filter   The circuitry in a DVD player that reduces the vertical size of anamorphic widescreen video (combining every four lines into three) and adds black mattes at the top and bottom. *See* filter.

level   In MPEG-2, levels specify parameters such as resolution, bit rate, and frame rate. Compare this to profile.

linear PCM   A coded representation of digital data that is not compressed. Linear PCM spreads values evenly across the range from highest to lowest, as opposed to nonlinear (companded) PCM that allocates more values to more important frequency ranges.

line doubler   A video processor that doubles the number of lines in the scanning system in order to create a display with scan lines that are less visible. Some line doublers convert from an interlaced to a progressive scan.

lines of horizontal resolution Sometimes abbreviated as TVL (TV lines) or LoHR, this is a common but subjective measurement of the visually resolvable horizontal detail of an analog video system, measured in half-cycles per picture height. Each cycle is a pair of vertical lines, one black and one white. The measurement is usually made by viewing a test pattern to determine where the black and white lines blur into gray. The resolution of VHS video is commonly gauged at 240 lines of horizontal resolution, broadcast video at 330, laserdisc at 425, and DVD at 500 to 540. Because the measurement is relative to picture height, the aspect ratio must be taken into account when determining the number of vertical units (roughly equivalent to pixels) that can be displayed across the width of the display. For example, an aspect ratio of 1.33 multiplied by 540 gives 720 pixels.

locale *See* regional code.

logical An artificial structure or organization of information created for convenience of access or reference, usually different from the physical structure or organization. For example, the application specifications of DVD (the way information is organized and stored) are logical formats.

logical unit A physical or virtual peripheral device, such as a DVD-ROM drive.

L_o/R_o Left only/right only. A stereo signal with no matrixed surround information in which optional downmixing is output in Dolby Digital decoders. It does not change the phase but simply folds surround channels forward into L_f and R_f.

lossless compression Compression techniques that enable the original data to be recreated without loss. Contrast with lossy compression.

lossy compression Compression techniques that achieve very high compression ratios by permanently removing data while preserving as much significant information as possible. Lossy compression includes perceptual coding techniques that attempt to limit the data loss so that it is least likely to be noticed by human perception.

LP Long-playing record. An audio recording on a plastic platter turning at 33 1/3 rpm and read by a stylus.

LPCM *See* linear PCM.

L_t/R_t Left total/right total. Four surround channels matrixed into two channels. The mandatory downmixing method in Dolby Digital decoders.

luma (Y′) The brightness component of a color video image (also called the grayscale, monochrome, or black-and-white component) with nonlinear luminance. The standard luma signal is computed from nonlinear RGB as $Y' = 0.299\ R' + 0.587\ G' + 0.114\ B'$.

luminance (Y)   Loosely, the sum of RGB tristimulus values corresponding to brightness. This may refer to a linear signal or (incorrectly) a nonlinear signal.

M   Mega. An SI prefix for denominations of one million (10^6).

Mac OS   The operating system used by Apple Macintosh computers.

macroblock   In MPEG MP@ML, the four 8×8 blocks of luma information and two 8×8 blocks of chroma information that form a 16×16 arae of a video frame.

macroblocking   An MPEG artifact. *See* blocking.

Macrovision   An antitaping process that modifies a signal so that it appears unchanged on most televisions but is distorted and unwatchable when played back from a videotape recording. Macrovision takes advantage of the characteristics of AGC circuits and burst decoder circuits in VCRs to interfere with the recording process.

magneto-optical   A recordable disc technology using a laser to heat spots that are altered by a magnetic field. Other formats include dye-sublimation and phase-change.

main level (ML)   A range of proscribed picture parameters defined by the MPEG-2 video standard, with a maximum resolution equivalent to ITU-R BT.601 ($720 \times 576 \times 30$). *See* level.

main profile (MP)   A subset of the syntax of the MPEG-2 video standard designed to be supported over a large range of mainstream applications such as digital cable TV, DVD, and digital satellite transmission. *See* profile.

mark   The non-reflective area of a writable optical disc. Equivalent to a pit.

master   The metal disc used to stamp replicas of optical discs, or the tape used to make additional recordings.

mastering   The process of replicating optical discs by injecting liquid plastic into a mold containing a master. This is often used inaccurately to refer to premastering.

matrix encoding   The technique of combining additional surround-sound channels into a conventional stereo signal. *See* Dolby Surround.

matte   An area of a video display or motion picture that is covered (usually in black) or omitted in order to create a differently shaped area within the picture frame.

MB   Megabyte.

Mbps   Megabits/second. Millions (10^6) of bits per second.

M byte   One million (10^6) bytes. Not to be confused with MB or megabyte (2^{20} bytes).

megabyte   1,048,576 (2^{20}) bytes. Refer to Chapter 1, "Introduction," for more information.

megapixel   An image or display format with a resolution of approximately one million pixels.

memory   Data storage used by computers or other digital electronics systems. Read-only memory (ROM) permanently stores data or software program instructions. New data cannot be written to ROM. Random-access memory (RAM) temporarily stores data, including digital audio and video, while it is being manipulated and holds software application programs while they are being executed. Data can be read from and written to RAM. Other long-term memory includes hard disks, floppy disks, digital CD formats (CD-ROM, CD-R, and CD-RW), and DVD formats (DVD-ROM, DVD-R, and DVD-RAM).

MHz   One million (10^6) Hz.

Microsoft Windows   The leading operating system for Intel CPU-based computers developed by Microsoft.

middle area   On a dual-layer OTP disc, the physical area 1.0 mm or wider on both layers, adjacent to the outside of the data area.

Millennium Group   The group of companies proposing the Millennium watermarking format that includes Macrovision, Philips, and Digimarc.

mixed mode   A type of CD containing both Red Book audio and Yellow Book computer data tracks.

MKB (Media Key Block)   A set of keys used in CPPM and CPRM for authenticating players.

MLP (Meridian Lossless Packing)   A lossless compression technique (used by DVD-Audio) that removes redundancy from PCM audio signals to achieve a compression ratio of about 2:1 while allowing the signal to be perfectly recreated by the MLP decoder.

MO   Magneto-optical rewritable discs.

modulation   Replacing patterns of bits with different (usually larger) patterns designed to control the characteristics of the data signal. DVD uses 8/16 modulation, where each set of eight bits is replaced by 16 bits before being written onto the disc.

mosquitoes   A term referring to the fuzzy dots that can appear around sharp edges (high spatial frequencies) after video compression. Also known as the Gibbs Effect.

mother   The metal discs produced from mirror images of the father disc in the replication process. Mothers are used to make stampers, often called sons.

motion compensation   In video decoding, the application of motion vectors to already-decoded blocks in order to construct a new picture.

motion estimation   In video encoding, the process of analyzing previous or future frames to identify blocks that have not changed or have changed only their location. Motion vectors are then stored in place of the blocks. This is very computation-intensive and can cause visual artifacts when subject to errors.

motion vector   A two-dimensional spatial displacement vector used for MPEG motion compensation to provide an offset from the encoded position of a block in a reference (I or P) picture to the predicted position (in a P or B picture).

MP@ML   Main profile at main level. The common MPEG-2 format used by DVD (along with SP@SL).

MP3   MPEG-1 Layer III audio. A perceptual audio coding algorithm. Not supported in DVD-Video or DVD-Audio formats.

MPEG   Moving Picture Experts Group. An international committee that developed the MPEG family of audio and video compression systems.

MPEG audio   Audio compressed according to the MPEG perceptual encoding system. MPEG-1 audio provides two channels, which can be in Dolby Surround format. MPEG-2 audio adds data to provide discrete multichannel audio. Stereo MPEG audio is one of two mandatory audio compression system for 625/50 (PAL/SECAM) DVD-Video.

MPEG video   Video compressed according to the MPEG encoding system. MPEG-1 is typically used for low data rate video such as on a Video CD. MPEG-2 is used for higher-quality video, especially interlaced video, such as on DVD or HDTV.

MTBF   Mean time between failure. A measure of reliability for electronic equipment, usually determined in benchmark testing. The higher the MTBF, the more reliable the hardware.

Mt. Fuji   *See* SFF 8090.

multiangle   A DVD-Video program containing multiple angles, allowing different views of a scene to be selected during playback.

multichannel   Multiple channels of audio, usually containing different signals for different speakers in order to create a surround-sound effect.

multilanguage   A DVD-Video program containing sound tracks or subtitle tracks for more than one language.

multimedia   Information in more than one form, such as text, still images, sound, animation, and video. Usually implies that the information is presented by a computer.

multiplexing   Combining multiple signals or data streams into a single signal or stream. This is usually achieved by interleaving at a low level.

MultiRead   A standard developed by the Yokohama group, a consortium of companies attempting to ensure that new CD and DVD hardware can read all CD formats (refer to "Innovations of CD" in Chapter 2, "The World Before and After DVD," for a discussion of CD variations).

multisession   A technique in write-once recording technology that enables additional data to be appended after data is written in an earlier session.

mux   Short for multiplex.

mux_rate   In MPEG, the combined rate of all packetized elementary streams (PES) of one program. The mux_rate of DVD is 10.08 Mbps.

NAB   National Association of Broadcasters.

NCTA   National Cable Television Association.

nighttime mode   A Dolby Digital dynamic range compression feature that enables low-volume nighttime listening without losing dialog legibility.

noise   Irrelevant, meaningless, or erroneous information added to a signal by the recording or transmission medium or by an encoding/decoding process. An advantage of digital formats over analog formats is that noise can be completely eliminated (although new noise can be introduced by compression).

noise floor   The level of background noise in a signal or the level of noise introduced by equipment or storage media, below which the signal can't be isolated from the noise.

NRZI   Non-return to zero, inverted. A method of coding binary data as waveform pulses. Each transition represents a one, while a lack of a transition represents a run of zeros.

NTSC   National Television Systems Committee. A committee organized by the Electronic Industries Association (EIA) that developed commercial television broadcast standards for the U.S. The group first established black-and-white TV standards in 1941, using a scanning system of 525 lines at 60 fields per second. The second committee standardized color enhancements using 525 lines at 59.94 fields per second. NTSC refers to the composite color-encoding system. The 525/59.94 scanning system (with a 3.58-MHz color subcarrier) is identified by the letter M and is often incorrectly referred to as NTSC. The NTSC standard is also used in Canada, Japan, and other parts of the world. NTSC is facetiously referred to as meaning never the same color because of the system's difficulty in maintaining color consistency.

NTSC-4.43 A variation of NTSC in which a 525/59.94 signal is encoded using the PAL subcarrier frequency and chroma modulation. Also called 60-Hz PAL.

numerical aperture (NA) A unitless measure of the capability of a lens to gather and focus light. NA = n sin 1, where 1 is the angle of the light as it narrows to the focal point. A numerical aperture of 1 implies no change in parallel light beams. The higher the number, the greater the focusing power and the smaller the spot.

OEM Original equipment manufacturer. A computer maker.

operating system The primary software in a computer, containing general instructions for managing applications, communications, input/output, memory, and other low-level tasks. DOS, Windows, Mac OS, and Unix are examples of operating systems.

opposite path *See* OTP.

Orange Book The document begun in 1990 that specifies the format of recordable CD. Its three parts define magneto-optical erasable (MO) and write-once (WO) discs, dye-sublimation write-once (CD-R) discs, and phase-change rewritable (CD-RW) discs. Orange Book also added multisession capabilities to the CD-ROM XA format.

OS Operating system.

OSTA Optical Storage Technology Association. Refer to Appendix C, "References and Information Sources."

OTP Opposite track path. A variation of DVD dual-layer disc layout where readout begins at the center of the disc on the first layer, travels to the outer edge of the disc, then switches to the second layer, and travels back toward the center. Designed for long, continuous-play programs. Also called RSDL. Contrast this with PTP.

out of band In a place not normally accessible.

overscan The area at the edges of a television tube that is covered to hide possible video distortion. Overscan typically covers about four or five percent of the picture.

pack A group of MPEG packets in a DVD-Video program stream. Each DVD sector (2,048 bytes) contains one pack.

packet A low-level unit of DVD-Video (MPEG) data storage containing contiguous bytes of data belonging to a single elementary stream such as video, audio, control, and so forth. Packets are grouped into packs.

packetized elementary stream (PES) The low-level stream of MPEG packets containing an elementary stream, such as audio or video.

PAL Phase alternate line. A video standard used in Europe and other parts of the world for composite color encoding. Various versions of PAL use different scanning systems and color subcarrier frequencies (identified with letters B, D, G, H, I, M, and N), the most common being 625 lines at 50 fields per second, with a color subcarrier of 4.43 MHz. PAL is also said to mean "picture always lousy" or "perfect at last," depending on which side of the ocean the speaker comes from.

palette A table of colors that identifies a subset from a larger range of colors. The small number of colors in the palette enables fewer bits to be used for each pixel. Also called a color look-up table (CLUT).

pan and scan The technique of reframing a picture to conform to a different aspect ratio by cropping parts of the picture. DVD-Video players can automatically create a 4:3 pan and scan version from widescreen anamorphic video by using a horizontal offset encoded with the video.

parallel path *See* PTP.

parental management An optional feature of DVD-Video that prohibits programs from being viewed or substitutes different scenes within a program depending on the parental level set in the player. Parental control requires that parental levels and additional material (if necessary) be encoded on the disc.

part of title In DVD-Video, a division of a title representing a scene. Also called a chapter. Parts of titles are numbered 1 to 99.

PCI Presentation control information. A DVD-Video data stream containing details of the timing and presentation of a program (aspect ratio, angle change, menu highlight and selection information, and so on). PCI and DSI together make up an overhead of about one Mbps.

PCM An uncompressed, digitally coded representation of an analog signal. The waveform is sampled at regular intervals, and a series of pulses in coded form (usually quantized) are generated to represent the amplitude.

PC-TV The merger of television and computers. A personal computer capable of displaying video as a television.

pel *See* pixel.

perceived resolution The apparent resolution of a display from the observer's point of view, based on viewing distance, viewing conditions, and physical resolution of the display.

perceptual coding Lossy compression techniques based on the study of human perception. Perceptual coding systems identify and remove information that is least likely to be missed by the average human observer.

PES   Packetized elementary stream. A single video or audio stream in MPEG format.

PGCI   Program chain information. Data describing a chain of cells (grouped into programs) and their sector locations, thus composing a sequential program. PGCI data is contained in the PCI stream.

phase-change   A technology for rewritable optical discs using a physical effect in which a laser beam heats a recording material to reversibly change an area from an amorphous state to a crystalline state, or vice versa. Continuous heat just above the melting point creates the crystalline state (an erasure), while high heat followed by rapid cooling creates the amorphous state (a mark). Other recording technologies include dye-sublimation and magneto-optical.

physical format   The low-level characteristics of the DVD-ROM and DVD-Video standards, including pits on the disc, the location of data, and the organization of data according to physical position.

picture   In video terms, a single still image or a sequence of moving images. Picture generally refers to a frame, but for interlaced frames, it may refer instead to a field of the frame. In a more general sense, picture refers to the entire image shown on a video display.

picture stop   A function of DVD-Video where a code indicates that video playback should stop and a still picture be displayed.

PIP   Picture in picture. A feature of some televisions that shows another channel or video source in a small window superimposed in a corner of the screen.

pit   A microscopic depression in the recording layer of a optical disc. Pits are usually 1/4 of the laser wavelength in order to cause cancellation of the beam by diffraction.

pit art   A pattern of pits to be stamped onto a disc to provide visual art rather than data. A cheaper alternative to a printed label.

pixel   The smallest picture element of an image (one sample of each color component). A single dot of the array of dots that make up a picture. Sometimes abbreviated to pel. The resolution of a digital display is typically specified in terms of pixels (width by height) and color depth (the number of bits required to represent each pixel).

pixel aspect ratio   The ratio of width to height of a single pixel. This often means the sample pitch aspect ratio (when referring to sampled digital video). Pixel aspect ratio for a given raster can be calculated as $y/x \times w/h$ (where x and y are the raster horizontal pixel count and vertical pixel count, and w and h are the display aspect ratio width and height). Pixel aspect ratios are also confusingly calculated as $x/y \times w/h$, giving a height-to-width ratio. Refer to Table 6.22.

pixel depth   *See* color depth.

PMMA   Polymethylmethacrylate. A clear acrylic compound used in laserdiscs and as an intermediary in the surface transfer process (STP) for dual-layer DVDs. PMMA is also sometimes used for DVD substrates.

POP   Picture outside picture. A feature of some widescreen displays that uses the unused area around a 4:3 picture to show additional pictures.

P picture (or P frame)   In MPEG video, a "predicted" picture based on the difference from previous pictures. P pictures (along with I pictures) provide a reference for following P pictures or B pictures.

premastering   The process of preparing data in the final format to create a DVD disc image for mastering. This includes creating DVD control and navigation data, multiplexing data streams together, generating error-correction codes, and performing channel modulation. This often includes the process of encoding video, audio, and subpictures.

presentation data   DVD-Video information such as video, menus, and audio that is presented to the viewer. *See* PCI.

profile   In MPEG-2, profiles specify syntax and processes such as picture types, scalability, and extensions. Compare this to level.

program   In a general sense, a sequence of audio or video. In a technical sense for DVD-Video, a group of cells within a program chain (PGC).

program chain   In DVD-Video, a collection of programs, or groups of cells, linked together to create a sequential presentation.

progressive scan   A video scanning system that displays all lines of a frame in one pass. Contrast this with interlaced scan. Refer to Chapter 3, "DVD Technology Primer," for more information.

psychoacoustic   *See* perceptual encoding.

PTP   Parallel track path. A variation of DVD dual-layer disc layout where readout begins at the center of the disc for both layers. This is designed for separate programs (such as a widescreen and a pan and scan version on the same disc side) or programs with a variation on the second layer. PTP is most efficient for DVD-ROM random-access application. Contrast this with OTP.

PUH   Pickup head. The assembly of optics and electronics that reads data from a disc.

QCIF   Quarter common intermediate format. Video resolution of 176×144.

quantization levels   The predetermined levels at which an analog signal can be sampled as determined by the resolution of the analog-to-digital converter (in bits per sample), or the number of bits stored for the sampled signal.

quantize   To convert a value or range of values into a smaller value or smaller range by integer division. Quantized values are converted back (by multiplying) to a value that is close to the original but may not be exactly the same. Quantization is a primary technique of lossless encoding.

QuickTime   A digital video software standard developed by Apple Computer for Macintosh (Mac OS) and Windows operating systems. Quick-Time is used to support audio and video from a DVD.

QXGA   A video graphics resolution of 2,048 × 1,536.

RAM   Random-access memory. This generally refers to solid-state chips. In the case of DVD-RAM, the term was borrowed to indicate the capability to read and write at any point on the disc.

RAMbo drive   A DVD-RAM drive capable of reading and writing CD-R and CD-RW media (a play on the word "combo").

random access   The capability to jump to a point on a storage medium.

raster   The pattern of parallel horizontal scan lines that makes up a video picture.

read-modify-write   An operation used in writing to DVD-RAM discs. Because data can be written by the host computer in blocks as small as two KB, while the DVD format uses ECC blocks of 32 KB, an entire ECC block is read from the data buffer or disc, modified to include the new data and new ECC data, and then written back to the data buffer and disc.

Red Book   The document first published in 1982 that specifies the original compact disc digital audio format developed by Philips and Sony.

Reed-Solomon   An error-correction encoding system that cycles data multiple times through a mathematical transformation in order to increase the effectiveness of the error correction, especially for burst errors (errors concentrated closely together, as from a scratch or physical defect). DVD uses rows and columns of Reed-Solomon encoding in a two-dimensional lattice, called Reed-Solomon product code (RS-PC).

reference picture (or reference frame)   An encoded frame that is used as a reference point from which to build dependent frames. In MPEG-2, I pictures and P pictures are used as references.

reference player   A DVD player that defines the ideal behavior as specified by the DVD-Video standard.

regional code   A code identifying one of the world regions for restricting DVD-Video playback. Refer to Table A.25.

regional management   A mandatory feature of DVD-Video to restrict the playback of a disc to a specific geographical region. Each player and DVD-ROM drive include a single regional code, and each disc side can specify in which regions it is allowed to be played. Regional coding is optional; a disc without regional codes will play in all players in all regions.

replication   1) The reproduction of media such as optical discs by stamping (contrast this with duplication); 2) A process used to increase the size of an image by repeating pixels (to increase the horizontal size) and/or lines (to increase the vertical size) or to increase the display rate of a video stream by repeating frames. For example, a 360×240 pixel image can be displayed at 720×480 size by duplicating each pixel on each line and then duplicating each line. In this case, the resulting image contains blocks of four identical pixels. Obviously, image replication can cause blockiness. A 24-fps video signal can be displayed at 72 fps by repeating each frame three times. Frame replication can cause jerkiness of motion. Contrast this with decimation. *See* interpolate.

resampling   The process of converting between different spatial resolutions or different temporal resolutions. This can be based on a sample of the source information at a higher or lower resolution or it can include interpolation to correct for the differences in pixel aspect ratios or to adjust for differences in display rates.

resolution   1) A measurement of the relative detail of a digital display, typically given in pixels of width and height; 2) The capability of an imaging system to make the details of an image clearly distinguishable or resolvable. This includes spatial resolution (the clarity of a single image), temporal resolution (the clarity of a moving image or moving object), and perceived resolution (the apparent resolution of a display from the observer's point of view). Analog video is often measured as a number of lines of horizontal resolution over the number of scan lines. Digital video is typically measured as a number of horizontal pixels by vertical pixels. Film is typically measured as a number of line pairs per millimeter; 3) The relative detail of any signal, such as an audio or video signal. *See* lines of horizontal resolution.

RGB   Video information in the form of red, green, and blue tristimulus values. The combination of three values representing the intensity of each of the three colors can represent the entire range of visible light.

ROM   Read-only memory.

rpm   Revolutions per minute. A measure of rotational speed.

RS   Reed-Solomon. An error-correction encoding system that cycles data multiple times through a mathematical transformation in order to increase the effectiveness of the error correction. DVD uses rows and columns of Reed-Solomon encoding in a two-dimensional lattice, called Reed-Solomon product code (RS-PC).

RS-CIRC   *See* CIRC.

RSDL   Reverse-spiral dual-layer. *See* OTP.

RS-PC   Reed-Solomon product code. An error-correction encoding system used by DVD employing rows and columns of Reed-Solomon encoding to increase error-correction effectiveness.

R-Y, B-Y   The general term for color-difference video signals carrying red and blue color information, where the brightness (Y) has been subtracted from the red and blue RGB signals to create R-Y and B-Y color-difference signals. Refer to Chapter 3, "DVD Technology Primer."

sample   A single digital measurement of analog information or a snapshot in time of a continuous analog waveform. *See* sampling.

sample rate   The number of times a digital sample is taken, measured in samples per second, or Hertz. The more often samples are taken, the better a digital signal can represent the original analog signal. The sampling theory states that the sampling frequency must be more than twice the signal frequency in order to reproduce the signal without aliasing. DVD PCM audio enables sampling rates of 48 and 96 kHz.

sample size   The number of bits used to store a sample. Also called resolution. In general, the more bits are allocated per sample, the better the reproduction of the original analog information. The audio sample size determines the dynamic range. DVD PCM audio uses sample sizes of 16, 20, or 24 bits.

sampling   Converting analog information into a digital representation by measuring the value of the analog signal at regular intervals, called samples, and encoding these numerical values in digital form. Sampling is often based on specified quantization levels. Sampling can also be used to adjust for differences between different digital systems. *See* resampling and subsampling.

saturation   The intensity or vividness of a color.

scaling   Altering the spatial resolution of a single image to increase or reduce the size, or altering the temporal resolution of an image sequence to increase or decrease the rate of display. Techniques include decimation, interpolation, motion compensation, replication, resampling, and subsampling. Most scaling methods introduce artifacts.

scan line   A single horizontal line traced out by the scanning system of a video display unit. 525/60 (NTSC) video has 525 scan lines, about 480 of which contain the actual picture. 625/50 (PAL/SECAM) video has 625 scan lines, about 576 of which contain the actual picture.

scanning velocity   The speed at which the laser pickup head travels along the spiral track of a disc.

SCMS   The serial copy management system used by DAT, MiniDisc, and other digital recording systems to control copying and limit the number of copies that can be made from copies.

SCSI   Small Computer Systems Interface. An electronic interface and command set for attaching and controlling internal or external peripherals, such as a DVD-ROM drive, to a computer. The command set of SCSI was extended for DVD-ROM devices by the SFF 8090 specification.

SDI   *See* Serial Digital Interface. Also Strategic Defense Initiative, a.k.a. *Star Wars*, which as of 2000 was still not available on DVD other than as bootleg copies.

SDDI   Serial Digital Data Interface. A digital video interconnect designed for serial digital information to be carried over a standard SDI connection.

SDDS   Sony Dynamic Digital Sound. A perceptual audio-coding system developed by Sony for multichannel audio in theaters. A competitor to Dolby Digital and an optional audio track format for DVD.

SDMI   Secure Digital Music Initiative. Efforts and specifications for protecting digital music.

SDTV   Standard-definition television. A term applied to traditional 4:3 television (in digital or analog form) with a resolution of about 700×480 (about 1/3 megapixel). Contrast this with HDTV.

seamless playback   A feature of DVD-Video where a program can jump from place to place on the disc without any interruption of the video. This enables different versions of a program to be put on a single disc by sharing common parts.

SECAM   Séquential couleur avec mémoire/sequential color with memory. A composite color standard similar to PAL but currently used only as a transmission standard in France and a few other countries. Video is produced using the 625/50 PAL standard and is then transcoded to SECAM by the player or transmitter.

sector   A logical or physical group of bytes recorded on the disc, the smallest addressable unit. A DVD sector contains 38,688 bits of channel data and 2,048 bytes of user data.

seek time   The time it takes for the head in a drive to move to a data track.

Serial Digital Interface (SDI)   The professional digital video connection format using a 270-Mbps transfer rate. A 10-bit, scrambled, polarity-independent interface, with common scrambling for both component ITU-R 601 and composite digital video and four groups each of four channels of embedded digital audio. SDI uses standard 75-ohm BNC connectors and coax cable.

SFF 8090 The specification number 8090 of the Small Form Factor Committee, an ad hoc group formed to promptly address disk industry needs and to develop recommendations to be passed on to standards organizations. SFF 8090 (also known as the Mt. Fuji specification) defines a command set for CD-ROM- and DVD-ROM-type devices, including implementation notes for ATAPI and SCSI.

SI Système International (d'Unités)/International System (of Units). A complete system of standardized units and prefixes for fundamental quantities of length, time, volume, mass, and so on.

signal-to-noise ratio The ratio of pure signal to extraneous noise, such as tape hiss or video interference. Signal-to-noise ratio is measured in decibels (dB). Analog recordings almost always have noise. Digital recordings, when properly prefiltered and not compressed, have no noise.

simple profile (SP) A subset of the syntax of the MPEG-2 video standard designed for simple and inexpensive applications such as software. SP does not enable B pictures. *See* profile.

simulate To test the function of a DVD disc in the authoring system without actually formatting an image.

SMPTE The Society of Motion Picture and Television Engineers. An international research and standards organization. This group developed the SMPTE time code, used for marking the position of audio or video in time. Refer to Appendix C, "References and Information Sources."

S/N Signal-to-noise ratio. Also called SNR.

son The metal discs produced from mother discs in the replication process. Fathers or sons are used in molds to stamp discs.

space The reflective area of a writable optical disc. Equivalent to a land.

spatial resolution The clarity of a single image or the measure of detail in an image. *See* resolution.

spatial Relating to space, usually two-dimensional. Video can be defined by its spatial characteristics (information from the horizontal plane and vertical plane) and its temporal characteristics (information at different instances in time).

S/P DIF Sony/Philips digital interface. A consumer version of the AES/EBU digital audio transmission standard. Most DVD players include S/P DIF coaxial digital audio connectors providing PCM and encoded digital audio output.

SP@ML   Simple profile at main level. The simplest MPEG-2 format used by DVD. Most discs use MP@ML. SP does not allow B pictures.

squeezed video   *See* anamorphic.

stamping   The process of replicating optical discs by injecting liquid plastic into a mold containing a stamper (father or son). Also (inaccurately) called mastering.

STP   Surface transfer process. A method of producing dual-layer DVDs that sputters the reflective (aluminum) layer onto a temporary substrate of PMMA, and then transfers the metalized layer to the already-molded layer 0.

stream   A continuous flow of data, usually digitally encoded, designed to be processed sequentially. Also called a bitstream.

subpicture   Graphic bitmap overlays used in DVD-Video to create subtitles, captions, karaoke lyrics, menu highlighting effects, and so on.

subsampling   The process of reducing spatial resolution by taking samples that cover areas larger than the original samples, or the process of reducing temporal resolutions by taking samples that cover more time than the original samples. This is also called downsampling. *See* chroma subsampling.

substrate   The clear polycarbonate disc onto which data layers are stamped or deposited.

subtitle   A textual representation of the spoken audio in a video program. Subtitles are often used with foreign languages and do not serve the same purpose as captions for the hearing impaired. *See* subpicture.

surround sound   A multichannel audio system with speakers in front of and behind the listener to create a surrounding envelope of sound and to simulate directional audio sources.

SVCD   Super Video Compact Disc. MPEG-2 video on CD. Used primarily in Asia.

SVGA   A video graphics resolution of 800×600 pixels.

S-VHS   Super VHS (Video Home System). An enhancement of the VHS videotape standard using better recording techniques and Y/C signals. The term S-VHS is often used incorrectly to refer to s-video signals and connectors.

s-video   A video interface standard that carries separate luma and chroma signals, usually on a four-pin mini-DIN connector. Also called Y/C. The quality of s-video is significantly better than composite video because it does not require a comb filter to separate the signals, but it's not quite as good as component video. Most high-end televisions have s-video inputs. S-video is often erroneously called S-VHS.

SXGA   A video graphics resolution of 1280 × 1024 pixels.

sync   A video signal (or component of a video signal) containing information necessary to synchronize the picture horizontally and vertically. Also, sync is specially formatted data on a disc that helps the readout system identify location and specific data structures.

syntax   The rules governing the construction or formation of an orderly system of information. For example, the syntax of the MPEG video encoding specification defines how data and associated instructions are used by a decoder to create video pictures.

system menu   The main menu of a DVD-Video disc, from which titles are selected. Also called the title selection menu or disc menu.

T   Tera. An SI prefix for denominations of one trillion (10^{12}).

telecine   The process (and the equipment) used to transfer film to video. The telecine machine performs 2-3 pulldown by projecting film frames in the proper sequence to be captured by a video camera.

telecine artist   The operator of a telecine machine. Also called a colorist.

temporal   Relating to time. The temporal component of motion video is broken into individual still pictures. Because motion video can contain images (such as backgrounds) that do not change much over time, typical video has large amounts of temporal redundancy.

temporal resolution   The clarity of a moving image or moving object, or the measurement of the rate of information change in motion video. *See* resolution.

tilt   A mechanical measurement of the warp of a disc. This is usually expressed in radial and tangential components, with radial indicating dishing and tangential indicating ripples in the perpendicular direction.

time code   Information recorded with audio or video to indicate a position in time. This usually consists of values for hours, minutes, seconds, and frames. It is also called SMPTE time code. Some DVD-Video material includes information to enable the player to search to a specific time code position.

title   The largest unit of a DVD-Video disc (other than the entire volume or side). A title is usually a movie, TV program, music album, or so on. A disc can hold up to 99 titles, which can be selected from the disc menu. Entire DVD volumes are also commonly called titles.

title key   A value used to encrypt and decrypt (scramble) user data on DVD-Video discs.

track   1) A distinct element of audiovisual information, such as the picture, a sound track for a specific language, or the like. DVD-Video enables one track of video (with multiple angles), up to eight tracks of audio, and up to 32 tracks of subpicture; 2) One revolution of the continuous spiral channel of information recorded on a disc.

track buffer   The circuitry (including memory) in a DVD player that provides a variable stream of data (up to 10.08 Mbps) to the system decoders of data coming from the disc at a constant rate of 11.08 Mbps (except for breaks when a different part of the disc is accessed).

track pitch   The distance (in the radial direction) between the centers of two adjacent tracks on a disc. The DVD-ROM standard track pitch is 0.74 μm.

transfer rate   The speed at which a certain volume of data is transferred from a device such as a DVD-ROM drive to a host such as a personal computer. This is usually measured in bits per second or bytes per second. It is sometimes confusingly used to refer to the data rate, which is independent of the actual transfer system.

transform   The process or result of replacing a set of values with another set of values. It can also be a mapping of one information space onto another.

trim   *See* crop.

tristimulus   A three-valued signal that can match nearly all the colors of visible light in human vision. This is possible because of the three types of photoreceptors in the eye. RGB, YC_bC_r, and similar signals are tristimulus and can be interchanged by using mathematical transformations (subject to a possible loss of information).

TVL   Television line. *See* lines of horizontal resolution.

TWG   Technical Working Group. A general term for an industry working group. Specifically, the predecessor to the CPTWG. It is usually ad hoc group of representatives working together for a period of time to make recommendations or define standards.

UDF   Universal Disc Format. A standard developed by the Optical Storage Technology Association designed to create a practical and usable subset of the ISO/IEC 13346 recordable, random-access file system and volume structure format.

UDF Bridge   A combination of UDF and ISO 9660 file system formats that provides backward-compatibility with ISO 9660 readers while allowing the full use of the UDF standard.

universal DVD   A DVD designed to play in DVD-Audio and DVD-Video players (by carrying a Dolby Digital audio track in the DVD-Video zone).

universal DVD player   A DVD player that can play both DVD-Video and DVD-Audio discs.

user data The data recorded on a disc independent of formatting and error-correction overhead. Each DVD sector contains 2,048 bytes of user data.

UXGA A video graphics resolution of 1600 x 1200.

VBI Vertical blanking interval. The scan lines in a television signal that do not contain picture information. These lines are present to enable the electron scanning beam to return to the top, and they are used to contain auxiliary information such as closed captions.

VBR Variable bit rate. Data that can be read and processed at a volume that varies over time. A data compression technique that produces a data stream between a fixed minimum and maximum rate. A constant level of compression is generally maintained, with the required bandwidth increasing or decreasing depending on the complexity (the amount of spatial and temporal energy) of the data being encoded. In other words, a data rate is held constant while quality is allowed to vary. Compare this to CBR.

VBV Video buffering verifier. A hypothetical decoder that is conceptually connected to the output of an MPEG video encoder. It provides a constraint on the variability of the data rate that an encoder can produce.

VCAP Video capable audio player. An audio player that can read the limited subset of video features defined for the DVD-Audio format. Constrast this with a universal DVD player.

VCD Video Compact Disc. Near-VHS-quality MPEG-1 video on CD. Used primarily in Asia.

VfW *See* Video for Windows.

VGA (Video Graphics Array) A standard analog monitor interface for computers. It is also a video graphics resolution of 640 × 480 pixels.

VHS Video Home System. The most popular system of videotape for home use. Developed by JCV.

Video CD An CD extension based on MPEG-1 video and audio that enables the playback of near-VHS-quality video on a Video CD player, CD-i player, or computer with MPEG decoding capability.

Video for Windows The system software additions used for motion video playback in Microsoft Windows. Replaced in newer versions of Windows by DirectShow (formerly called ActiveMovie).

Video manager (VMG) The disc menu. Also called the title selection menu.

Video title set (VTS) A set of one to 10 files holding the contents of a title.

videophile   Someone with an avid interest in watching videos or in making video recordings. Videophiles are often very particular about audio quality, picture quality, and aspect ratio to the point of snobbishness.

VLC   Variable length coding. *See* Huffman coding.

VOB   Video object. A small physical unit of DVD-Video data storage, usually a GOP.

volume   A logical unit representing all the data on one side of a disc.

VSDA   Video Software Dealers Association. Refer to Appendix C, "References and Information Sources."

WAEA   World Airline Entertainment Association. Discs produced for use in airplanes contain extra information in a WAEA directory. The in-flight entertainment working group of the WAEA petitioned the DVD Forum to assign region 8 to discs intended for in-flight use.

watermark   Information hidden as invisible noise or inaudible noise in a video or audio signal.

White Book   The document from Sony, Philips, and JVC begun in 1993 that extended the Red Book CD format to include digital video in MPEG-1 format. It is commonly called Video CD.

widescreen   A video image wider than the standard 1.33 (4:3) aspect ratio. When referring to DVD or HDTV, widescreen usually indicates a 1.78 (16:9) aspect ratio.

window   A usually rectangular section within an entire screen or picture.

Windows   *See* Microsoft Windows.

XA   *See* CD-ROM XA.

XDS   Line 21.

XGA   A video graphics resolution of 1024×768 pixels.

XVCD   A non-standard variation of VCD.

Y   The luma or luminance component of video, which is the brightness independent of color.

Y/C   A video signal in which the brightness (luma, Y) and color (chroma, C) signals are separated. This is also called s-video.

YC_bC_r   A component digital video signal containing one luma and two chroma components. The chroma components are usually adjusted for digital transmission according to ITU-R BT.601. DVD-Video's MPEG-2 encoding is based on 4:2:0 YC_bC_r signals. YC_bC_r applies only to digital video, but it is often incorrectly used in reference to the YP_bP_r analog component outputs of DVD players.

Yellow Book   The document produced in 1985 by Sony and Philips that extended the Red Book CD format to include digital data for use by a computer. It is commonly called CD-ROM.

YP$_b$P$_r$ A component analog video signal containing one luma and two chroma components. It is often referred to loosely as YUV or Y, B-Y, R-Y.

YUV   In the general sense, any form of color-difference video signal containing one luma and two chroma components. Technically, YUV is applicable only to the process of encoding component video into composite video. *See* YC$_b$C$_r$ and YP$_b$P$_r$.

ZCLV   Zoned constant linear velocity. This consists of concentric rings on a disc within which all sectors are the same size. It is a combination of CLV and CAV.

INDEX

E

F

J-K

M

T

U

ABOUT THE AUTHOR

Jim Taylor has been hip deep in DVD since 1995. Called a "minor tech legend" by *E! Online*, Jim created the official Internet DVD FAQ, writes articles and columns about DVD, serves as President of the DVD Association, was named one of the 21 most influential DVD executives by *DVD Report*, and received the 2000 DVD Pro Discus Award for Outstanding Contribution to the Industry. Jim has worked with interactive media for over 20 years, developing educational software, laserdiscs, CD-ROMS, Web sites, and DVDs; and teaching workshops and courses on multimedia, computer-based education, computer applications, and DVD. Formerly VP of Information Technology at Videodiscovery, an educational multimedia publishing company, Jim championed the format as Microsoft's DVD Evangelist from 1998 to 2000. Jim Taylor is based in Seattle, Washington.